Proceedings in Information and Communications Technology 4

Jong-Hyun Kim Kangsun Lee Satoshi Tanaka
Soo-Hyun Park (Eds.)

Advanced Methods, Techniques, and Applications in Modeling and Simulation

Asia Simulation Conference 2011
Seoul, Korea, November 2011
Proceedings

Springer

Volume Editors

Jong-Hyun Kim
Yonsei University
Korea
E-mail: jhkim34@yonsei.ac.kr

Satoshi Tanaka
Ritsumeikan University
Japan
E-mail: stanaka@media.ritsumei.ac.jp

Kangsun Lee
MyongJi University
Korea
E-mail: ksl@mju.ac.kr

Soo-Hyun Park
Kookmin University
Korea
E-mail: shpark21@kookmin.ac.kr

ISSN 1867-2914 e-ISSN 1867-2922
ISBN 978-4-431-54215-5 e-ISBN 978-4-431-54216-2
DOI 10.1007/978-4-431-54216-2
Springer Tokyo Heidelberg New York Dordrecht London

Library of Congress Control Number: 2012946923

CR Subject Classification (1998): I.6

Typesetting: Camera ready by author and data conversion by Scientific Publishing Services, Chennai, India.

Printed on acid-free paper

Springer is part of Springer Science+Business Media (www.springer.com)

Preface

The AsiaSim2011 (Asia Simulation Conference 2011) was held November 16–18, 2011, in Seoul, Korea. The Asia Simulation Conference, held annually, is organized by ASIASIM (Federation of Asian Simulation Societies), KSS (Korea Society for Simulation), CASS (Chinese Association for System Simulation), and JSST (Japan Society for Simulation Technology) and provides a forum for scientists, academicians, and professionals from the Asia-Pacific region and from other parts of the world. Participants present their latest exciting research findings in various fields of modeling, simulation, and their applications.

This volume is the proceedings of AsiaSim2011. At the conference, 61 papers, stringently selected, were presented and three keynote speeches introduced new simulation technologies and research trends. Full-length versions of all submitted papers were refereed by the international program committee, with each paper receiving at least two independent reviews. This volume publishes papers selected from among those presented at the conference.

In addition to the scientific papers presented, the conference featured keynote talks by three invited speakers: Axel Lehmann (Bundeswehr University, Munich, Germany), Ichiro Hagiwara (Tokyo Institute of Technology, Japan), and Bo Hu Li (Beijing University of Aeronautics and Astronautics, China). We are grateful to them for accepting our invitation and for their talks. We also would like to express our gratitude to all contributors, reviewers, and program committee and organizing committee members who made the conference very successful. Special thanks are due to Soo-Hyun Park, Yun-Bae Kim, and Kangsun Lee, Chairs of the Program Committee, Organization Committee, and Publication Committee of AsiaSim2011, for their hard work in various aspects of conference organization.

Finally, I would like to acknowledge the partial financial support of the Korea Federation of Science and Technology Societies. I also would like to express appreciation for the publication support from Springer Japan.

November 2011

Jong-Hyun Kim
General Chair of AsiaSim2011

Organization

Honorary Chairs

Chan-Mo Park	POSTECH, Korea
Seong-Joo Park	KAIST, Korea
Doo-Kwon Baik	Korea University, Korea
Bo Hu Li	Beijing University of Aeronautics and Astronautics, China
Toshiharu Kagawa	Tokyo Institute of Technology, Japan

General Chair

Jong-Hyun Kim	Yonsei University, Korea

General Co-chairs

Jin-Woo Park	Seoul National University, Korea
Tag Gon Kim	Korea Advanced Institute of Science and Technology, Korea
Myoung Hee Kim	Ewha Womans University, Korea
Axel Lehmann	Bundeswehr University, Munich, Germany
Osamu Ono	Meiji University, Japan
Koji Koyamada	Kyoto University, Japan
Qinping Zhao	Beihang University, China
Kitti Tirasesth	KMITL, Thailand
Marzuki Khalid	UTM, Malaysia
Bernard P. Zeigler	University of Arizona, USA

Organizing Committee

Chair

Yun Bae Kim	Sungkyunkwan University, Korea
Dong-Won Seo	Kyung Hee University, Korea
Zhang Lin	Beihang University, China

International Program Committee

Chairs

Sung-Do Chi	Hankuk Aviation University, Korea
Soo-Hyun Park	Kookmin University, Korea
Kangsun Lee	Myongji University, Korea
Satoshi Tanaka	Ritsumeikan University, Japan

Hyung-Jong Kim	Seoul Women's University, Korea
Tianyuan Xiao	Tsinghua University, China
Noriyuki Komine	Tokai University, Japan
Jung Gil Choi	Korea Institute of Industrial Technology, Korea
Norman Foo	University of New South Wales, Australia
Sumit Ghosh	University of Texas at Tyler, USA
David Goldsman	Georgia Institute of Technology, USA
Alemdar Hasanoglu	Kocaeli University , Turkey
Farnando J. Barros	University of Coimbra, Portugal
V. Kachitvichyanukul	Asian Institute of Technology, Thailand
Yukio Kagawa	Nihon University, Japan
Ki-hyung Kim	Ajou University, Korea
Haklin Kimm	East Stroudsburg University, USA
Hessam S. Sarjoughian	University of Arizona, USA
Roy Sato	University of Tsukuba, Japan
Mi Ae Shon	Sungkyunkwan University, Korea
Shingo Takahashi	Waseda University, Japan
Xiao Tianyuan	Tsinghua University, China
Peng Xiaoyuan	Beijing University, China
Guang Leng Xiong	Tsinghua University, China
Mohd Nazmi Yosoff	International Islamic University, Malaysia

Publication Committee

Chair

Seong Yong Jang	Seoul National University of Science and Technology, Korea
Hee Seok Seo	Korea University of Technology and Education, Korea

Industrial Committee

Chair

Dug Hee Moon	Changwon National University, Korea

Table of Contents

Distributed Simulation

Numerical Simulation

Modeling Methodology (I)

Modeling Methodology (II)

Communication Networks and Simulations

Simulation in General Engineering (I)

Simulation in General Engineering (II)

Simulation in General Engineering (III)

Military Simulation (I)

Military Simulation (II)

Military Simulation (III)

Simulation Methods and Tools

Simulation and Visualization

Flight Simulation on Tiled Displays with Distributed Computing Scheme

Sangwon Chae[1], Soon-Bin Yim[2], and Youngshin Han[2,*]

[1] University of California, Irvine,
Irvine, CA 92697, U.S.A.
chaes@uci.edu

[2] Sungkyunkwan University
School of Information and Communication Engineering,
Suwon, Gyeonggi-do 440-746, South Korea
{sbyim,hanys}@skku.edu

Abstract. Flight simulator is based on real aircrafts. The main purpose of flight simulator is to prevent accidents when people control real flight through the training with various unexpected scenarios during flight. It is difficult to handle large scale geographical data on ordinary personal computer. Therefore, we propose distributed flight simulation which provides wider screen view on tiled displays and high system performance with distributed computing environment.

1 Introduction

Simulation is based on the imitation of real things, state of affairs or process. It shows similar characteristics of real objects such as aircraft, generator, analog/digital circuit and so on. Purposes of those simulators are various such as research, drill, and entertainment[1]. Applications of simulators which show the simulation of real objects are war game, aero system, factory automation and so on. For educational purposes, we can simulate on personal computers, it is more efficient to run simulators on tiled display which consists of multiple displays and computing nodes. Tiled displays use a diversity of implementation immersed virtual reality(VR). VR which is implemented on tiled displays provides wide display environments, so it is possible for developers to co-work. Tiled displays are efficient to not only face to face VR which multiple users share one system to communicate and work together, but also remote VR which small group users can share their works through network. Moreover, if we run simulators on distributed computing environments, we can achieve scalability to handle large data and stability to run on real time. It is an important issue to show the characteristics of real objects on real time.

In this paper, we target a flight simulator which a user can control virtual flight objects in order to get used to handling real ones. Indeed, we run a flight simulator on tiled displays which provide distributed computing environments. That is, the

* Corresponding author.

J.-H. Kim et al. (Eds.): AsiaSim2011, PICT 4, pp. 1–6, 2012.

purpose of our research is for users/pilots to understand various scenarios of battle, interceptions, and evasion more effectively through our flight simulations. In addition, we can provide wider and larger screen view to users on tiled displays and handle large dataset through distributed environments.

This paper describes our proposed algorithm in Section 2. Section 3 shows our system overview and Section 4 indicates the results of our system. Section 5, which concludes, summarizes our paper and discusses the future works of our system.

2 Algorithm

In this section, we describe our proposed algorithms. This simulator consists of three parts, a simulation engine, geographical database, and a rendering engine. We generate random geographical data and implement simulation engine and rendering algorithms.

2.1 Simulation Engine

Both simulation and rendering engine of our system have functions to save and load both of input and output data. Simulation engine provides various input methods such as joysticks, keyboards, and mouse for users to control virtual aircraft and saves data of their flights. In addition, we can see pre-saved data to see educational scenarios without any control.

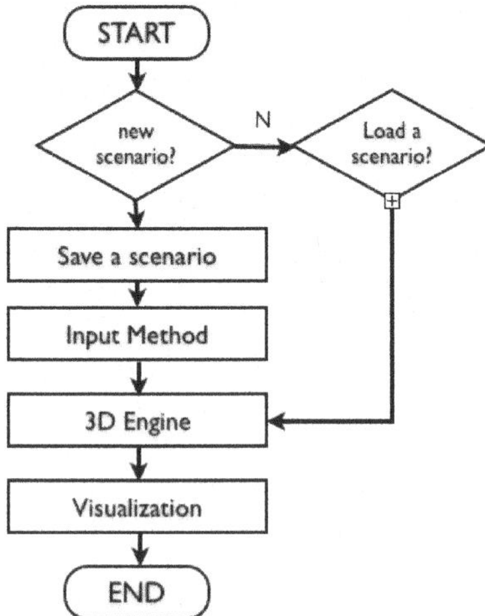

Fig. 1. It shows a simple simulation engine flowchart. We can choose whether it is new scenario to create or old scenario to load. We can control virtual crafts through input method and see the result on displays.

2.2 Distributed Rendering Algorithm

Display wall system is shown recently[4], these walls consist of one or multiple displays which are connected multiple computing nodes. In this environment, it is important how to distribute data to each computing node for rendering.

As you can see, it consists of two parts, one is a control node and the other is computing display node. A main purpose of control node is to collect and distribute all data to each computing display node. It is difficult to handle large data on one node, so it divides an entire data into pieces of data. Since geographical data and aircrafts' information are based on positions, it is available to divide a whole data into sub-dataset. In addition, a main purpose of computing displays nodes is to visualize all data on their displays after rendering. Each computing display node communicates to a control node for sending and receiving dataset after processing its procedures.

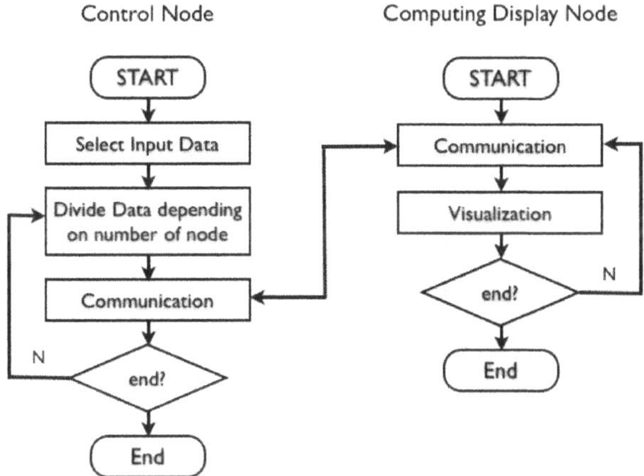

Fig. 2. It shows a simple distributed rendering engine flowchart. It consists of two parts, a control node which control geographical data and flight information and computing display nodes which display all data on each display.

3 System Overview

Our system consists of two parts, a control node and compute-displaying nodes. Figure 3 shows the system overview of our research we discuss in Section 2. Figure 4 shows Highly Interactive Parallel Display Wall(HIPerWall)[5]. Our system consists of 50 displays and 25 compute-displaying nodes. Its resolution is 200M pixel at once. It supports to display very large picture without any zoom or compression of data. When we simulate virtual aircrafts on HIPerWall, it is possible to understand scenarios more effectively and detail.

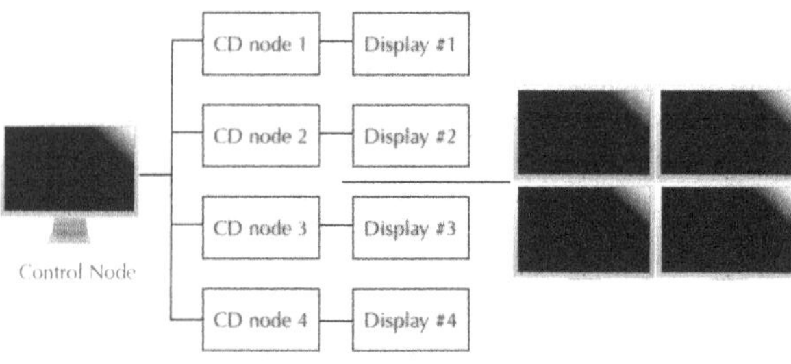

Fig. 3. There is a control node to collect and distribute data to each compute-displaying node. Each compute-displaying node is connected to one or two display(s). They are visualizing data set for users to see a scenario of simulation.

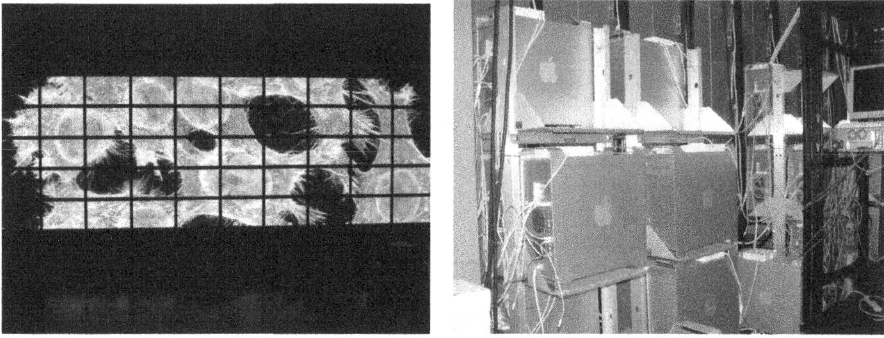

Fig. 4. Highly Interactive Parallel Display Wall(HIPerWall) consists of 25 compute-displaying nodes and 50 mac cinema displays. Left picture shows an example of visualizing medical information and right picture shows the back of HIPerWall.

4 Result

In this section, we show the simulation results on a single display and tiled displays. In Figure 5, the result on tiled displays consists of four displays and four compute-displaying nodes(lower picture) shows more details and higher quality rather than the result on a single node and a single display(upper picture). When practical data is applied, it is possible to have more complicated and larger geographical data to visualize. It compromises the overall system performance such as processing time and visualization. Tiled displays provide users to have wider views with higher resolution to display large geographical data at once.

Fig. 5. There are the results on a single display and compute-displaying node(upper) and tiled displays which consists of four displays and four compute-displaying nodes(lower). Tiled display provides more detailed and higher quality results to users.

5 Conclusion and Future Works

We implemented distributed flight simulation on tiled displays. It provides more detailed results to users with higher resolutions and it is scalable on tiled display with distributed computing scheme.

However, there are several issues of tiled displays. Because of the frames of each display, it is difficult to discern objects when multiple objects go through panels. Moreover, it is expensive to build entire display walls. We can build projector based display wall instead of displays, but the cons of projector based wall are lower resolution and a keystone issue.

Major component of our future work is the implementation of the module to provide practical geographical data and more detailed aircraft. Also, we will implement the module to control multiple aircrafts from multiple input methods in real-time environment. It can show educational scenarios of large scale battles or training on tiled displays.

Acknowledgement. This work is supported by Basic Research Program through the National Research Foundation of Korea(NRF) funded by Ministry of Education, Science and Technology(2011-0379-000).

References

1. Choi, S., Chae, S., Han, Y., Lee, C.: 3D Flight simulator for flying tactics. In: Korean Simulation Conference, pp. 37–42 (2003)
2. Park, K., Park, C.: A research of flight simulator' components. In: Korean Aeronautical and Space Sciences Conference, pp. 239–245 (November 2001)
3. Kim, J., You, C.: Light aircraft simulator on personal computer. In: Korean Aeronautical and Space Sciences Conference, pp. 392–397 (April 1994)
4. Mueller, C., Gregor, D., Lumsdaine, A.: Distributed Force-Directed Graph Layout and Visualization. In: Eurographics Symposium on Parallel Graphics and Visualization (EGPVG 2006) (May 2006)
5. HIPerWall, `http://hiperwall.calit2.uci.edu/`
6. Ratan, K., Jeff, L.: Flight Simulator Evaluation of a Cockpit Display of Traffic Information (CDTI) With and Without Indications & Alerts (IA). Honeywell Contract Report to FAA, Redmond (2010)

A Novel Federation Development and Execution Process Based on Collaboration Ontology

Hongbo Sun, Wenhui Fan, Tianyuan Xiao, Yanbing Liu, and Yueting Chai

National CIMS Engineering Research Centre, Tsinghua University, Beijing 100084,China
hsun@ieee.org

Abstract. This paper presents a novel Federation Development and Execution Process aiming to employ collaboration ontology as mutual understanding of High Level Architecture. The proposed approach includes three main stages: Conceptual Analysis, Collaboration Preparation and Federation Execution. It uses collaboration ontology to represent collaboration knowledge such as conceptual model, domain knowledge, SOM, FOM, FED/FDD files, class RootObject in a consistent way. This knowledge representation method improves semantic completeness and transformation efficiency of collaboration knowledge and reduces workload without the loss of accuracy and consistency. This paper also discusses some ontology based collaboration knowledge modeling, manipulating, organization and implementation issues on the basis of collaboration ontology, which includes FCA modeling method, extending UML sequence diagram, ontology fusion, ontology maintenance and ontology template. The proposed approach has great potential to improve efficiency of FEDEP, reduce the work load for adaptive adjustment of ever-existing platforms, and enhance the applicability and flexibility of HLA systems.

1 Introduction

As a well-known modeling and integration standard of distributed simulation, HLA (High level Architecture [1]) has been successfully adopted in various simulation systems, and extended to some other research areas [2]. Since simulation is an important part of collaborative product development, some CPD (Collaborative Product Development) systems use HLA as their basic architecture [3,4].

However, when applying HLA in this research area, there raise some new challenges. Within these challenges, mutual understanding (collaboration) knowledge modeling and manipulating among several subsystems is an essential issue to support solutions of other issues.

Fortunately, ontology in knowledge engineering is the semantic basis of communication among domain entities. It is applicable to automatic reasoning, knowledge representation and reuse [5].

Sun H. has proposed the concept of collaboration ontology to solve mutual understanding among several independent systems [6]. Because the algebraic system defined on the concept set of HLA collaboration and the partial order relations of these

J.-H. Kim et al. (Eds.): AsiaSim2011, PICT 4, pp. 7–14, 2012.

concepts have the same upper bound and lower bound, it can be deemed as a concept lattice [7]. The collaboration ontology is formally defined as follows:

$$O ::= (C, H_C, R_C, H_R, M, R_M, A)$$

Collaboration ontology O is defined as a seven tuple. C denotes collaboration concept set of HLA collaboration. H_C defines a set of partial orders on concept set C, which give the inherit relations among the concepts involved. The concepts set and inherit relations defined on that set form a Directed Acyclic Graph (DAG) whose source is the given model of collaborative product and whose sink is binary fragments. R_C denotes a set of non-inherit partial order relations on concept set C, corresponding to concept attributes. H_R defines inherit relations on partial order relation set R_C . M is a series of meta ontology concepts, which give a series inheritable instances of R_C. R_M denotes a set of partial order relations under M, which describe the relations among elements in meta ontology set, and are also the basis for ontology reasoning. A defines a set of axioms among ontology concept set and meta ontology relation set, which provide the major premises of ontology reasoning.

This paper proposes a collaboration ontology based FEDEP (FEderation Development and Execution Process) to give an outline of how to use, organize and manipulate collaboration ontology to meet various requirements. This paper is organized as follows: the motivation of this research is described in Section 2; collaboration ontology based FEDEP is proposed in Section 3; discussions and conclusions are given in the last section.

2 Motivation

In HLA-based federation development, collaboration knowledge is mainly manipulated as Fig. 1 shows [8], and can be divided into 11 steps: Analyze Scenario, Generate Federation Conceptual Model, Analyze Collaboration Requirements, Generate FOM, Negotiate Federation Agreements, Realize Federate, Realize RTI, Create Federation, Start Federation Execution, Execute Federation and Terminate Federation Execution.

In this process, the input is federate SOMs, Existing Conceptual Models and collaboration Aim. The output is Scenario Instances, Federate Modification, Supporting Database, RTI & its initial data, interaction/object class instance, class subscriber, class publisher, synchronize point and results. Resources involved include Scenario Lib, Domain Resources, Data Dictionary, OM Lib and Other Resources. Scenario Lib supports *Analyze Scenario* by *Existing Scenarios.* Domain Resources provides *Domain Knowledge* to analyze scenario and generate collaboration conceptual model. Data Dictionary stores Meta OM models for Generate *FOM.* When *Generate FOM*, related *Existing FOMs and BOMs* also need to be retrieved from OM Lib. Other Resources include tools and knowledge related to *Generate FOM.*

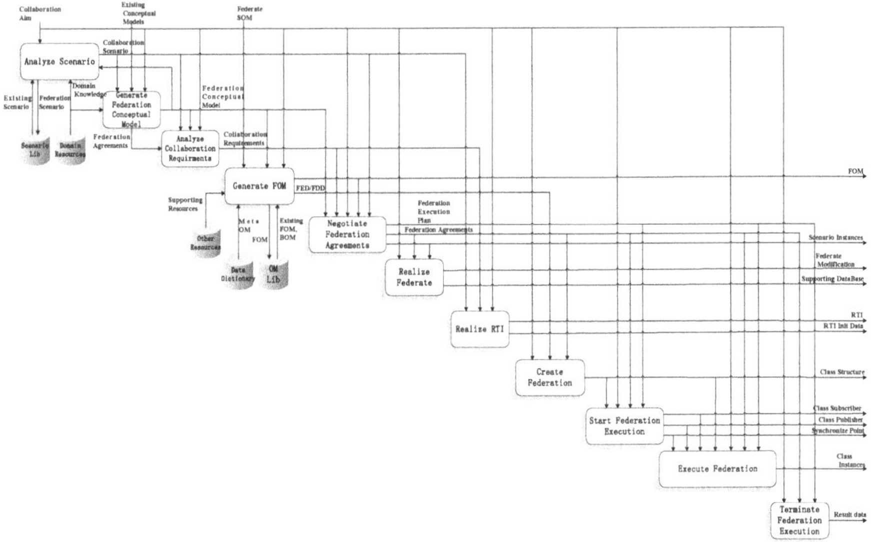

Fig. 1. Collaboration knowledge modeling and usage in FEDEP

Collaboration knowledge also represents itself as Collaboration Scenario, Federation Conceptual Model, Federation Execution Plan, FED/FDD, FOM, Collaboration Requirements, Class Subscriber, Class Publisher, and Class Structure. Collaboration Scenario describes environment and desired process of collaborations. Federation Conceptual Model is the description of concepts and their relations in collaborations. Federation Execution Plan gives operation order of participant federates. FOM is the standard description of federation concepts. FED/FDD stores FOM information. Collaboration Requirements describes *Federation Execution Plan* and collaboration requirements of federates. Interaction/object Class Structure is inheritance relations among these classes. Class Subscriber is federates that subscribe interaction or object classes. Class Publisher is federates that publish interaction or object classes, which supervise data distribution during collaborations.

From the view of collaboration knowledge, HLA semantic representation, transformation, and organize have several shortcomings which limits its consistency and efficiency.

First, collaboration knowledge takes different forms according to separated steps, which include conceptual models, domain knowledge, SOM, FOM, FED/FDD files and class RootObject. This will lead to semantic loss and incorrect delivery in the process of transformation. Moreover, it is a time-consuming and cost task to transform collaborative knowledge among different steps, and semantic accuracy and consistency are also hard to guarantee.

Second, SOM generation is not mentioned in FEDEP, which is the basis for FOM construction.

Third, class RootObject is the only one form of collaboration knowledge which can be used during collaboration execution. However, it is loaded only once before *Start Federation Execution*, so it cannot satisfy collaboration requirement changing in the process of federation execution.

In this paper, a collaboration ontology based FEDEP is proposed to give a consistent way of collaboration knowledge representation, organization and manipulating so that collaboration knowledge can be used in a more consistent, complete, flexible and efficient way.

3 Collaboration Ontology Based FEDEP

In collaboration ontology based FEDEP, corresponding to SOM and FOM, Federate collaboration ontologies describe collaboration capability and requirements of federates, federation collaboration ontology describes collaboration knowledge of federation respectively. Modeling collaboration knowledge is reflected in the creation of collaboration ontologies, manipulating collaboration knowledge corresponds to dynamical maintenance of federation collaboration ontology. Federation collaboration ontology is an important identification and data exchange basis of given federation.

The process of modeling and utilizing collaboration knowledge can be separated into three stages: Conceptual Analysis, Collaboration Preparation and Federation Execution (Fig. 2).

In stage Conceptual Analysis, scenario and domain knowledge is converted to federation collaboration scenarios which are represented by extended UML sequence graphs under the control of *Collaboration Aim*. *Collaboration Requirement* is analyzed and used as the input of collaboration preparation.

In stage Collaboration Preparation, federation agreements and federation collaboration ontology are obtained according to federate SOMs, existing collaboration ontologies, ontology template(including common collaboration ontology schema, basic information, data type transform rules), related bridge equivalence(mutual exclusion) axioms, collaboration requirements and federation collaboration scenario.

In stage Federation Execution, collaboration is executed according to federation agreements, federation collaboration ontology, collaboration requirements and collaboration aim.

3.1 Conceptual Analysis

The input of Conceptual Analysis is *Collaboration Aim*, the output is Federate Collaboration Scenario Described by Extended UML Sequence Diagram, and the resources involved include Scenario Lib, Domain Resources and Collaboration Ontology Lib. Scenario Lib provides the existing scenarios supporting scenario analysis and stores analysis results, Collaboration Ontology Lib supports collaboration requirements analysis by existing federate collaboration ontologies, and Domain Resources provide domain knowledge. This stage can be divided into two activities: Analyze Scenario and Analyze Collaboration Requirements (Fig. 3).

Analyze Scenario generates federate collaboration scenarios described by extended UML sequence diagram according to existing scenarios and domain knowledge under control of the collaboration target.

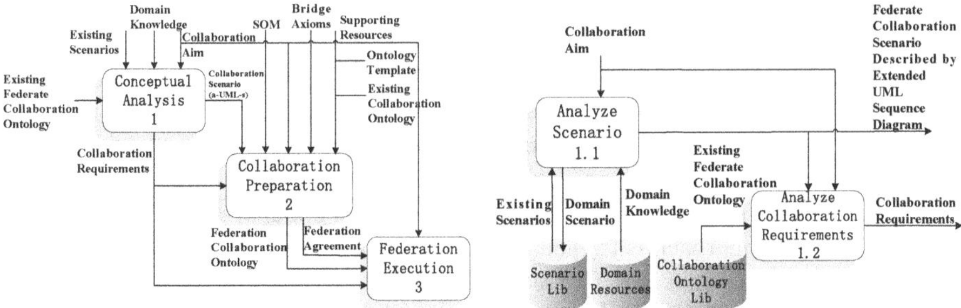

Fig. 2. Top-Level View

Fig. 3. Concept Analysis

Analyze Collaboration Requirement analyzes existing federate collaboration scenarios, existing federate collaboration ontology and collaboration aim, and provides collaboration requirement for future use .

3.2 Collaboration Preparation

In this stage, the input is Federate Collaboration Scenario Described by Extended UML Sequence Diagram, SOM, Collaboration Aim, Bridge Axioms, and Collaboration Requirements. The output is Federation Collaboration Ontology, Federation Agreements, Federate Modifications, Supporting Database, ORTI (Ontology-based RunTime Infrastructure, in this research ORTI is reached by add Protégé container to TH_RTI), and ORTI initial Data. And the reusable resources involve Domain Resources, Ontology Template and Collaboration Ontology Lib (Fig. 4).

In collaboration ontology-based FEDEP, federate collaboration ontology corresponds to SOM, federation collaboration ontology corresponds to FOM. Federate collaboration ontology can be established by either automatic transformation from SOM files or FCA-like method with the help of Extended UML Sequence Diagram. Federation collaboration ontology is automatically generated by federate collaboration ontologies fusion [6].

The process of Collaboration Preparation can be separated into 5 activities: Generate Federate Collaboration Ontology, Generate Federation Ontology, Negotiate Federation Agreements, Realize Federates, and Realize RTI.

Federate collaboration ontology is generated by reference basic schema, axioms and basic elements of ontology template, domain axioms of domain resources, and existing federate collaboration ontologies.

Based on federate collaboration ontologies, bridge axioms, basic schema and axioms and basic elements in ontology template, the federation collaboration ontology is fused under the control of *Collaborative Aim.*

According to federation collaboration ontology and collaboration requirement, federation agreements are generated under the guidance of *Collaboration Aim.*

Under the control of *Collaboration Aim* and *Requirements*, according to *Federation Agreements* and federate collaboration ontologies, federates and RTI can be realized properly.

3.3 Federation Execution

The input of Federation Execution is Collaboration Aim, Federation Collaboration Ontology, Federation Agreements, and Collaboration Requirements. The output is concept subscriber, publisher, synchronize points, concept instances and result data. In this process, class RootObject is replaced by federation collaboration ontology, class instances substituted by its concept instances. When collaboration requirement is changed, an online approach is trigged to fit the changes by so-called ontology maintenance. This process involves five activities: Create Federation, Start Federation Execution, Modify Collaboration Requirements, and Terminate Federation Execution (Fig. 5).

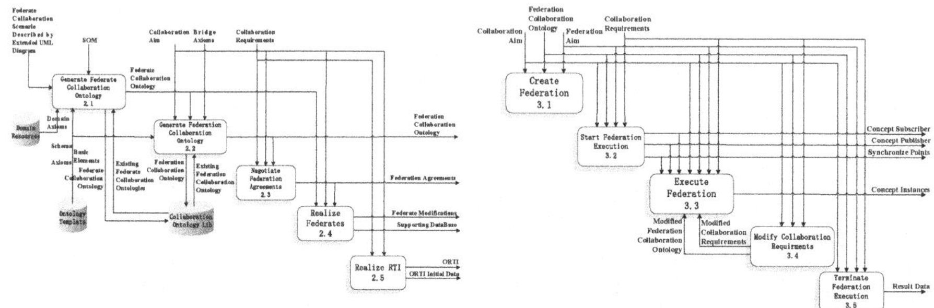

Fig. 4. Collaboration Preparation

Fig. 5. Federation Preparation

Federation is created by one federate and other federates will join in one by one. The main work is to specify federation name, federation execution name, related federates and federation collaboration ontology handle.

The purpose of Start Federation Execution is to create the conceptual instances of synchronization points, register publishers and subscribers of concept instances which correspond to interaction and object classes.

In Execute Federation, data distribution is controlled by instances of publishers and subscribers. Execution process is guided by execution plan, federation agreements and collaborative target. The outputs in this stage are concept instances, update and reflection of property values.

In the process of the federation execution, collaboration requirements can be changed because of uncertainty of collaborative product development. For example, before knowing the detail of optimization requirements, optimization federate cannot tell whether it needs to ask help from simulation federates. Under this circumstance, a dynamic federation execution plan adjustment mechanism is of great importance. Its main task includes collaboration requirement modification and federation collaboration ontology maintenance.

Terminating Federation Execution stops execution of given federation and complete results collection according to Federation Execution Plan, Federation Agreements and Collaboration Aim.

From the processed described above, collaboration ontology based FEDEP can be described as Fig. 6. In this process, collaboration ontology represents collaboration knowledge such as conceptual model, domain knowledge, SOM, FOM, FED/FDD files, class RootObject in a consistent way. This knowledge representation method improves semantic completeness and transformation efficiency of collaboration knowledge and reduces workload without the loss of accuracy and consistency.

Started with analyzing federates collaboration background by extended UML sequence graph, then add appropriate semantic referring to ontology template and domain knowledge, federate collaboration ontologies could be well established by automatic transformation from SOMs or its formal background.

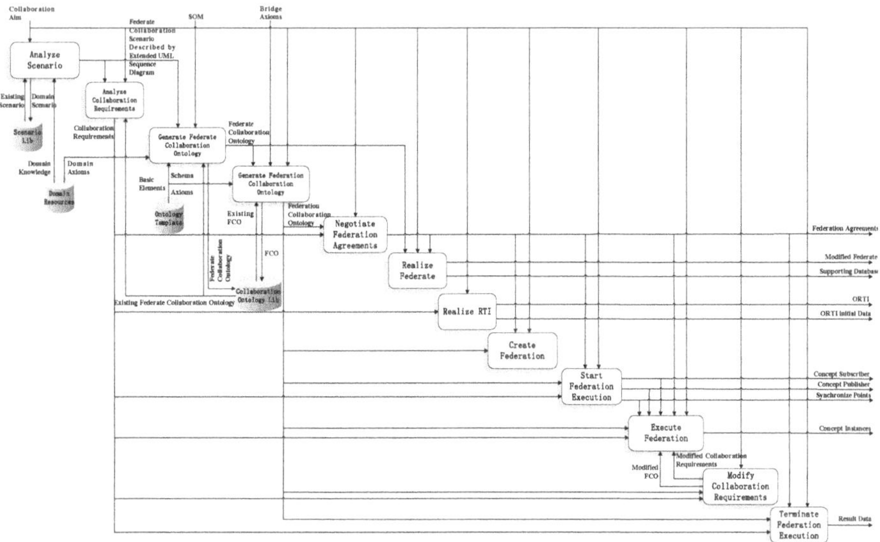

Fig. 6. Global view of collaboration ontology based FEDEP

In federation execution, collaboration knowledge is stored in federation collaboration ontology and its instances. When collaboration requirement changing, collaboration requirement modification could solve changes in an on-line manner.

4 Conclusion

Since collaboration knowledge of HLA federations adopts different representations like conceptual model, domain knowledge, SOM, FOM, FED/FDD files, and RootObject class, semantics always loses or misunderstands in their translations. In this paper, collaboration ontology describes the collaboration knowledge, and supports HLA federation development in a consistent way.

Other than collaboration knowledge is transformed ambiguously and inconsistently by negotiations which is time-consuming and cost much, this paper analyses and creates federate ontology on the basis of ontology template by extended UML sequence diagram, or uses an automata to transform SOM files to federate collaboration ontologies, constructs federation collaboration ontology by fusion of federate ontologies. These construction methods contain more semantic information, improve collaboration modeling efficiency.

Compare to HLA keep a blind eye on collaboration requirements changes, this paper introduces a novel ontology modification mechanism which employs federation ontology to substitute RootObject class, and concept instances of federation ontology to replace class instances in HLA federation executions. This approach can respond to collaboration requirements change in an "online" manner.

To support the methods mentioned above, this research also adds Protégé container to TH_RTI, changes some services, and realizes the key parts of collaboration ontology based runtime infrastructure, ORTI.

Verified by industrial application, the collaboration ontology based federated development process can keep semantic consistency of collaboration knowledge, improve preparation efficiency of collaborative simulation, and enhance flexibility of HLA federations.

Acknowledgment. This work is supported by Chinese National High-tech Research and Development Program (863 Program, Grant No. 2009AA110302) and Chinese Nature Science Foundation (Grant No. 60874066).

References

1. Simulation Interoperability Standards Committee (SISC) of the IEEE Computer Society, IEEE Standard for Modeling and Simulation (M&S) High Level Archi-tecture (HLA)-Framework and Rules. The Institute of Electrical and Electronics Engineers, Inc., New York (2000)
2. Zhang, H., Wang, H., Chen, D.: Integrating web services technology to HLA-based multidisciplinary collaborative simulation system for complex product development. In: 12th International Conference on Computer Supported Cooperative Work in Design, CSCWD 2008, Xi'an, China, pp. 420–426 (April 2008)
3. Xu, Y., Xiao, T., Liang, C., Zhang, L.: Federated integration of networked man-ufacturing service platforms. Advanced Engineering Informatics 22, 317–327 (2008)
4. Tang, S., Xiao, T., Fan, W.: A collaborative platform for complex product de-sign with an extended HLA integration architecture. Simulation Modelling Practice and Theory 18(8), 1048–1068 (2010)
5. Zhou, W., Liu, Z., Chen, H.: A Survey of the Research about Both FCA and Ontology. Computer Science 33(2), 8–12 (2006)
6. Sun, H., Fan, W., Shen, W., Xiao, T.: Ontology Fusion in HLA-based Collabora-tive Product Development. In: Proceedings of 2010 IEEE International Conference on Systems, Man, and Cybernetics, SMC 2010, Istanbul, Turkey, pp. 2526–2532 (October 2010)
7. Qu, K., Liang, J., Wang, J., Shi, Z.: The algebraic properties of Concept Lattice. Journal of Systems Science and Information 2(2), 271–277 (2004)
8. IEEE Computer Society. IEEE Recommended Practice for High Level Architecture (HLA) Federation Development and Execution Process (FEDEP, IEEE Std 1516.3-2003). The Institute of Electrical and Engineers, NewYork (2003)

PASENS: Parallel Sensor Network Simulator

Banghyun Kim[1] and Jong-Hyun Kim[2]

[1] Ocean Engineering Research Department, Korea Ocean Research & Development Institute, 1312gil 32,Yuseongdaero, Yuseong-gu, Daejeon, 305-343, Republic of Korea
bhkim@moeri.re.kr

[2] Computer and Telecommunication Engineering Division, Yonsei University, Maeji 234, Heungup, Wonju, Kangwon, 220-710, Republic of Korea
Jhkim34@yonsei.ac.kr

Abstract. This paper presents a Parallel Sensor Network Simulator (PASENS) to shorten the time in a large-scale wireless sensor network simulation. The degree of details of the simulation must be high to verify the behavior of the network and to estimate its power consumption and execution time of an application program as accurately as possible. Instruction-level simulation can provide those functions. But, when the degree of details is higher, the simulation time becomes longer. We propose an optimal-synchronous parallel discrete-event simulation method to shorten the simulation time. In this method, sensor nodes are partitioned into subsets, and PCs interconnected through a network are in charge of simulating one of the subsets. Results of experiments using PASENS show, in the case that the number of sensor nodes is large, the speedup tends to approach the square of the number of PCs participating in a simulation. We verified that the simulator provides high speedup and scalability enough to simulate maximum 20,000 sensor nodes.

1 Introduction

Wireless Sensor Network (WSN), which is an infrastructure of ubiquitous computing, consists of a number of sensor nodes of which the hardware is very small and simple. The network topology and routing scheme of the network should be determined according to its purpose. Its hardware and software may have to be changed as needed from time to time. Thus, a WSN simulator, which is capable of verifying its behavior and predicting its performance, is required to improve the design [1].

The WSN Simulator should be able to verify the behavior of WSN, estimate execution time and power consumption, and simulate a large-scale WSN. To satisfy the first two requirements, we use an instruction-level discrete-event simulation (DES), which simulates the behavior of WSN programs at the machine code level with cycle-accuracy. Instruction-level simulation, that is both language and operating system independent, provides the highest behavioral and timing accuracy. But it trades scalability for accuracy. As the degree of details increases, the simulation time becomes longer. Moreover, when the number of sensor nodes increases, the time becomes

J.-H. Kim et al. (Eds.): AsiaSim2011, PICT 4, pp. 15–24, 2012.

extremely long. Thus, we propose an optimal-synchronous Parallel DES (PDES) method to satisfy the last requirement without sacrificing cycle accuracy.

In optimal-synchronous PDES method, sensor nodes are partitioned into a number of subsets, and PCs interconnected through a network are in charge of simulating one of the subsets. The parallel simulation advances with the increment of the Global Virtual Time (GVT) by a time interval between a transmit operation and a receive operation of the RF communication. And each PC performs the simulation with increasing a Local Virtual Time (LVT) asynchronously within a time interval of the GVT.

We have implemented a Parallel Sensor Network Simulator (PASENS) using instruction-level DES and optimal-synchronous PDES. In PASENS, Execution times of most event routines are much shorter than the handling time of an event queue. This means that the handling time is the most significant factor in the simulation time. In this paper we present a speedup formula derived from our simulation model, which shows that the speedup approaches the square of the number of PCs participating in a simulation as the number of sensor nodes increases. The results of our experiments using PASENS prove validity of the formula in predicting the speedup achievable. Instruction traces used as a workload for simulation are executable images produced by the cross-compiler for an Atmega128L microcontroller unit (MCU).

The remainder of this paper is organized as follows: Section 2 reviews related work in PDES. Section 3 describes an instruction-level DES, and Section 4 proposes an optimal-synchronous PDES. Predicting speedup is shown in Section 5. The results of the experimental study with PASENS are reported in Section 6. Finally, Section 7 concludes the paper.

2 Related Work

PDES can solve the problem of slow simulation speed. It is based on partitioning the simulation model into a set of Logical Processes (LPs), which simulate distinct parts of the system under investigation. There are three major PDES techniques; synchronous simulation, conservative simulation, and optimistic simulation. Conservative simulation and optimistic simulation are both asynchronous simulations, where every LP maintains a LVT with generally different clock values at a given point in real time [2].

Synchronous simulation [3] is the simplest simulation technique known. The advantages of synchronous simulation include its low overhead, ease of implementation and performance predictability [4]. However, it is more prone to poor load balancing and high communication cost caused by synchronization steps between cycles. These problems get worse when the number of processors is increased. Synchronous simulation is suited for simulation that has very small computational granularity and reasonably large cycle parallelism [5]. Conservative simulation algorithm dates back to the original works of Chandy and Misra [6] and Bryant [7], and are often referred to as the Chandy-Misra-Bryant (CMB) protocols. Each generated event is accompanied with the LVT of the sending LP. The receiver LP knows that it will not receive any event in future operations with smaller time stamp than the LVT from the sender. Each LP can safely execute the events with a time stamp smaller than the LVT value of all input channels. Time Warp [8] algorithm allows optimistic asynchronous simulation of discrete-event systems. Optimistic mechanisms allow each LP to execute events whenever they are available, thus performing no preventive verification on whether the execution itself meets the correct criterion. On the other hand, if a time stamp order violation is detected, a rollback procedure recovers the LP state to a previous correct value.

GloMoSim [9], QualNet [10], SWAN [11], and SNAP [12] used conservative simulation techniques for WSN simulation. GloMoSim provides a scalable simulation environment for wireless and wired network systems. It is designed using PDES capability provided by Parsec [13]. QualNet is a commercial product of GloMoSim. SWAN is a high performance framework for wireless ad-hoc network simulation, and SNAP achieves conservative simulation using Time-Based Synchronization (TBS) protocol. These simulators can simulate only wireless network simulation, but cannot simulate the activity of each sensor node.

3 Instruction-Level DES

In PASENS, instruction traces used as workloads for simulations are executable images produced by the cross-compiler for an Atmega128L MCU. Intel's hex-record (.hex), Motorola's S-record (.srec), and ROM image (.rom) format are acceptable as workloads. An executable image is loaded onto a Virtual Sensor Node (VSN), which simulates a real sensor board. VSN is an abstract module of the MICAz sensor board including CrossBow MPR2400 [14] and Octacomm NANO-24 [15] . MICAz platform contains an Atmel ATmega128L MCU [16] and a Chipcon CC2420 RF transceiver [17]. Only difference between MPR2400 and NANO-24 is pin assignments.

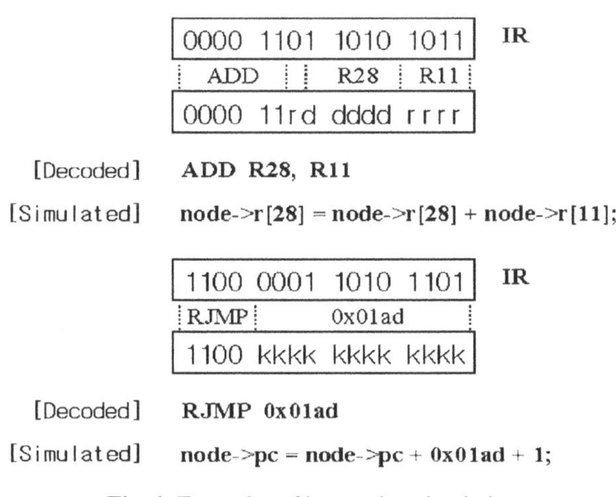

Fig. 1. Examples of instruction simulation

Examples of two instructions decoded and executed are shown in Fig. 1. In the first instruction, "0000 11" is translated into "ADD" instruction by opcode rules, and "01 1010 1011" is translated into "R28, R11" by operand rules of "ADD". Thus simulation of execution "R28 = R28 + R11" is performed by VSN, and LVT is increased by one clock cycle. In the second instruction, "1100" is translated into "RJMP" instruction, and "0001 1010 1101" into "0x01ad" operand. Thus simulation of "PC = PC + 0x01ad + 1" is performed, and LVT is increased by two clock cycles. These instruction-level simulations allow estimating program execution time and predicting the amount of power consumption based on the estimated execution times.

4 Optimal-Synchronous PDES

Instruction-level DES can estimate execution time and power consumption, but cannot simulate a large-scale WSN because it takes a lot of time to obtain a high degree of details. Parallel simulation can solve the problem of slow simulation speed, and overcome limitation of the number of sensor nodes to be simulated. Traditional PDES techniques may be used for instruction-level simulation, but they are not appropriate. Synchronous simulation will lead to a high communication cost caused by synchronization steps between cycles. Because instruction-level simulation breeds events frequently, conservative simulation has to check too many events for the synchronization between LPs, and optimistic simulation has to save too much information for a rollback. In order to overcome this, we introduce optimal-synchronous PDES for the instruction-level WSN simulation.

Optimal-synchronous PDES acts in a Central Process (CP) and several LPs interconnected through a network. Sensor nodes are partitioned into a number of subsets, and LPs (PCs) are in charge of simulating one of the subsets. Each LP progresses a simulation with increasing its LVT until GVT is increased by CP. This centralized method can provide good load balancing and easy implementation with low cost. Optimal-synchronous PDES has characteristics of not only synchronous but also asynchronous simulation. It is based on synchronous simulation because instruction-level simulation has very small computational granularity like logic simulation. But, the increment interval of GVT in optimal-synchronous PDES is much longer than that of logic simulation. Each LP increases its LVT independently in the time interval, and then progresses a simulation asynchronously.

Fig. 2 shows the optimal-synchronous PDES algorithm in PASENS. CP increases GVT by a time interval, and broadcasts it to all LPs until GVT reaches at the end of simulated time as shown in Fig. 2(a). The next increment of GVT occurs when CP receives completion messages from all LPs. When an LP receives GVT from CP, the LP progresses a simulation and in-creases its LVT until it reaches at the current GVT as shown in Fig. 2(b). LVT is updated to the occurrence time of an event that has the earliest LVT in the event queue. If the updated LVT is greater than GVT, the event is returned to the event queue and LVT is set to GVT. While LP executes the event routine, if an RF transmission occurs, it broadcasts the RF message to other LPs. LP that receives the message inserts an event to process the RF reception into the event queue. Each LP sends a completion message to CP after the simulation ends for the time interval and waits for the next GVT broadcast by CP.

```
/* Δt : A time interval of GVT
   End_Time : End of simulated time */
GVT = 0;
do {
   GVT += Δt;
   if (GVT > End_Time)
     GVT = End_Time;
   Broadcast GVT to all LPs;
   Wait until receiving completion messages from all LPs;
} while (GVT < End_Time);
```

(a)

```
/* E_i : ith event
   LVT_i : The occurrence time of E_i */
while (1) {
   Wait for an GVT from CP;
   while (1) {
     Remove an event E_i with the earliest LVT_i from the event queue;
     LVT = LVT_i;
     if (LVT > GVT) {
       Insert E_i into an event queue;
       LVT = GVT;
       break;
     }
     Execute the event routine of E_i;
     if (RF transmission occurs by E_i)
       Broadcast the RF message to other LPs;
   }
   Insert an event to process the received message into the event queue;
   Send a completion message to CP;
}
```

(b)

Fig. 2. Optimal-synchronous PDES algorithm (a)CP's operation. (b) LP's operation.

5 Predicting Speedup

In DES, when a new event is inserted into an event queue, searching a proper position in time sequence must be performed. In the case of instruction-level discrete-event simulation, the queue operation time (including times to search and insert an entry into the event queue) is an important factor in predicting speedup because it may be much longer than the time for an event routine execution. For a unit time the total time of sequential simulation is:

$$T_S = T_{queue} + T_{execute}, \tag{1}$$

where T_{queue} is the queue operation time and $T_{execute}$ is the time for an event routine execution. We assume that the number of incoming events to the event queue is the same as the number of outgoing events from the event queue for a unit time. That is, an executed event creates a new event on an average. The queue operation time can be described as follows:

$$T_{queue} = aseN, \tag{2}$$

where a is the access time of the event queue, s is the average number of search operations, e is the average number of events executed by a sensor node, and N is the number of sensor nodes. The total number of events executed during a unit time is eN. PASENS is implemented by using *smpl* [18] adopting sequential searching as a simulation engine. Since the average length of the event-queue is assumed by *kN*, the average number of search operations *s* can be represented as *kN/2*, where *k* is the average number of events generated by a sensor node. Thus, T_{queue} becomes:

$$T_{queue} = \frac{ake}{2} N^2. \tag{3}$$

The time for an event routine execution $T_{execute}$ is *beN*, where *b* is the average execution time of event routines. Therefore, the total time of a sequential simulation is:

$$T_S = \frac{ake}{2} N^2 + beN, \tag{4}$$

and the total time of a parallel simulation can be described as:

$$T_P = \frac{ake}{2} (\frac{N}{P})^2 + be\frac{N}{P} + v, \tag{5}$$

where *P* is the number of LPs and *v* is the time due to synchronization overhead for parallel simulation.

The speedup is the ratio of the sequential simulation time to the parallel simulation time. The following formula presents the predicted speedup (Sp) of an optimal-synchronous PDES.

When the number of sensor nodes *N* is large, each value including *N* is close to 0 because other parameters without *N* are constant values. Therefore, the speedup

$$Sp = \frac{T_S}{T_P} = \frac{\frac{ake}{2} N^2 + beN}{\frac{ake}{2} (\frac{N}{P})^2 + be\frac{N}{P} + v} = \frac{P^2 + \frac{2b}{ak}\frac{P^2}{N}}{1 + \frac{2b}{ak}\frac{P}{N} + \frac{2v}{ake}\frac{P^2}{N^2}}. \tag{6}$$

approaches P^2 and we will observe close to the square speedup. That is, the maximum speedup achievable is the square of the number of LPs participating in a simulation.

6 Experiments

In our experiments, the sensor board model is a Cross-Bow MPR2400 board with MICAz type, which includes a 7.3728 MHz MCU. We assume that all sensor nodes can communicate with each other. Experiments were performed for one virtual second by using 41 PCs interconnected through 100 Mbps Fast Ethernet. One of PCs is used as a CP and others as LPs. Each PC includes a 1.7 GHz Pentium-4 CPU and 256 MB RAM running MS windows XP. We used executable images of CntToRfm and RfmToLeds examples also used for TinyOS [19] 1.1.7 as workloads for simulations. CntToRfm maintains a counter on a 4 Hz timer and sends out the value of the counter in a packet on each increment. RfmToLeds listens to messages. When it receives one, it sets its LEDs to the lowest three bits of the received value. Combined with CntToRfm, this is a simple test of networking. In our experiments, one node runs CntToRfm and other nodes run RfmToLeds. This situation is similar to that of broadcasting from a head sensor node to others periodically. The increment interval of GVT is 352 us.

The first experiment is conducted to verify the effect of workload size on the simulation time in a sequential (one LP) simulation when the number of sensor nodes is in-creased from 100 to 2,000. In WSN simulation, the number of sensor nodes determines the workload size. As shown in Fig. 3(a), when the number of sensor nodes increases, the simulation time becomes much longer in a square relation. For example, the simulation time was 32 seconds for 200 sensor nodes. While, the simulation time is 2,864 seconds for 2,000 sensor nodes: about 89 times slower when compared to 200 sensor nodes. It is due to the longer queue operation time that is caused by the increased number of accesses to the event queue as shown in Fig. 3(b). Such a phenomenon is already predicted by formulas in Section 5.

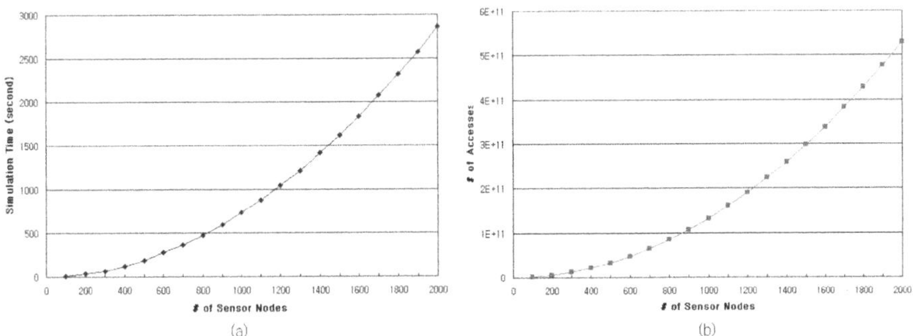

Fig. 3. The simulation results according to the number of sensor nodes in single simulation. (a) Simulation time. (b) The number of accesses to event queue.

The second experiment is performed to measure speedups as the number of LPs is increased from two to ten when the number of sensor nodes is fixed to 2,000. Fig. 4(a) shows, as the number of LPs increases, the simulation times are shortened. For example, the simulation time with two LPs is 750 seconds while that of the sequential simulation is 2,864 seconds. Fig. 4(b) represents corresponding speedups. Results prove that speedups are close to the square of the number of LPs, while speedups of most parallel processing are below the number of processors.

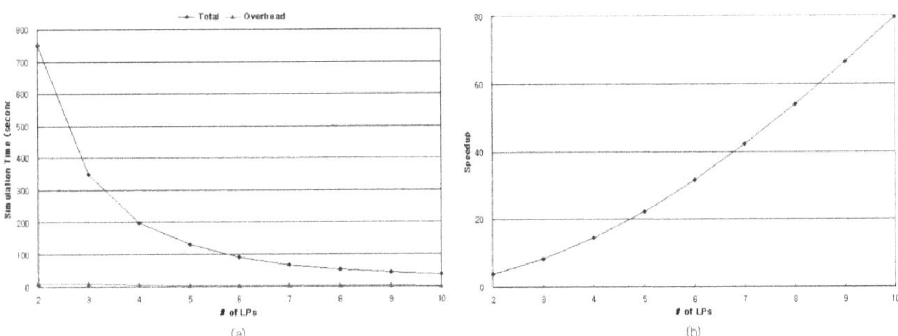

Fig. 4. The simulation results according to the number of sensor nodes in parallel simulation. (a) Simulation time. (b) Speedup.

We carry out the third experiment to verify the speedup formula in Section 5. The formula predicts that the speedup is close to the square of the number of LPs as the number of sensor nodes increases. In this experiment, speedups are measured when the number of sensor nodes is increased from 100 to 2,000 and the number of LPs is two or ten. Fig. 5 shows that the speedup approaches the square of the number of LPs. The maximum speedup achieved with two LPs is about four times. When the number of LPs is ten, the speedup approaches 100 times. This confirms our formula in predicting the maximum speedup achievable, which is the square of the number of LPs.

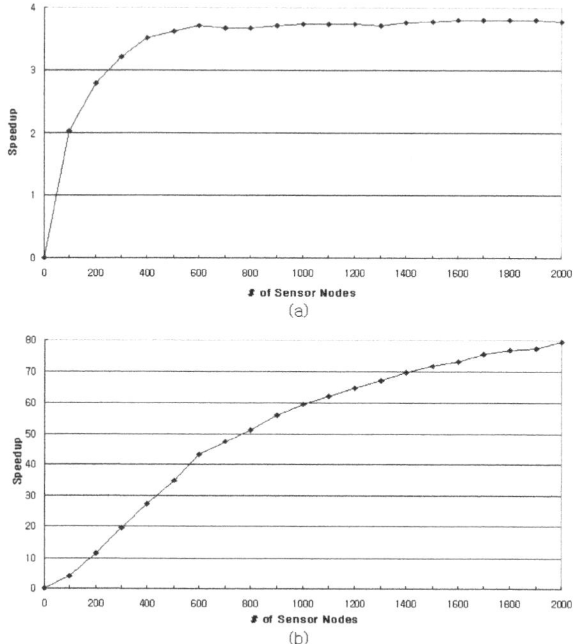

Fig. 5. The speedup approaching the square of the number of LPs. (a) The number of LPs = 2. (b) The number of LPs = 10.

We conducted the next experiment to check the scalability of the number of LPs with a fixed workload. The numbers of sensor nodes in each LP are fixed to 50, 250, and 500. The experiments are progressed while increasing the number of LPs from 5 to 40. In this case, the maximum number of simulated sensor nodes is 20,000. Although the number of LPs is increased, the simulation time is not increased as shown in Fig. 6. The reason for this result is that the overhead by the increment of LPs is too small to affect on the simulation time. Therefore, if the number of LPs is enough, PASENS can simulate the infinite number of sensor nodes.

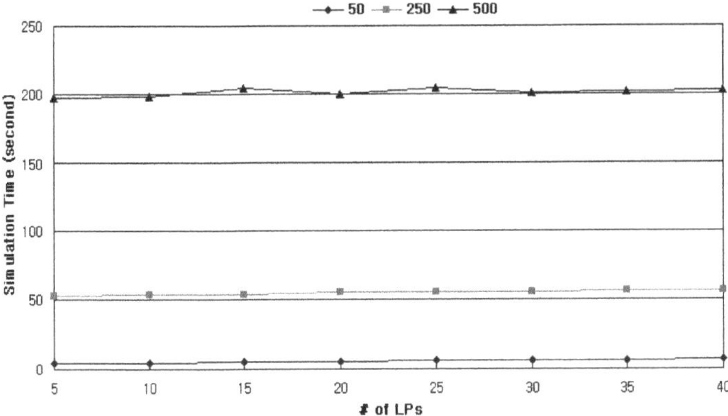

Fig. 6. The scalability of the number of LPs

7 Conclusion

We have developed a parallel simulator for large-scale wireless sensor network using instruction-level discrete-event simulation method and optimal-synchronous parallel discrete-event simulation method. The speedup formula derived from our simulation model shows that the speedup approaches the square of the number of LPs as the number of sensor nodes increases. The results of our experiments proved validity of the formula in predicting the speedup achievable.

Instruction-level simulation is useful to verify the activities of sensor nodes and to estimate program execution time and power consumption. The simulator based on optimal-synchronous PDES method can easily and cheaply be implemented and provide high speedup and scalability. If the number of LPs is enough, there is no limit to the number of sensor nodes to be simulated. In this paper, we simulated maximum 20,000 sensor nodes. The executable files and sources programs of PASENS are available from http://user.chol.com/~legnamai/pasens/.

This simulation method also can be applied to complex simulations, such as computational fluid dynamics or walking algorithm simulation, for a multi-legged walking subsea robot because their characteristics are similar to large-scale WAN.

Acknowledgement. The authors would like to thank the support by Ministry of Land, Transportation and Maritime Affairs (MLTM) of Korea for the "Development of a Multi-legged Walking Flying Subsea Robot".

References

1. Kim, B.H., Kim, T.K., Jung, Y.D., Kim, J.H.: Development of Sensor Network Simulator for Estimating Power Consumption and Execution Time. Journal of The Korea Society for Simulation, 35–42 (2006)

2. Ferscha, A.: Parallel and Distributed Simulation of Discrete Event Systems. In: Handbook of Parallel and Distributed Computing. McGraw-Hill (1995)
3. Peacock, J.K., Wong, J.W., Manning, E.G.: Distributed Simulation using a Network of Processors. Computer Networks, 44–56 (1979)
4. Xu, J., Chung, M.J.: Predicting the Performance of Synchronous Discrete Event Simulation. IEEE Transaction on Parallel and Distributed Systems, 1130–1137 (2004)
5. Xu, J., Zhang, J.: Efficiently Unifying Parallel Simulation Techniques. In: Proceedings of the 44th Annual Southeast Regional Conference ACM-SE 2006 (2006)
6. Chandy, K.M., Misra, J.: Distributed Simulation: A Case Study in Design and Verification of Distributed Programs. IEEE Transactions on Software Engineering, 440–452 (1979)
7. Bryant, R.E.: A Switch-Level Model and Simulator for MOS Digital Systems. IEEE Transactions on Computers, 160–177 (1984)
8. Jeerson, D.A.: Virtual Time. ACM Transactions on Programming Languages and Systems, 404–425 (1985)
9. Zeng, X., Bagrodia, R., Gerla, M.: GloMoSim: A Library for Parallel Simulation of Large-scale Wireless Networks. In: Proceedings of the 12th Workshop on Parallel and Distributed Simulation (1998)
10. Scalable Network Technologies, Qualnet, `http://www.scalable-networks.com/`
11. Hill, J., Szewczyk, R., Woo, A., Hollar, S., Culler, D.E., Pister, K.S.: System Architecture Directions for Networked Sensors. In: Proceedings of International Conference on Architectural Support for Programming Languages and Operating Systems (2000)
12. Kelly IV, C., Manohar, R.: An Event-Synchronization Protocol for Parallel Simulation of Large-Scale Wireless Networks. In: Seventh IEEE International Symposium on Distributed Simulation and Real Time Applications (2003)
13. Bagrodia, R., Meyerr, R.: PARSEC: A Parallel Simulation Environment for Complex System. UCLA Technical Report (1997)
14. CrossBow: MPR/MIB Users Manual (2005)
15. Ocracomm, `http://www.octacommm.net/`
16. Atmel: ATmega128(L) Complete (2006)
17. Chipcon: SmartRF CC2420 Preliminary Datasheet 1.2 (2004)
18. MacDougall, M.H.: Simulating Computer System: Techniques and Tools. MIT Press (1987)
19. Simon, G., Volgyesi, P., Maroti, M., Ledeczi, A.: Simulation-based Optimization of Communication Protocols for Large-scale Wireless Sensor Networks. In: Proceedings of the IEEE Aerospace Conference (2003)

A Study on Measurements System for Evaluating DA-C2's Effectiveness

Zhao Fang[1,2], Zhao Xinjun[2], and Lu Xiaojun[3]

[1] College of Automation Science and Electrical Engineering,
Beijing University of Aeronautics and Astronautics, Beijing 100083, China
zhaofang219@163.com
[2] Unit 63963 PLA, Beijing 100072, China
[3] College of Information Systems and Management,
National University of Defense Technology, Changsha, 410073, China

Abstract. Effectiveness evaluation of the command & control system (C2, central to C4ISR) directly influences the success of system building and development. Critical to the evaluation is the selection of a proper measurements system, which can also provide data support for the optimization of DA's C2 system. Based on the idea of system model decomposition and the principle of hierarchical modeling to deal with the characteristics of DA-C2, this paper built a measurements system for evaluating DA-C2's effectiveness on each tier's measurements structure, definition and evaluation approaches. It also analyzes and compares the various approaches for acquiring the evaluation measurements.

Keywords: C2, measurements system.

1 Introduction

With the increasingly big role the C4ISR system is playing in the modern war, the issue of effectiveness evaluation becomes increasingly important to the extent of directly influencing the whole process of requirement demonstration, design, development and application. The quality of the advance evaluation will critically influence the resultant system's being put into combat and other applications. As C2 is the core of C4ISR, we should establish a proper and complete measurements system for evaluating C2's effectiveness, so that it can be used to conduct comprehensive and in-depth analyses of the mass data of simulation results acquired from the DA-C2's simulation, for the purpose of evaluating the DA-C2's effectiveness and providing data support for optimizing the DA-C2 system.

2 Building C2 Measurements System

The DA C2 effectiveness-measurements system is established on the idea of system-model decomposition and the principle of hierarchical modeling, and on the

J.-H. Kim et al. (Eds.): AsiaSim2011, PICT 4, pp. 25–31, 2012.

basis of an analysis of such factors as DA establishment plans, simulation scenario, and composition of army units according to the features of the DA architecture and its combat.

2.1 Combat System's Architecture

A DA's operational system architecture is a 3-tier structure (see Fig. 1): the first tier is for combined commanding and decision making, and it is also the tier for IT-based combined-forces combat's command decision and action control, physically matching the combined commanding mechanism, which commands its combat units through its command sites. The second tier is for troop-specific commanding, physically matching the special-troops' regiment / battalion / group commanding mechanisms. Each troop's commanding mechanism commands its own troop's combat units via its own commanding seats. The third tier is for combat execution, physically matching the company-level commanding mechanisms, individual combatants, and weapon units. Through the C2 unit, the combat units realize direct commanding and action control of the weapon platforms in accordance with the troop-specific weapon platforms' combat application characteristics.

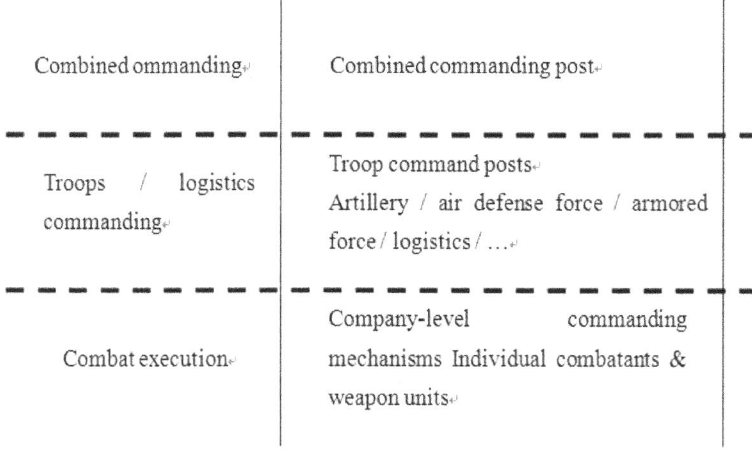

Fig. 1. DA's 3-tier operational command-system architecture

2.2 Identifying Principles for Measurements System

(1) The principle of objectivity. The chosen measurements should be able to objectively reflect the changes of C2 system's internal states in such a manner as to precisely link problems under research to system-related uncertainties, so as to avoid deviations caused by too many subjective factors involved.

(2) The principle of scientificalness. The evaluation measurements are scientific in the following two ways: ① With corresponding scientific justifications and support

by scientific rationales, the chosen measurements as well as the evaluation mode and process can reflect the reality of the C2 system; ② The evaluation measurements system can cover all the major factors affecting the C2 system. The scientificalness of the C2 effectiveness' evaluation measurements, in a word, is determined by the rationality of the measurements and the integrity of the measurements system.

(3) The principle of flexibility. System effectiveness should be able to have change corresponding to the change of the system's measurement parameters. If the change of measurement parameters in the system's measurements structure leads to no or little change of the system's effectiveness value, then such measurements are useless to effectiveness evaluation of the entire C2 system.

(4) The principle of operability. The evaluation measurements are operable in the following two ways: ① The measurements should be quantifiable, or at lease convenient for qualitative judgment; ② Efforts should be made for the realization of evaluation-measurements targets to be operable in such a manner as to break down, tier by tier, the target process into managerial organizations' units, even into individuals.

(5) The principle of the biggest simplicity. As long as virtually meeting evaluation demands and combat needs, the evaluation should be conducted by choosing critical measurements as reasonably few as possible. It should be noted that in effectiveness evaluation, the number of measurements is not "the greater, the better." What matters is whether the measurements have played a big role in evaluating the system's effectiveness. Too many measurements would not only increase the complexity of the evaluation results, but may also affect the objectivity of the results. Therefore, we should screen out those measurements with little influence on the evaluation targets and focus our attention on those measurements that reflect the system's essential characteristics.

2.3 Building the Measurements System

The measurements system should be a complete set of measurements that fully reflects the C2 system's characteristics based on the hierarchicality of the system of systems' operational system architecture, on the DA system-of-systems C2 effectiveness evaluation's evaluation targets and evaluation needs, and on system effectiveness evaluation's scientificalness and completeness. A particular effectiveness measurement can only reflect a particular aspect of the evaluation object; only the measurements system can reflect the whole picture of the object and form a complete and objective perception of the C2 system, so as to essentially ensure the evaluation's scientificalness.

The measurements system for DA C2's effectiveness evaluation is shown in Fig. 2, comprised of level-1 measurements and groups of level-2 measurements:

(1) Situation sharing effectiveness

The situation sharing effectiveness mainly includes the enemy situation updating time and our positioning information updating time.

- The command-post enemy-situation updating period is the time difference for the change between two enemy situations (unit: second), and the situation change refers to the process in which there are changes in enemy situations facing the combat units one or two levels under this command unit, the combat units processes the information and report upward, and this command unit receives the information and updates the situations.
- The command post's own positioning information updating period refers to the time difference for the change of positioning information between two times (unit: second), and the positioning-information change is the process in which the combat units one or two levels under this command unit report their positioning information to this command unit, and the latter receives the information and updates the situations.

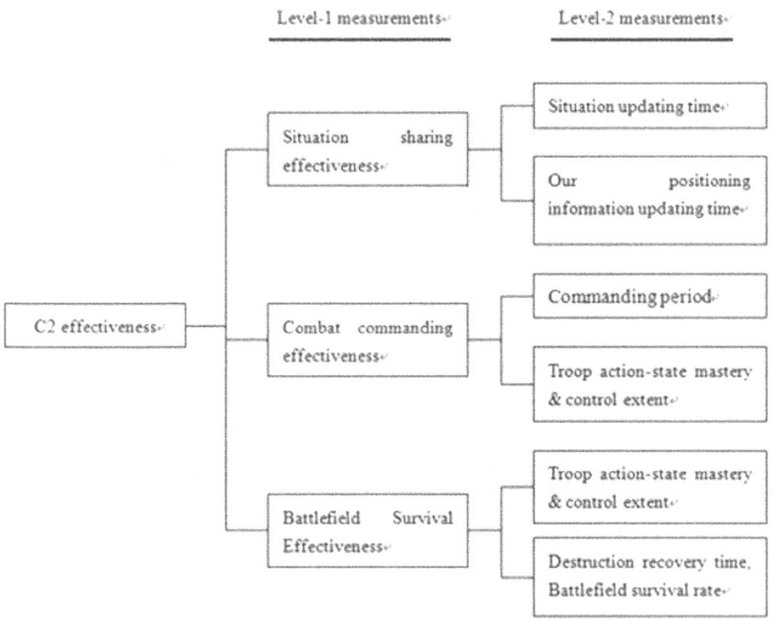

Fig. 2. DA C2 effectiveness-evaluation's measurements system

(2) Combat commanding effectiveness

The combat commanding effectiveness includes the commanding period and the troop actions' mastery and control extent.

- The commanding period refers to the whole time from the command post's receiving a command instruction from a higher-level commander, through finishing the processing of the instruction, to the lower-level units' receiving its command (unit: minute) ;

- The troop actions' mastery and control extent concerns the number of information to the command post from its lower-level combat units: It is the number actually received out of the number the lower-level units should have sent (unit: 1).

(3) Battlefield Survival Effectiveness

The battlefield survival effectiveness mainly includes the destruction recovery time and the battlefield survival rate:

- The destruction recovery time is the time between a command post's destruction and its re-opening (unit: minute);
- The battlefield survival rate concerns the number of our army's combat units. It refers to its survival number out of its total number (unit: 1).

3 Approaches to Acquiring Measurements

The C2 system's effectiveness measurements are mainly acquired through the Expert Assessment Approach, the Test Statistics Approach, the Analytical Approach and the Combat Simulation Approach, each covering specific application areas with pros and cons.

(1)Expert Assessment Approach
To evaluate some measurements that cannot be quantitatively described, this approach is resorted to, generally by selecting those characteristic measurements that can best reflect system effectiveness for experts to score, and then process the scores to get system effectiveness. Usually involved are such methods as the Scoring Method, the Grading Method, the Weighted Scoring Method and the Optimum Order Method, which are relatively simple and popular. The Expert Assessment Approach is relatively effective in assessing unquantifiable measurements, two of its challenges being how to select experts and what measurements to choose for experts to evaluate. The drawback of this approach is big subjectivity: in evaluation, experts tend to show much personal propensity.

(2) Test Statistics Approach
The Test Statistics Approach is applied in designated sites or accurately simulated environments, where researchers observe performance characteristics of command automation systems and collect data, so as to evaluate system effectiveness. This approach is characterized by acquiring plenty of statistics from real combats, exercises/maneuvers and/or tests to assess the effectiveness measurements, on condition that the statistics' stochastic characteristics can clearly be modeled and correspondingly utilized. Usually involving such methods as sampling survey, parameter estimation, hypothesis testing, regression analysis and correlation analysis, the statistics-based evaluation approach can not only give effectiveness measurements' evaluated values, but also show the influence of such factors as command automation system's performance and combat rules on effectiveness measurements, providing the basis of quantitative analysis for the improvement of command automation system's

performance and combat rules. The results obtained are relatively accurate, but the approach needs to use a great number of command automation systems as the physical basis of the tests, a task that is relatively difficult in the availability of command automation systems, too costly in testing, and time consuming.

(3) Analytical Approach
The Analytical Approach is the computation of measurements according to the analytical expression showing the functional relations between effectiveness measurements and the given conditions. This approach is strong in that the formula is transparent enough to understand, that the computation is relatively simple, and that relations among the variables can be analyzed, a plus for its application. It is weak in that the factors considered are few, and that it is only effective under assumed conditions strictly set. It is fit, therefore, for the acquisition of macro effectiveness measurements in simplified conditions.

(4) Combat Simulation Approach
Also called the Combat Mimicking Approach, the Combat Simulation Approach essentially conducts the experiment of combat simulation based on the computer simulation model. From the experimental data of the combat process and results, the assessed values of the effectiveness measurements can be given, directly or through statistical processing. Conducting the evaluation in the antagonistic conditions against the background of specific combat environments and certain compositions of troops, the Combat Simulation Approach is vivid enough to demonstrate the combat process, but needs to be supported by plenty of reliable basic data and original materials, the acquisition of complete information depending on collecting tons of data according to long-term plans. The approach tends to be challenged by the simulation of the combat environment: factors such as environment-interfering uncertainties will directly affect the results. Generally speaking, however, combat simulation is indispensable for effectiveness evaluation of the command automation system. With such advantages as short in time and low in costs, this approach can, in a way, reflect antagonistic conditions and combat objects, taking into consideration the influence on the entire process of combat by the cooperation among the sub-systems of the command automation system, and by the attributes of the command automation system, and particularly fit for predicting and evaluating the effectiveness measurements of the command automation system or the combat scheme.

This study acquired measurements mainly through the Expert Assessment Approach, complemented by the Test Statistics Approach and the Analytical Approach. Examining a DA's C2 process covering intelligence, planning, combat control and cooperation, this study identified the C2 system's effectiveness measurements, and use them to evaluate the effectiveness of a certain DA to be tested, by having the measurements assessed with the five grades of "very good", "good", "not bad", "bad" and "very bad", to which 1 to 5 points were assigned, respectively. The reviewing panelists' scored measurements of the DA to be tested were translated, through aggregation, into the troop's construction level of digitalization as well as its advantages and disadvantages.

4 Summary

Regarding DA C2's effectiveness evaluation, the establishment of its measurements system is critical for its success. The DA C2 effectiveness-measurements system is established in this study on the idea of system-model decomposition and the principle of hierarchical modeling, and on the basis of an analysis of such factors as DA establishment plans, simulation scenario, and composition of army units according to the features of the DA C2 system. It has been proved through practice that this measurements system is feasible for evaluating DA C2's effectiveness.

References

1. Zhu, M.-J., Zhang, W.: Modeling and Simulation on Work flow of Combat Command and Control. Fire Control and Command Control 33(12) (December 2008)
2. Li, P., Liu, S.-F., Zhang, X.-J.: A Distributed Command and Control Simulation System Framework Based on Design Pattern. Journal of Ji Lin University (Science Edition)
3. Lu, H.-T., Fu, H., Hou, D.-L.: Study on Simulation of Command & Control System for command Tank. Fire Control and Command Control 34(7) (July 2009)

Modeling and Simulation of the Acoustics of a Steelpan, a Percussion Instrument

Yukio Kagawa[1], Kohei Tanaka[2], Ken Yamazaki[3], and Lingyun Chai[4]

[1] Okayama Univ., Okayama, Japan
y2kagawa@gmail.com
[2] Nihon Univ. College of Industrial Technology, Chiba, Japan
[3] Nihon University College of Industrial Technology, Chiba, Japan
[4] Solar Frontier, Kanagawa, Japan

Abstract. The paper discusses the acoustics and vibration of a steelpan. It is a shallow curved circular plate on which many spherical shells are formed. Each shell sounds a musical scale frequency. Each shell vibrates almost independently. Shell's vibration is numerically analyzed for guiding the design of the instrument.

1 Introduction

In the era of the electronics technology, the introduction of new acoustic instruments is not so frequent.

The steelpan, a kind of the percussion instruments, is a very rare example, which can play not only rhythm but also melody. It was originally developed in the Carribian Islands for necessity.

It was first made from a part of the oil drum cask for Carribian folk music and then evolved. As it has not only beautiful clear sound but also attractive and mild tones, the instrument becomes increasingly popular, which is now played from popular music to classic music. Examples of the performances can be viewed through the Internet YouTube, in which it is also interesting to compare with the sound of a glockenspiel. The glockenspiel and the steelpan are both made of a steel plate which is played hummered with mallets. The glockenspiel is a metallophone which consists of lined steel plate bars, in which each corresponds to each musical scale, and each bar vibrates in bending, whose resonant frequency depends on the length and thickness. On the other hand, the steelpan is a slightly concaved circular plate, on which multiple smaller shallow spherical shells are formed. The plate has the multiple modes of vibrations, in which each inner shallow spherical shell corresponds to each musical scale. The inner shells' regions vibrate almost independently with week coupling each other through their edges. The physics of many kinds of musical instruments have been studied [1], in which however, the steelpan is excluded. Present paper is to reveal the acoustics of the steelpan by investigating the vibrations with experimental modal analysis and the mathematical model of the spherical shells. The work is undertaken to search the steelpan sound mechanism and to guide for the design and

J.-H. Kim et al. (Eds.): AsiaSim2011, PICT 4, pp. 32–40, 2012.

production. Until now, the instrument has been made by craftsmanship. The experimental and finite element modal analysis has been reported in reference [2]. We are rather concerned with the spectral analysis and reverberation of the sound, and the eigenvalues of the shallow spherical shells for the musical scale.

2 Fourier Analysis

First we investigate the frequency spectrum of the sound for each musical scale when each scale shell is hammered by the mallets. Our instrument under test is a tenor steelpan produced by SONOBE Studio Co., whose front surface and cross-section are shown in Fig.1 (a) and (b).

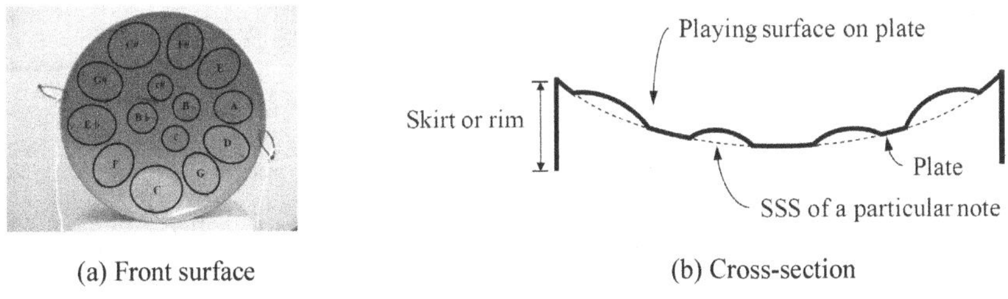

(a) Front surface (b) Cross-section

Fig. 1. A steelpan under test(Diameter:35cm)

The steelpan has the edge rim or side skirt which helps separating the sounds emitted from the both surfaces of the plate. These two sounds are out of phase. Fig 2 shows the experimental set-up, in which PULSE is a platform for FFT analysis by B&K.[3]. The sounds emitted from the steelpan and the analyzed results are transmitted to the computer.

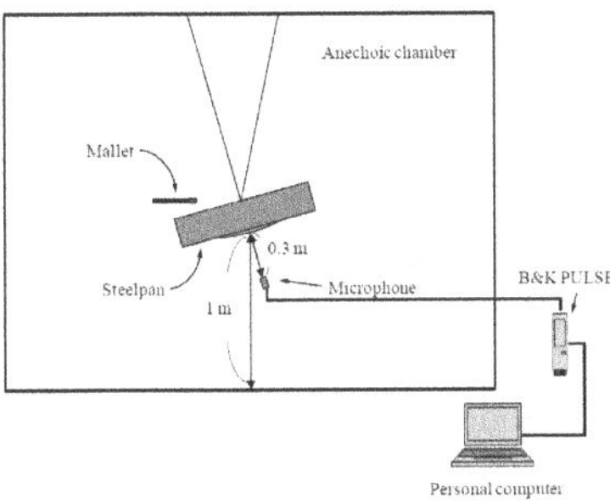

Fig. 2. Experimental set-up

Fig.3 shows an example of the decay sound waveform and the frequency spectra corresponding to the note E. As shown in Fig.(b), the lowest dominant frequency 664Hz corresponds to the frequency of the note E, and there appear many overtone (or higher harmonics) spectra equally spaced. The abscissa is given in terms of a log scale. The decay of the sound shown in Fig.(a) can be characterized by the reverberation time as defined in the room acoustics. The definition is the time required so that the sound pressure diminishes to the one-thousandth (-60dB). If the ordinate is given in terms of a log scale, the reverberation time is easily calculated from the inclination of the delay line. Fig.4 shows the reverberation times. The values are about the same for all notes.

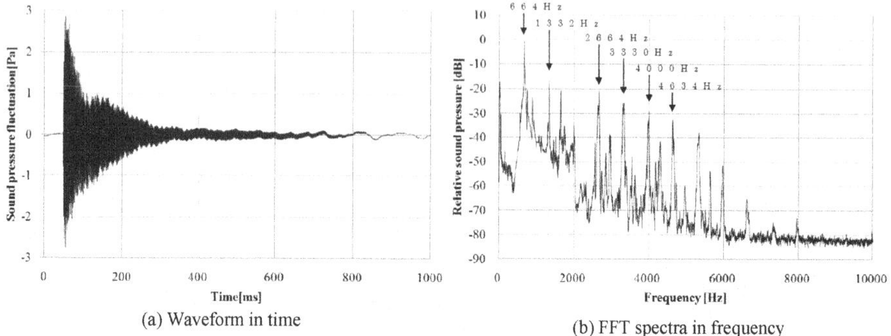

(a) Waveform in time

(b) FFT spectra in frequency

Fig. 3. Sound radiated when E shell is hammered by the mallets

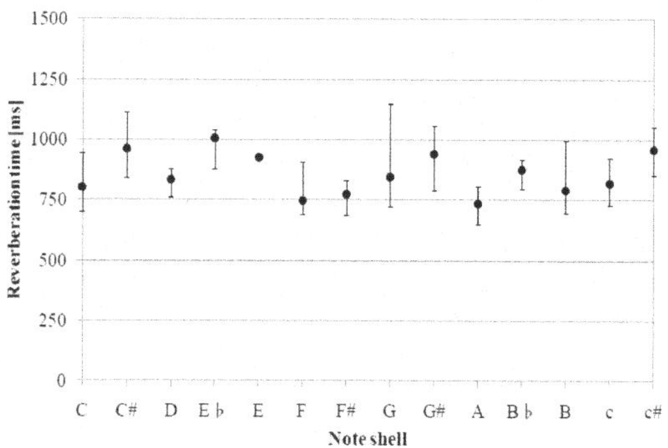

Fig. 4. Reverberation time of each note

3 Modal Models

Modal models identify the modes of vibrations through the modal analysis. Fig.5 shows the experimental set-up for the measurement of the transmission characteristics between arbitrary two points over the steelpan, from which modal models are made. The procedure has been well established and practiced as a standard [4]. The experiments are carried out with B&K measuring apparatuses. The deriving ports (hammering points) are taken within the note shell and the receiving points are placed outside of the shell to avoid the effect of the sensor's weight.

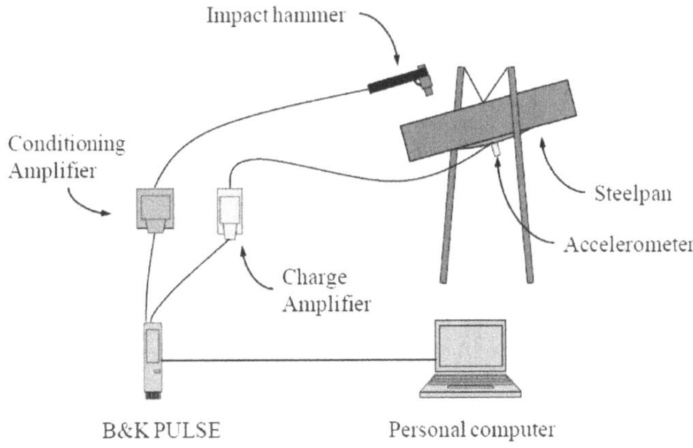

Fig. 5. Experimental set-up for the experimental modal analysis

Fig.6 shows note shells and the driving points. Fig.7 shows an example of the transfer function which shows many peaks and dips. The lowest peak frequency corresponds to the note frequency of the corresponding shell. Form these data, eigenmodes can be deduced. Fig.8 shows the examples of the eigenmodes for C and C# note shells. Each mode is localized for each note shell. The lateral displacement is mainly as the result of the extensional vibration of the shell which contributes to the volume displacement. The couplings between the shells look very small. There characteristic is similar for other modes of vibration. Modal localization agrees with the reference (2)'s result.

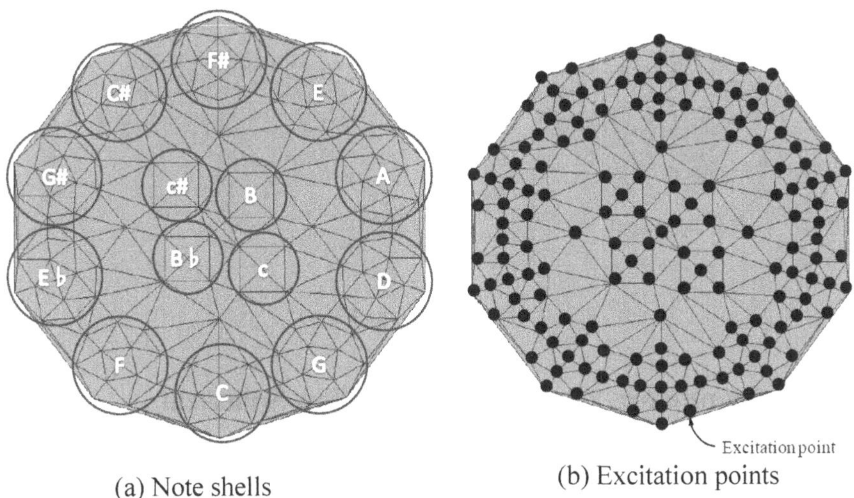

Fig. 6. Steelpan's top views

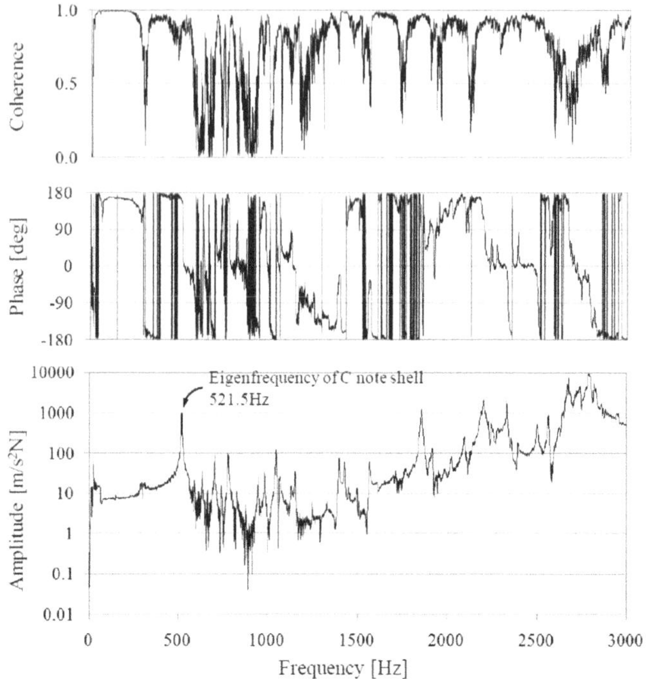

Fig. 7. Transfer frequency response function

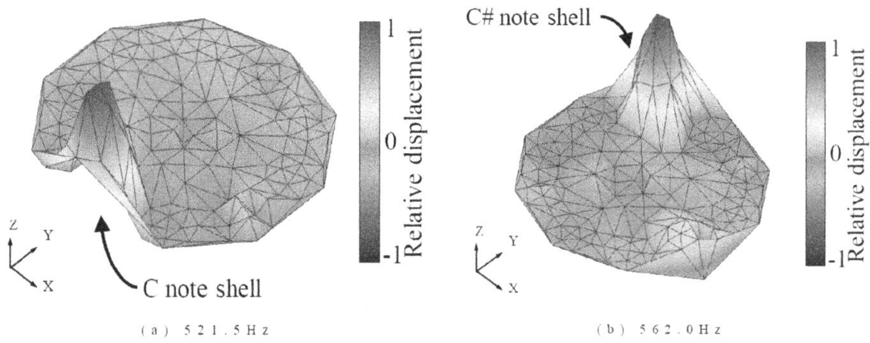

Fig. 8. Eigenmodes

4 Axisymmetric Vibrations of a Shallow Spherical Shell-Numerical Simulation

In the previous section, the modal analysis showed that the modes of vibrations are local and almost independent. Here we examine the vibration of each note shell. It can be modeled as a shallow spherical shell. We investigate the contribution of the shell's physical dimensions to the natural frequency. In the real situation, the craftsmen form the shape of the shell by their experiences. We consider its extensional vibration, which causes the lateral displacement and contributes to the total volume displacement.

4.1 Equations of Motion

We consider a shallow spherical shell as shown in Fig 9. The governing equations for the extensional vibration are given as

$$\frac{d^2u}{d\theta^2}+\cot\theta\frac{du}{d\theta}-\frac{dw}{d\theta}+(K^2-\cot^2\theta)u=0 \tag{1}$$

$$\frac{du}{d\theta}+u\cot\theta+(K^2-2)w=0 \tag{2}$$

$$K=\frac{2\pi R}{c_p}f \tag{3}$$

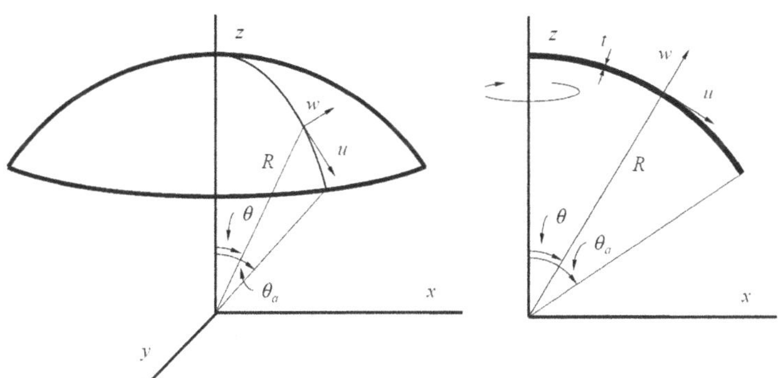

Fig. 9. A spherical shell

where u and w are the extensional and lateral displacements on the neutral surface of the shell. Other definition can be given in the Figure. These equations can be easily derived from the Love's equations of motion for the shell of general coordinates [5]. The flexural vibrations are ignored assuming that the shell is very thin. K is the wave number to be determined as the eigenvalue. f is the eigenfrequency and $c_p=\sqrt{(1-\sigma^2)\rho/E}$ is the sound speed for the longitudinal wave.

4.2 Boundary Conditions

At the top of the shell, the extensional displacement must be zero so that

$$u=0 \text{ at } \theta=0 \tag{4}$$

There are two possibilities to be considered at the edge.

(a) Assuming that the edge is fixed for the extension, $u=0$ at $\theta=\theta_a$ (5-a)

(b) Assuming that the edge is free, $\dfrac{du}{d\theta}-w=0$ at $\theta=\theta_a$ (5-b)

The fixed edge must be closer to the real situation.

4.3 Numerical Solution

The simplest way to solve numerically is to introduce the finite difference method [6]. Equations (1) and (2) are dicretized with the central difference scheme for the step h, so that

$$-\left(\frac{1}{h^2}-\frac{\cot\theta_n}{2h}\right)u_{n-1}+\left(\frac{2}{h^2}+\cot^2\theta_n\right)u_n-\left(\frac{1}{h^2}+\frac{\cot\theta_n}{2h}\right)u_{n+1}$$
$$-\frac{1}{2h}w_{n-1}+\frac{1}{2h}w_{n+1}=K^2u_n \tag{6}$$

$$\frac{1}{2h}u_{n-1}-u_n\cot\theta_n-\frac{1}{2h}u_{n+1}+2w_n=K^2w_n \tag{7}$$

The boundary conditions are

$$u_0=0 \text{ at } n=0 \tag{8}$$

(a) For the fixed boundary at the edge

$$u_N=0 \quad n=N \tag{9-a}$$

(b) For the free boundary at the edge,

$$-\frac{1}{h}u_{N-1}+\frac{1}{h}u_N-w_N=0 \tag{9-b}$$

where $h=\theta_a/N$ and $\theta_n=nh$ $(0\le n\le N)$ and u_n and w_n are the displacements at the node n, as the shells curvature is divided into N segments with step h and the each node is numbered with n from 0 to N. There are therefore $N+1$ node points.

We have $2(N+1)\times 2(N+1)$ linear algribraic equations for $N+1$ nodal points. In those days, the equations can be easily solved as an eigenvalue problem by Eigensystem, Mathematica[7].

4.4 Eigenmodes and Corresponding Eigenvalues

We here show an example of the solution for the C note shell. For the calculation, N=1000 for $\theta_a=0.02[rad]$. Fig 10 shows the lower three modes when the edge is assumed "fixed". At the tip of the shell, shell is laterally deflected. It is interesting that eigenfrequencies are very close.

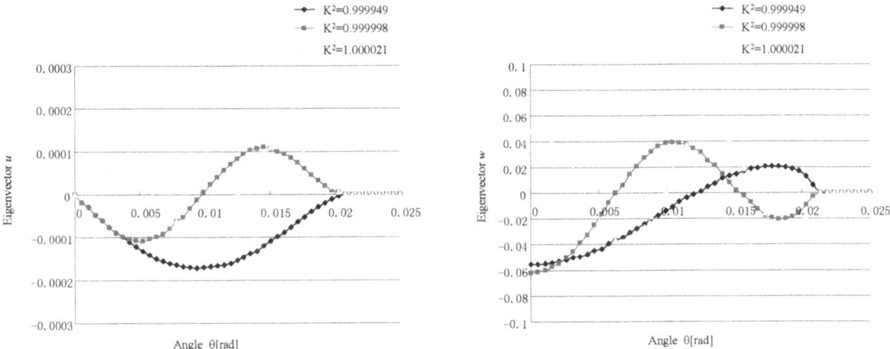

Fig. 10. Eigenvector u and w for fixed boundary at the edge

Fig.11 shows the eigenvalue against the meridional length θ_a. The change of the eigenvalue is small and independent of the boundary condition at the edge for small θ_a. The natural frequency is thus inversely proportional to the radius of the curvature R. The R is measured for the several directions in the shell diameter and the averaged value is used for the calculation. The comparisons are given in Table 1 for the each scale shell. The frequency here is the frequency of the note. The frequencies of the sound and the shell's vibration are very close, which shows that the airloading to the vibration is very light. The calculated eigenvalues can predict very well for all the notes.

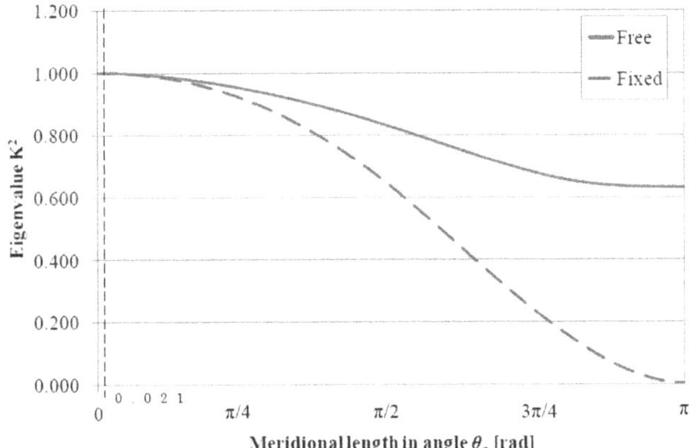

Fig. 11. Eigenvalue

Table 1. Frequency of the fundamental mode, measured and calculated

Note	Frequency [Hz]			
	Reference	Acoustic	Modal	Calculated
C	523.3	522.0	521.5	519.7
C#	554.4	558.0	562.0	531.3
D	587.3	590.0	592.5	586.7
E b	622.3	634.0	635.5	638.3
E	659.3	664.0	665.5	649.0
F	698.5	702.0	706.0	708.0
F#	740.0	742.0	745.5	796.5
G	784.0	784.0	783.0	715.5
G#	830.6	832.0	831.5	861.2
A	880.0	886.0	883.5	837.6
B b	932.3	932.0	950.5	900.3
B	987.8	986.0	986.0	963.9
c	1046.5	1064.0	1067.5	1010.1
c#	1108.7	1118.0	1131.5	1101.6

5 Concluding Remarks

Here we discussed the physics of the steelpan, a percussion instrument for the sound and vibrations. It is quite interesting that in spite of the fact that the instrument is naturally developed among common people, sheelpans are well designed so that the modes of vibrations are localized and each scale region vibrates almost independently corresponding to the musical scale frequency. We modeled each scale region as a shallow spherical shell, whose eigenfrequency is calculated for its axisymmetric vibrations. In the spherical and conical shells, the extensional vibrations play an important role for volume displacement modes. We revealed the relation between the shell's dimension and the natural frequency. This can be a good guide for the steelpan design.

References

1. Flecher, N.H., Rossing, T.D.: The physics of Musical Instruments. Springer (1981)
2. Iwatsuki, N., Morikawa, K., Ito, A.: A research on musical scale generating mechanism of a steel drum. In: Japan Society of Mechanical Egineers, Dynamics and Design Conference No. 08-14, Yokohama (September 2-5, 2008)
3. Guide Book for PULSE operation. Bruel & Kjaer (2005)
4. Nagamatsu, A.: Introduction to Mode Analysis. Carona-sha, Tokyo (2009)
5. Love, A.E.H.: The Mathematical Theory of Elasticity. Cambridge Univ. Press (1934)
6. Kagawa, Y., Yamabuchi, T., Kato, S.: On the axisymmertric vibration of conical shells and their natural frequencies. J. Acoutic. Soc. Japan 37(10) (1981)
7. Mathematics & Documentation. Wolform Research Inc. (2011)

Derivation of 2-D Flow Lines from Water Table Data in Heterogeneous Aquifer by Developing Specific Finite Difference Method Programs

Samuel Sangwon Lee

Federal Energy Regulatory Commission,
San Francisco Regional Office,
100 First Street, Suite 2300,
San Francisco, CA 94105
Samuel.Lee@ferc.gov

Abstract. This study makes it possible to draw conclusions about the configuration of regional groundwater systems by using a numerical model. The model uses finite difference numerical method commonly used to investigate saturated groundwater flow system. A two-dimensional (2-D) vertical cross-section is used in the model because it is useful to conceptualize the flow system, determine reasonable ranges of aquifer parameters, assets model boundaries, and to determine the most influential parameters in the system. Details of steady-state flow in regional groundwater basins can be investigated using specific numerical model simulations of finite difference method (FDM). The developed programs were run to evaluate the factors that control the interaction of bay and groundwater. The study concerns only with bay encircled by water table mounds that are at a higher altitude in vertical section show that for typical hydrogelogic settings, the line (divide) separating the local from the regional groundwater flow systems in continuous beneath bay. Bay bottom sediments are considered this study flow net. Two cases of aquifer hydraulic conductivity (K_{a1} = 100 ft/day, K_{a2} = 50 ft/day) and bay sediments hydraulic conductivity (K_b = 5 ft/day) are assumed in this study.

Keywords: Modeling Methodology, Numerical Simulation, General Engineering, Groundwater).

1 Introduction

The objectives of this study were: (1) to calculate the potential lines using water table potential values from field measurements, (2) to calculate the flow rate and flow lines from the potential lines obtained above, (3) to perform the above calculations assuming heterogeneity in hydraulic conductivity for bay sediments to simulate a coastal area, and (4) to develop original programs to perform the above tasks if existing programs are not appropriate.

A vertical 2-D steady state model for particular regional groundwater system involving a bay basin and coastal area was developed by using the finite differences method (FDM). This model is expressed by a set of computer codes for simulating a

J.-H. Kim et al. (Eds.): AsiaSim2011, PICT 4, pp. 41–49, 2012.

saturated groundwater flow system where the water table level is above the mean sea level (MSL). A 2-D finite difference approximation to Laplace's equation was employed for calculating the potential lines, flow rate, and flow lines. The model can be applied to any aquifer system including in regional and local basin.

Several computer codes (*e.g.*, MODFLOW, FLOWPATH) are available which can incorporate 2- or 3-D groundwater systems using detailed field-measured aquifer properties [1],[2],[3]. Because none of these programs was found appropriate. Some original programs have been simply designed to fulfill the objectives by using only water table values (potential values at MSL) and hydraulic conductivity of each pixel.

2 Description of the Problem and Assumptions

Figure 1 represents a cross section through a regional watershed bounded on one side by a topographic high and on the other side by a MSL. This aquifer is assumed to be isotropic in a steady state. The left and right groundwater boundaries assume to represent groundwater divide and thus to be impermeable, no-flow boundaries.

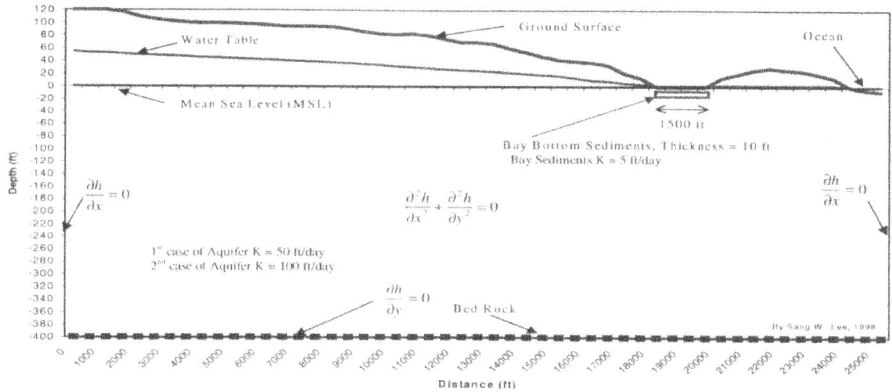

Fig. 1. Vertical section view and boundary conditions of 2-D regional aquifer area

A no-flow boundary also can be used to approximate the freshwater-saltwater interface in right side coastal aquifer. Groundwater in costal aquifers discharges to the ocean through a zone of dispersion that forms the saltwater interface. Rigorous treatment of the saltwater interface requires a model that allows for density effects as well as diffusion and dispersion of saltwater [4].

The lower boundary is also a no-flow boundary because the impermeable bedrock forms a physical barrier to the flow. The upper boundary of the mathematical model is the horizontal line at the MSL even though the water table of the physical system lies above the MSL. Thus the effect of aquifer the MSL is simulated through the measured water table values which are regarded as the potential values at the MSL.

The 2-D model domain is shown in Figure 1. Bay sediments section is located in a region where x = 18,000 to 19,500 ft and y = -10 to 0 ft. In this region, a different hydraulic conductivity from the rest of the aquifer may be assumed.

3 Theoretical Background

Toth (1962, 1963) found that the numerical model is a realistic representation of the general configuration of the flow system where the topography is subdued and the water table slope is gentle [5],[6]. Toth (1962, 1963) also used a more general expression for the configuration of the water table in a region of gently rolling topography.

3.1 Governing Equation

The 2-D generalization of Darcy's law requires that the 1-D form be true for each of the x and y components of flow:

$$\begin{cases} q_x = -K\dfrac{\partial h}{\partial x} \\ q_y = -K\dfrac{\partial h}{\partial y} \end{cases} \tag{1}$$

where $h(x, y)$ is the head potential, $q_x(x, y)$ and $q_y(x, y)$ are the specific flow rates in the x and y direction, respectively, and $K(x, y)$ is the hydraulic conductivity.

The 2-D continuity equation for steady-state conditions is

$$\frac{\partial q_x}{\partial x} + \frac{\partial q_y}{\partial y} = 0 \tag{2}$$

Laplace's equation combines Darcy's law and the continuity equation into a single second-order partial differential equation. Darcy's law, equation (1), is substituted component by component into equation (2) to give

$$\frac{\partial}{\partial x}\left(-K\frac{\partial h}{\partial x}\right) + \frac{\partial}{\partial y}\left(-K\frac{\partial h}{\partial y}\right) = 0 \tag{3}$$

If we assume hydraulic conductivity K is constant, equation (3) becomes

$$\frac{\partial^2 h}{\partial x^2} + \frac{\partial^2 h}{\partial y^2} = 0 \tag{4}$$

Toth (1962) assumed that, in an undeveloped watershed, the fluctuations of the water table on an annual basis were small. That is, he used an average water table position and assumed that the system was at a steady state on an annual basis. The water table position at the beginning of the year was the same as the position at the end of the year; there was no net accumulation or loss of water from the system. Therefore, with this idealization, the 2-D Laplace equation (4) is the required governing equation. The numerical model, consisting of the governing equation together with the four boundary conditions, is summarized in Figure 1.

Because q_x and q_y satisfy equation (2) and thus (4), they can be expressed in the following forms:

$$\begin{cases} q_x = -\dfrac{\partial V}{\partial y} \\ q_y = -\dfrac{\partial V}{\partial x} \end{cases} \tag{5}$$

where $V(x, y)$ is called the flow function.

For visualization of the flow patterns, we have to define the flow rate vectors in x- and y-directions. If h function is given flow rate vector (q_x, q_y) can be calculated from equation (1). Then, $V(x, y)$ can be calculated from equation (5).

3.2 Boundary Conditions

The upper boundary is located at $y = 0$ ft on MSL for x rating from 0 ft to 25,000 ft. The distribution of head along this boundary was determined at 500-ft interval based on the field measurements from the topographic high to coastal MSL (Table 1). The other three boundary conditions are for no-flow boundaries.

Darcy's law relates flow to gradient of head according to equation (1). Along a vertical, no-flow boundary, $q_x = 0$ implies $\partial h/\partial x = 0$, and along a horizontal, no-flow boundary, $q_y = 0$ implies $\partial h/\partial y = 0$. Specification of flow across these three boundaries makes them Neumann boundary conditions.

Table 1. Initial heads in regional aquifer area

Distance (ft) from Topo-graphic high	Head (ft) from MSL	*Distance (ft) from Topo-graphic high*	Head (ft) from MSL	*Distance (ft) from Topo-graphic high*	Head (ft) from MSL
0	55.00	8500	37.50	17000	6.00
500	54.00	9000	36.00	17500	3.50
1000	53.00	9500	34.80	18000	2.00
1500	51.00	10000	33.50	18500	2.00
2000	50.00	10500	31.80	19000	2.00
2500	49.00	11000	30.00	19500	2.00
3000	48.20	11500	28.00	20000	2.40
3500	47.50	12000	27.50	20500	2.60
4000	47.00	12500	25.50	21000	2.80
4500	46.00	13000	24.00	21500	3.00
5000	45.00	13500	22.10	22000	2.80
5500	44.00	14000	20.00	22500	2.55
6000	43.00	14500	17.80	23000	2.40
6500	42.00	15000	16.00	23500	1.80
7000	41.00	15500	13.00	24000	0.00
7500	40.00	16000	10.00	24500	0.00
8000	38.50	16500	8.00	25000	0.00

Top	$h(x, 0)$ in Table 1	$0 \le x \le 25{,}000$ (ft)
Bottom	$\partial h/\partial y \big\rvert_{y=-400} = 0$	$0 \le x \le 25{,}000$ (ft)
Left	$\partial h/\partial x \big\rvert_{x=0} = 0$	$-400 \le y \le 0$ (ft)
Right	$\partial h/\partial x \big\rvert_{x=25{,}000} = 0$	$-400 \le y \le 0$ (ft)

4 Numerical Implementation

4.1 Finite Differences Method

For many problems, the assumptions that must be made to obtain an analytical solution such as Toth's (1962) will not be realistic. In this case, we must resort to approximate methods using numerical techniques in order to solve the equations in a model. The finite differences technique is a good approximation to transform differential equations that make up a model into a set of algebraic equations.

Numerical solutions yield values for only a predetermined, finite number of points in the problem domain. By limiting our need to know the head to a reasonable number of points *N*, we can convert a partial differential equation into a set of *N* algebraic equations involving *N* unknown potentials.

Consider a finite set of points on a regularly spaced grid (Figure 2). Lattice points are spaced horizontally by a distance Δx =500 ft and vertically by a distance Δy = 10 ft. To locate any point in the grid, we specify an integer ordered pair (i, j). Relative to the origin (0, 0) located in the upper left-hand corner, (i, j) is located at a distance of $i\Delta x$ in the position x direction and $j\Delta y$ in the negative y direction. This indexing convention inverts rows and columns when compared with standard matrix notation.

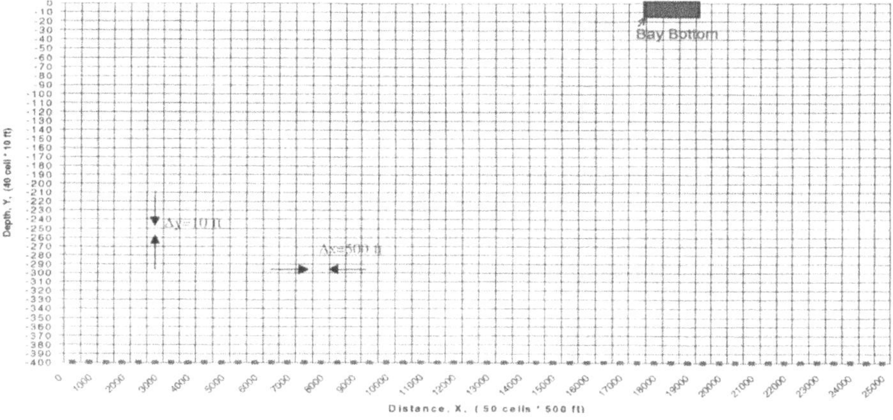

Fig. 2. Finite difference grid of regional aquifer area, where the spacing in the *x* direction is 500 ft, the spacing in the *y* direction is 10 ft, and the aquifer thickness is the unit length

4.2 Calculation of the Head Values – Gauss-Seidel Iteration Method

The value of the head at the point represented by the indices (i, j) is $h_{i,j}$. Let the Cartesian coordinates (x_0, y_0) be represented by (i, j). Along the horizontal line $y = y_0$, consider a profile of head which has in succession the values $h_{i-1,j}$, $h_{i+1,j}$. In the finite difference approximation, derivatives are replaced by differences taken between nodal points.

A central approximation to $\partial^2 h/\partial x^2$ (x_0, y_0) is obtained by approximating the first derivative at [(x_0+Δx)/2, y_0] and at [(x_0+Δx)/2, y_0], and then obtaining the second derivative by taking a difference between the first derivatives at those points. That is,

$$\frac{\partial^2 h}{\partial x^2} \cong \frac{\dfrac{h_{i+1,j}-h_{i,j}}{\Delta x} - \dfrac{h_{i,j}-h_{i-1,j}}{\Delta x}}{\Delta x} = \frac{h_{i+1,j}-2h_{i,j}+h_{i-1,j}}{(\Delta x)^2} \tag{6}$$

Similarly

$$\frac{\partial^2 h}{\partial y^2} \cong \frac{h_{i,j-1}-2h_{i,j}+h_{i,j+1}}{(\Delta y)^2} \tag{7}$$

According to Laplace's equation (4), the finite difference approximation for Laplace's equation at the point (i, j) simplifies to

$$\frac{h_{i-1,j}+h_{i+1,j}}{(\Delta x)^2} + \frac{h_{i,j-1}+h_{i,j+1}}{(\Delta y)^2} - 4\left(\frac{1}{(\Delta x)^2} + \frac{1}{(\Delta y)^2}\right)h_{i,j} = 0 \tag{8}$$

If the finite difference equation (9) were solved for $h_{i,j}$, then

$$h_{i,j} = \frac{(h_{i,j-1}+h_{i,j+1}) + R(h_{i-1,j}+h_{i+1,j})}{2(1+R)}, \quad R = \left(\frac{\Delta y}{\Delta x}\right)^2 \tag{9}$$

That is, the value of $h_{i,j}$ at any point is an weighted average value of head computed from its four nearest neighbors in the nodal array. Equation (9) is often called the five-point operator because the algebraic equations which approximate Laplace's equation are created one after another by moving the star of five points throughout the domain of the problem. Suppose we guess a set of trial answers and then successively improve our guesses until we get the right answer. In short, iterative methods consist of guessing and adjusting.

In Gauss-Seidel iteration method, we work through the grid in an orderly way; that is, we start at $i = 0$ and $j = 0$ and sweep across from left to right and down line-by-line as if we were reading a page. In this way, we can always use two newly computed values in the iteration formula. Thus the Gauss-Seidel iteration formula is

$$h_{i,j}^{m+1} = \frac{(h_{i,j+1}^{m+1}+h_{i,j-1}^{m+1}) + R(h_{i-1,j}^{m+1}+h_{i+1,j}^{m+1})}{2(1+R)} \tag{10}$$

The change between two successive Gauss-Seidel iterations is called the residual c. The residual is then defined by

$$c = h_{i,j}^{m+1} + h_{i,j}^{m} \tag{11}$$

By replacing $h^{m}{}_{i,j}$ with $h^{m+1}{}_{i,j}$ after each calculation, the Gauss-Seidel procedure liquidates or relaxes the residuals at every node and thus leads to a solution of each algebraic equation.

Let us consider the left boundary. The finite difference approximation to $\partial h/\partial x$ is

$$\frac{\partial h}{\partial x} \cong \frac{h_{i+1,j} - h_{i-1,j}}{2\Delta x} \tag{12}$$

For the left boundary, the point referred to by indices (i+1, j) is inside the problem domain, but the point referred to by indices (i-1, j) is outside. Therefore, we expand the finite difference problem domain by one additional column to the left by putting in a column of so-called imaginary nodes. The boundary condition $\partial h/\partial x = 0$ translates in finite difference from to $h_{i+1,j} = h_{i-1,j}$. That is, the value of head along the imaginary column must reflect across the left boundary. Thus, the left boundary is a symmetry line. The right and bottom no-flow boundaries are also handled by creating imaginary nodes. The top boundary has specified heads (Table 1) and, therefore, no imaginary nodes are needed along this boundary.

4.3 Calculation of the Flow Rates and Flow Lines

The finite difference expression of equation (1) becomes

$$\begin{cases} q_x = -K\dfrac{h_{i+1,j} - h_{i-1,j}}{2\Delta x} \\ q_y = -K\dfrac{h_{i,j-1} - h_{i,j+1}}{2\Delta y} \end{cases} \tag{13}$$

which is used to calculate flow rate q_x and q_y at each grid point from head values h of neighboring grid points.

The finite differences expression of equation (5) becomes

$$\begin{cases} q_x = -\dfrac{V_{i,j-1} - V_{i,j+1}}{\partial \Delta y} \\ q_y = -\dfrac{V_{i+1,j} - V_{i-1,j}}{\partial \Delta x} \end{cases} \tag{14}$$

In order to obtain flow function, V values from q_x and q_y based on equation (14), we use Gauss-Seidel iteration method as we did in Section 4.2 to calculate the heads. The difference from Section 4.2 is the boundary condition. Because we assume there is no-flow across left, right, and bottom boundaries, the flow function V should have is arbitrary, we put $V = 0$ on those boundaries. Therefore, we initially put 0 at all the grid points and performed Gauss-Seidel iterations using equation (14) until residual error becomes below threshold. In the iteration process, the V values at the left, right, and bottom boundaries were kept 0, and the V values at all other grid points were optimized.

5 Computer Programs

The numerical implementation described in the last Chapter was realized by three FORTRAN programs. The first program, HEADS.f90, was developed for simulating the potential heads in the flow system. The second FFLOW.f90 was developed for calculating the flow rate vectors from the heads. The last VFLOW.f90 was developed for calculating the flow functions from the flow rate vectors.

6 Calculation and Results

Three computer programs were applied to the field-measured table heights in Table 1. The grid internals Δx and Δy are set to 500 and 10 ft, respectively. The calculated heads and flow functions are plotted as contours in Figures 3 and 4.

Plots of flow lines and equipotential lines are called flow-nets. Flow-nets are useful in depicting groundwater flow paths and calculating flux through the system. Two adjacent flow lines form a stream tube; for steady-state flow, the flux through the stream tube is constant. In a homogeneous and isotropic medium, flow lines and equipotential lines intersect at right angles and form curvilinear squares.

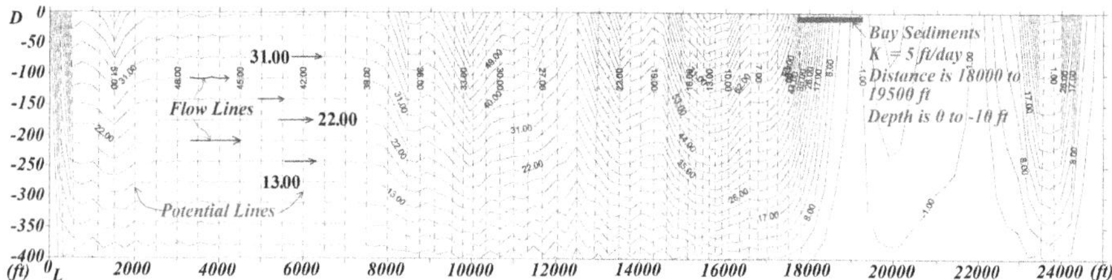

Fig. 3. Hydrologic section showing aquifer area Flow-Net of groundwater flow in aquifer K_1 = 50 ft/day and K_2 = 5 ft/day of bay sediments

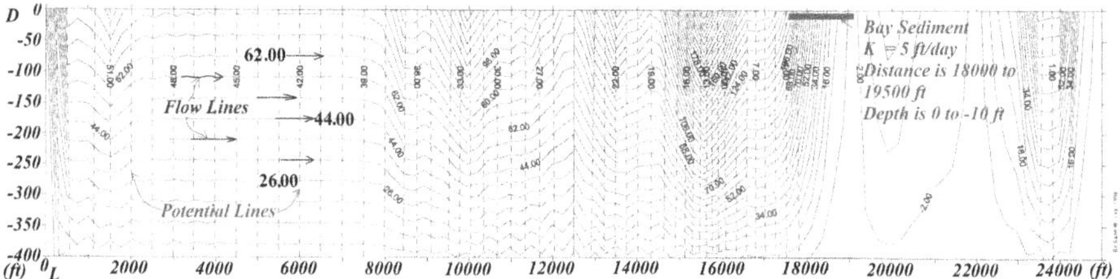

Fig. 4. Hydrologic section showing aquifer area Flow-Net of groundwater flow in aquifer K_1 = 100 ft/day and K_2 = 5 ft/day of bay sediments

7 Summary

In this study, it has become possible to estimate the configuration of regional groundwater system from the table heights by computer programs using the FDM. A 2-D vertical cross-section was used to conceptualize the flow system. This study concerns only with bay encircled by water table mounds that are at a higher altitude in vertical section show that for typical hydrogelogic settings, the line (divide) separating local from regional groundwater flow system in continuous beneath bay. Bay bottom sediments are considered this study by assuming two cases of aquifer hydraulic conductivity (K_{a1} = 100 ft/day, K_{a2} = 50 ft/day) and bay sediments hydraulic conductivity (K_b = 5 ft/day). The programs correctly produced head value contours and flow lines intercepting at right angles. The calculation results of above two case also correctly reflected differences in the aquifer hydraulic conductivity.

References

1. Franz, T., Guiguer, N.: FLOWPATH, two-dimensional horizontal aquifer simulation model. In: Waterloo Hydrogeologic Software, Waterloo, Ontario, p. 74 (1990)
2. Hill, M.C.: MODFLOWP: A computer program for estimating parameters of a transient, three-dimensional groundwater flow model using nonlinear regression. USGS, Open-File Report, p. 317 (1990)
3. Carrera, J., Neuman, S.P.: Estimation of aquifer parameters under transient and steady state conditions. 1, 2, 3, Water Resources Research 22(2), 199–242 (1986a, 1986b, 1986c)
4. Urish, D.W., Ozbilgin, M.M.: The coastal groundwater boundary. Ground Water 27(3), 310–315 (1988)
5. Toth, J.: A theory of groundwater motion in small drainage basins in cebtral Alberta. J. Geophys. Res. 67, 4375–4387 (1962)
6. Toth, J.: A theoretical analysis of groundwater flow in small drainage basins. J. Geophys. Res. 67, 4795–4812 (1963)

On a Relationship between Typical Behavior and Initial Conditions of Lagrange's Top

Kazumasa Miyamoto

Takasaki University of Health and Welfare, Faculty of Health and Welfare,
Department of Healthcare Informatics, 37-1 Nakaooruimachi, Takasaki, Gunma, Japan
miyamoto@takasaki-u.ac.jp

Abstract. The parameter α , which at first was introduced to modify the coordinate of the equilibrium point of Euler's equations[1], is used to represent the canonical equations of (θ, p_θ) for Lagrange's top. We then take a perspective approach to the top by doing an analysis of the potential function $U(\theta,\alpha)$, and settle on a mapping between typical motions of the top and the values of α . This approach makes it clear that there are two different modes of stationary precession and that a sleeping top mode can be reduced from a Lagrange's top.

1 Introduction

I have clarified that the typical motions of a Lagrange's top such as precession can be generated by a parameter α which modify the coordinate of the equilibrium point of Euler's equations[1]. In addition, I have proposed a new integration method in which the deviation of the Hamiltonian of a Lagrange's top is controlled to be zero on solving the canonical equations of (θ, p_θ) by the symplectic rule [2].

In this paper I will develop the above argument in more detail and in a systematic manner with a numerical simulation and a potential analysis method. The systematic manner involves mainly two steps as follows: Firstly, we specify the values of α which generate precession or nutation. Secondly, for each section decided by the adjacent values of α , we will analyze the potential function and simulate the canonical equations. Furthermore, a relationship between sleeping top and Lagrange's top shall be clarified.

2 Euler Angles

Euler angles are a means of representing the spatial orientation of the rotated system xyz as a composition of rotations from the fixed system XYZ. The composition of these rotations is given by Eq. (1).

$$\begin{bmatrix} X \\ Y \\ Z \end{bmatrix} = \begin{bmatrix} \cos\varphi & -\sin\varphi & 0 \\ \sin\varphi & \cos\varphi & 0 \\ 0 & 0 & 1 \end{bmatrix} \begin{bmatrix} \cos\theta & 0 & \sin\theta \\ 0 & 1 & 0 \\ -\sin\theta & 0 & \cos\theta \end{bmatrix} \begin{bmatrix} \cos\psi & -\sin\psi & 0 \\ \sin\psi & \cos\psi & 0 \\ 0 & 0 & 1 \end{bmatrix} \begin{bmatrix} x \\ y \\ z \end{bmatrix} = \Gamma(\varphi,\theta,\psi) \begin{bmatrix} x \\ y \\ z \end{bmatrix}. \tag{1}$$

J.-H. Kim et al. (Eds.): AsiaSim2011, PICT 4, pp. 50–61, 2012.

The derivative of Euler angles with respect to time is defined by Eq. (2) with the angular velocity vector ${}^t[\omega_x \quad \omega_y \quad \omega_z]$ represented by the rotated system.

$$ {}^t[\omega_x \quad \omega_y \quad \omega_z] = \Sigma \ {}^t[d\psi/dt \quad d\theta/dt \quad d\varphi/dt] \tag{2} $$

Similarly, the derivative of Euler angles with respect to time is defined by Eq. (3) with the angular velocity vector ${}^t[\omega_X \quad \omega_Y \quad \omega_Z]$ represented by the fixed system.

$$ {}^t[\omega_X \quad \omega_Y \quad \omega_Z] = \Phi \ {}^t[d\psi/dt \quad d\theta/dt \quad d\varphi/dt] \tag{3} $$

From equations (2) and (3), the relation between ${}^t[\omega_X \quad \omega_Y \quad \omega_Z]$ and ${}^t[\omega_x \quad \omega_y \quad \omega_z]$ is given by Eq. (4), whose relation is the same as Eq. (1).

$$ {}^t[\omega_X \quad \omega_Y \quad \omega_Z] = \Phi\Sigma^{-1} \ {}^t[\omega_x \quad \omega_y \quad \omega_z] = \Gamma^t[\omega_x \quad \omega_y \quad \omega_z] \tag{4} $$

3 ND(Non-Dimensional) Euler's Equations and First Integrals

Euler's equations of a symmetric top are given as follows

$$ A\dot{\omega}_x - (A - C)\omega_y\omega_z = M_x \ , A\dot{\omega}_y - (C - A)\omega_z\omega_x = M_y \ , C\dot{\omega}_z = M_z \, , $$

where C is the moment of inertia relative to the symmetric axis of the top, and A is the moment of inertia around an axis through the tip, directed perpendicular to the symmetry axis. The angular velocity of the rotation around the symmetry axis is denoted by ω_z while ω_x and ω_y are the angular velocities around two axes going through the tip and perpendicular both to the symmetry axis and to each other. With the origin of the coordinate system placed at the tip and the Euler angles defined by Eq. (1), the components of the applied torques (M_x, M_y, M_z) are given as follows

$$ M_x = mgl\sin\theta\sin\psi \ , \ M_y = mgl\sin\theta\cos\psi \ , \ M_z = 0 \, , $$

where m is the mass of the top, g is the gravitational acceleration, and ℓ is the position of the center of mass along the symmetric axis. The equation of $\omega_z = \omega_0 = const.$ is a direct result of $M_z = 0$. Here we will introduce new variables and new constants as follows.

$$ \hat{t} = t\omega_0 \, , \hat{\omega}_x = \omega_x / \omega_0 \, , \hat{\omega}_y = \omega_y / \omega_0 \ , \hat{\omega}_z = \omega_z / \omega_0 = 1, $$

$$ k = \{mgl / A(\omega_0)^2\}, r = C / A - 1 \, . $$

Changing $t, \omega_x, \omega_y, \omega_z$ to these new variables and using new constants allows us to rewrite Euler's equations as follows.

$$ d\hat{\omega}_x / d\hat{t} = -r\hat{\omega}_y + k\sin\theta\sin\psi \quad (5) \qquad d\hat{\omega}_y / d\hat{t} = r\hat{\omega}_x + k\sin\theta\cos\psi \ (6) $$

$$ d\psi / d\hat{t} = (\hat{\omega}_x\cos\psi - \hat{\omega}_y\sin\psi)\cos\theta / \sin\theta + 1 \tag{7} $$

$$ d\theta / d\hat{t} = \hat{\omega}_x\sin\psi + \hat{\omega}_y\cos\psi \tag{8} $$

$$ d\varphi / d\hat{t} = (-\hat{\omega}_x\cos\psi + \hat{\omega}_y\sin\psi) / \sin\theta \tag{9} $$

$$\hat{\omega}_z = d\psi / d\hat{t} + (d\varphi / d\hat{t})\cos\theta = 1 \tag{10}$$

Here, Equation (10) is derived directly from equations (7) and (9), and the unit vector of z axis is described by $(\sin\theta\cos\varphi', \sin\theta\sin\varphi', \cos\theta)$ in the fixed system, where φ' is defined by equation (11).

$$d\varphi' / d\hat{t} = d\varphi / d\hat{t}\,(1-\cos^2\theta) = (-\hat{\omega}_x \cos\psi + \hat{\omega}_y \sin\psi)\sin\theta \tag{11}$$

Note that $\hat{\omega}_x$, $\hat{\omega}_y$, ψ and θ are independent variables while φ and φ' are dependant variables of those. First integrals of Lagrange's top are total energy E_0 and the angular momentum in the Z-direction M_{Z0}, which are given by equations. (12) and (13) respectively.

$$(\hat{\omega}_x{}^2 + \hat{\omega}_y{}^2)/2 + k\cos\theta = E_0 / A\omega_0{}^2 \equiv \hat{E} \tag{12}$$

$$(-\hat{\omega}_x \cos\psi + \hat{\omega}_y \sin\psi)\sin\theta + (1+r)\cos\theta = M_{Z0} / A\omega_0 \equiv \hat{M} \tag{13}$$

3.1 Equilibrium Point and Initial Values of Euler's Equations

The equilibrium point of equations (5) - (8) is determined under the condition (14), which is also the stable condition of the system. The coordinates of the point are given by Eq. (16), where θ_0 is given by Eq. (15) and ψ is arbitrary. In addition, proper angular frequencies in the neighborhood of the point are given by

$$\omega^2 = 0, (r-1)^2 + (k^2/r^2 - 1)\ [1].$$

$$|\, r/k \,| < 1 \quad (14) \qquad \theta_0 = \cos^{-1}(r/k) \tag{15}$$

$$(-\tan\theta_0 \cos\psi,\ \tan\theta_0 \sin\psi,\ \psi,\ \theta_0 = \cos^{-1}(r/k)) \tag{16}$$

Since ψ is arbitrary, we may set $\psi \equiv 0$ in Eq. (16). Then the equilibrium point is expressed by $(-\tan\theta_0, 0, 0, \theta_0)$. Now I will introduce a new initial value defined by Eq. (17) with a new parameter α so that it may generate various typical motions such as precession and nutation.

$$(\hat{\omega}_{x0}, \hat{\omega}_{y0}, \psi_0, \theta_0) = (-\alpha\tan\theta_0,\ 0\ ,\ 0\ ,\ \theta_0 = \cos^{-1}(r/k)) \tag{17}$$

4 ND Canonical Equation and Its Behaviors (In Case of $\theta_0 \neq 0$)

Lagrangian of the system is given by Eq. (18). Then generalized momentum $p_\theta, p_\varphi, p_\psi$ are defined in equations (19) - (21)

$$L = A\dot{\theta}^2/2 + A\dot{\varphi}^2\sin^2\theta/2 + C(\dot{\psi} + \dot{\varphi}\cos\theta)^2/2 - mgl\cos\theta \tag{18}$$

$$p_\theta = \partial L / \partial\dot{\theta} = A\dot{\theta} \tag{19}$$

$$p_\varphi = \partial L / \partial\dot{\varphi} = A\dot{\varphi}\sin^2\theta + C(\dot{\psi} + \dot{\varphi}\cos\theta)\cos\theta \equiv a \tag{20}$$

$$p_\psi = \partial L/\partial\dot{\psi} = C(\dot{\psi}+\dot{\varphi}\cos\theta) = C\omega_0 \equiv b, \tag{21}$$

where "a" and "b" are constants of motion, because φ, ψ are cyclic. Therefore, Hamiltonian H, the total energy of the system conserved, is given by Eq. (22').

$$H = {p_\theta}^2/2A + (a-b\cos\theta)^2/2A\sin^2\theta + b^2/2C + mgl\cos\theta \tag{22}$$

$$H' \equiv H - b^2/2C = {p_\theta}^2/2A + (a-b\cos\theta)^2/2A\sin^2\theta + mgl\cos\theta \tag{22'}$$

Then canonical equations are given by equations (23) and (24).

$$\dot{\theta} = \partial H'/\partial p_\theta = p_\theta/A \tag{23}$$

$$\dot{p}_\theta = -\partial H'/\partial\theta = mgl\sin\theta - (a-b\cos\theta)(b-a\cos\theta)/A\sin^3\theta \tag{24}$$

Here we newly introduce a variable and constants as follows.

$$\hat{p}_\theta = p_\theta/A\omega_0 \quad \hat{H} = H'/A{\omega_0}^2, \hat{a} = a/A\omega_0, \ \hat{b} = b/A\omega_0$$

Changing p_θ, H', a, b into those allows us to rewrite equations (22'), (23) and (24) as follows

$$\hat{H} = {\hat{p}_\theta}^2/2 + (\hat{a}-\hat{b}\cos\theta)^2/2\sin^2\theta + k\cos\theta = {\hat{p}_\theta}^2/2 + U(\theta) \tag{25}$$

$$d\theta/d\hat{t} = \partial\hat{H}/\partial\hat{p}_\theta = \hat{p}_\theta \tag{26}$$

$$\begin{aligned} d\hat{p}_\theta/d\hat{t} &= -\partial\hat{H}/\partial\theta = -\partial U(\theta)/\partial\theta \\ &= k\sin\theta - (\hat{a}-\hat{b}\cos\theta)(\hat{b}-\hat{a}\cos\theta)/\sin^3\theta \equiv F(\theta) \end{aligned} \tag{27}$$

where the potential function $U(\theta)$ is defined by Eq. (25).

4.1 Initial Values and ND Constant of Canonical Equation

On the basis of Eq. (17), the initial values of (θ, p_θ) at $\hat{t}=0$ is given by

$$\theta_0 = \cos^{-1}(r/k), \hat{p}_{\theta 0} = A\dot{\theta}/A\omega_0 = d\theta/d\hat{t} = \hat{\omega}_{x0}\sin\psi_0 + \hat{\omega}_{y0}\cos\psi_0 = 0 \,.$$

Then the constants of motion $\hat{a}, \hat{b}$ and the first integral $\hat{H}$ are given as follows.

$$\hat{a} = a/A\omega_0 = \hat{M}_0 = \alpha\tan\theta_0\sin\theta_0 + (r+1)\cos\theta_0, \ \hat{b} = C\omega_0/A\omega_0 = r+1$$

$$\hat{H}_0 = \hat{E}_0 = (\hat{a}-\hat{b}\cos\theta_0)^2/2\sin^2\theta_0 + r = \alpha^2\tan^2\theta_0/2 + r$$

Considering equations (20) and (21), equations of (9), (7) and (11) are transformed into equations (28) - (30) respectively.

$$d\varphi/d\hat{t} = (\hat{a}-\hat{b}\cos\theta)/\sin^2\theta \equiv P(\theta,\alpha) \tag{28}$$

$$d\psi/d\hat{t} = 1 - (\hat{a}-\hat{b}\cos\theta)\cos\theta/\sin^2\theta \equiv Q(\theta,\alpha) \tag{29}$$

$$d\varphi'/d\hat{t} = \hat{a}-\hat{b}\cos\theta \equiv R(\theta,\alpha) \tag{30}$$

In the above equations, since the values of $\hat{a}$ and $\hat{b}$ are decided by α, the expressions of $P(\theta,\alpha), Q(\theta,\alpha)$ *and* $R(\theta,\alpha)$ are adopted to make it clear. Similarly $F(\theta), U(\theta)$ in equations (25) and (27) may be replaced with $F(\theta,\alpha), U(\theta,\alpha)$ respectively.

4.2 Equilibrium Points of Canonical Equations

Under the condition of $\theta_0 = \cos^{-1}(r/k)$, equilibrium points of canonical equations (26) and (27) must satisfy the conditions $d\theta/d\hat{t} = \hat{p}_\theta \equiv 0, d\hat{p}_\theta/d\hat{t} \equiv d^2\theta/d\hat{t}^2 \equiv 0$. These requirements are satisfied only if $\alpha = 1, r$. If $\alpha = 1$, then the equations of $d\hat{\omega}_x/d\hat{t} = d\hat{\omega}_y/d\hat{t} = d\psi/d\hat{t} = d\theta/d\hat{t} \equiv 0$ hold; if $\alpha = r$, then only $d\theta/d\hat{t} \equiv 0$ holds. In the latter case, $(\hat{\omega}_x, \hat{\omega}_y)$ are represented by equation (31), which means simple harmonic oscillation with constant angular velocity $d\psi/d\hat{t} = Q(\theta_0, r) = 1 - r$.

$$^t[\hat{\omega}_x \quad \hat{\omega}_y] = P(\theta_0, r)\sin\theta_0 \, {}^t[\cos(\pi - \psi) \quad \sin(\pi - \psi)] \tag{31}$$

In the case of $k = 0.5$, $r = 0.4$; for example, when α is 1 or 0.4, both of which generate precession. For both cases, there exist two different potential functions; consequently the shapes of them are also different from each other. But they take the minimum at the same point of $\theta_0 = \cos^{-1}(r/k) = 0.64351$. Table 1 lists the values of typical variables in those two cases for comparison.

Table 1. Values of typical variables at precession

	condition of precession: $\dot{\theta} = \ddot{\theta} = 0, \theta_0 = \cos^{-1}(r/k)$	
α	1	r
ω_x	$-\tan\theta_0$	$-r\tan\theta_0\cos\psi$
ω_y	0	$r\tan\theta_0\sin\psi$
ψ	0	$(1-r)\hat{t}$
$\dot{\psi}$	0	$(1-r)$
$\dot{\varphi}$	k/r	k
$\Omega_{prec.}$	$\sin\theta_0\tan\theta_0 = k/r - r/k$	$r\sin\theta_0\tan\theta_0 = r(k/r - r/k)$
a=M	$k(1/r + k/r)$	$k(1+k/r) + r(1-r)/k$
H=E	$(k^2 - r^2)/(2r^2) + r$	$(k^2 - r^2)/2 + r$

4.3 Dependence of Dynamic Range of θ and $d\varphi'/d\hat{t}$ on α

For a given value of α, the potential function $U(\theta, \alpha)$ and the constant of motion $\hat{a}$ are determined and then the dynamic range of θ can be calculated by the analysis of the $U(\theta, \alpha)$. And furthermore, it is determined whether $\hat{a}$ is greater than $\hat{b}$ or not. As a result, the sign of $d\varphi'/d\hat{t}$, which represents the direction of the axis revolution around Z-axis, is determined by taking consideration of Eq. (30). Note that $U(\theta, \alpha)$ takes the minimum at some value α under the condition of value of fixed θ, because it is a positive quadratic function of α.

Here, we assume $0<r<k<1, 0<\hat{a}$ for a Lagrange's top. The condition of $0<r<k<1$ means that a top is flat and can have stable precession at $\theta_0 = \cos^{-1}(r/k)$, $0<\theta_0<\pi/2$, and that the kinetic energy is sufficiently larger than the potential energy. Since the Z-component of angular momentum is a constant of motion, we may assume $\hat{a}>0$ without any loss of generality. Now we will examine the three cases as follows.

[CASE $\hat{a}>\hat{b}$]: If $\hat{a}>\hat{b}$, then $d\varphi'/d\hat{t}>0$ and $\alpha>\alpha^* = (r+r^2)/(r+k)$, which satisfies $1>\alpha^*>r$ under the assumed condition of $0<r<k<1$. So, it is sufficient to examine $U(\theta,\alpha)$ for the three disjoint regions of $\alpha\,;\alpha>1, \alpha=1, \alpha>\alpha^*$ separately. As a result, if $\alpha>1$, then dynamic range of θ is estimated $\pi/2>\theta\geq\theta_0$; if $\alpha=1$, then $\theta\equiv\theta_0$; if $1>\alpha>\alpha^*$, then $0<\theta\leq\theta_0$.

[CASE $\hat{a}=\hat{b}$]: If $\hat{a}=\hat{b}$, then $\alpha=\alpha^*$, under which condition equations (25) - (30) are expressed by equations (32) - (37) respectively.

$$\hat{H}_s = \hat{p}_\theta{}^2/2+(1+r)^2\tan^2(\theta/2)/2+k\cos\theta \equiv \hat{p}_\theta{}^2/2+U_s(\theta) \tag{32}$$

$$d\theta/d\hat{t} = \hat{p}_\theta \tag{33}$$

$$d\hat{p}_\theta/d\hat{t} = -2\sin(\theta/2)((r+1)^2/4-k\cos^4(\theta/2))/\cos^3(\theta/2) \equiv F_s(\theta) \tag{34}$$

$$d\varphi/d\hat{t} = (1+r)/2\cos^2(\theta/2) \equiv P_s(\theta) \tag{35}$$

$$d\psi/d\hat{t} = 1-(1+r)\cos\theta/2\cos^2(\theta/2) \equiv Q_s(\theta) \tag{36}$$

$$d\varphi'/d\hat{t} = (1+r)(1-\cos\theta) \equiv R_s(\theta) \tag{37}$$

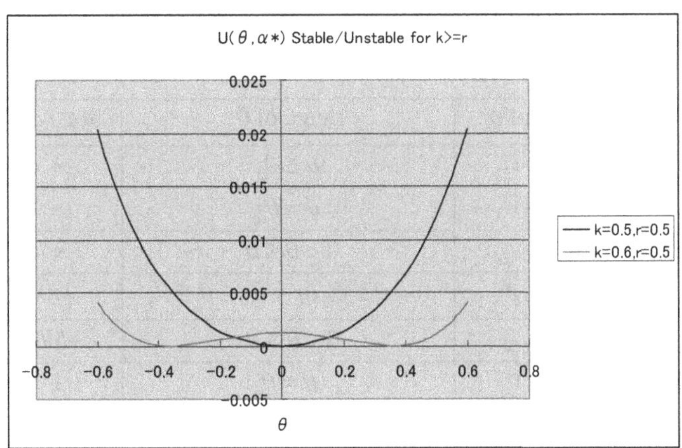

Fig. 1. The Potential Function stable/unstable at $\theta=0$ (the case of $\alpha=\alpha^*$)

Note that $U_s(\theta)$ is symmetric about $\theta=0$, and that $d\varphi'/d\hat{t} \geq 0$. If $(r+1)^2/4 \geq k\cos^4(\theta/2)$ for $\pi/2>\theta\geq 0$, namely $(r+1)^2/4 \geq k$, then $\theta=0$ is a unique stable point. Otherwise, $\theta=0$ is an unstable point and there are two stable points at $\theta \neq 0$. Graphs of stable/unstable $U(\theta,\alpha^*)$ are shown in Fig. 1. Thus the stable condition at $\theta=0$ is as follows.

$$(r+1)^2/4 \geq k \cdot \tag{38}$$

From the view of Fig. 1, it is concluded that if the system is stable at $\theta=0$, then the dynamic range of θ is $-\theta_0 \leq \theta \leq \theta_0$. Otherwise, it depends on the values of $U(0,\alpha^*)$ and $U(\theta_0,\alpha^*)$. That is to say, if $U(\theta_0,\alpha^*)$ is greater than $U(0,\alpha^*)$, then it is $-\theta_0 \leq \theta \leq \theta_0$, and if not, it is $0<\theta\leq\theta_0$.

[CASE $\hat{a}<\hat{b}$]:If $\hat{a}<\hat{b}$, then it is sufficient to examine the $U(\theta,\alpha)$ for the five disjoint regions of α:

$$\alpha^* > \alpha > r, \alpha = r, r > \alpha > 0, \alpha = 0, \alpha < 0.$$

If $\alpha^* > \alpha$ and α is in the neighborhood of α^*, then θ can take values close to 0. In such a situation, the value of $d\varphi'/d\hat{t}$ may become negative, which generates the trajectory of loops. If $r \geq \alpha \geq 0$, then $\theta \geq \theta_0$, in which the condition is sufficient for $\hat{a}-\hat{b}\cos\theta \geq 0$, that is, $d\varphi'/d\hat{t} \geq 0$. If $\alpha=0$, then $d\theta(0)/d\hat{t} = d\varphi(0)/d\hat{t} = 0$, where the trajectory has cusps. If $\alpha<0$ and α is in the neighborhood of 0, then θ can take values close to θ_0. In such a situation, the value of $d\varphi'/d\hat{t}$ may become negative, which also generates the trajectory of loops [1]. The summary of the discussion stated above is shown in Table 2. Fig. 2-1 shows the graphs of $U(\theta,\alpha)$ for $\alpha = 1.1, 1., 0.8, 0.65$, and Fig. 2-2 for $\alpha = 0.6, 0.4, 0.2, 0.0, -0.1$. The dynamical simulation results on some values of α will be described in section 6.

Table 2. Typical mode of movement with respect to α

$\hat{a}:\hat{b}$	range of α	range of θ	$d\varphi'/dt$	mode
$\hat{a}>\hat{b}$	$\alpha>1$	$\theta\geq\theta_0$	+	n
	$\alpha=1$	$\theta=\theta_0$	+	p
	$1>\alpha>\alpha^*$	$0<\theta\leq\theta_0$	+	n
$\hat{a}=\hat{b}$	$\alpha=\alpha^*$	$0<\theta\leq\theta_0 \; or -\theta_0\leq\theta\leq\theta_0$	+/0	z/n
$0<\hat{a}<\hat{b}$	$\alpha^*>\alpha>r$	$0<\theta\leq\theta_0$	+/0/-	loop/n
	$\alpha=r$	$\theta=\theta_0$	+	p
	$r>\alpha>0$	$\theta\geq\theta_0$	+	n
	$\alpha=0$	$\theta\geq\theta_0$	+/0	cusp

$\theta_0 = \cos^{-1}(r/k), \alpha^* = (r+r^2)/(r+k)$, n=nutation, p=precession, z=zero cross

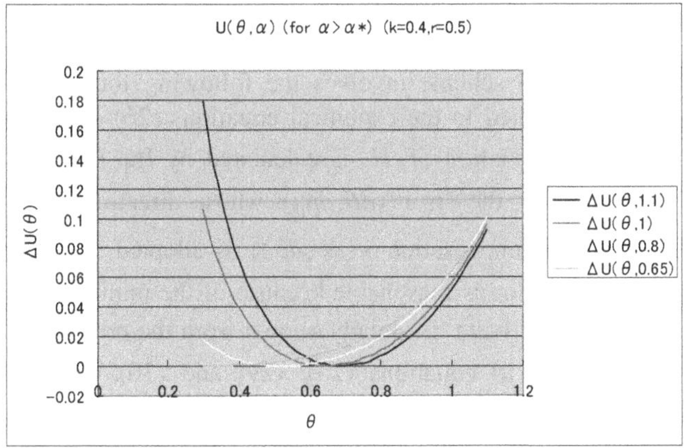

Fig. 2-1. Relative values of the Potential Functions (α>α*, $\theta_0 = 0.64351$)

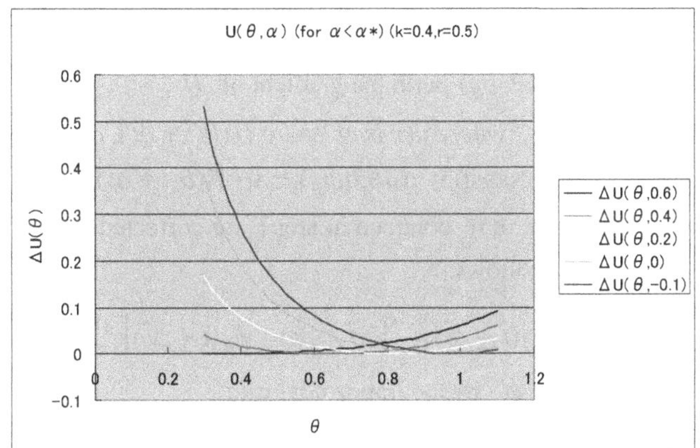

Fig. 2-2. Relative values of the Potential Functions (α<α*, $\theta_0 = 0.64351$)

5 ND Canonical Equation of a Sleeping Top (In Case of $\theta_0 = 0$)

If $\theta_0 = 0$, then $\hat{a} = r + 1$. On the condition, equations (25)-(30) are reduced to equations (32)-(37). Thus the stability condition of a sleeping top is the same as the equation (38). This condition can be expressed by two disjoint conditions: $(r+1)^2/4 \geq k > r$ and $r \geq k$. The former satisfies the equation (14), which implies that stationary precession can be realized in the neighborhood of $\theta_0 = 0$; the latter doesn't. Strictly speaking, the latter only can be the stability condition of a sleeping top. The shapes of potential functions of both stable and unstable cases at $(\theta, \hat{p}_\theta) = (0,0)$ are similar as those described in Fig. 1.

6 Time Integral Scheme and Simulations

A numerical time integral scheme involves the following four steps. Firstly, the symplectic method is applied to the canonical equations (26) and (27) with respect to $\theta, \hat{p}_\theta$. Secondly, the deviation of $\hat{H}$ is calculated by Eq. (25). Supposing that the deviation derives from the error $(d\theta, d\hat{p}_\theta)$ whose direction is the same as the gradient of $\hat{H}$, reverse compensation $-(d\theta, d\hat{p}_\theta)$ is adopted in order to cancel the deviation. This compensation is reasonable because of the minimum quantity of correction. Thirdly, since $\theta(\hat{t}+\Delta\hat{t})$ is already known from the previous step, the trapezoidal rule is applicable to equations (28), (29) and (30). Finally, we update $(\hat{\omega}_x, \hat{\omega}_y)$ from Eq. (2) with $\theta(\hat{t}+\Delta\hat{t}), \hat{p}_\theta(\hat{t}+\Delta\hat{t}), and\ \psi(\hat{t}+\Delta\hat{t})$ decided in the previous steps. The practical procedure of those four steps is summarized as follows.

(step1) Integration of $(\theta, \hat{p}_\theta)$ by the symplectic rule

$$\theta(\hat{t}+\Delta\hat{t}) = \theta(\hat{t}) + \hat{p}_\theta(\hat{t})\Delta\hat{t},\ \ \hat{p}_\theta(\hat{t}+\Delta\hat{t}) = \hat{p}_\theta(\hat{t}) + F(\theta(\hat{t}+\Delta\hat{t}))\Delta\hat{t}$$

(step2) Compensation of $(\theta, \hat{p}_\theta)$ with the gradient of $\hat{H}$

For $d\hat{H} = -F(\theta)d\theta + \hat{p}_\theta d\hat{p}_\theta,\ where\ d\hat{H}\ is\ given\ by\ \hat{H}(\theta(\hat{t}+\Delta\hat{t}), \hat{p}_\theta(\hat{t}+\Delta\hat{t})) - \hat{H}_0$, you should decide λ that satisfies $(d\theta,\ d\hat{p}_\theta) = \lambda(-F(\theta(t+\Delta\hat{t}), \hat{p}_\theta(t+\Delta\hat{t}))$. The values of $(\theta(t+\Delta\hat{t}), \hat{p}_\theta(t+\Delta\hat{t}))$ obtained at step1 are corrected by the reverse compensation $-(d\theta,\ d\hat{p}_\theta)$ as follows.

$${}^t\left[\theta(\hat{t}+\Delta\hat{t}) \quad \hat{p}_\theta(\hat{t}+\Delta\hat{t})\right] \leftarrow {}^t\left[\theta(\hat{t}+\Delta\hat{t}) \quad \hat{p}_\theta(\hat{t}+\Delta\hat{t})\right] - {}^t\left[d\theta \quad d\hat{p}_\theta\right]$$

(step3) Integration of ψ, φ, φ' by the trapezoidal rule

$$\begin{bmatrix} \varphi(\hat{t}+\Delta\hat{t}) \\ \psi(\hat{t}+\Delta\hat{t}) \\ \varphi'(\hat{t}+\Delta\hat{t}) \end{bmatrix} = \begin{bmatrix} \varphi(\hat{t}) \\ \psi(\hat{t}) \\ \varphi'(\hat{t}) \end{bmatrix} + \begin{bmatrix} P(\theta(\hat{t})) + P(\theta(\hat{t}+\Delta\hat{t})) \\ Q(\theta(\hat{t})) + Q(\theta(\hat{t}+\Delta\hat{t})) \\ R(\theta(\hat{t})) + R(\theta(\hat{t}+\Delta\hat{t})) \end{bmatrix} \Delta\hat{t}/2$$

(step4)Update of $(\hat{\omega}_x, \hat{\omega}_y)$

$$\begin{bmatrix} \hat{\omega}_x(\hat{t}+\Delta\hat{t}) \\ \hat{\omega}_y(\hat{t}+\Delta\hat{t}) \end{bmatrix} = \begin{bmatrix} \sin\psi & -\cos\psi \\ \cos\psi & \sin\psi \end{bmatrix} \begin{bmatrix} \hat{p}_\theta \\ P(\theta)\sin\theta \end{bmatrix}$$

6.1 Dynamical Simulations Result

[In Case of $\theta_0 \neq 0$: Lagrange's Top]

Under the condition of $k = 0.5, r = 0.4$, since the simulation results have been shown for $\alpha = 0.8, 0.0, -0.2$ [1], I will show the result for $\alpha = 0.6$, which is near to

$\alpha^* = 0.622$, in Fig. 3-1,3-2 and 3-3. If $\alpha = 0.6 < \alpha^*$, then $\hat{a} = 1.39, \hat{b} = 1.4$ and the value of θ can be under 0.1, which is determined from the potential function of Fig. 3-1, then $d\varphi'/d\hat{t} = 1.39 - 1.4\cos\theta$ can be negative. Both Fig. 3-2 and 3-3 show $d\varphi'/d\hat{t} < 0$ also. This result coincides with the result of the potential analysis shown in section 5. Since the condition $\alpha = \alpha^*$ corresponds to a sleeping top, we will examine the case of $\alpha = \alpha^*$ next. (By the way, the condition of $k = 0.5, r = 0.4$ does not satisfy the stable condition of a sleeping top.)

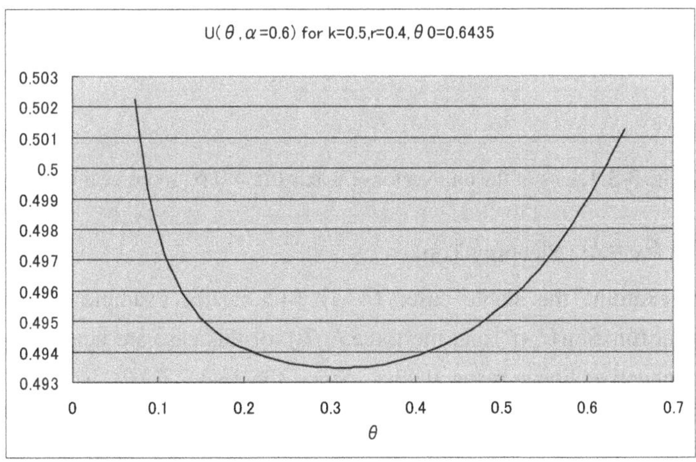

Fig. 3-1. Relative values of the Potential Function ($\alpha = 0.6$)

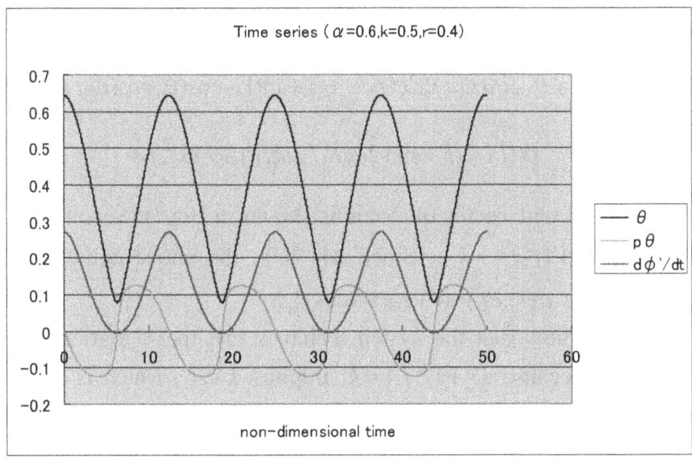

Fig. 3-2. Time series ($\alpha = 0.6$, $d\phi'/d\hat{t}$ becomes negative, when θ is small)

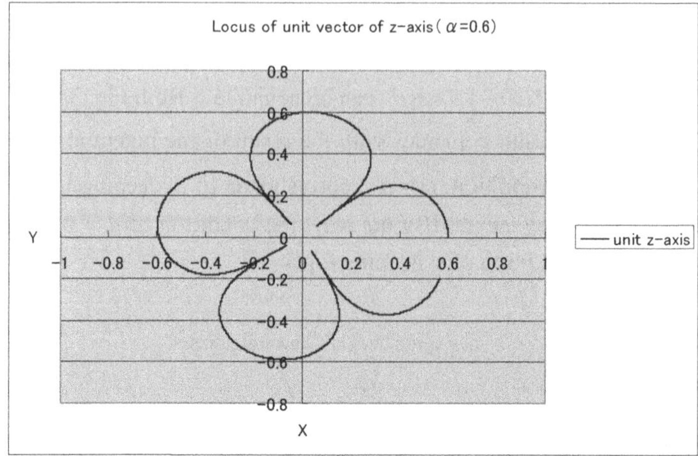

Fig. 3-3. Locus of the unit vector of z-axis ($\alpha = 0.6$, loops near the origin)

[In Case of $\theta_0 = 0$: Sleeping Top]

Firstly we examine the stable case: $(r+1)^2/4 > k$, for example $k = 0.4, r = 0.5$. The potential function $U_s(\theta)$ and the force $F_s(\theta)$ of this case are nearly positive quadratic and negative linear respectively. Since the force $F_s(\theta)$ is a monotone decreasing function, and has a unique zero point at $\theta_{zero} = 0$, it has a negative slope. Consequently, dynamical behaviors around $\theta_{zero} = 0$ is described by the harmonic oscillation with a proper angular frequency $\omega_0{}^2 = (r+1)^2/4 - k = 0.4301^2$.

Now suppose $(\theta_0, \hat{p}_{\theta 0}) = (0.1, 0), \varphi'(0) = 0$, then $\theta(\hat{t}), \varphi'(\hat{t})$ are approximated as follows.

$$\theta(\hat{t}) \approx \theta_0 \cos(\omega_0 \hat{t}), d\varphi'/d\hat{t} \approx (1+r)(1-\cos(\theta_0 \cos(\omega_0 \hat{t})) ,$$

$$\varphi'(\hat{t}) \approx (\hat{t} + \sin(2\omega_0 \hat{t})/2\omega_0)(1+r)\theta_0{}^2/4 .$$

Thus the locus of the unit vector of z-axis is shown in Fig. 4. Secondly we examine the critical case: $(r+1)^2/4 = k$. Since $\omega_0{}^2 = 0$, the above equations are further reduced to $\theta(t) \approx \theta_0$, $d\varphi'/d\hat{t} \approx (1+r)(1-\cos\theta_0)$.

Those equations show that the system behaves just like a stationary precession. That is reasonable, because $(r+1)^2/4 = k$ implies $k \geq r$, which is also a sufficient condition for a stationary precession. Thirdly we examine the unstable case: $(r+1)^2/4 < k$. Since $k \geq r$ is satisfied automatically, θ moves from $\theta_0 = 0$ to some $\theta*$ periodically.

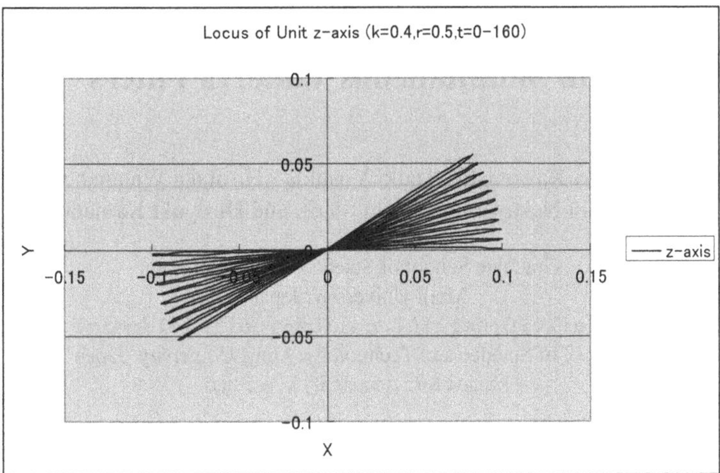

Fig. 4. Locus of a sleeping top

7 Result

The parameter α, which at first was introduced to modify initial values of Euler's equations, is used in the expression of the canonical equations of (θ, p_θ) for Lagrange's top in this paper. Then we can take a perspective approach to the top by the analysis of the potential function $U(\theta,\alpha)$, and can decide the dynamic behavior of a Lagrange's top for a given value of α. This approach makes it clear that there are two different modes of stationary precession and that the dynamical model of a sleeping top can be reduced from that of a Lagrange's top. The precision of Dynamical simulations is kept by the new integral scheme also.

Acknowledgements. I wish to thank Mr. Y. Kojima for helping me to translate Japanese into English.

References

1. Miyamoto, K.: Dynamical Simulation of the Symmetric Lagrange's Top Precession and Nutation. JSST 1(1), 11–15 (2009) (in Japanese)
2. Miyamoto, K.: On the Numerical Integration of Lagrange's Top Keeping the First Integral Value Fixed. JSST 3(2), 19–23 (2011) (in Japanese)

Chaotic Modem System Using Nonlinear Map with Simultaneous Volterra Filters

Yoshiaki Kawanishi[1], Daiki Yoshida[1], Hirotaka Watanabe[1], Ariyoshi Nakayama[1], Taichi Sato[2], and Hiroyuki Kamata[2]

[1] Graduate School of Science and Technology, Meiji University, Japan
{ce01030,ce11096,ce11101,ce11066,ce11044}@meiji.ac.jp
[2] School of Science and Technology, Meiji University, Japan
kamata@isc.meiji.ac.jp

Abstract. This paper proposes the chaotic modem system using nonlinear map and simultaneous Volterra filters for cipher and secure communications. By using the proposed structure, the sensitivity of the parameter mismatching between the transiver and receiver sides improved, and the randomness of the generated chaotic signal is obtained in various parameters combination.

1 Introduction

The chaotic modem system is one of the stream cipher, therefore, the time required for the modulation should be originally short. However, in the conventional study, when it is got the robustness as the cipher, the expressions become complicated, as the result, the rapidity of the processing becomes poor.

This study, a novel structure of the chaotic modem system is proposed. The chaotic modulator is comprised of the nonlinear map with volterra filter and another volterra filter for synchronization between the modulator and the demodulator.

Besides, many parameters are included in the proposed modulator, however, all parameters may not be effective as an encrypting key. The parameters that are not effective as encrypting keys are removed from the system, as the result, a rapid chaotic modem system having only effective parameters can be constituted.

In this paper, the proposed chaotic system is explained, and the affectivity and robustness are evaluated.

2 Chaotic Modulator

In this section, we investigate a chaotic system. This system use 16-bit fixed-point digital computation.

2.1 Proposed Chaotic Modulator

The chaotic system adopted here show in (1)-(6) taking both linear and nonlinear digital filter structure.

J.-H. Kim et al. (Eds.): AsiaSim2011, PICT 4, pp. 62–70, 2012.

<Encryption part>

$$x_1(n) = s(n) + \sum_{i=0}^{3}\sum_{j=0}^{3} g_{ij}\, x_i(n-1)x_j(n-1) + g_{123}\prod_{k=0}^{3} x_k(n-1) + g\big(g_0 + x_3(n-1)\big)$$

$$x_2(n) = \sum_{i=0}^{3}\sum_{j=0}^{3} h_{ij}\, x_i(n-1)x_j(n-1).$$

$$x_3(n-1) = x_2(n-1)$$

< Decryption part >

$$r(n) = x_4(n) - \sum_{i=3}^{6}\sum_{j=3}^{6} g_{ij}\, x_i(n-1)x_j(n-1) - g_{123}\prod_{k=3}^{6} g_{123}x_k(n-1) - g\big(g_0 + x_3(n-1)\big)$$

$$x_5(n) = \sum_{i=0}^{3}\sum_{j=0}^{3} h_{ij}\, x_i(n-1)x_j(n-1)$$

$$x_6(n-1) = x_5(n-1)$$

Where Eps. (1), (2) and (3) present the encryption part (3), (4) and (5) show the decryption part, s(n) is the denotes the information signal, *r(n)* is the recover signal, $x_1(n), x_2(n)$,,, are the internal state variables of the system, and g_i, g_{ij}, h_{ij},,, are coefficients that are the private keys of the cipher. The nonlinear function *g(x)* is given by the following equation.

$$g(x) = \begin{cases} \dfrac{(Min - Max)x + (Max - a_1)Min}{Min - a_1} : (x_{min} \le x \le a_1) \\ \dfrac{(Max - Min)x + (a_1 Min - a_2 Max)}{a_1 - a_2} : (a_1 \le x \le a_2) \\ \dfrac{(Min - Max)x + (a_2 Max - a_3 Max)}{a_2 - a_3} : (a_2 \le x \le a_3) \\ \dfrac{(Max - Min)x + (a_3 Min - Max^2)}{a_3 - Max} : (a_3 \le x \le x_{max}) \end{cases}$$

We assume that a chaotic variable $x_1(n)$ issues form the modulator is transmitted exactly to the demodulator, namely,

$$x_4(n) = x_1(n).$$

2.2 Conventional Chaotic Modulatyor

The chaotic system adopted here show in (7)-(13) taking both linear and nonlinear digital filter structure.

<Encryption part>

$$x_1(n) = s(n) - g\big(x_1(n-1)\big) + \alpha x_3(n-1) + \theta$$

$$x_2(n) = h_0 + \sum_{i=0}^{3} h_i x_i(n-1) + \sum_{i=0}^{3}\sum_{j=0}^{3} h_{ij}\, x_i(n-1)x_j(n-1) + h_{123}\prod_{k=0}^{3} x_k(n-1)$$

$$x_3(n-1) = x_2(n-1)$$

< Decryption part >

$$r(n) = x_4(n) + g\big(x_1(n-1)\big) - \alpha x_6(n-1) - \theta$$

$$x_5(n) = h_0 + \sum_{i=0}^{3} h_i x_i(n-1) + \sum_{i=0}^{3}\sum_{j=0}^{3} h_{ij}\, x_i(n-1)x_j(n-1) + h_{123}\prod_{k=0}^{3} x_k(n-1)$$

$$x_6(n-1) = x_5(n-1)$$

Where Eps. (7), (8) and (9) present the encryption part (10), (11) and (12) show the decryption part, s(n) is the denotes the information signal, r(n) is the recover signal, $x_1(n), x_2(n)$,,, are the internal state variables of the system, and h_i, h_{ij},,,are coefficients that are the private keys of the cipher. The nonlinear function g(x) is same equation of proposed chaotic modulator, therefore explanation is omitted.

We assume that a chaotic variable $x_1(n)$ issues form the modulator is transmitted exactly to the demodulator, namely,

$$x_4(n) = x_1(n).$$

3 Fixed Point Calculation

In this section, we investigate fixed point calculation. 16 bit fixed point calculation called Q10 format was adopted in the chaotic modem system. Fixed point calculation has the advantage of faster than floating-point arithmetic calculations. But fixed point calculation interferes with accurate calculation often caused overflow. However the chaotic modem system incorporating overflow as a nonlinear function. In conclusion, the chaotic modem system gets more randomness.

3.1 Q10 Format

The chaotic system use 16-bit fixed-point digital computation called Q_{10}format that suppose lower 10 bits to be a decimal point. Therefore Q_{10}format's effective real number range is -32 to 31.99.

When 16-bit fixed-pint computation is adopted, chaotic properties of the modulated signal should be checked carefully as well as calculation error problems (especially, overflow problems). However, the chaotic system uses overflow problems as a non-linear. Therefore the chaotic encryption system gets more complexity.

The figure of Q10 format image show Fig 1.

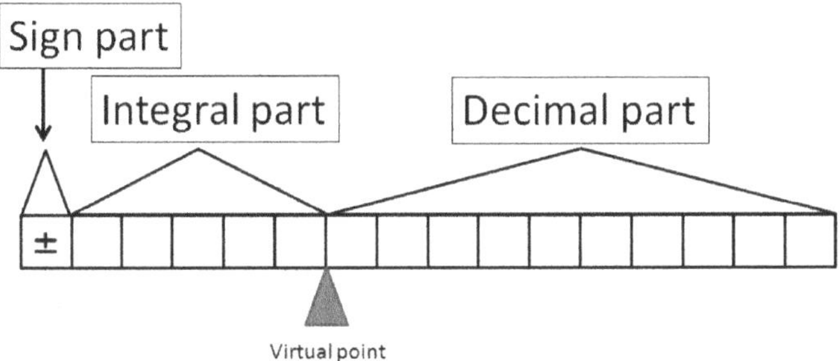

Fig. 1. Q10 format

4 Lyapunov Exponent of the Proposed Modulator

In this section, we investigate cipher Lyapunov exponent. The Lyapunov exponent is one of the quantitative indexes of chaos. If the maximum of the Lyapunov exponent is a positive number, the corollary shows chaotic. The value of Lyapunov exponent is higher, the system is more chaoric. The results of Lyapunov exponent of one of the parameters result by the chaotic modulator system show Fig 2.

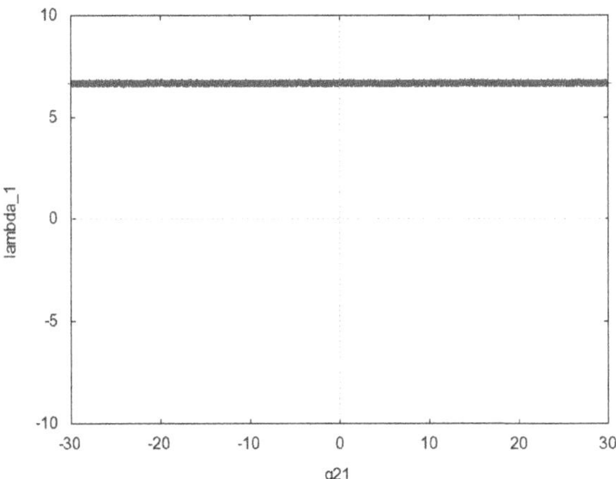

Fig. 2. Lyapunov exponent

5 Randomness Tests

In this section, we investigate randomness Tests.

The chaotic system takes two randomness tests of DIEHARD TESTS and NIST SP 800-22.

5.1 DIEHARD TESTS

The DIEHARD TESTS are randomness test developed by George Marsaglia. DIEHARD TESTS are built on 18 kinds of statistics tests. The tests in DIEHARD TESTS are output one or more p-value. DIEHARD TESTS output many p-value as a test result but explicit criteria of p-value is not provided. Therefore, the test is regarded as pass if the range of 0.001 to 0.009 the value of p-vlue.

5.2 NIST SP 800-22

The NIST SP 800-22 is randomness test developed by NIST. DIEHARD TESTS is built on 15 kinds of statistics tests. The paper shows best result of NIST SP 800-22.

The tests in NIST SP 800-22 are output one or more p-value same DIEHARD TESTS. The test is regarded as pass if the range of 0.001 to 0.009 the value of p-value.

5.3 Mersenne Twister

Mersenne Twister is one of the random number generator developed by Makoto Matumoto and Takuji Nishimura[4]. Mersenne Twister has the characteristic of long period and high-dimensional uniform distribution. In this paper, Mersenne Twister used to define parameters in Randomness Tests when the parameters are random changes.

6 Result of Randomness Tests (Regular Changes in the Parameters)

In this section, we investigate randomness Test that parameters of chaotic modulator are Regular changes.

An initial value of parameters of the chaotic system that play a key part is 2.4. When the chaotic system takes two randomness tests, one of the parameters of the chaos moves to -32 from 31.99 of Q_10format in units of 1. The purpose of this test determines any behavior that indicates when the parameters change.

6.1 Result of DIEHARD TESTS

The results of DIEHARD TESTS of one of the parameters result by the proposed chaotic system and conventional show Table 1.

Table 1. Result of DIEHARD TESTS

Test	Result
Birthday Spacing Test	100%
Overlapping 5-Permutation Test	92%
Binary Rank Test for 31x31 matrices	100%
Binary Rank Test for 32x32 matrices	100%
Binary Rank Test for 6x8 matrices	100%
Bit stream Test	93%
OPSO Test	95%
OQSO Test	92%
DNA Test	95%
Count-The-1's Test on a stream of bytes	100%
Count-The-1's Test for specific bytes	92%
Parking Lot Test	100%
Minimum Distance Tes	100%
3DSpheres Test	98%
Sqeezed Test	98%
Overlapping Sum Test	100%
Runs Test	100%
Craps Test	100%

6.2 Result of NIST SP 800-22

The results of NIST SP 800-22 of chaotic modulator with simultaneous Volterra filters show Table 2.

Table 2. Result of NIST SP 800-22

Test	Result
FREQUENCY TEST	100%
BLOCK FREQUENCY TEST	100%
CUMULATIVE SUMS TEST	100%
RUNS TEST	96%
LONGEST RUNS OF ONES TEST	100%
RANK TEST	100%
DFT TEST	100%
NONPERIODIC TEMPLATES TEST	30%
OVERLAPPING TEMPLATE OF ALL ONES TEST	96%
UNIVERSAL STATISTICAL TEST	100%
APPROXIMATE ENTROPY TEST	100%
RANDOM EXCURSIONS TEST	70%
RANDOM EXCURSIONS VARIANT TEST	66%
SERIAL TEST	100%
LINEAR COMPLEXITY TEST	98%

7 Result of Randomness Tests (Random Changes in the Parameters)

In this section, we investigate randomness Test that parameters of chaotic modulator are Random changes.

An initial value of parameters of the chaotic system that play a key part is random provided by Mersenne Twister. The sample number of tests result is 1000. The proposed of this test that check the chaotic modulator's behaviour intended to use in practice.

7.1 Result of DIEHARD TESTS

We show the result of DIEHARD TESTS of proposed chaotic modem system and conventional chaotic modem system Table 3.

Comparing the two results reveals that the improvement of the proposed method is more up to 5 percent.

Table 3. Result of DIEHARD TESTS (Result of proposed Chaotic Modem System)

	proposed	conventional
Birthday Spacing Test	99%	98%
Overlapping 5-Permutation Test	88%	87%
Binary Rank Test for 31x31 matrices	99%	99%
Binary Rank Test for 32x32 matrices	99%	99%
Binary Rank Test for 6x8 matrices	99%	99%
Bit stream Test	99%	92%
OPSO Test	94%	89%
OQSO Test	93%	89%
DNA Test	91%	91%
Count-The-1's Test on a stream of bytes	99%	95%
Count-The-1's Test for specific bytes	95%	93%
Parking Lot Test	99%	98%
Minimum Distance Test	99%	98%
3DSpheres Test	99%	99%
Sqeezed Test	98%	97%
Overlapping Sum Test	99%	99%
Runs Test	99%	98%
Craps Test	99%	97%

7.2 Result of NIST SP 800-22

We show the result of NIST SP 800-22 of proposed chaotic modem system and conventional chaotic modem system Table 4. Comparing the two results, the proposed method has improved slightly. But, the result of NONPERIODIC TEMPLATES TEST, RANDOM EXCURSIONS TEST and RANDOM EXCURSIONS VARIANT TEST is down.

Table 4. Result of NIST SP 800-22 (Result of proposed Chaotic Modem System)

	proposed	conventional
FREQUENCY TEST	99%	98%
BLOCK FREQUENCY TEST	98%	99%
CUMULATIVE SUMS TEST	98%	98%
RUNS TEST	99%	99%
LONGEST RUNS OF ONES TEST	99%	99%
RANK TEST	98%	98%
DFT TEST	98%	98%
NONPERIODIC TEMPLATES TEST	21%	23%
OVERLAPPING TEMPLATE OF ALL ONES TEST	99%	98%
UNIVERSAL STATISTICAL TEST	98%	98%
APPROXIMATE ENTROPY TEST	99%	98%
RANDOM EXCURSIONS TEST	57%	60%
RANDOM EXCURSIONS VARIANT TEST	57%	60%
SERIAL TEST	98%	96%
LINEAR COMPLEXITY TEST	99%	98%

8 Cipher Speed

In this section, we show the cipher speed test.

The test of cipher speed measure the time that encrypt a random number 500000 times. The result of this test is average 100 times of cipher speed tests.

The result of test shows Table 5.

Table 5. Cipher Speed

Result of proposed chaotic modulater	0.09261[sec]
Result of conventional chaotic modulater	0.07902[sec]

9 Cipher Speed

The novel chaotic modulator for chaotic cipher and secure communications has been proposed in this study. The proposed system includes simultaneous Volterra Filters. Therefore, the proposed system gets more randomness than conventional chaotic systems in DIEHARD TESTS. However, the proposed system has lower results of the tests in NIST SP 800-22 than conventional chaotic systems. And the proposed system's cipher speed is slower than conventional chaotic systems. Therefore, we will study of the chaotic system for speedup and get more randomness in NIST SP 800-22. And report this research continuously in this future.

References

1. Kamata, H., Umezawa, Y., Dobashi, M., Endo, T., Ishida, Y.: Private communications with chaos based on the fixed-point computation. IEICE Trans. E83-A(6), 1238–1246 (2000)
2. Iwata, K., Nakamura, T., Ikeue, I., Irikura, H., Kamata, H.: Chaotic modulator with volterra filter for cipger. In: NOLTA 2007 (2007)
3. Akihara, K.: Chaotic neural Network. Bifurcation Phenomena in Nonlinear Systems and Theory of Dynamical System, 143–161 (1990)
4. Mersenne Twister Home Page, `http://www.math.sci.hiroshima-u.ac.jp/~m-mat/MT/mt.html`
5. NIST.gov - Computer Security Division - Computer Security Resource Center, `http://csrc.nist.gov/groups/ST/toolkit/rng/documentation_software.html`

Research on Maneuver Model of the Parameterized Tank Entity in CGF

Qian Lidong[1], Song Xiao[1], Shi Xuecheng[1], Zhao Xinjun[2], and Zhao Fang[2]

[1] School of Automation Science and Electronic Engineering, Beihang University, 100191 Beijing, China
qianlidong2006@126.com
[2] China North Vehicle Research Institute, 100072 Beijing, China

Abstract. Starting with demonstrational parameters, this paper comes up with a practical parameterized maneuver model of tank entities in CGF system, which solves the problem of real-time simulation in large-scale virtual environment. This paper introduces the path planning of tank entities during the simulation including global and local path planning. With the limit of maneuver demonstrational parameters, it plans the optimal path in the global and local scope. Besides, with the accord of some demonstrational parameters, it analyses the motion state of tank modeling on the base of optimal path. The analysis shows that the parameterized maneuver modeling can improve the capability of real-time in CGF system. It is a practical method of simulating.

1 Introduction

CGF(Computer Generated Force) is diveded into two types of platform-level and aggregate-level. The platform-form CGF is used for mainly used for tactial exercises of small-scale units, such as battalion units. And the aggregate level CGF conducts adequate operational model by a certain amount of combat units such as the division units, which can complete simulately the same task as the real combat units without the human control in the virtual environment.

In recent years, the simuation of maneuver model of tank entities in large-scale virtual battlefield has become more and more important to domestic and foreign reseach institutes. Literature [1] indicates that the comprehensive project of virtual simulation in the U.S Central Folride, IST(Institute for Simulation and Training) solves the connections and conversions between aggregate-level tanks and platform-level tank effectively, which gives a clear description of the complex relationships among various factors. Literature [2] comes up with an Agent-Oriented CGF system. It designs the simulation system architecture of a tank company and discusses its behavior of intelligent mobility in virtual environment. Literature [3] evaluates the maneuver as one of the most important characteristics for tank simulation

J.-H. Kim et al. (Eds.): AsiaSim2011, PICT 4, pp. 71–79, 2012.

and finds that its advantages and disadvantages of the combat capability of the tank playing an extremely important impact.

Taking the above factors into account, this paper presents a practical parameterized maneuver model of tank entities to CGF system. The parameterized maneuver model of tank entities can both reflect the relationship of tank itself and combat environment and describe the internal complex force behavior by limited state. It can improve the capability of real-time in CGF system.

2 Global Path Planning before Simulation

The movement of tank entities is not random clutter movement. It takes the planned path to the list as support. From the macro level, it is to obey the order from the leader in the battlefield and march in accordance with the planned path list. The determination of the path list must be obtained through the path planning module. Therefore, the path planning module is a motorized model of force generation system. The path planning of tank is divided in two parts: global planning and local planning. Global planning reflects the route planning from the superior and local planning reflects the autonomy of the tank platform resulted in the road information with a list consisting of serises of points.

2.1 Planning Points Based on Military Plans

Global path planning begins before the simulation of tank foces. In the first phase of the global planning, it executes the military plans. Assume that there is seven points in the path and the tank entity moves from point 1 to point 2 during the time pushing. It begins the second response 30 minutes later from the point 2 to point 3 and the rest can be done in the same manner.

2.2 Path Planning among the Planned Points

In the second phase of the global planning, it conducts the path planning during the planned points with genetic algorithms [6]. There is a collection U combined with all points during the path including all vertices of obstacles. Assume that the list L consists of all planned points and then the path planning based on obstacle ploygon can be expressed as to find an entity in the list L, which can make the maximum fitness value of the entity.

Assuming that tank entities march along with less than 90 degree from the starting point to ending point. Suppose that there are three polygon obstacles(forest, swamp, lake) as shown in Figure 1:

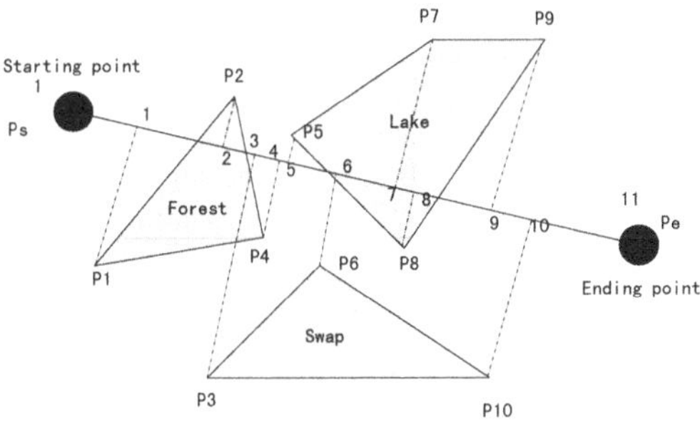

Fig. 1. Polygon obstacles

In Figure 1, the numbers 1-10 correspond to the vetices of the polygons $P_1, \cdots, P_{10}$, which are the array of all vertices along the direction tank marches. So $P_s \to P_1 \to P_2 \to \cdots \to P_9 \to P_{10} \to P_e$ is the total sequence of the points noted as U . So the path planning from starting point P_s to ending point P_e can be described as selecting m points from the sequence U and forming the new sequence L, which makes the minimum cost of L and the maximum fitness value of the entity.

3 Local Path Planning during Simulation

Local planning is mean to search the route in a small-scale and unknown space, such as obstacle avoidance, which shows the autonomy of the tank platform. In the basis of global path planning, it makes a path plan in the small-scale space. Assume that tanks move forward 100 meters each time, which can be adjusted with the actual simulation system, and set a path point in each 100-meter. Finding an element from the collection L to calculate its cost function and making the cost of collection L minimum and the fitness value maximum. Besides, Local path planning should determine whether the vehicle is outside the scope of the global path, that is, to determine whether the vehicle is driving side, in which case, it needs to determine whether the vehicle is left or right side of the rode, the distance between vehicles and road.

Parameterized maneuver model of tank entities corresponds to the analytical model of formation instruction when calculating whether it can move from the current point to the next point. According to the demonstational parameters, it can be determined whether the tank can continue the next formation changing. For example, when climbing the slope, it should be determined whether the slope is larger than the tank entity gradeability. If the slope is larger than the tank entity gradeability, it stops and conducts the local path planning again. Other maneuver parameters are similar to the slope. The tank maneuver parameter table is shown in Table 1.

Table 1. Maneuver parameter table

number	Maneuver parameter	variable	Influence range
1	Gradeability	a	the current position + visual distance
2	Maximum roll slope	b	the current position + visual distance
3	trench width	c	the current position + visual distance
4	vertical wall height	d	the current position + visual distance

The value of tank maneuver parameters is taken based on the actual model.

4 Kinematic Model in Tank Simulation

The maneuver model is to deal with the change of tank's velocity and acceleration in specific battlefield environment, including natural geographical environment, the enemy situation and entities' movement mode. The most important is the physical kinematic model as the rest of the environment can be translated into the influence factors to physical kinematic model. As a result, the basic part of maneuver model is the kinematic model. It is the basis for entities in CGF system and a variety of tactics and strategy is reflected through the movement.

Tank kinematic simulation model consists of two sub-models:

(1) Analytical model of action instruction. This model is mainly responsible for the commands in relevant to movement, which are mapped to the action that the driver needs to execute. The commands which are in relevant to movement correspond to the acceleration, uniform motion and deceleration. And the three actions correspond to three different throttle open.

(2) Motion state computing model. This model is mainly responsible for that the throttle actions the driver executes which correspond to power value and the current velocity and acceleration values.

4.1 Analytical Model of Action Instruction

General action commands can be parsed into the following actions that the driver needs to execute.It is shown in Table 2:

Table 2. Analytical table of action instruction

Action command	Action that the driver needs to execute
Accelerated foward	Accelerate
Decelerated foward	Decelerate
short stop shooting	Stop
passing the mine field	Uniform
entered	Uniform
Moving shooting	Uniform
Cover	Stop
Right bypass	Uniform and turning

4.2 Motion State Computing Model

4.2.1 Analytical Model of Action

The main function of the analytical model of action is to parse the actions the driver needs to execute into power value of different state during the process. It is shown in Table 3.

Table 3. Analytical table of action

Actions dirver needs to execute	Power value
startup	$p = p_S$
Accelerate	$p = p_A$
Decelerate	$p = p_D$
Uniform	$p = p_U$
Stop	p= 0

As for turning movements, turn the yaw angle to some degrees at each moment. The equations are as follows:

$$\begin{cases} yaw_t = yaw_{t-1} + \Delta yaw, yaw_t \le yaw_{max} \\ yaw_t = yaw_{max}, yaw_t > yaw_{max} \end{cases} \tag{2}$$

4.2.2 Analytical Model of Velocity

Analytical model of velocity is mainly responsible for calculating the velocity of tank entity for the current moment. It can be calculated as follows:

$$\begin{cases} v_t = v_{t-1} + a_{t-1}\Delta t, v_t \le v_{max} \\ v_t = v_{max}, v_t > v_{max} \end{cases} \tag{3}$$

v_t is the velocity of the tank entity for current moment and v_{t-1} is the velocity of the tank entity for previous moment. a_{t-1} is the acceleration of tank entity for previous moment. Δt is the time interval which also means simulation step. v_{max} is the maximum velocity of the tank entity.

4.2.3 Analytical Model of Acceleration

Analytical model of acceleration is mainly responsible for calculating the acceleration of tank entity for the current moment.

Analytical model of acceleration is mainly responsible for calculating the acceleration of tank entity for the current moment.

It calculates the power of the tank entity by analytical model of action and the velocity for the current moment by analytical model of velocity. With these values, the traction for the current moment can be achieved. It can be calculated as follows:

$$F_T = \frac{p}{v_t} \tag{4}$$

F_T is the traction of the tank entity. p is the power and v_t is the velocity of the tank entity for the current moment.

Therefore the resultant force can be calculated as follows:

$$F_R = F_T - (F_w + F_f + G \cdot \sin\alpha) \tag{5}$$

F_R is the resultant force that the tank entity. F_w is the force by air resistance. F_f is the force by rolling resistance. G is the gravity for the tank entity. α is the slope angle of the location the entity stays.

Finally, the acceleration of the tank entity for the current moment can be calculated as follows:

$$a_t = \frac{F_R}{m} \tag{6}$$

a_t is the acceration of the tank entity for the current moment. m is the quality of the entity.

4.3 Corresponding Relationship with Maneuver Parameters

Parameterized maneuver model of tank entity corresponds to the CGF simulation system during the calculation of the state of motion. The demonstration table of parameterized maneuver model is shown in Table 4.

Table 4. Demonstration table of parameterized maneuver model

Number	Demonstration parameter	variable	Influence range
1	Maximum power	A	throttle open
2	Maximum velocity	B	current topographic map
3	Average velocity of highway	C	current topographic map with current terrain
4	Average velocity of cross-country	D	current topographic map with current terrain
5	Maximum acceleration	E	current topographic map
6	Maximum deceleration	F	current topographic map

Each value of maneuver parameter is determined by the type of tank entity.

Tank entities start from the stationary state and determine whether it needs to execute the startup command. If there is no startup command, then continue to remain stationary, otherwise accelerating with startup power p_s. According to the formulas $p = F_T * v$ and $a = F_R / m$, calculating the current velocity and determining whether the average velocity reaches the average velocity of highway C or cross-country D. If the average velocity is not reached, then continue to accelerate to the average velocity and go on marching forward. If it has reached the average velocity, then determine whether it reaches the maximum velocity B. If there is a command from the superior that requires accelerating to the maximum velocity, then continue to accelerate to the maximum velocity B with maximum acceleration E. Go on marching forward with the average velocity if there is no commands from the superior. According to the formulas $p = F_T * v$ and $a = F_R / m$ determining whether it reaches the maximum velocity. If it has reached the maximum velocity B, determining whether it needs to decelerate. Decelerate with the power $p = p_D$ if it needs decelerating. According to the formulas $p = F_T * v$ and $a = F_R / m$ determining whether it has decelerated to the average velocity of highway C or cross-country D. When it decelerates to the average velocity, determining where it needs to continue decelerating. Decelerate with the power $p = p_D$ if there is a command of deceleration from the superior and calculate its velocity with the formulas and $a = F_R / m$ until it reaches the stationary state. At that moment, the power of the entity is 0. Kinematic state transition is shown in Figure 2.

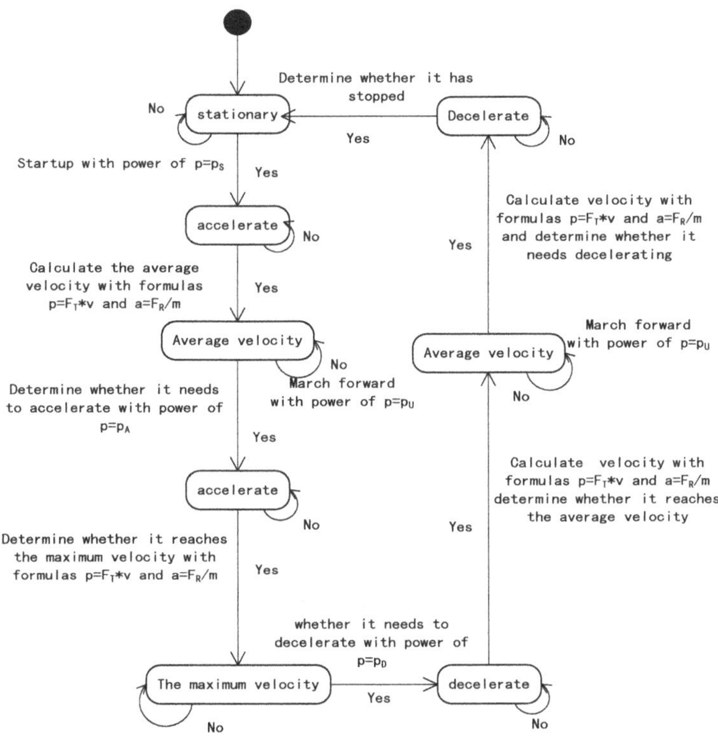

Fig. 2. Kinematic state transition

5 Simulation Analysis

With the path planning model and kinematic model, it takes an experiment of 5 tanks of the regiment level at the ZZSIM simulation system. It makes a statistics of maneuver time for these tanks and the result is as follows:

Table 5. Cycle time of the maneuver process/μs

	Cycle time 1	Cycle time 2	Cycle time 3	Average cycle time
Tank 1	339	336	330	335
Tank 2	277	253	254	261
Tank 3	269	266	251	262
Tank 4	266	265	250	260
Tank 5	266	266	251	261

The simulation results show that it can improve the real-time effectively with parameterized maneuver model of tank entities. The average cycle time can reach 0.3ms, which shows that this method can simulate the maneuver model in large-scale virtual battlefield fast.

6 Conclusion

This paper makes a discussion about the maneuver model of tank entities in the large-scale virtual battlefield. It comes up with a parameterized maneuver model of tank to CGF system with the charateristics of the virtual battlefield. With maneuver parameter, planning the optimal path in the global and local space and calculating the motion state of the tank during the movement. After preliminary testing, the cycle time of the maneuver model is about 0.3ms, which meets the requirement of real-time.

References

1. Yang, L., Guo, Q.: The Development of the Research of Computer Generated Force. Computer Simulation 17(3), 4–7 (2000)
2. Pei, H., Song, L., Liu, L.: Research on Intelligent Maneuver Modeling of Tank Company CGF System Based on Agent. Journal of Academy of Armored Force Engineering 21(2), 20–23 (2007)
3. Liu, X., Gao, H., Wang, L.: An Application of Grey Relational Analysis in Mobilityevaluation of Main Battle Tanks. Acta Armamentarii the Volume of Tank, Armored Vehicle and Engine (3), 43–49 (2000)

4. McDowell, P., Darken, R., Sullivan, J., Johnson, E.: Delta3D: A Complete Open Source Game and Simulation Engine for Building Military Training System. The Journal of Defense Modeling and Simulation: Applications, Methodology, Technology 3(3), 143–154 (2006)
5. Xing, L., Chen, Y., Wang, Z.: Evaluation of Overlang Mobility of Tank Based on Simulation. Computer Simulation 21(7), 4–7 (2004)
6. Lu, M.: Research of Path Planning Based on Virtual Battefield Environment. Beihang University, Beijing (2010)

Extended RUSES

Suk-Hoon Shin, Eun-Bog Lee, and Sung-Do Chi

Intelligent System Research Laboratory, Korea Aerospace University,
Dept. of Computer Engineering, Goyang-si, Kyunggi-do, Korea
{ev4shin,danalee,sdchi}@kau.ac.kr

Abstract. This paper proposes an Extended RUSES for improving the conventional rule-based SES(RUSES). The Extended RUSES is maintaining the advantages of RUSES for supporting the configuration automation, capability as well as providing context sensitive pruning by adding the new property setting rules to entity based on this idea, the methodology of Extended RUSES has been suitably proposed.

1 Introduction

The SES(System Entity Structure, Zeigler, 1992) is a representation scheme that contains the decomposition, coupling and taxonomy information of a system. The SES contains three types of nodes-entities, aspect, and specialization-which represent three types of knowledge about systems, the entity mode, which may have several aspects and/or specializations, corresponds to a model component that represents a real-world object. One application of the SES framework relates to the design of a system. Here the SES serves as a compact knowledge representation scheme of organizing and generating the possible configurations of a system to be designed. To generation a candidate design, we can use pruning that reduces the SES to a PES(Pruned Entity Structure)[1,2,3].

The pruning process restricts the space of possibilities for selection of components and coupling that can be used to realize the system being designed. Thus we can assume that design may now be reduced to the synthesis problem. Synthesis involves putting together a system from a known and fixed set of components in a fairly well prescribed manner. In the synthesis problem, we are modeling a rather restricted design process, one amenable to automation by extracting concepts and procedures from expert's knowledge and experience, augmenting them and modeling them into a coherent set of rules had proposed the rule-based SES(RUSES)[1,2,3,4,5].

This rule-based approach can be easily realized using the generic frame concept, i.e., the objectives, requirements, and constraints of the design are expressed in an object called a generic frame. At any point in an SES where more than one choice is presented, e.g., a selection between two or more different special types of a component, attributes and rules are attached to the components to guide in the pruning process, A pruning procedure checks these attributes and rules against those in a generic frame to select the appropriate entities, the selected entities are used to construct

J.-H. Kim et al. (Eds.): AsiaSim2011, PICT 4, pp. 80–85, 2012.

design models. The pruning algorithms generate all design model structures that conform to the design objectives and constraints [1,5]. This methodology was differentiated from others in that it can support configuration automation by utilizing the knowledge of a system design expert and it is established on the basis of the well formalized framework that can support hierarchical and modular system design[1,2,3,5]. However, but RUSES cannot provide the for fully supporting the context sensitive pruning introduced by Zeigler(1992)[1,5].

In this paper, we propose the Extended RUSES methodology through the complement and extension of RUSES. Zeigler(1992) presented that getting context sensitive pruning process and result to define the SES expressed sufficient knowledge for synthesis problem solving and proposed pruning methodology and constraint rule(See Fig.1)[1].

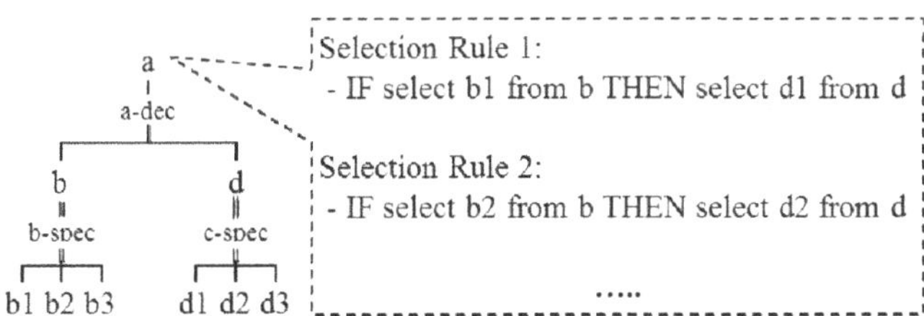

(a) Context Sensitive Constraint Rules

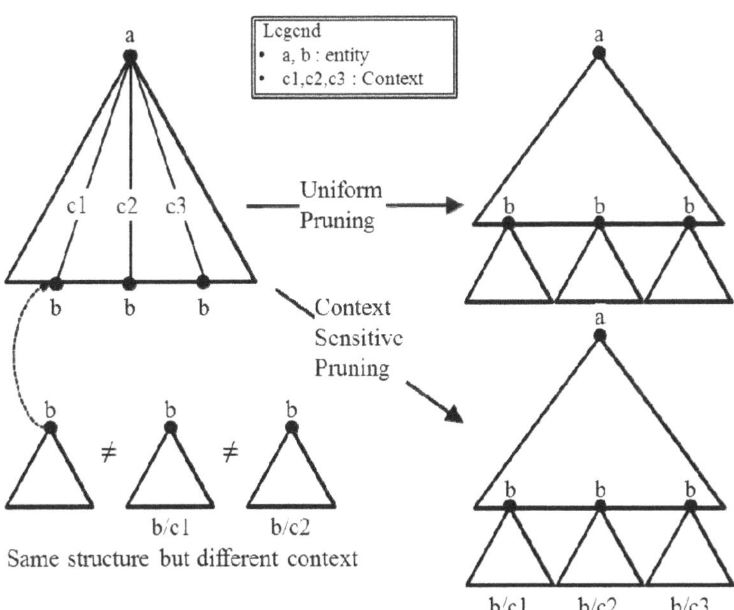

(b) Context Sensitive Pruning

Fig. 1. Context Sensitive Rules and Pruning[1]

First, context sensitive pruning is a pruning based on the specification by semantic difference for each specialization when entity has same substructure is chosen. Next, context sensitive constraint rule is the rule of suggest choices that pruning result to be a synthesis result meet SES design intent[1,2,5].

In context sensitive pruning, copies of **b** are made for each occurrence of entity **b**(See Fig.1(b)). Each copy is separately pruned, renamed and pasted back in the appropriate context. Recall that due to the valid brothers axiom each occurrence of **b** is reached by a uniquely labeled path from the root. This path (actually the minimal part needed for proper discrimination) provides the context needed by the user for pruning. Each item in a pruned copy of **b** is renamed to be distinct from occurrences in other copies of **b**. In renaming, couplings and priority lists are altered appropriately. Note that **b** may itself have multiple occurrences if entities. Thus context sensitive pruning is a recursive procedure[1]. In this approach there is no formal way of recognizing that models of the same underlying entity are being referenced. The various models of bottle are dispersed throughout the structure, we can greatly improve coherence by using an SES in which the generic entity **b** replaces each of its special cases and by collecting them together as specializations in an SES, **b**, as in Fig.1(b). Organizing models by the entity they model, rather than the context they are used in, facilitates evolvability. Recall that models of an entity may be related by abstraction relationships so that when one is changes other must be amended to retain consistency[1].

Context sensitive constraint rule is the rule of suggest choices that pruning result to be a synthesis result meet SES design intent. A corresponding class of rules can be defined to assist context sensitive pruning. As shown Fig.1(a), we can have the selection of an entity from a specialization be governed by context. Moreover as shown in **b** of Fig.1(a), imposing syntactic patterns, we can state a generalized rile to replace an indefinite number of special cases of the same form[1].

RUSES produces the context sensitive result by limited the possible combinations of synthetic applying constraint rules. But it cannot clearly express semantic differences of the pruned result by specialization that entity has the same sub-structure.

2 Extended RUSES

The purpose of the Extended RUSES is maintaining the advantages of RUSES and the providing fully context sensitive. Extended RUSES is maintaining the automated methodology. When entered the user's requirements, the pruning engine automatically generates a corresponding PES. During the pruning, the pruning engine checks the requirement from the user by utilizing the inference engine. Inference engine retrieves the corresponding entity rules to find the goal entity as well as proper properties by firing the rules based on the mechanism to inference. In this way, the pruning(selection) process is proceeded until to create a PES.

Overall methodology for developing the Extended RUSES is in Fig.2. Extended RUSES uses generic frame for define the entity property and rules like conventional RUSES. Conventional RUSES only deals with the synthesis rules, selection rules as its generic frame. However, the Extended RUSES can provide the property setting rules in addition to conventional rules such as synthesis rules, selection rules(See Fig. 2). Entity **b** and **c** in Fig.2 are specialized entity that is decomposed into the entity **a1**, **a2** and **a3**'s property (initial value of inner variable, e.g.,) in entity **a**'s structure.

When SES a in Fig. 3 does context sensitive pruning, generated a structure having **b** or **c**'s special property, at this time property rule is defined as rule for setting **a1**, **a2**, **a3** by chosen **b** or **c**'s property value.

The context sensitive pruning result of applying Extended RUSES is under Fig.3. Entity **b.a** and **c.a** have same substructure model. But context sensitive pruning results do not have property of substructure. In this way, the pruning of Extended RUSES can be satisfied given context sensitivity.

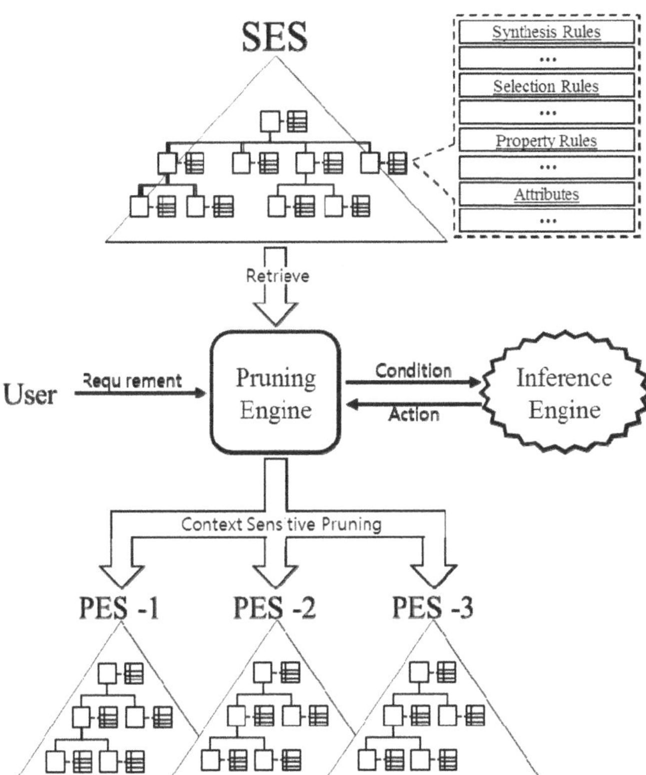

Fig. 2. The Methodology of Extended RUSES

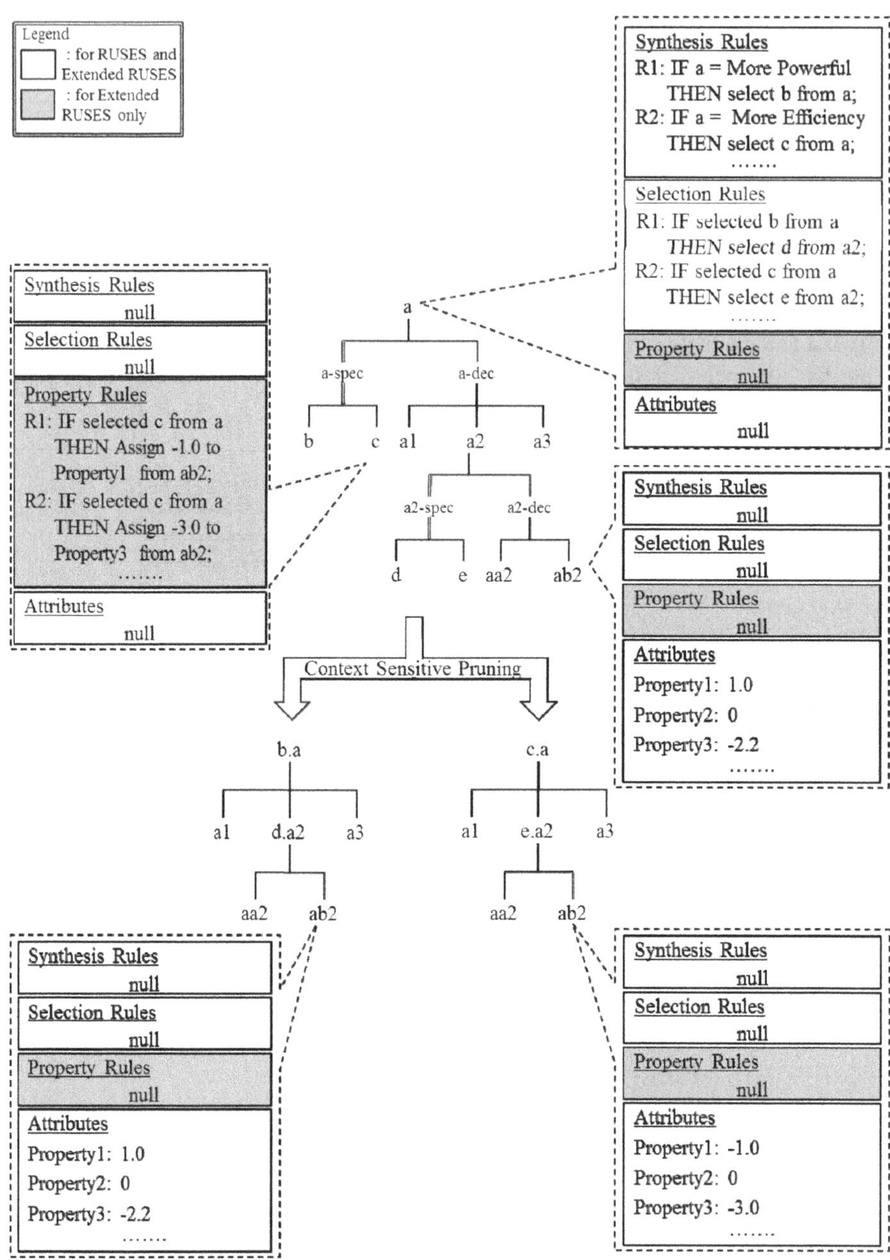

Fig. 3. An Example of Extended RUSES

3 Conclusions

The extended version of RUSES has been successfully proposed by adopting the context sensitive pruning concept introduced by Zeigler[1]. The major advantage of our research is:

(1) Supporting high level of context sensitivity of SES
(2) Rule-based context sensitive pruning

In near future, the research should be continued to verify our idea.

Acknowledgment. This paper with the support of underwater vehicle research center specializing in the SM-11 challenge "A system of underwater vehicle / sub system features and performance simulation models for the network-based structural studies of the virtual com-plex system" was carried out.

References

1. Zeigler, B.P.: Object-oriented Simulation with Hierarchical, Modular Models: Intelligent Agent and Endomorphic Systems. Academic Press (1990)
2. Zeigler, B.P.: System Representation From Newton to Minsky and Beyond. In: Applied Artificial Intelligence, vol. 1, pp. 87–107. Hemisphere Pub. (1987)
3. Zeigler, B.P.: System-theoretic Representation of Simulation Models. IIT Transactions, 19–34 (1984)
4. Lee, H.J., Zeigler, B.P.: SES-based Ontological Process for High Level Information Fusion. In: Proceeding of the 2010 Spring Simulation Multiconference, SpringSim 2010 (2010)
5. Chi, S.D., Lee, J.K.: Evaluating Alternative System Designs, TRANSACTIONS of the Society for Computer Simulation International. The Society for Computer Simulation International 17(3), 148–160 (2000)

Application of Symbiotic Decision Support to Managed Evacuation Studies Using a Perennial Framework

Seth Normand Hetu and Gary Tan

Department of Computer Science, School of Computing,
National University of Singapore , Computing 1, 13 Computing Drive,
Singapore 117417
seth.hetu@gmail.com, gtan@comp.nus.edu.sg

Abstract. A perennial simulation is used to optimize a managed evacuation through the use of symbiotic simulation in conjunction with the Indriya sensor network. To the best of our knowledge, this represents the first full-scale application of perennial simulation to the building of a computer simulation system. We show several key benefits from the use of this modeling methodology technique, and present components of the designed system.

1 Introduction

Crisis management in urban spaces often calls for effective strategies for evacuating buildings in a safe and orderly manner. One common technique is called a "managed" (alt. "guided") evacuation, in which teams of evacuation personnel provide guidance to large groups of occupants as they evacuate in real-time. The use of intercommunicating teams, combined with a central dispatch, improves both evacuee safety and total evacuation time. It should come as no surprise that this conjoint task is a suitable candidate for simulation-based optimization. In particular, a promising new approach called *symbiotic simulation* seems capable of fulfilling these particular real-time needs. Early limitations of symbiotic simulation prevented it from being used in the domain of crisis management. In this paper, we will present the use of our previously-detailed "perennial" simulation framework to represent and reason about this problem. We will show that the perennial framework is a good tool for construction of these types of simulations.

Our previous work showed that symbiotic simulation possessed a promising potential benefit to crisis management tasks such as building egress, as well as confirming the observation that people tend to react positively to advice under conditions of low to no stress. This paper expands on that work by offering the first fully-featured example of a perennial simulation designed from the ground up according to our framework. It is of immense value, from a modeling and methodology point of view, to observe the benefits gained through an appeal to perennial simulation.

The remainder of this paper will proceed as follows. Section 2 will explain the problem with respect to related work in the field. Section 3 will introduce our existing work on symbiotic crisis management, and will frame the problem within this

J.-H. Kim et al. (Eds.): AsiaSim2011, PICT 4, pp. 86–94, 2012.

domain. Section 4 will explain the design and setup of our experiments, and Section 5 will provide preliminary results from this system. Section 6 will conclude the paper.

2 Problem Description and Related Work

This study builds on existing work in the fields of crisis management and symbiotic simulation, both of which will be discussed next. Following that, a brief overview of our previous work on perennial simulation is will be given.

2.1 Work in Crisis Management

Significant existing work abounds in the domain of crisis management. Basic movement simulation was first enabled through Fruin's (and later Togawa's) observations on crowd movement speed [6][21], which allowed crisis planners to predict movement throughout a crisis. These simple speed/density relationships were soon replaced by reverse-engineered behavioral models such as Helbing's "Social Forces" model, which treated each pedestrian's goal as a virtual force analogous to physical forces like gravity [9][10]. Later work improved behavioral models to handle cases where people did not simply move from source to sink node, such as "window shopping", wandering, or route choice in the face of multiple, equally-distanced paths. The *Legion* simulator, for example, uses computer software agents to model human movement to very high degrees of precision in areas such as subway stations and sports arenas [18].

This slow evolution from simple equation-based models to computer simulation is a trend common in many fields. Crisis management simulation makes an appeal to simulation for many tasks such as earthquake and fire spread models [5][2]. In terms of actively saving human lives, crisis-related tasks include things like simulating hospital operating conditions at peak capacity [17], training communities through the use of realistic training simulations [16], and allocating resources or guiding evacuations at the time of crisis itself [14]. Crisis management exhibits a clear trend from simple models to more powerful simulations, and a natural extension of this trend is an appeal to real-time simulation-generated feedback.

2.2 Work in Symbiotic Simulation

Symbiotic simulation is a technique introduced in 2002 to deal with the complexities of monitoring and responding to real-time systems via simulation [7]. Conceptually speaking, symbiotic simulation consists of a feedback loop based on "What If?" simulation analysis. Multiple "What If?" analyses are performed either in reaction to a threshold being crossed or in real-time to prevent it from being breached. Output data from these simulation runs are captured and analyzed to determine how to optimize the system and maximize the amount of time it is functioning at a level below the threshold. Once this decision has been made, the system is altered using *effectors*, which are analogous to *sensors* except that they modify their target instead of measuring it. Finally, the system continues running in its altered state, and output analysis is performed to determine how accurate the chosen "What If?" scenario actually was. This can be used for auto-validation, or to initially calibrate the system.

Despite its promising potential, symbiotic simulation as originally detailed had several key shortcomings. Introduced for "rigid" domains such as manufacturing and shipping [15][1], the technique has no directly obvious way of dealing with more flexible domains such as those with large numbers of human participants. Our previous work on perennial simulation attempts to alleviate these concerns, and introduces a notion of utility to help capture the myriad performance parameters of crisis.

2.3 The Perennial Simulation Framework

We observed that the primary stumbling block in using symbiotic simulation for real-time, human-saturated applications was the lack of a suitable design framework. As such, in [12] we introduced the notion of a *perennial* simulation framework, which was partly tested with a traffic simulation study in [13]. The full details of a perennial simulation are outlined in these papers, and relevant sections will be expounded in Section 3 as we describe our sample study. Figure 1, below, provides a UML diagram of the latest version of the Java reference implementation of the perennial framework. Figure 2 shows a high-level overview of the framework.

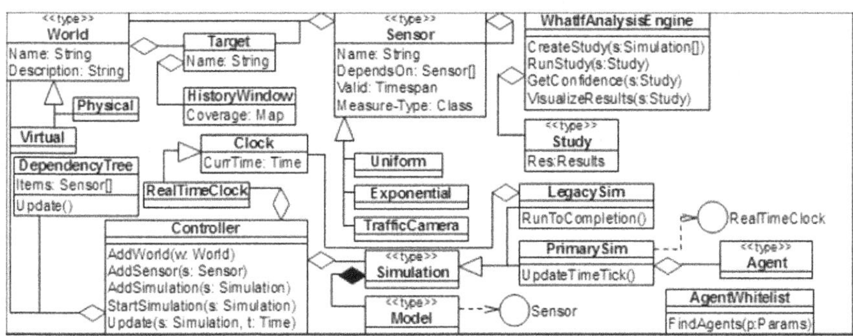

Fig. 1. UML diagram of the Java implementation of a perennial simulation system

Several key benefits are immediately obvious from these diagrams. First, the framework allows for a more systematic construction of a long-term simulation than was previously possible. Second, it does so by placing a very small number of restrictions on the design stage of the system, and although the reference implementation is in Java, the conceptual diagram (Figure 2) is clearly generic enough to be realized in any general purpose programming language. Third, by complying with our framework, it becomes significantly easier to perform *foresight* and *hindsight* studies, as well as *symbiotic decision support*. Some examples of the former will be given in Section 6.

3 A Symbiotic-Guided Managed Egress System

Our system involves managed evacuation exercises, in which small teams proceed through a building room by room and notify each room's occupants that the building must be evacuated. This kind of methodical and thorough evacuation is necessary when one must ensure that all rooms are emptied while minimizing panic. The crisis event in question could be a bomb threat, a recent epidemic outbreak, or an imminent earthquake.

We intend to abstract the occupants into a virtual setting through the use of our perennial framework. Figure 2 presents the high-level overview of this framework. We will now detail each component as it pertains to our study.

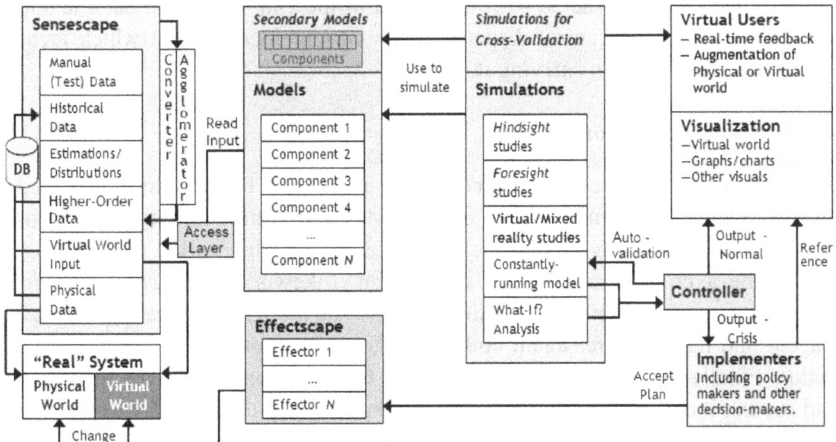

Fig. 2. High-level diagram of a perennial simulation framework

3.1 Worlds, Sensors and Effectors

Our system consists of two **worlds**: the physical building that teams navigate and a virtual room-for-room copy of that world within which occupants roam. Based on [12], we can say that these two worlds are *virtually established*, which is the most common setup for virtual world simulations. The building being studied is the COM1 building at the National University of Singapore. Its floor plan is displayed in Figure 3, and is typical of university complexes.

The worlds mentioned contain a variety of sensors. The physical location contains WiEye, SBT80, and SBT30 sensors, which detect human presence through long-range passive infrared scanning in combination with visual light and acoustic sensor streams. The latter two sensors can also detect temperature, magnetic fields, and acceleration. All of these sensors communicate through the use of TelosB motes, as part of the Indriya testbed at NUS. [4] The virtual world has access to all of these sensors in the form of historical data, which are automatically saved by our perennial system. Finally, both worlds make use of the perennial system's agglomerator to combine these myriad and noisy data points into a higher-order sensor that detects human presence in a room. The combinatorial function used to achieve this is built on typical acoustical tracking approaches such as [19], combined with some considerations for multi-target environments [3]. Since the TelosB motes are located at ceiling level, accurately detecting room counts was not possible, so we restricted our measurements to several "levels" of room occupancy: Empty, Sparse, Full, and Overfilled.

Our system contains two effectors, both of which affect the virtual world in real-time. The first informs teams of their notification routes, and is updated by the "What If?" simulation to choose the best notification strategy for all teams combined. In reality, the physical effector may take the form of a PDA or cell phone, by which the simulation can communicate with the teams and pass them information. The second effector informs occupants how to leave the building. As the occupants are virtual agents, they can then be instructed to find a route to the exit. Note that each team's PDA regularly informs the central dispatch of its current location and which areas have been notified by the team carrying it.

3.2 Models and Simulations

The perennial system leverages several models, in addition to the pseudo-model used by the agglomerator to determine where users are located in the building. Several of these are simple. The *building model* tracks each team's location within the building, and contains the average population of each room. Likewise, the *occupant model* tracks all occupants in the building and their status. The *motion model* assigns a size to each agent and handles inter-agent collisions. Team locations are estimated using interpolation. Finally, the *team movement model* is the most complex: given the list of teams and rooms, it determines how long it will take to notify the rest of the building, and what alternative notification strategies are available. This model uses a weighted exponential backoff algorithm to model how likely a team is to respond to a new plan, and in turn use this to throttle notification updates.

The system only contains two simulations: the Constantly-Running set of models which track user locations and sensor levels, and the "What If?" analyses which are triggered when either a previously-notified room detects activity, or when the system detects that teams are moving too slowly for the global notification strategy to remain effective.

3.3 Implementers and Visualizations

The implementers have a small but far-reaching role in our managed evacuation study. They can select which notification strategy to follow when several different strategies are all considered optimal. They can also choose to blacklist sensors (in case some are repeatedly reporting erroneous results) and may also manually inject data into a running simulation for the purpose of stressing the system or testing the teams.

Given the above description of sensors and agents, our system defines several visualizations and tools. First, the *building visualization* shows a floorplan of the world with team positions and room population densities at any given time. Given the nature of our historical sensors, it is easy to "rewind" and play back existing studies. The *strategy visualization* lists the most promising plans generated by "What If?" analysis, categorized by their estimated effectiveness and potential disruptiveness to the teams' current plans. Finally, the *statistics visualization* is a more classical simulation visualization component, and shows the estimated time left to evacuate, average time spent by a team in a room, and other relevant data relating to the progress of the evacuation.

3.4 MMOHILS Users

One final element of our perennial simulation study remains to be explained: MMOHILS users. First introduced in related work on virtual reality [11], a Massively Multiplayer Online Human-In-the-Loop Simulation –or MMOHILS– is a form of virtual world within which real humans act as agents in a simulation. In our guided evacuation exercise, the notification teams are exploring a mixed-reality type of environment, which qualifies them as MMOHILS agents. Unlike the study introducing the idea, however, these human agents have very limited capabilities within the virtual world.

4 Experiment Setup

Two experiments are proposed, both of which start the system in some normative state and then disrupt it by sending notification teams to guide the evacuation. The first of these assumes an even distribution of occupants across rooms, something which is common in egress studies [22][20]. The second experiment tries to generate a "point of interest" in the building around which a majority of the occupants have gathered.

We make several assumptions about our system. First, it is assumed that agents will follow the evacuation instructions given to them. This has been shown in our previous work to be valid under low stress levels. Second, there is no external announcement of the evacuation; all occupants hear the announcement from the evacuation team first. Third, it is assumed that agents do not re-enter the building once they have evacuated (though they may delay their egress while still inside the building).

4.1 First Experiment

Our first experiment observes the egress procedure on a "normal" day with no events. 300 agents are distributed throughout the building's rooms uniformly with respect to each room's relative size. After a small warm-up period, 5 teams enter the building and proceed to inform each room. All occupants and teams are instructed to move at the speed of a moderate walk.

This experiment uses a utility function to optimize the entire exercise, placing the priority on total evacuation time. Each symbiotic decision is recorded, and later validated (along with the utility points) with respect to a Pareto front estimation overall performance variables. The purpose of the first experiment is to obtain a baseline value for egress under normal conditions with a fairly distributed group of occupants.

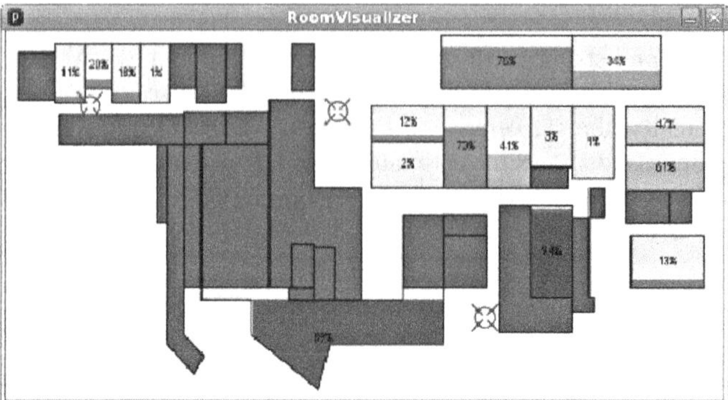

Fig. 3. Screenshot of our building visualization using example data. Evacuation progress is charted on a per-room basis. Team locations are represented by bulls-eyes.

4.2 Second Experiment

Our second experiment also initializes 300 agents, but unlike the first they are not uniformly distributed. Instead, 75% of the agents will all be gathered in one room, representing an "event" such as a presentation or an exam. Dense groups are usually more difficult to quickly and safely egress from a building due to non-optimal behaviors like herding [8].

The same utility function will be used for both experiments, and the same validation effort will be performed. Although the primary purpose of this study is to examine a real-world application of our perennial framework, it is expected that symbiotic feedback will nonetheless improve evacuation time.

5 Preliminary Results

Our results fall into two categories. The primary results of this study (and its primary research value) pertain to the methodology of the perennial framework and its applicability to real-world simulation studies. Thus, we will first discuss the results of implementing the building egress study using our framework. Following that, we will present some preliminary results obtained from the completed system.

Of paramount importance, we observed that designing our system through the use of the perennial framework alleviated a large number of otherwise-alarming concerns. The automatic historical sensors made replaying specific simulation runs trivial, while the agglomerator allowed us to cleanly specify combinational sensors (whose histories were nonetheless automatically recorded). The separation of our implementers from the simulation models allowed *hindsight* and *foresight* studies to be carried out easily, and a clean separation of visualization tools allowed us to cater to non-crisis needs (e.g., by estimating study room occupancy) without taxing the system for this trivial data at the time of crisis. Leveraging MMOHILS for our teams and occupants made large real-time studies possible without disrupting the normal operation of the building, and representing entities as agents left us the option of running purely-virtual or purely-physical world studies if desired. Using our distinction of physical versus virtual world overlap allowed us to tightly couple parts of these two worlds together while leaving other areas completely untethered. Finally, using the Indriya sensor test-bed allowed us to apply symbiotic simulation to a crisis management task in real-time.

In addition to these primary benefits, we produced several tools to help understand building egress better. First, implementers can replay simulations, and can "step through" the virtual world as an agent in an attempt to detect flaws in the teams' notification strategies. Second, the plug-and-play nature of our models makes it easy to test different sensing technologies within the context of the wired building. For example, sensing cell phone signals would help to accurately count the number of people in a room, and would be fairly simple to add to our existing conglomerate data. Finally, the use of Pareto-estimation might be used at runtime (instead of after-the-fact) to help implementors make decisions about performance trade-offs.

6 Conclusions

We have described a narrow, highly practical application of our previously-introduced perennial system. By focusing on evacuation exercises, we create a system which can be used directly at the time of crisis with no modification to the system whatsoever. Moreover, we can test the notification capability of our system in similar conditions without bringing its validity into question. The system is also geared to hindsight studies, and has possible applications to foresight studies which can, again, be implemented without changing the core system.

By leveraging our perennial system, we gained the benefits of large amounts of functionality with a minimal amount of system configuration. Real-time simulation results are notoriously expensive to collect, but the perennial framework allowed us to keep our options open without limiting our design. The framework is lithe, agile, and powerful.

References

1. Aydt, H., Turner, S.J., Cai, W.T., Low, M.Y.H.: Research Issues in Symbiotic Simulation. In: Rossetti, M.D., Hill, R.R., Johansson, B., Dunkin, A., Ingalls, R.G. (eds.) Proc. of the 2009 Winter Simulation Conference (2009)
2. Belkhouche, R.B.B., Al-Saleh, S.: An expert decision support system for crisis management. EECS Department, Tulane University (2002)
3. Braunling, R., Jensen, R.M., Gallo, M.A.: Acoustic target detection, tracking, classification, and location in a multiple-target environment. In: Proc. SPIE - the International Society for Optical Engineering (1997)
4. Doddavenkatappa, M., Chan, M.C., Ananda, A.L.: Indriya: A Low-Cost, 3D Wireless Sensor Network Testbed. Developed at School of Computing. National University of Singapore, Singapore (2009)
5. Fiedrich, F.: An hla-based multiagent system for optimized resource allocation after strong earthquakes. In: Winter Simulation Conference (2006)

6. Fruin, J.J.: Pedestrian Planning and Design. Metropolitan Association of Urban Designers and Environmental, New York (1971)
7. Fujimoto, R., Lunceford, E., Uhrmacher, A.M. (eds.): Grand Challenges for Modeling and Simulation: Dagstuhl Report. Technical Report 350, Schloss Dagstuhl, Seminar No. 02351 (2002)
8. Han, C.S., Pan, X.S., Law, K.H.: A multi-agent based simulation framework for the study of human and social behavior in egress analysis. Center for Integrated Facility Engineering, Standford University (2004)
9. Helbing, D., Molnar, P.: Social Force Model for Pedestrian Dynamics. American Physical Society (1998)
10. Helbing, D., Farkas, I., Vicsek, T.: Simulating Dynamical Features of Escape Panic. Nature, 407, 487–490 (2000)
11. Hetu, S., Tan, G.: MMOHILS: A Simpler Approach to Valid Agents in Human Simulation Studies. In: Mason, S.J., Hill, R.R., Mönch, L., Rose, O., Jefferson, T., Fowler, J.W. (eds.) Proceedings of the 2008 Winter Simulation Conference, pp. 909–913 (2008)
12. Hetu, S., Tan, G.: The Big Picture of Symbiotic Decision Support: Designing a "What-If" Simulation Framework for Crisis Management. In: FISAT: Second International Conference on Advanced Computing and Communications Technologies for High Performance Applications, Keynote (2010)
13. Hetu, S., Tan, G.: Perennial Simulation of a Legacy Traffic Model: Implementation, Considerations, and Ramifications. In: Jain, S., Creasey, R.R., Himmelspach, J., White, K.P., Fu, M. (eds.) Proc. of the 2011 Winter Simulation Conference (Appearing) (2011)
14. Jain, S., McLean, C.R.: A concept prototype for integrated gaming and simulation for incident management. In: WSC 2006: Proceedings of the 38th Conference on Winter Simulation, Winter Simulation Conference, pp. 493–500 (2006) ISBN 1-4244-0501-7
15. Low, M.Y.H., Lye, K.W., Lendermann, P., Turner, S.J., Chim, R.T.W., Leo, S.H: An agent-based approach for managing symbiotic simulation of semiconductor assembly and test operation, Utrecht, The Netherlands, pp. 85–92 (2006)
16. McGrath, D., Hunt, A., Bates, M.: A simple distributed simulation architecture for emergency response exercises. In: DS-RT 2005: Proceedings of the9th IEEE International Symposium on Distributed Simulation and Real-Time Applications, pp. 221–228. IEEE Computer Society, Washington, DC (2005) ISBN 0-7695-2462-1
17. Medeiros, D.J., Swenson, E., DeFlitch, C.: Improving patient flow in a hospital emergency department. In: WSC 2008: Proceedings of the 40th Conference on Winter Simulation, pp. 1526–1531 (2008) ISBN 978-1-4244-2708-6
18. Still, G.K.: Crowd Dynamics. PhD thesis, Warwick University (2000)
19. Succi, G., Pedersen, T.K., Gampert, R., Prado, G.: Acoustic target tracking and target identification: recent results. In: Proc. SPIE – The International Society for Optical Engineering (1999)
20. Taaffe, K., Johnson, M., Steinmann, D.: Improving Hospital Evacuation Planning Using Simulation. In: Perrone, L.F., Wieland, F.P., Liu, J., Lawson, B.G., Nicol, D.M., Fujimoto, R.M. (eds.) Proceedings of the 2006 Winter Simulation Conference (2010)
21. Togawa, K.: Study on fire escapes basing on the observation of multitude currents. Building Research Institute. Ministry of Construction, Tokyo (1955)
22. Yang, B.W., Wang, Y.G., Wang, C.: A GIS-Based Simulation for Occupant Evacuation in an Amusement Building. In: 2010 2nd International Asia Conference on Informatics in Control, Automation and Robotics (2010)

Distribution Analysis in Manufacturing 2.0 Using Agent Based Modeling and Simulation

Bodeum Choi, Yun Bae Kim, Jinsoo Park, Kiburm Song, and Chul Woo Jung

Department of Systems Management Engineering, Sungkyunkwan University
Cheoncheon-dong 300 Jang-an-gu Suwon, Korea
bd0725@nate.com, {kimyb,jsf001,idshadow,chulking05}@skku.edu

Abstract. The mechanism of past manufacturing system (manufacturing 1.0) causes long period of development, cost increment and risks, which cannot manage the rapid change of current business environment such as diversified customer needs. The manufacturing 2.0 companies which applied the long tail intend to create values through planning, development, production and marketing for satisfying the customers' specialized needs. Companies can secure the superiority in the market by analysis of distribution channel. Focusing accuracy of distribution channel analysis in manufacturing 1.0 causes it to take long to analyze. However, for the case of manufacturing 2.0, companies sacrifice accuracy for fast analysis of distribution channel. In this paper, we suggest a method to establish and evaluate alternatives with a single model using agent based modeling. Then, we provide the classical simulation results to verify our method.

1 Introduction

The Traditional manufacturing companies gave priority to main products satisfying simple customer needs, which account for 20 percent of total production. The situation that these products have guaranteed 80 percents of the sales is called Pareto rule. We call this manufacturing scheme as the manufacturing 1.0 corresponding to the manufacturing 2.0 scheme. To monopolize the market under manufacturing 1.0, companies have advanced their own technologies confidentially. This manufacturing scheme causes increment of duration and cost in product development, and relevant risks. Also, it cannot properly respond to the fast change of business environment such as reduction of product lifecycles or diversification of customer needs. Fundamentally, the companies under manufacturing 1.0 environment intend to achieve the economies of scale through mass production. The products or services which are made by interaction between customers and companies make the latest customer satisfaction maximized. Therefore, current manufacturing method and capability cannot answer to customers in these days having much importance of speed and participation. As a result, the limits of manufacturing 1.0 opened an avenue for the manufacturing 2.0 companies.

The long tail theory in web 2.0 is a rule that the unpopular items occupying 80 percents of total items make internet companies successful (Anderson, 2006). The manufacturing 2.0 companies indicate those companies which are applied the long tail

J.-H. Kim et al. (Eds.): AsiaSim2011, PICT 4, pp. 95–104, 2012.

theory (DeVol et al., 2009, Jacobson et al., 2010). After appearance of the manufacturing 2.0 scheme, many companies map out various business strategies to respond to the change of customer needs. These companies try to occupy competitive superiority through new paradigm of planning, development, and production, which are creative and timely, and satisfy personal customer needs. In general, to maximize the profits, it is important that analyze distribution channels. Under the manufacturing 2.0, to respond to change of market environment caused by diversity of customer needs and short market lifecycle, rapidity is more important than accuracy in distribution analysis. As a new market is opened up, it is almost impossible that the companies collect the information about it. In this circumstance, it is difficult to perform distribution analysis. The agent-based modeling and simulation (ABMS) can be a useful tool to help it possible in this situation.

ABMS is a modeling and simulation technique that the autonomous agents take actions individually and/or interact with one another to perform assigned missions under given system environments and rules (Bonabeau, 2002). These agents are intelligent and have their individual purpose, which are restricted by time and space, and follow their algorithms. That is, they make decisions on their actions via purpose and/or interaction with the others. Therefore, one can analyze the system with only observing the simulation process in progress after determining the attributes of agents, system environments, and so on. Applying the ABMS in the distribution channel analysis, the agent can be any item such as the part on production line, worker, equipment, product, or etc. One can design the system which can properly cope with changes of external environments such as factory layout or process alteration. In this paper, we propose a new method of distribution channel analysis which suits well for rapid under manufacturing 2.0.

The organization of this paper is as follows. Section 2 introduces the concepts of the manufacturing 2.0 and the ABMS. Section 3 provides alternative distribution system modeling and analysis method that we suggest. Section 4 illustrates the numerical results by implementation of ABMS and analyzes them to validate our method. Section 5 concludes our method and presents future works.

2 Two Basic Concepts

2.1 Manufacturing 2.0

The past companies as we mentioned by manufacturing 1.0 pursued the maximized efficiency through the mass production of main items which responded to customer needs on the ground of Pareto rule. Under the manufacturing 1.0 environment, companies tried to close their own technologies and monopolize the market values by utilizing them. This situation made the duration and the cost of product development increased, caused relative risks, and could not cope with the fast changes of business environment such as substantial decrease of product lifecycle or variations in customer needs. However, this business environment is changed by appearance of the web 2.0. At first, the means to secure the ability is shifted to innovation by opening and reunion from the past closing and monopolization. Also, while the decision makings of the past companies have been performed by unidirectional flows of information, they exchange opinions with interaction of the various fields on their own initiatives nowadays. At last, the customer needs become diversified and detailed. This manufacturing paradigm is called the manufacturing 2.0.

The manufacturing 2.0 companies should act in concert with variation of business environment. They obtain the market values laying emphasis to creativity, timeliness, and customer needs satisfaction in the planning, the development, and the production activities. The spreading of long tail theory makes the importance of collecting various opinions more increased. Also, the manufacturing 2.0 companies should import new production methods which can fulfill the innovation through customer participations. As the customers take part in production from the planning to the deciding quantity and design, the companies do not have to consider the complex problems such as customer taste, demand forecasting, and so on. As a result, instead of the production methods through downward and unidirectional decision making, they should build up the conditions which can exchange and feed back the opinions at the customer-planning-production parts in real time. Table 1 summarizes these differences between manufacturing 1.0 and 2.0.

Table 1. The Differences between Manufacturing 1.0 and 2.0

Differences	Manufacturing 1.0	Manufacturing 2.0
Basic economic theory	Pareto	Long tail
Target	efficiency of production	rapid change of market environment
Decision making	downward and unidirectional	customer participation
Distribution analysis	accuracy	rapidity
Securing capability	closing and monopoly	opening and reunion
Customer's rule	consumer	consumer and producer

2.2 Agent Based Modeling and Simulation

The system which is compounded by innumerable elements and shows a new state and order different from characteristics of the elements by their interactions is called the complex system. The agent implies the element of complex system, learns, evolutes, and adapt through interactions with other agents. The complex system is not passive system controlled by predefined assumptions and schedules, but active system utilizing the gained experiences. Under the manufacturing 2.0 paradigm, the customers aggressively take part in the production activities as well as consumption in their own areas as the units of society. Supposing that the society indicates a system, the manufacturing 2.0 system is a complex system.

The agent based modeling is one of the useful analysis methods of the complex systems (Bonabeau, 2002). An agent in ABMS interacts with others having individual knowledge of the rules without system information. The agent takes actions according to the given rules, and diverse agents make knowledge exchanges possible from communication and cooperation. The attributes of an agent are classified into cognitive faculty, response ability, achieving ability, and so on. The agent autonomously recognizes the purpose, responds to stimulus, and achieves tasks without intervening of the exterior. Figure 1 illustrates an outline of the agents.

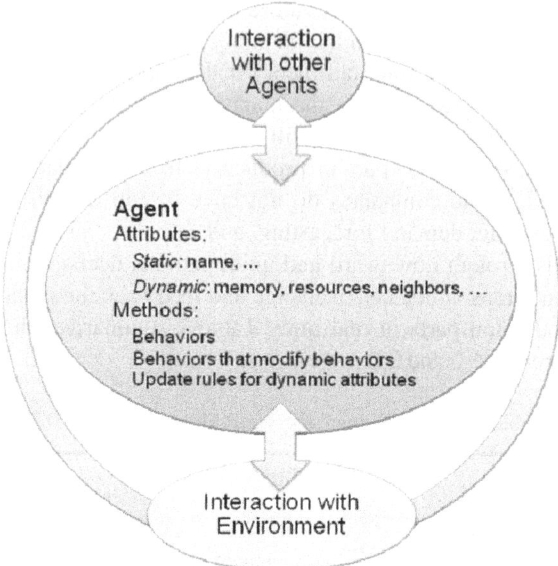

Fig. 1. Illustration of the Agents

3 Alternative Method

3.1 Introduction to a New Distribution Analysis Method

As the market lifecycle becomes short, it is almost impossible to apply the traditional method to analysis of distribution channel. Traditional method takes long time and high cost while the product items are replaced fast. That is, traditional method cannot obtain the reasonable results which can timely satisfy individual customer needs. This situation makes the companies give up or fail to analyze the distribution channel. The ABMS method is a useful tool that makes the distribution analysis possible under the rapid changes of market environment caused by the shortened lifecycle. In other words, the ABMS can overcome limits of time and cost in traditional method. Therefore, we suggest a new distribution analysis method using the ABMS.

Figure 2 presents the systematic structure of our alternative method. In the beginning, randomly spread agents in the waiting space move to the production line in their order. The agents on the line enter into the each processing area, wait in the queue until completion of preceding process, and then take the process. After process completion, the agents move into the line again. Passing through user-defined the number of process, they move into the shipping space and wait for shipping. In our system, these agents will be any item on the line such as parts on the production line, products on the packaging line, or transporters on the path.

This alternative method has two characteristics. One is reusability which means user can reuse predefined models, the other is flexibility. These characteristics bring the economic effects on time and cost, which satisfy the conditions of the manufacturing 2.0 paradigm. In the ABMS, the agents run the simulation behaving autonomously corresponding to the user-defined static attributes, dynamic attributes, behaviors, rules, and so on. Before the simulation run, user can modify the attributes, the behaviors, or rules to perform flexible distribution analysis with various scenarios. This makes it possible to get the rapidity of distribution analysis on the manufacturing 2.0 configurations. In the distribution system that the agents follow the given path, an important subject of the analysis is the bottleneck phenomenon. As this phenomenon increase moving time and cost in the distribution system, the final purpose of our method is solving the bottleneck problem.

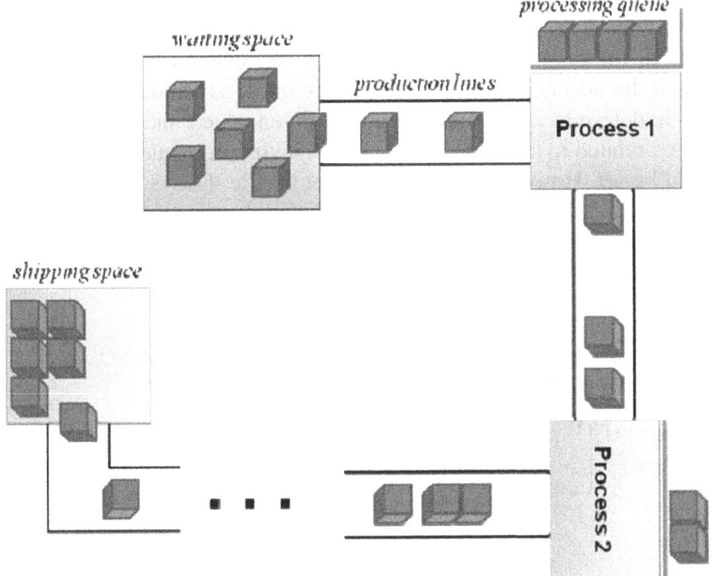

Fig. 2. Structure of Alternative system

3.2 Design of Modeling and Simulation

Figure 3 is the procedural flow chart of the agents on our alternative model. The user-defined agents created randomly wait in the queue; i.e., waiting space. Observing the production line flow, the agents estimate a favorable time and move into the line at that time. The agents follow the path performing their given processes. They increase or decrease their moving speed to maintain proper distance on the line. After the all given processes are performed, agents move into the shipping space. As the all agents pass through given processes, the simulation is terminated.

The alternative model can check the bottleneck phenomenon designing whole system layout or setting each process. In this situation, there are two reasons raise the possibility of the bottleneck. One is the short interval of successive agents on the path. The other is the high moving speed of the agents. Therefore, alternative model is designed to optimize the moving speed and interval of the agents. That is, finding optimal speed and interval is the final goal of our method.

Our model is organized into system layout, agent modeling, and process settings. The users can customize the system modifying each detail option. First, the users can fix the number of processes area and the system scale such as factory size at the modeling time. The length of production lines is changed according to the system scale. The number of processes also changes the system layout. In a system layout, the total distance of working path, the distance of processing area, and the distance between processing areas are fixed by the users. Secondly, the users should select characteristics of the agents such as the maximum number, moving speed, degree of acceleration or deceleration. The agents follow the production line with allocated speed passing through given processes. Each agent maintains the uniform distance from the previous agent increasing or decreasing the speed. The lower and upper limits of agents speed are defined in the model, so the speed of the agent is confined within these limits. Even though the agent speed are given by the users as an attribute, the real speed of the agents is determined by random sum of given speed and defined speeds. Lastly, the measures related to processes such as the maximum queue size and the processing time should be set. However, the users cannot handle the system with these measures as the reason of dependence nature on the system.

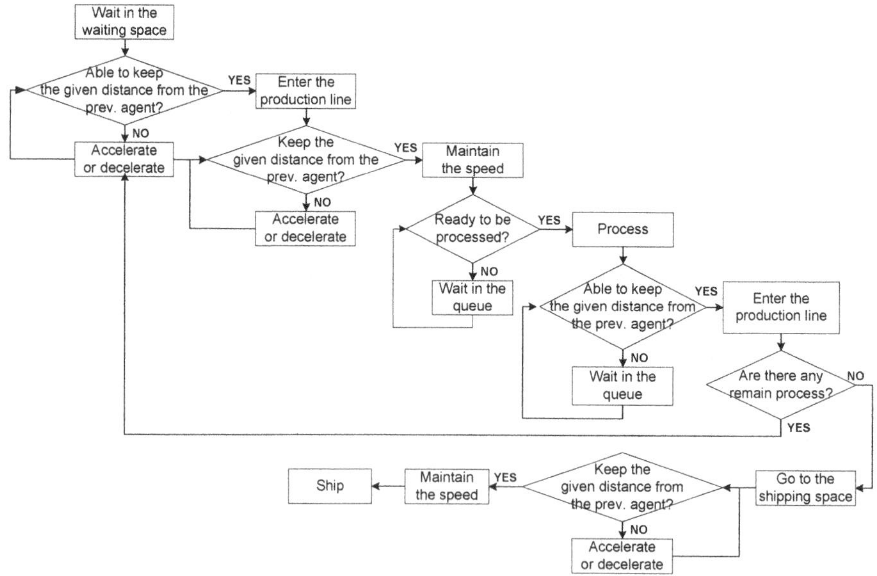

Fig. 3. Agent Flow of Alternative System

4 Example

4.1 Virtual Shipping Factory

To confirm the possibility of applying our method to actual distribution systems, we construct a virtual shipping factory system. In this system, there are the waiting space and the shipping space out of working space, and the packaging area with 5 packaging machines. The total length of working space is valued by 30, the length of each machine is 3, and the distance between machines is 2.5. The distances from the waiting space to the first machine and from the last machine to the shipping space are also 2.5. The products convey the given path from the waiting space to the shipping space passing through the machines. The mean processing time of every machine is 1 unit times except the third machine, and the third machine takes 2 unit times. We assume that the maximum speed of product is 1 and the minimum is 0. To understand this virtual factory, image the situation that there are 5 processing areas in the figure 2 of the previous section.

4.2 Simulation and Results

We use the NetLogo, the commercial software as the ABMS tool to materialize our model. As this software can model the natural and social phenomenon, it has the advantage of modeling the changes with the lapse of time in the complex system. It also allows the numerous kinds of agent in the system to active independently. These characteristics make it possible to discover the detailed features of individual agents or the large scaled patterns of whole system.

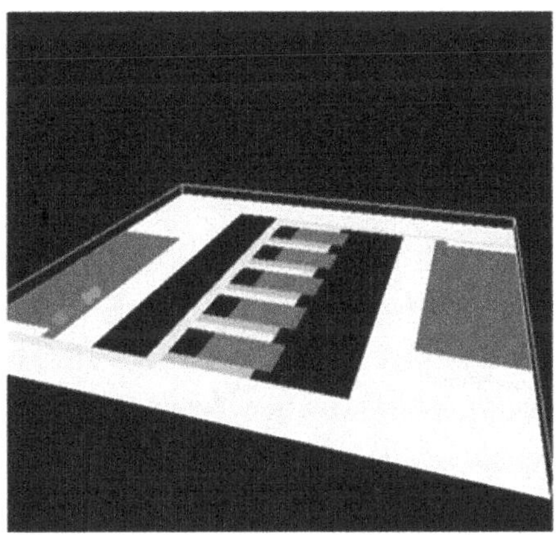

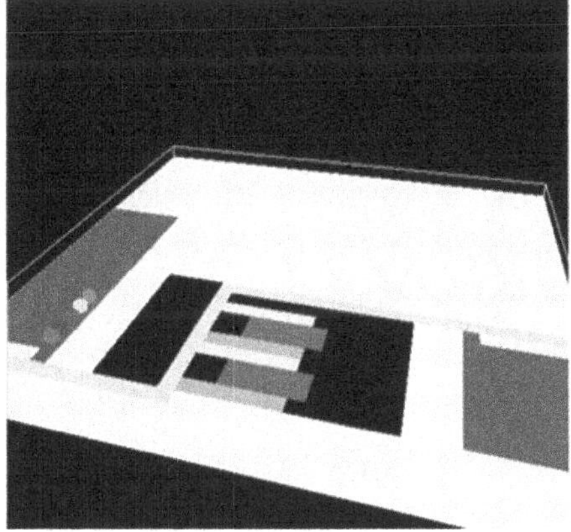

Fig. 4. Examples of System Layout

Materialized model of our proposed system can settle the values for 3 modeling categories in the section 3.2. In the system layout, the total length of working space, the lengths of processing areas, the defined agent speed, the number of processing areas, and the maximum number of agents in the line can be valued. We can select the values of agent characteristics, the maximum number of agents in the system, the moving speed, the degrees of acceleration and deceleration, and the interval of successive agents in the line. The processing times can be also fixed for each processor. Figure 4 presents layouts of two example systems captured from our NetLogo model by changing the number of processing areas. The left system has 5 processing areas and the right has 2. In this figure, the boxes imply the agents, and the left system is the implementation of our virtual shipping factory.

The first goal of our simulation is to find the range of moving speed that does not cause the bottleneck without exceeding the upper limit of pre-defined speed. The second is the lowest interval of successive agents in the line with the same role. However, as the interval of agents depends on the feature of agents in general, we use the fixed value of the interval in simulation. We replicated the simulation models changing the degrees of acceleration and deceleration which changes the actual speed of agents. Undoubtedly, while the third machine can make the bottleneck, this situation is discovered graphically. Figure 5 shows the examples of our simulation results observing the tagged agent (a red box). The left one is the result by setting the acceleration degree to 0.005 and the deceleration to 0.025. From this result, we can confirm that the actual speed of agent converges to 1, which increase the possibility of the bottleneck. The right is resulted by setting the acceleration to 0.006 and the deceleration to 0.035. The actual speed of the tagged agent is cycled from 0.3 to 0.8, which means this system decreases the probability of the bottleneck. In this case, by calculating the mean of the actual speed, we can find a visible solution. The maximum value of the visible solutions from iteration can be a local optimum. As our model uses the heuristic method which needs the iteration and the replication, there is some weak point such that the optimum may not be global. However, the advantage of rapidity and economical efficiency in our method makes it possible to analyze the distribution system on a newly created market.

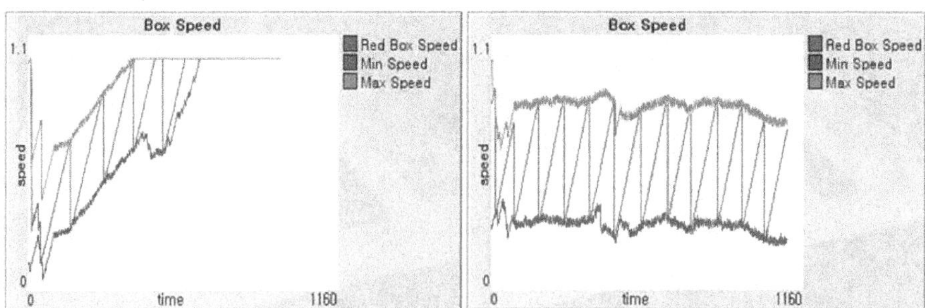

Fig. 5. Examples of Simulation Results

To validate our optimal solution, we simulated the identical model with Arena software. We confirmed that the optimal speed does not cause bottlenecks. In addition, we iterated the systems with changing the value of agent's speed to verify that our optimum is global. If we increased the speed from the optimal value we found, the number and magnitude of bottlenecks are increased. Decreasing the speed from the optimal value did not generate any bottlenecks. However, the process utilization was decreased. This explains that our optimum may be the global and our method performs as we intended. As our simulated system is virtual model, we do not present the numerical results.

5 Conclusion

We proposed a new method of the distribution system analysis in manufacturing 2.0 with the ABMS to apply the rapid changes of customer needs, instead of past method which can get accurate solution but takes long time and high costs. The reason of the speedy analysis required is that the changes of customer needs shorten the lifecycle of products. Therefore the manufacturing 2.0 companies have to determine the number of machines, spending time of each process, and speed of movement between processes depending on the layout and procedural processes. We can get the solution improving the process in various situations using the ABMS.

There are three categories in our proposed model, and the measures related to these categories can be provided prior to running simulation experiment. Then we can find the appropriate moving speed of items and specific processes causing the bottlenecks using the ABMS. These solutions we obtained by our heuristic method may not be the global optimum. However, in a situation that we have no base knowledge of new production processes, we use them as a basis of decision making such as establishing the manufacturing strategy. The advantages of our model are the rapidity and the flexibility in analyzing the distribution system under the manufacturing 2.0. We can reuse the established agent based model by updating configurations or conditions; i.e., work spaces, equipments, processing speed, and so on.

Our heuristic method needs macroscopic and detailed design of systems to get reliable results. Our proposed system simplifies configurations to moving speed for single working space. However, more configurations such as scheduling, breakdown and repair of equipments, and so on, should be treated for multiple spaces. Also, constructing algorithm which can select the optimal condition of configurations increases the accuracy of results and analysis.

References

1. Anderson, C.: The Long Tail: Why the future of business is selling less of more. Hyperion Books (2006)
2. Axelrod, R.M.: The Complexity of Cooperation: Agent-based Models of Competition and Collaboration. Princeton University Press (1997)
3. Bonabeau, E.: Agent-based Modeling: Methods and Techniques for Simulating Human Systems. Proceeding of the National Academy of Sciences 99, 7280–7287 (2002)
4. Borshchev, A., Filippov, A.: From System Dynamics and Discrete Event to Practical Agent Based Modeling: Reasons, Techniques, Tools. In: The 22nd International Conference of the System Dynamics Society, pp. 1–23 (2004)

5. De Vol, R.C., Wong, P., Bedroussian, A., Hynek, C.F., Rice, D.: Manufacturing 2.0: A More Prosperous California. Milken Institute (2009)
6. Jacobson, S.F., Eriksen, L., Kim, P.: Manufacturing 2.0: A Fresh Approach to Integrating Manufacturing Operations with DDVN. Gartner, Inc. (2010)
7. Jennings, N.R.: An Agent-based Approach for Building Complex Software Systems. Communications of the ACM 44(4), 35–41 (2001)
8. Mainzer, K.: Thinking in Complexity: The Computational Dynamics of Matter, Mind, and Mankind. Springer (2007)

Modeling and Simulation Framework for Cyber Physical Systems

Tianyuan Xiao and Wenhui Fan

Department of Automation, Tsinghua University, Beijing, China, 100084
xty-dau@tsinghua.edu.cn

Abstract. A cyber-physical system (CPS) is a system featuring a tight combination of, and coordination between the system's computational and physical elements. Although much research has addressed modeling and simulation of such systems, most solutions usually do not cover all requirements from different domains. Rather, these solutions restrict the development environment and reduce its flexibility by enforcing the use of a specific software platform and tightly-coupled tools. This paper presents problems and challenges of CPS and discusses a MDA based Modeling & Simulation framework, which has a loosely coupled architecture, to enable design, modeling & simulation of large-scale, heterogeneous CPS systems in an integrated manner. Meta-model based Model Driven Architecture (MDA) is realized for dynamic reuse of simulation model across platforms in this paper at first. Then the paper introduces an Extended HLA architecture based on multi-layers and multi-federation and its implementation, which is realized based on Resource Management Federate (RMF) technique. Finally, Federate Integration Architecture (FIA) and its Federate Executive Infrastructure (FEI) are described, which are realized based on Web service and OGSI.

Keywords: Cyber Physical System (CPS), extended HLA, Model Driven Architecture (MDA), Resource Management Federate (RMF), Federate Integration Architecture (FIA), Federate Execution Infrastructure (FEI).

1 Introduction

Cyber Physical Systems (CPS) is a new paradigm combining computation, communication and control to construct real-time systems. It is envisioned as heterogeneous systems of systems, which involve communication, computation, sensing, and actuating through heterogeneous and widely distributed physical devices and computation components.

Today, a pre-cursor generation of cyber-physical systems can be found in areas as diverse as aerospace, automotive, chemical processes, civil infrastructure, energy, healthcare, manufacturing, transportation, entertainment, and consumer appliances. This generation is often referred to as embedded systems.

Unlike more traditional embedded systems, a full-fledged CPS is typically designed as a network of interacting elements instead of as standalone devices. The components

J.-H. Kim et al. (Eds.): AsiaSim2011, PICT 4, pp. 105–115, 2012.

of a CPS are connected through wired and wireless networks in a large scale and orchestrated together as a whole. Moreover, CPS introduces several challenges for system design: (1) to support high system flexibility such that the CPS components in the system are free to join or leave dynamically, (2) to support various Quality of Services (QoS) requirements through out every level of CPSs.

The analysis of CPS should be quite different from conventional embedded system because of many reasons, including that a CPS usually consists of many sensors and actuators communicating with each other, and CPS analysis requires the modeling of external resources such as light source, temperature, and signals. Furthermore, most CPSs are resource constrained in terms of memory, processing power, and battery life. Because of that, the analysis environment should provide software architectural and design solutions to minimize resources, and an accurate simulation environment to support battery life and timing related communication. In addition, the nature of heterogeneity and scalability introduce a big challenge to deploy, test and maintain CPS. Thus, accurate modeling and simulation environment can reduce the effort and cost of deployment, testing and maintenance. Lastly, the domains using CPS often require mobile nodes that must interact in a scalable manner with the various nodes in the system.

2 M&S Features for CPS

The following features are essential in Modeling & Simulation framework for CPS.

-Heterogeneous applications support: CPS usually consists of non-homogeneous applications. Thus, it should be able to simulate heterogeneous application logics simultaneously.
-Various physical modeling environments: the physical modeling environment should support mathematical expressions and incorporate domain specific physical modeling descriptions (e.g. floor plan of buildings) by extracting relevant information from them.
-Scalability support: support for the development and simulation ranging from small scale (tens) to large scale (thousands of) sensors and actuators.
-Integration of existing simulation tools: easy-to-use support to link to existing simulation tools is required.
-Integration of proprietary solutions and open standards support: proprietary solutions and open standards including protocols, infrastructures and existing software should be able to be easily incorporated into a generic framework.
-Software reuse: a generic framework should support software reuse either by exploiting code generation techniques (which can also use proprietary infrastructure), linking libraries or using configurable components.

3 Challenges

CPS modeling & simulation environment should be distributed interactive framework, which has loosely coupled architecture. Currently, the majority of the international distributed interactive simulation frameworks for complex system are built according to High Level Architecture (HLA). But in many ways, there are still some shortcomings in HLA itself.

For instance, there is not a unified modeling framework on how to use the simulation model to construct a federate in HLA, and much work needs to be done when coding a federate, which leads to a low efficiency, thus the independency and reusability of the code are both influenced.

To solve the problems above, much work has been done, especially in the field of MDA based HLA. For example, Calytrix Technologies Pty Ltd in Australia implements the core concept of MDA in SIMplicity IDE, and M&S components can be exported from different platforms and middleware. The researchers in National Science Foundation, George Mason University and SAIC Company put forward Extensible Modeling and Simulation Framework (XMSF) together. Europe Space Administration proposed simulation model portability (SMP) specifications referenced by MDA presented by OMG, and the simulation model design information and operating information of phase separation.

However, the researches mentioned above are carried out mainly from the point of modeling environment, and the architecture of HLA is not improved. Essentially, the configuration of OMT in HLA must be completed before the start of federation, and it cannot join different FOM dynamically. So the model reuse on federate level is classified as merely static reuse.

Essentially, all federates in HLA are located in same level, thus HLA can be considered as planar architecture. A model is impossible to join different FOM federation dynamically as federate in this architecture.

To meet the need of heterogeneity of CPS and reuse of available software and simulation tools, this paper introduces meta-model based Model Driven Architecture (MDA) first to solve the problem of dynamic reuse of simulation model and cross-platform. Then, this paper presents extended HLA architecture based on multi-layers and multi-federation to meet the demand of CPS, the concept "resource management federation" is specified to configure federate OMT of federation in HLA dynamically. Finally, based on Web service and OGSI, the Federate Integration Architecture (FIA) and its Federate Executive Infrastructure (FEI) for CPS are proposed and realized.

4 MDA Used in HLA

4.1 MDA Architecture

Figure 1 shows the implementation process of typical MDA. This paper proposes the implementation process of MDA as shown in Fig. 2, in which a modeling process can be divided into four levels, called as M0, M1, M2 and M3, respectively.

The logical relations from layer M3 to layer M1 are: the lower layer is the base of the upper layer, and the upper layer is the application of the lower layer. Layer M3 is meta-meta model, and MOF is included. MOF provides more abstract modeling support to define the meta-model in layer M2. It defines the uniform semantic basis for all model/meta-model in MDA, and provides with common standard for the definition of model language (meta-model). The layer M2 is meta-model layer, which offers the specific model language corresponding to different field and modeling symbol for layer M1. Layer M1 is model layer, which is the link with modeling worker, including CIM, PIM and PSM defined by MDA. PIM defines the interaction between the simulation logical function and model, using UML; PSM is relational model on HLA standard, using UML Profile. Layer M0 is instance layer. It is the implementation of the model in Layer M1, and it is the final goal of model description.

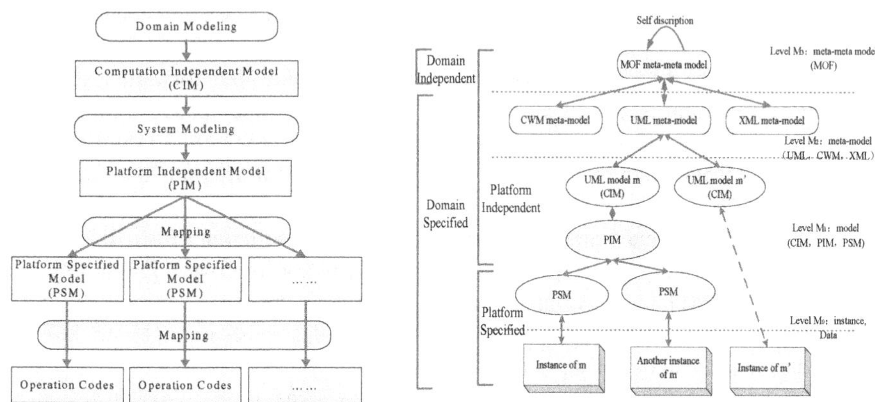

Fig. 1. The implementation process of MDA

Fig. 2. MDA model hierarchy

4.2 Transforming PIM to PSM

One important feature of MDA is it can transform PIM to PSM automatically by mapping rules and assistant tools, as well as finish the conversion from PSM to code. The way of conversion in MDA between models in different layers is: defining a conversion between different meta-models first, and according to the definition, utilizing tools to finish the conversion from meta-model to target model automatically.

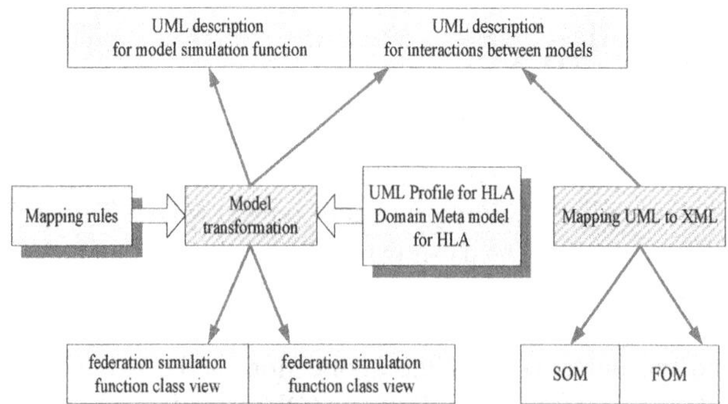

Fig. 3. Transforming PIM to PSM

The conversion between the models in PIM and PSM based on meta-model mainly refers to the conversion between UML of simulation modeling logical function and UML of interaction among models; federation simulation function class view, federate integrated function class view, FOM lists and SOM lists. This paper defines the set of mapping rules for the conversion, and utilizes XMI (XML Metadata Interchange) to implement the mapping UML to XML, and generates FOM and SOM files needed when the federation development platform is running, as shown in Fig. 3.

4.3 Transforming PSM Model to Code Model

When the software development theory of MDA is introduced into the architecture of HLA collaborative simulation, the problem of automatic code generation of federate is transformed to how to convert PSM model to code layer in MDA software development framework.

Many researches have been done on automatic code generation tools of different federates, but these tools are all based on member level to implement the development of federation, to achieve the push-pull operation of federate dynamically for a specific FOM. When facing different federation based on different simulation application, it is impossible to switch flexibly between the same FOM. So it is hard to reuse and interoperate between different federations.

The most important points are platform independent model PIM and platform specific model PSM, and the mapping between them, when applying MDA to describe and transform models in the application of HLA collaborative simulation system.

When HLA is mapping to PSM as the simulation integration middleware, based on the division of federate members, PSM layer models can be divided into simulation logical model of federate and interactive integrated functional model of federate. It can be mapping from PIM. The simulation functional model of federate can be described by the class view in UML, and integrated model can be described by SOM (object model) in OMT (object model template), and composed of FOM (federate object model) and object interactive class in class view of UML.

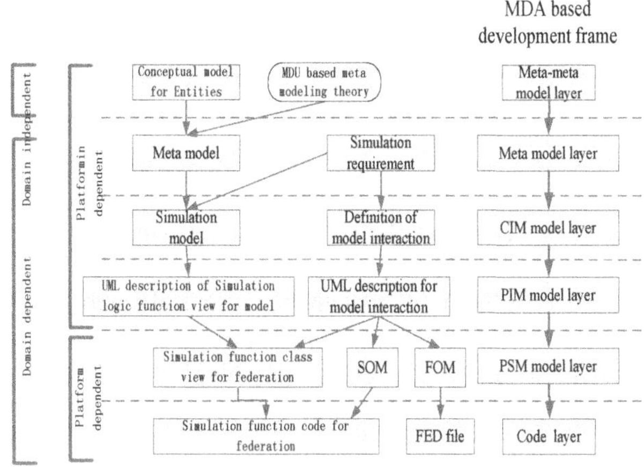

Fig. 4. MDA based HLA simulation system development framework

Figure 4 shows the federate development framework that we propose for HLA simulation, which supports automatic code generation from simulation model in PSM layer to the federate.

A complete federate usually covers two parts of code. One part is the code to build simulation model and make simulation running, and the other part is the code of some interfaces to communicate with RTI. The previous is closely related to model, as different models are simulated based on various simulation software, and the latter needs to implement a fixed function, so it can be encapsulated into the federate software development framework. Therefore, the fixed function part in model and federate can be separated. The model part exists in independent module, and SOM definition in model and federate uses dynamic mapping. Doing like this, PNP of model in federate framework architecture can be implemented.

Federate development based on the framework architecture mentioned before can get rid of restrict of specific SOM information, thereby the goal of reuse and interoperation of federate in different federations can be achieved

Comparing to the development pattern of federate existing, the new one bring the following benefits to the federate developer:

- The model of federate can exists as independent module, which is developed and debugged by simulation model developer. The simulation developer does not need to know the detailed information of RTI, nor does federate developer need to know the model information in simulation field.
- The interface function module of RTI is only related to simulation software. It has no relation with the model detail or the purpose of simulation, so it will not change even if the model or SOM information changes, which enhance the reusability of the code.
- Different federates can be composed of different modules of different function models, and the models themselves in federate can support PNP, which enhance the utilization of the model.

After generating platform specific model PSM, the model should be mapped to code. The mapping of code is mainly from the federate layer PSM to federate code, and is also from simulation function UML, integrated function SUML and SOM, to generate logical code of federate and HLA integrated code. Finally the simulation integration and test should run based on federate.

5 Two Layers HLA Architecture Based on MDA

The federate in HLA is hard to implement reuse and interoperation now, mainly because it is restricted by specific SOM information and the user model is hard coded. To enhance the reusability of simulation model, SOM information should be departed from simulation model, and when building a federate according to the simulation task, configure SOM and FOM dynamically.

5.1 Resource Management Federation

This paper proposes the concept of "resource management federation", and its main ides is: every computer (node) that joins collaborative simulation runs a resource management federation program which represents itself, and every node's resource is set up as the member of "resource management federation", and this resource management federation program is responsible for simulation resources, such as local models and so on.

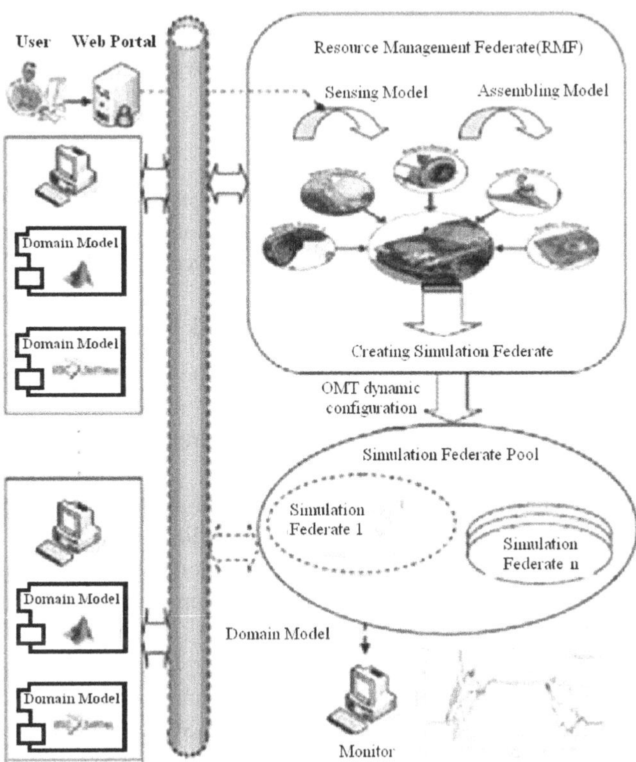

Fig. 5. Resource management federation

Based on resource management federation, users can build simulation application federation. During the running of the platform, resource management federation always exists, and all of the simulation application federations are under its control. This changes the planar architecture of HLA, and it makes resource management federation and simulation application federation constitute a hierarchy between the federal structures. Through the dynamical configuration of simulation federation OMT by "Resource Management Federation" in HLA, the share and maintenance of distributed and heterogeneous simulation model can be implemented, and the simulation application federation can generate dynamically. The simulation application federations which are active form the "simulation application federation pool", as shown in Fig. 5.

Resource management federation does not manage the federate code that is related to the implementation technique, but the meta-model that has closely mapping relation with simulation model in PIM, including the meta-modeling description of simulation model, and the concept model of interaction between CIM. This part can be generated by the user's designation. This method can separate simulation model from FOM, SOM information naturally. SOM and FOM information, and even simulation member program, all can be generated dynamically under the management and control of resource management federation. So the reusability on federate level evolves to the simulation model level. At the same time, aiming to the users' different requirement and different simulation purpose, the federation can build up and re-configured quickly and flexibly, and the interoperability and reusability are also assured.

5.2 Mapping from Meta-model to Federate

MDA realizes the process that simulation model and meta-model transform into each other. Considering what resource management federation manages is the meta-model of simulation model, so the dynamic mapping between SOM and FOM through meta-model is convenient.

The establishment steps of a specific simulation task list as follows:

(1) According to the simulation application purpose, choose the meta-model for collaborative simulation;
(2) Let the users establish the model mapping relation on conceptual semantic level through the visual interface.
(3) Decide the input and output variables of meta-model by the mapping relation;
(4) For the meta-model obtained by meta-building, transform it into the field simulation model that can be run in simulation software; for the meta-model obtained by domain simulation model, map it into the original domain simulation model by corresponding relation.
(5) By way of meta-model and concept mapping relation on semantic level, model simulation logical function view UML and interaction among models UML can be generated automatically. And through SOM and FOM;
(6) By way of automatic mapping relation from PIM to PSM, generate simulation function class and integrated function class of federate, SOM and FOM.
(7) Generate FED file automatically by FOM list;
(8) Generate object class and interaction class of simulation model atomically from SOM;
(9) Generate federate that meets HLA norm by the federate automatic generation framework.
(10) Under the control of resource management federation, federate creates federation, joins federation and starts simulation.

6 MDA Based HLA Extension

The standard HLA does not provide with modeling environment, so it cannot manage federation in two layers architecture. Referring to HLA/RTI/web service and OGSI, this paper presents FIA based on MDA and the extension of HLA/RTI, and its federation running environment FEI.

FIA is the architecture of two levels, including system federation and application federation. The system federation is resource management federation, which offers modeling environment that is in accord with MDA modeling framework, and transform user model to federate model that conforms to HLA.

System federation services the whole modeling/simulation process, and manages all resources that simulation needs. Thus it clears away the obstacles for the cross-platform of resources.

Application federation is built on federate model which is accord with HLA based on resource management federation. The architecture enables multi-federation to run at the same time, and they are managed by application federation pool.

To achieve the above requirements, this paper researches on FEI which supports FIA to manage every process in system federation, including: FEI interface management, federation application management, system OMT management, objects management, declaration management and federation user management. Its function shows in Figure 6. FEI uses the ternary structure of logic, view and data, which separates the application logic, user view and backend data, and promises the smooth communication among them.

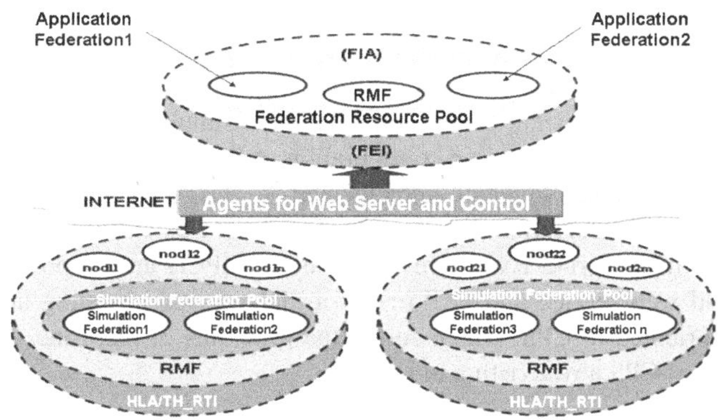

Fig. 6. Implementation architecture of FEI

FIA extends the function of HLA/RTI which is popular now, and implements the function of unified management of simulation models and dynamic sharing based on resource management federation. FEI offers a common, relatively independent support service program, which separates platform application from underlying support environment, thereby all parts can be developed relatively independently, and it can makes full use of advanced technologies in their respective fields.

7 Conclusion

This paper discusses problems and challenges in modeling and simulation of CPS.

The method of MDA is introduced to the HLA modeling, and MDA and HLA are given way to integration; there are different hardware, software and network condition in each node, so the heterogeneity is more obvious than the computer application before. The import of meta-model idea implements the process that model transform to federate well, as well as the using of MDA implements the rapid modeling from model to federate.

The paper gives the definition of PIM layer and PSM layer based on UML; based on XML technology, we study on the transformation from PIM to PSM, and give the mapping rules and the corresponding convert tools; the framework that separates SOM information from federate, which implement the dynamic mapping between SOM information and the interaction class code of federate. Simultaneously, it supports the PNP of simulation model, and constructs support for the rapid development from PSM model to federate code.

The HLA architecture consisted of multi-layer, multi-federation facing CPS is proposed, which breaks through the planer construction. Because every federate has different demand on real-time, so multi-layers federation architecture is organized according to different bandwidth. The faster system uses LAN, which forms the lowest level federation; the slower forms the middle level federation, and the slowest forms the highest level federation. In the interactive logic, Multi member in multi federation is a big union essentially. The upper level and lower level not only have data interaction, but also have a uniform time.

The idea of resource management federation is proposed. Resource management federation manages which bandwidth the federate in CPS can unified, and makes the transformation between different bandwidth when system is running. By supporting the management of meta-model, dynamic combination of simulation application federation based on meta-model and dynamic configuration of FOM information, resource reuse and share based on meta-model are implemented.

Finally, Federate Integration Architecture (FIA) as the extension of HLA/RTI based on MDA and its Federate Executive Infrastructure (FEI) are proposed referring to HLA/RTI and web service and OGSI. It is a MDA based modeling and simulation architecture oriented to CPS, which solves the problems of real-time, extension and heterogeneity of CPS now existing in HLA.

References

1. Lee, E.A.: Department of EECS, UC Berkeley, Cyber-Physical Systems - Are Computing Foundations Adequate? Position Paper for NSF Workshop On Cyber-Physical Systems: Research Motivation, Techniques and Roadmap, October 16 - 17 (2006)
2. Cárdenas, A.A., Amin, S., Sastry, S.: Secure Control:Towards Survivable Cyber-Physical Systems. In: First International Workshop on Cyber-Physical Systems (WCPS 2008) (2008), http://www.pdfgeni.com/book/intrusion-detection-kalman-pdf.html
3. Gill, H.: Co-Chair, CISE/CNS, National Science Foundation NSF, Perspective and Status On Cyber-Physical Systems. In: National Workshop on Cyber-Physical Systems, NITRD High Confidence Software and Systems Coordinating Group, October 16-17 (2006)
4. Branicky, M.S., et al.: Multi-Disciplinary Challenges and Directions in Networked Cyber-Physical Systems, http://varma.ece.cmu.edu/cps/Position-Papers/Branicky.pdf

5. Song, Z., Sastry, C.R.: Optimal Observation for Cyber-physical Systems. Springer-Verlag London Limited (2009)
6. Brutzman, D., Zyda, M., Pullen, M., et al.: Extensible Modeling and Simulation Framework(XMSF) Challenges for Web-Based Modeling and Simulation. In: NPS, GMU and SAIC (October 22, 2002)
7. Steinman, J.S.: Evolution of the Standard Simulation Architecture. In: Proceedings of the 2004 Spring Simulation Interoperability Workshop (2004)
8. OMG Staff Strategy Group, Model Driven Architecture, `http://www.omg.org/2000`
9. Calytrix Technologies, Welcome to SIMplicity (EB/OL) (2005), `http://www.simplicity.calytrix.com`
10. Brutzman, D., Tolk, A.: JSB Composability and Web Services Interoperability via Extensible Modeling & Simulation Framework (XMSF), Model Driven Architecture (MDA), Component Repositories, and Web-based Visualization, Study Report (November 2003)
11. European Space Agency, SMP 2.0 Handbook. ESOC (October 28, 2005)

Intelligent Agent-Based Mobility Prediction Method Using Velocity Inference

Yong Beom Ma and Jong Sik Lee

School of Information Engineering, Inha University, #253, YongHyun-Dong, Nam-Ku, 402-751 Incheon, South Korea
myb112@hanmail.net, jslee@inha.ac.kr

Abstract. In wireless network environment, to detect the location of a mobile node (MN) is an important factor for providing high-quality service to a MN. However, a limited communication range, disconnection, location error caused by various environmental influence of moving paths can be occurred in wireless network environment. In this paper, we propose an Intelligent Agent-based Mobility Prediction (IAMP) method that predicts the mobility of a MN by assigning an intelligent agent (IA) to each base station (BS) in a hybrid wireless network environment and provides suitable services according to change of location. To predict the mobility of a MN we build an ontology and carry out an inference based on some defined rules. To demonstrate superiority of the IAMP, we measured the average location error as time. Experimental results comparing the proposed method with established methods verify effectiveness and efficiency of the IAMP.

Keywords: Intelligent Agent, Mobility Prediction, Velocity Inference, Hybrid Wireless Network.

1 Introduction

In mobile ad hoc networks, a MN communicates with other nodes through a routing and a transmitter-receiver that each node has. Each node needs a more intelligence because it must find a path with the minimum overhead and perform a route reconstruction of broken routes. Recently, a hybrid wireless network [1], [2] that combines the cellular network with the mobile ad hoc network through a gateway has been researched. In a hybrid wireless network, MNs enable to communicate with each other and much information of the network is obtained by the communication between a BS and a central control station (CS). Therefore, the network enables to improve quality of service and to perform more intelligent works by the obtained information. It is one of the important factors that the mobility of a MN and the connectivity between each node are supported in hybrid wireless networks.

In wireless network environments, we need to detect accurate location of a MN which is known as location management, in order to guarantee the connection among nodes. And it is used for service area-based location management because of the limited power of a MN. The service area is called a cell. A BS detects location of MNs and maintains the topology graph of the MNs in its service area. Moreover, a BS provides resources for service execution to MNs in its service area. Another feature of

J.-H. Kim et al. (Eds.): AsiaSim2011, PICT 4, pp. 116–125, 2012.

wireless network environments is that links are frequently disconnected. Handoff is conducted in order to avoid call termination when a MN gets outside the service range of a BS. In a wireless computing environment such as a hybrid wireless network, it is necessary to predict the mobility of a MN and provide resources according to change of location by the location management and the handoff management.

In this paper we propose an Intelligent Agent-based Mobility Prediction (IAMP) method using velocity inference to predict the mobility of a MN accurately within a wireless computing environment such as a hybrid wireless network. The method builds an ontology for hybrid wireless network environment and provides an accurate prediction of mobility through an inference based on some defined rules for the acquirement of knowledge. We compare performances of the proposed method with those of established prediction methods by measuring average location error as time progresses.

2 Related Work

The location of a MN is detected and updated periodically in wired and wireless network environment. But an intelligent and efficient mobility management is necessary due to the feature of that environment such as the lack of communication resources and frequent connection failures. There have been some researches on mobility management based on prediction [3], [4], [5], [6].

In earliest distance-based location prediction scheme, the location of a MN is based on historical data and the location information is updated whenever its distance from the last updated location exceeds a specified value. But this scheme is too simple and the moving distance is made since the last updated. Hence, the scheme is difficult to predict accurately. Ben Liang and et al [3] proposed a predictive distance-based mobility management scheme. In this scheme, the MN reports both its location and velocity when updating the location information. And the MN checks the location periodically and performs location update whenever it reaches some threshold distance away from the predicted location [3]. In the adaptive distance-based scheme [4], authors proposed a stochastic model which enables to compute the optimal update boundary. And their numerical results indicate that the proposed scheme gives a more accurate update boundary in real wireless cellular environments [4].

In [5], two thresholds, time threshold T and distance threshold D, is used for representing the velocity of a MN. Authors represent the velocity of a MN by the combination of two threshold values. And the values do not need to observe a certain period in order to set them and to perform complicated computation. In [6], authors use a dynamic filtering phase in order to predict the velocity of a MN. The current velocity of a MN is predicted based on the location of computed in previous step and that of computed in the current step. And the computed value is used the velocity for the next step.

However, these studies not only do not consider various environmental factors which enable to be affected during the movement of an actual MN but also almost need to a complicated computation. Therefore, it is difficult to predict the mobility of an actual MN accurately. And additional power consumption for computation is needed. It causes us to increase the error between the location information of a BS and that of a MN. This paper proposes an intelligent agent-based mobility prediction to solve problems of existing methods.

3 Intelligent Agent-Based Mobility Prediction Using Velocity Inference

In a hybrid wireless network, a mobile node (MN) moves constantly with an arbitrary or specific mobility. As mentioned before, it is necessary to predict accurately the mobility of a MN in order to detect accurately the location of a MN. And a base station (BS) has to detect the location of a MN in order to communicate with a MN and provide resources for service execution to a MN. However, too much communication for location detection causes overheads and performance degradation. Therefore, it is very important to predict the MN mobility accurately in a hybrid wireless network. In this paper we propose an Intelligent Agent-based Mobility Prediction (IAMP) method using velocity inference as shown in Fig. 1.

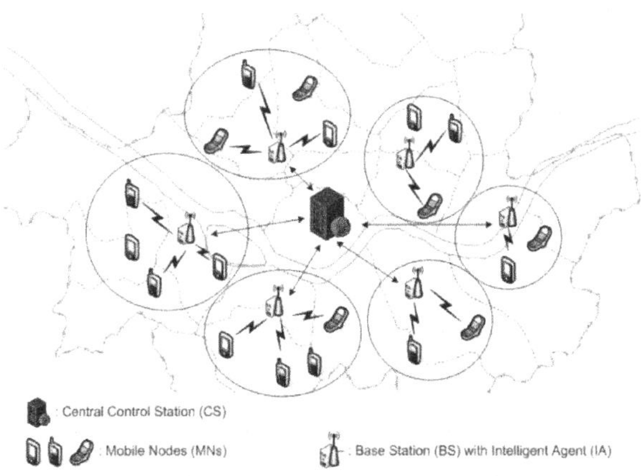

Fig. 1. High level view of the IAMP

A hybrid wireless network comprises a central control station (CS), several BSs and massive terminals [2]. We identify the terminal with a MN. The CS communicates with each BS and receives various information from BSs. MNs desire to receive a high quality service and it is greatly affected to the connectivity with a BS. Moreover, MNs move constantly with a specific mobility and need to use resources continuously during movement for the execution of an arbitrary service. If a MN cannot be guaranteed the connectivity with a BS, communication resources cannot be allocated to the MN as well as the BS is difficult to detect the location of the MN. Hence, we design an intelligent agent (IA) which supports the function of the established BS to predict mobility of a MN and provide adaptive resources according to change in the location of a MN. Arrows in Fig. 1 indicate that the communication between a BS with IA and the CS and the transmitted information. Links between BSs and the CS indicate the dynamic network topology within a cell. Fig. 2 shows an architecture of the IA for mobility prediction. The IA comprises some components that include the Mobility Predictor.

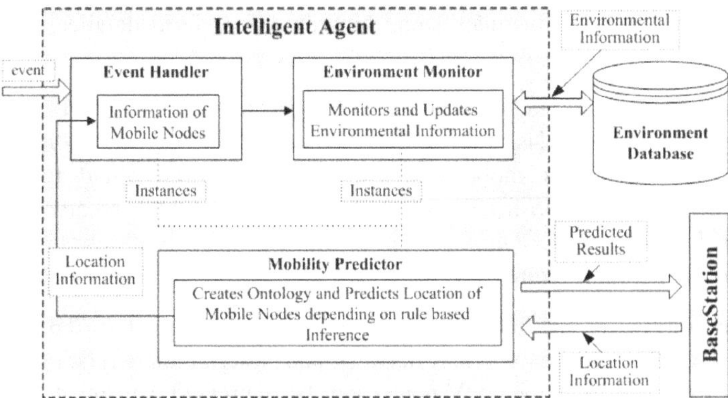

Fig. 2. Architecture of the IA

IA is composed of three module-based components; *Event Handler*, *Environment Monitor*, and *Mobility Predictor*. The *Event Handler* perceives an event to be occurred and manages the information about occurred events. This module obtains the current information of MNs when an event is occurred from an external component. Even though an event is not occurred, this module receives information about the precise location of MNs from the *Mobility Predictor* and updates the information. As shown in Fig. 2, The *Event Handler* proceeds with the next task by transmitting a message to the *Environment Monitor*. The *Environment Monitor* module monitors the environmental information managed by an external component and updates the information whenever a task is requested from the *Event Handler*. In order to obtain the environmental information, the *Environment Monitor* located in the IA communicates with other external components by transmitting messages. And the *Environment Monitor* makes instances of the information of MNs and updated environmental information. This module requests to predict mobility of MNs by transmitting instances to the *Mobility Predictor*. The *Mobility Predictor* uses an ontology and a rule-based inference in order to predict the next location of MNs. As shown in Fig. 3, internal modules of the *Mobility Predictor* are Information Manager, IAMP Ontology, Inference Engine, and Location Manager.

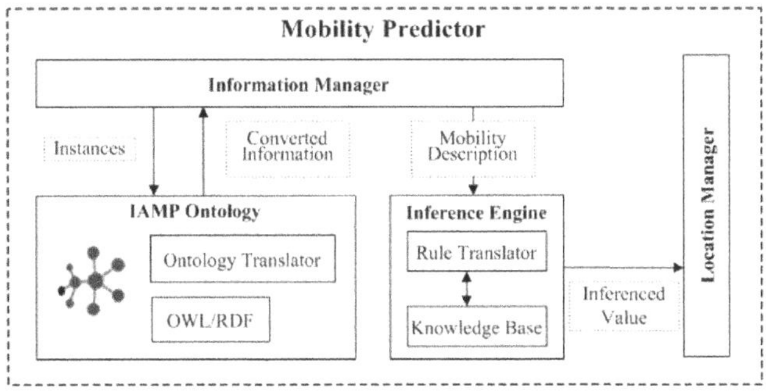

Fig. 3. Composition of the Mobility Predictor

The *Mobility Predictor* includes some modules to perform detailed functions as follows.

Information Manager: The Information Manager receives information about the mobile environment or MNs from upper modules of the *Mobility Predictor*. The information is needed for mobility prediction and is transmitted to the IAMP Ontology in order to construct an ontology. This module receives converted information from the IAMP Ontology and sends the mobility description converted into an ontology to the Inference Engine.

IAMP Ontology: This module constructs an ontology in order to define concepts underlying IAMP for hybrid wireless network and describe their relations. Instances transmitted from the Information Manager can be expressed a form of OWL [7] or RDF [8] by the Ontology Translator. The converted information is returned to the Information Manager.

Inference Engine: The Inference Engine carries out an inference based on the mobility description converted into an ontology by the Information Manager. Rules for an inference are defined. Results obtained by the rule-based inference are transmitted to the Location Manager in order to predict the mobility of a MN and are stored in the Knowledge Base.

Location Manager: The Location Manager predicts the next location of a MN based on results obtained from the Inference Engine and transmits predicted results to BS. And this module receives precise location information of MNs from BS every period and sends the information to the Event Handler.

We use some basic information about MN in order to represent the mobility of a MN accurately and extract some features of the mobility [10]. To represent the MN mobility, we use a velocity and acceleration. And we use x and y co-ordinates to represent the location of a MN. The Euclidean distance is used for representing the relation between two MNs. Basic information used in this paper are as follows.

1) $\vec{V}_i(t)$: Velocity vector of node i at time t
2) $\vec{a}_i(t)$: Acceleration vector of node i at time t
3) T : Simulation time
4) $random()$: Returns a value uniformly distributed in the interval [-1, 1]

On the basis of this information, we represent a change of the MN velocity as shown in equation (1). $\vec{V}_i(t+1)$ represents the next velocity of the MN and N_{env} is the number of environmental factors affecting a change of the MN velocity. β_j is an influence of j th predictable factor on the change of the MN velocity for a rate and $condition_j$ represents the condition value of the j th predictable factor. The value is normalized from zero to one and varies according to the current time and location of a MN. The next velocity of a MN is affected by the acceleration and depends on the current velocity. The current velocity is considerably affected by the weight value affecting environmental factors and the conditions of factors. Therefore, we can predict the next velocity of a MN accurately by taking the environmental factors related to the MN into consideration.

$$\left|\vec{V}_i(t+1)\right| = \left|\vec{V}_i(t)\right| \times \sum_{j=1}^{N_{env}} (\beta_j \times condition_j) + random() \times \left|\vec{\alpha}_i(t)\right| \quad (1)$$

4 IAMP Ontology for Hybrid Wireless Network

This section describes an ontology which predicts the mobility of a mobile node (MN) and provides adaptive resources to MNs in hybrid wireless network environment and an inference based on some rules. To describe concepts and form relations among concepts, we create an ontology which provides details of a MN and predicts the mobility of a MN. And we used Protégé [11], one of ontology editors, for developing an ontology. The Protégé supports not only OWL but also RDF and provides an integrated environment for editing an ontology and instances.

The ontology infers the velocity of a MN based on environmental factors which alter according to the current time and area. To infer the velocity, the ontology reflects concepts underlying hybrid wireless network environment and their relations among concepts. Fig. 4 presents the IAMP ontology which supports the prediction of mobility.

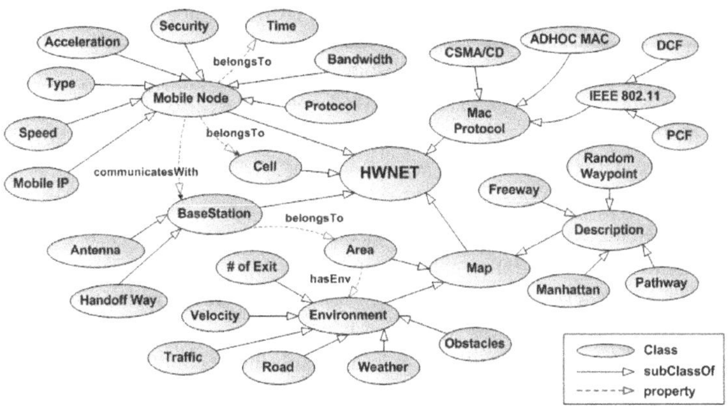

Fig. 4. IAMP ontology for hybrid wireless network

To build the IAMP ontology, we define the class HWNET first. The HWNET has the class Mobile Node that is one of members composing hybrid wireless network, the class Map that reflects the location information of MNs, the class Cell regarded as a coverage area, and the class Environment according to the current time and area. The IAMP ontology defines each class and describes the relation among classes. The Mobile Node can specify the information about its own location through the property *belongsTo* defined by the Cell. And the Mobile Node has the *communicatesWith* relation to the Base Station class in order to communicate with the base station in each cell and two sub class, Speed and Acceleration. The Mobile Node has the *belongsTo* relation to the class Time, because the location of MNs is changed as time progresses.

This paper focuses on allowing MNs obtain adaptive resources according to change of location and BSs predict accurately the mobility of MNs in order to provide adaptive resources. And we use the Base Station in order to detect the location of MNs and provide resources because the proposed method is based on the hybrid wireless network environment. The Base Station is a subclass of the Cell class and has the *communicatesWith* relation to the Mobile Node. The Area obtains information about environmental conditions which change according to the location of MNs. The information is obtained by the Environment class connected with the *hasEnv* relation. And the Area can refer the Base Station and the Cell including the relevant Area through the property *belongsTo* defined by the Base Station. The IAMP ontology creates instances of each class by loading the stored value in the relational database connected with the ontology.

MNs move on a map and the map is divided into cells in order to represent coverage areas. Moreover, the map is divided into areas and each area has different environmental conditions. Therefore, the relation between the mobility of MNs and environmental conditions according to change of location discovers through the inference engine. For an inference, we need to make a kind of rules and use Sematic Web Rule Language (SWRL) [9]. Table 1 shows some rules defined in this paper.

Table 1. Inference rules of the IAMP ontology

No	Domain rules
1	Mobile Node(?x) ∧ belongsTo(?x, ?y) ∧ Area(?y) ∧ areaNum(?y, ?z) → includedInArea(?x, ?z)
2	Mobile Node(?x) ∧ belongsTo(?x, ?y) ∧ Time(?y) ∧ timeNum(?y, ?z) → includedInTime(?x, ?z)
3	Mobile Node(?x)∧communicatesWith(?x, ?y) ∧ BaseStation(?y) ∧ bsNum(?y, ?z) → BSNumNode(?x, ?z)
4	Mobile Node(?x) ∧ includedInArea(?x, ?a) ∧ [?a=qArea] ∧ includedInTime(?x, ?t) ∧ [?t=qTime] → NodeInAreaTime(?x, ?a, ?t)
5	Mobile Node(?x) ∧ hasEnv(?x, ?y) ∧ Environment(?z) ∧ envNum(?z, ?e) ∧ NodeInAreaTime(?z, ?a, ?t) ∧ [?e=op:numeric-add(?a, ?t] → InferredNode(?x)
...	...

Defined rules are used for an inference. For instance, we identify the information about the cell, area, and time by the Rule 1, 2, and 3 in order to predict the mobility of a MN. The Mobile Node class in the Rule 5 can be replaced by the result class that has the same type with the Mobile Node class and obtained by the Rule 1, 2, and 3. If Mobile Node x has the property hasEnv y and the area and the time of the Environment z is the same as y, the Mobile Node is inferred to be included in specific area and time interval. Therefore, we can infer what environmental factors have an effect on a MN by the Rule 5.

5 Simulation and Results

We implemented a simulation model by applying the DEVS formalism [12] based on discrete events to IAMP method. We use Protégé [11] which enables to create and edit an ontology in order to conduct the velocity inference using an ontology and create an OWL file in order to import to JAVA API. To demonstrate the excellence of the proposed method, we measured the average location error of MNs with increasing simulation time. And we compared the IAMP method with Mobile Distance-based Mobility Prediction (MDMP) method and Mobile Velocity-based Mobility Prediction (MVMP) method in order to evaluate performances. In our experiments, we assumed that 100 MNs moves on a freeway map and the size of the map is 1000×1000. We conduct experiments on the basis of some additional assumptions because the area and the velocity do not have specific units in freeway mobility. In freeway mobility [10], [13], the velocity value of MNs is limited to 1.0. If a speed limit of a freeway is 100km/h, we assume that a velocity value of 1.0 corresponds to 100km/h and a map size value of 1.0 to 1km. And we can convert 100km/h to 27.8m/s because velocity multiplied by time is distance. As mentioned in section 3, the next velocity of a MN is affected by a weight value affecting environmental factors and current conditions of factors. If the speed limit of a freeway is 100km/h. Thus we assume that the condition value of the speed limit corresponds to 1.0 to be not affected by the current velocity. If we predict the next velocity of a MN by the speed limit, we need not the weight value affecting environmental factors. However, we set the weight value based on the relation of the velocity and each environmental factor in order to predict the next velocity of the MN consideration of various environmental factors. We set the weight value of the speed limit high because the velocity of a MN is directly related to the velocity of a MN. And we set the weight value of weather or road low because the condition of weather or road is not closely related to the velocity of a MN. For our experiments, we assumed that the weight value affecting environmental factors and set the values based on the condition of each environmental factor from 0.0 to 1.0.

5.1 Average Location Error

Generally MNs inform the BS about their location information every period and the BS updates the location information of MNs. If MNs inform the BS about their location information every time, there are few location errors of MNs. However, message transmission is needed in order to inform the BS and too much message transmission cause heavy traffic. Hence, we set the message transmission period of each method to be equal in total message transmission and then measure their average location error. The MDMP and the MVMP graph an almost similar form to each other. Each method has previous location information before it receives a message in order to update the location information. Therefore, the average location error of two methods increases gradually and become 0 when MNs transmit messages to each BS. Moreover, to predict the location of a MN in these two methods is inaccurate because they do not infer and predict the location of a MN through simple information such as the distance from previous node and the difference between the current velocity and the average velocity.

The Intelligent Agent-based Mobility Prediction (IAMP) performs an inference considering environmental factors which enable to affect the real mobility of a MN and predicts the velocity and the location of a MN based on the inferred result. The average location error of the IAMP is relatively few because the location of a MN predicted in the BS is not different from that of real MNs. And the location error does not increase continually but increases or decrease according to the inferred result. However, the location error becomes zero when a MN transmits the message for updating the location information of a BS. As mentioned before, the distance value of 1.0 corresponds to 27.8 m. Thus we can convert the average location error that is measured in this experiment into a real distance by multiplying the measured value and 27.8 m which is the moving distance per second of a MN. In our experiments, the average moving distance of 100 MNs recorded approximately 0.49 (=13.62 m). On average, the IAMP method provided a location error of about 9.45 m, which is lower than the average moving distance of MNs. Thus we consider that the value is not considerably out of the service range. The IAMP provided lower average location error than that of other methods because it infers the velocity of a MN by taking various environmental factors into consideration and predicts the location of a MN based on the inferred result. Fig. 5 shows the average location error of each method as mentioned above.

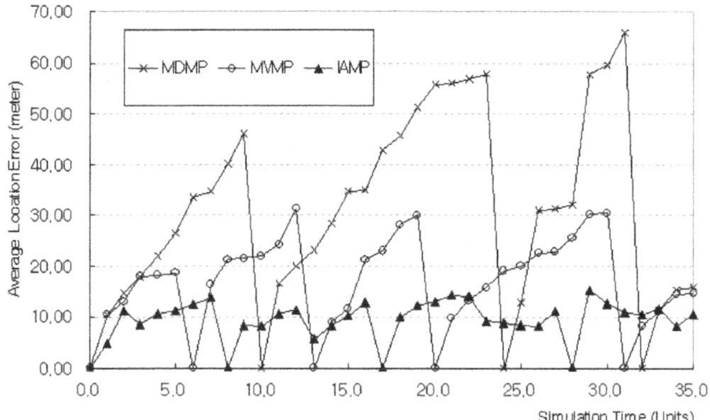

Fig. 5. Average location error

6 Conclusion

In this paper, we focused on mobility prediction in a wireless computing environment such as a hybrid wireless network and proposed an intelligent agent-based mobility prediction (IAMP) that provides adaptive resources to mobile nodes and improves the quality of services by performing a velocity inference. To provide services of high quality to a mobile node (MN) during the movement, we predict the mobility of a MN by assigning an intelligent agent (IA) to each BS. The IA predicts the mobility of a MN with greater accuracy than established mobility prediction methods by building an ontology for the velocity inference and taking various environmental factors in the path of movement into consideration. To evaluate performances, we conducted some experiments to compare the proposed IAMP method with the established two methods. Experimental results show that the IAMP provides the average location error 69.18% lower than that of MDMP and 43.22% lower than that of MVMP. This experimental result demonstrates that the IAMP can assist in reducing the location error of a MN.

References

1. Tchepnda, C., Moustafa, H., Laboid, H.: Hybrid Wireless Networks: Applications, Architectures and New Perspectives. In: 3rd Annual IEEE Communication and Networks (SECON), vol. 3, pp. 848–853 (2006)
2. Fujiwara, T.: A Multihop Wireless Network Architecture for Disaster Communications. Doctoral Thesis (2003)
3. Liang, B., Haas, Z.H.: Predictive Distance-based Mobility Management for PCS Networks. In: Proceddings of the Conference on Computer Communications (IEEE Infocom), New York, pp. 1390–1394 (1999)
4. Wong, V.W.S., Leung, V.C.M.: An Adaptive Distance-based Location Update Algorithm for Next Generation PCS Networks. IEEE Journal on Selected Areas in Communications (JSAC) 19(10), 9142–9152 (2001)
5. Hwang, S.H., Han, Y.H., Lee, B.K., Hwang, C.S.: An Adaptive Location Management Scheme Using The Velocity of Mobile Nodes. IEEE Wireless Communications and Networking (WCNC) 3, 1999–2004 (2003)
6. Su, C., Wan, J., Yu, N.: Dynamic Simulation Based Localization for Mobile Sensor Networks. In: Zhang, H., Olariu, S., Cao, J., Johnson, D.B. (eds.) MSN 2007. LNCS, vol. 4864, pp. 524–535. Springer, Heidelberg (2007)
7. McGuinness, D.L., Harmelen, F.: OWL Web Ontology Language Overview (2004), `http://www.w3.org/TR/owl-features/`
8. Beckett, D.: RDF/XML Syntax Specification (Revised) (2004), `http://www.w3.org/TR/rdf-syntax-grammar/`
9. Horrocks, I., Patel-Schneider, P.F., Boley, H., Tabet, S., Grodof, B., Dean, M.: SWRL: A Semantic Web Rule Language Combining OWL and RuleML (2004), `http://www.w3.org/Submission/SWRL/`
10. Bai, F., Narayanan, S., Helmy, A.: Wireless Ad Hoc and Sensor Networks: Chapter 1 A Survey of Mobility Models in Wireless Adhoc Networks, pp. 1–29. Kluwer Academic Publishers (2004)
11. Musen, M.: Protege (2010), `http://protege.stanford.edu/`
12. Zeigler, B.P., Moon, Y.K., Kim, D.H., Ball, G.: The DEVS Environment for High-Performance Modeling and Simulation, vol. 4(3), pp. 61–71. IEEE Computer Society Press (1997)
13. Bai, F., Sadagopan, N., Helmy, A.: IMPORTANT: A Framework to Systematically Analyze the Impact of Mobility on Performance of Routing Protocols for Adhoc Networks. In: 22th Annual Joint Conference of the IEEE Computer and Communications Societies, vol. 2, pp. 825–835 (2003)

Research on Global Path Planning of CGF Battlefield Terrain Based on Genetic Algorithms

Song Xiao, Teng Da, Qian Lidong, Zhang Lin, and Lu Ming

School of Automation Science and Electronic Engineering,
Beihang University 100191 Beijing, China
tengda0108@163.com

Abstract. Computer Generated Force (CGF) battlefield terrain is expressed in professional terrain database, For the information of polygonal obstacles of battlefield terrain obtained by the interface of CGF terrain database, we adopt the topology polygon to modeling circumstance, present a Genetic Algorithms coded with binary method based on the vertexes of polygonal obstacle, consider the battlefield terrain factors which affect the maneuverability of CGF entities, such as, soil types, terrain slope, into the design of adaptability function of Genetic Algorithms, research the method of global path planning which is suitable to CGF battlefield environment. The simulation results validate the validity and precision of this method.

Keywords: terrain database, genetic algorithms, path planning, battlefield terrain, computer generated force.

1 Introduction

CGF (Computer Generated Force) path planning is one of the hotspots of CGF research. CGF battlefield environment is characterized by specialized military terrain databases [1] [2], which include critical information of the battlefield terrain for the simulation. These descriptions of the terrain features are divided into physical features, abstract features and linear features. And abstract features are divided into surface-like features, line-like features and point-like features. The majority of obstacles in the battlefield environment are stored in the form of the surface-like features and the surface-like features take the polygon form of storage. We can get information of the polygonal obstacles through the interfaces of CGF terrain database. Progresses have been made in the field of CGF path planning home and abroad [3] [4].

This paper presents a genetic encoding approach for relatively large size polygonal obstacles based on the battlefield environment characterized by CGF terrain database, a design of the fitness function taking soil type, terrain slope etc. in consideration, and the process of the usage of GA to solve the path planning problem. At last, the efficiency of the binary encoding method is tested by simulation, and the results is compared with the ones of the typical real number encoding GA used for the robot path planning[5] [6]. With comparison the superiority of this path planning algorithm is proved.

J.-H. Kim et al. (Eds.): AsiaSim2011, PICT 4, pp. 126–133, 2012.

2 CGF Battlefield Environment Modeling Based on Polygonal Obstacles

For any two-dimensional battlefield terrain environment, there are a limited number of obstacles. Some of these obstacles can be passed through, such as grass and forest, and others can not, such as buildings, lakes, minefield, etc. CGF path planning is to find a path from the start point to the end point with the least cost, meaning the largest fitness value. The polygonal obstacle environment is shown in Figure 1:

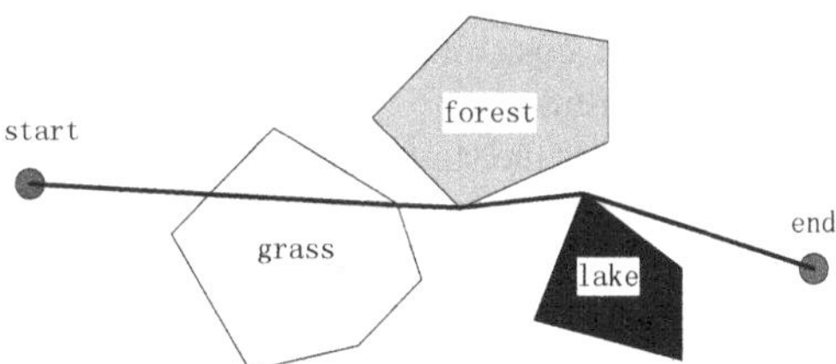

Fig. 1. CGF polygonal obstacle environment

In Figure 1, black polygons stand for the obstacles which can not be passed through, such as lakes, and gray polygons are barriers that can be went over. Light gray polygons can be passed easily for example grass, while dark gray ones take efforts to get over such as forest.

U : The set of path points, including the start point, the end point, and all the vertices of the polygonal obstacles. L: The set of all the paths from the start point to the end point, all the points of which are included in U .The optimal path must be one of the elements in L, so the path planning based on polygonal obstacles can be described as follows: finding an element in set L which has the largest fitness value.

In this paper we assume that, all battlefield entities go in the direction which makes an angle not larger than 90 degrees with the direction from the start point to the end point. There are three polygonal obstacles in Figure 2:

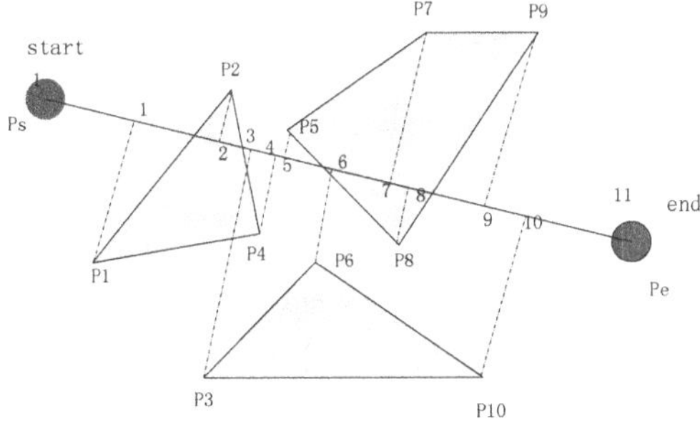

Fig. 2. CGF battlefield environment modeling

In Figure 2, Circle Ps represents the start point, and Circle Pe stands for the end point. Draw a straight line P_sP_e from the start point to the end point, and draw a vertical from each of the vertices of the polygons to the line P_sP_e, then sort the vertices by the distances their vertical from the start point. The figure 2 shows the serial numbers of the vertices of the 3 polygons. Add the start point and the end point to the series, and the whole path point sequence is: $P_s \to P_1 \to P_2 \to \cdots \to P_9 \to P_{10} \to P_e$. Mark the sequence as U, then the path planning problem form the start point to the end point can be described as : select m vertices from the sequence U to form a new sequence L to make L having the least cost, or the largest fitness value. The sequence L should meet the following requirements:

The first point of the sequence is the start point Ps, and the last is the end point Ps;

$m <= n$, n is the number of all the vertices.

3 Research on Genetic Algorithm of Global Path Planning in the Battlefield Environment

3.1 Path Encoding Method for Genetic Algorithm

(1) Encoding: Genetic information in DNA is arranged in a long chain in certain form, which is the genetic code. The coded string is called chromosome. Genetic code can be seen as a mapping from phenotypes to genotypes.

This paper presents a simple binary code for the polygon vertices. Taking the battlefield environment shown in Figure 2 as an example, the sequence of all path points $P_s \to P_1 \to P_2 \to \cdots \to P_9 \to P_{10} \to P_e$ is encoded in a corresponding binary code. Taking $s_1, \cdots, s_{10}$ as 10 binary bits corresponding to the 10 vertices $P_1, \cdots, P_{10}$, the path from the start point to the end point can be encoded as:

$$S_s S_1 S_2 S_3 S_4 S_5 S_6 S_7 S_8 S_9 S_{10} S_e$$

s_s : The binary bit for the start point P_s,

s_e: The binary bit for the end point P_e.

The meaning of the binary bit is:

1 — the path passes through the point,

0 — the path does not pass through the point.

The objective is to find the optimal path from the start point to the end point, so the start point and the end point must be passed by the path, which means $s_s = 1$ and $s_e = 1$.

(2) Decoding: A mapping form genotypes to phenotypes, or form chromosomes to phenotypes.

Taking the terrain shown in Figure 2 as an example, assuming there is a path encoded as 101001001001 by the method mentioned above, the path passes the vertices P_s, P_2, P_5, P_8, P_e, so the phenotypes of the path is $P_s \to P_2 \to P_5 \to P_8 \to P_e$, as the thick line shown in the following figure:

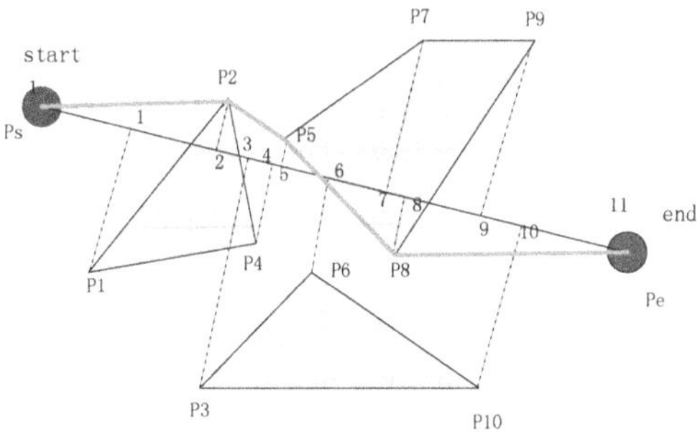

Fig. 3. The path decoding figure of GA

3.2 The Design of the Cost Function and the Fitness Function

(1) The design of the cost function

Assuming s is a path represented by a chromosome, $\text{cost}(path)$ stands for the cost function of the path. In accordance with the encoding approach described in Section 3.1, the cost function can be expressed as:

$$\text{cost}(path)=\text{Path}(s,e)+k\cdot\sum_{i=0}^{n}\lambda_i\cdot\text{Dis}(s_i,e_i) \tag{1}$$

In this expression, s is the start point, e is the end pint, $\text{Path}(s,e)$ is the length of the whole path from s to e, $\text{Dis}(s_i,e_i)$ is the distance between Vertex s_i and Vertex e_i, and n is the number of the polygonal obstacles. $\lambda_i, i=1,2,\cdots n$ are factors that represent the difficulty of passing though this area, so it reflects the terrain factors on the CGF path planning results. The bigger λ is, the harder this area is to be passed through. When the soil type is easy for passing through, for example plain, then $\lambda_i=0$, when it can not be got over $\lambda_i=1$, and when it is between the two situations above, λ_i is a real number between 0 and 1.

$Dis(s_i,e_i)$ stands for the length of the section where the path overlapped with the i th polygonal obstacle. According to the encoding approach mentioned in 3.1, each path can at most intersect the edge of a polygon at two pints, without consideration of concave polygons. s_i stands for one of intersection points, and e_i the other.

k stands for the cost weight of the section of path intersecting the obstacle, in this algorithm $k=2$.

It can be seen from expression (1) that the longer a path is and the longer it intersects the obstacles, the more it costs, where k is the weighting factor of the second cost.

(2) The design of the fitness function

In the genetic algorithm, a chromosome is better if it has smaller value of the cost function, and larger value of the fitness function. The cost function above is greater than zero, and then the fitness function can be the reciprocal of the cost function multiplied by a constant, as shown below:

$$\begin{aligned}\text{fitness}(path) &= \frac{a}{\text{cost}(path)} \\ &= \frac{a}{Path(s,e) + k \cdot \sum_{i=0}^{n} \lambda_i \cdot Dis(s_i, e_i)}\end{aligned} \tag{3}$$

In this expression, a is a constant. In order not to being too small, take

$$a = (ptNum / 2) \cdot Dis(s,e) \tag{4}$$

$ptNum$ is the number of all the vertices, including the start point and the end point, $Dis(s,e)$ is the distance from the start point to the end point.

3.3 The Improvement of Genetic Algorithm

In this paper, an improved Adaptive Genetic Algorithm (AGA) is used to search the optimal path. In the early generations of AGA evolution, genetic operators should ensure that the generation search in a wide space, in order to avoid convergence too early; in the late stages of AGA evolution, the generation should search in a smaller scope to improve the accuracy of the result. The convergence of GA mainly depends on its core operations as crossover and mutation operators. Because crossover offers global search capabilities, and mutation provides local search capabilities, AGA improved the standard GA from two aspects:

(1) Adjust the probabilities of the crossover and the mutation according to the value of fitness function. The adaptive probabilities of crossover and mutation of AGA is adjusted as follows:

$$p_c = \begin{cases} p_{c\_max} - \left(\dfrac{p_{c\_max} - p_{c\_min}}{it\max}\right) * iter, & f' > f_{avg} \\ p_{c\_max} & , f' \le f_{avg} \end{cases} \tag{7}$$

$$p_m = \begin{cases} p_{m\_max} - \left(\dfrac{p_{m\_max} - p_{m\_min}}{it\max}\right) * iter, & f > f_{avg} \\ p_{m\_min} & , f \le f_{avg} \end{cases} \tag{8}$$

In the expressions above, p_c is the probability of crossover, $p_{c\max}$ is the maximum probability of crossover, $p_{c\min}$ is the minimum probability of crossover, p_m is the probability of mutation, $p_{m\max}$ is the maximum probability of mutation, $p_{m\min}$ is the minimum probability of mutation, $iter$ is the number of the current generation, $it_{\max}$ is the maximum number of the generations, f_{avg} is the average value of the fitness of the generation, f' is the larger fitness value of the two chromosomes in the crossover operation, f is the fitness value of the chromosome in the mutation operation.

It can be seen from the expression (7) and (8) that if an individual is poor (its fitness value is less than the average), it would be given a large crossover probability and a small mutation probability, then if the individual is good (its fitness value is greater than the average), the probabilities of crossover and mutation will be determined according to its fitness value and its iteration state too. This adjustment method of crossover probability and mutation probability has a strong global search capability and a week local search capability in the early stage of evolution, while with the evolution the global search capability is weakened gradually and the local search capability increases. This improvement approach can prevent the "premature" phenomenon.

(2) Compare the fitness value after the crossover (or mutation), if the fitness value increased then keep this variation, otherwise keep the original individual. This approach guarantees the excellence of new-generated individuals, improves the efficiency of GA, and prevents the individual retrogression in the evolution process. The processes of the polygonal obstacle based binary encoding GA for path planning is shown below:

Step 1, use the method described in the Chapter 2 to model the battlefield environment of the polygonal obstacle, and get a list of the polygonal obstacles.

Step 2, set the parameter of GA, such as the number of the individuals in a generation N_p, the maximum probability of the crossover $p_{c\max}$, the minimum probability of the crossover $p_{c\min}$, the maximum probability of the mutation $p_{m\max}$, the minimum probability of the mutation $p_{m\min}$, and the maximum number of the iteration generations $it_{\max}$.

Step 3, initialize the generation using the encoding method described in Section 3.1.

Step 4, use the operations of selection, crossover and mutation to get a new generation, where roulette wheel selection method is used in the selection operation and single-point crossover is used as the crossover operator.

Step 5, if it reaches the maximum number of the iteration generations or meets the requirement of the accuracy, stop the evolution, and if not go to step 4.

4 Simulation Experiment

To verify the effect of the algorithm, simulation experiments have been done to the polygonal obstacle based binary encoding algorithm for path planning. For the sake of the convenience for verification and comparison, barriers that can be passed through are ignored.

In Environment 1, the number of the individuals in a generation is 48, and the maximum number of the iteration generations is 20. In Environment 2, the number of the individuals in a generation is 80, the maximum number of the iteration generations is 100, the maximum probability of the crossover is 0.9, the minimum probability of the crossover is 0.5, the maximum probability of the mutation is 0.1, and the minimum probability of the mutation is 0.05. Figure 4 shows the relationship between the fitness value and the ordinal number of the generation in Environment 1. From the curve in

Figure 4 we can see that the optimal path is found in the 6th generation, and the time is 0.203 second when it reaches the 20th generation, so the average time of finding the optimal path is 0.06 second in Environment 1.In Environment 2, it takes 4.125 seconds to reach the 100th generation, and the optimal path is found in the 15th generation, that is, 0.62 second on average.

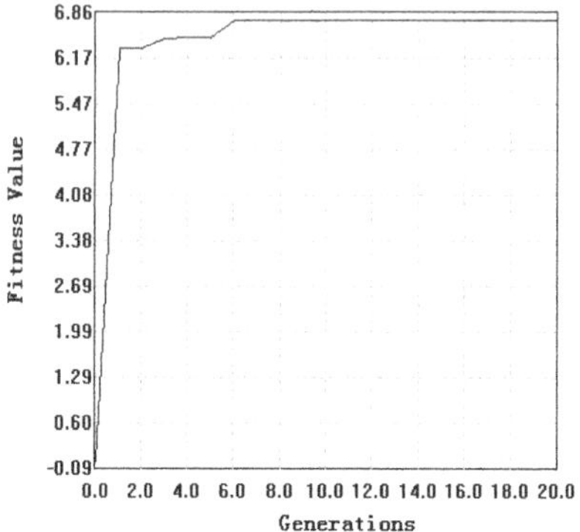

Fig. 4. The curve of the fitness value in Environment 1

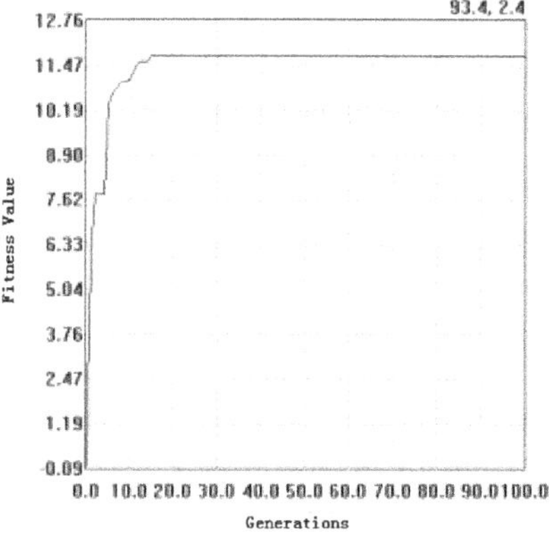

Fig. 5. The curve of the fitness value in Environment 2

Table 1 shows the comparison between the binary encoding GA proposed by this paper and the real number encoding GA used in [1] in the same polygonal obstacle environment.

Table 1. The comparison between two genetic algorithms

	Binary encoding GA		Real number encoding GA	
	Optimal path found	Average search time(second)	Optimal path found	Average search time(second)
Environment 1	Yes	0.06	Yes	5.17
Environment 2	Yes	0.62	Yes	7.76

Table 1 shows that the binary encoding GA proposed by this paper is far more effective than the real number encoding GA under the same situation.

5 Conclusion

This paper presents an obstacle modeling method for CGF battlefield path planning problem and a binary encoding genetic algorithm according to this obstacle modeling method. Based on the traditional genetic algorithm, this algorithm improves the genetic operators using the adaptive genetic algorithm. The results of the simulation experiments indicate that the binary encoding GA proposed by this paper greatly shortens the search time of the optimal path compared to the real encoding GA and it meets the needs of rapidness and accuracy for CGF simulation system. This paper provides a feasible and effective solution to the CGF path planning of complicated battlefields.

References

1. Wang, Z., Qiu, X., Li, G.: Research on CGF Terrain Database Format. Computer Simulation 21(9), 4–6 (2004)
2. Joe, L., Feldman, P.M.: Fundamental research policy for the digital battlefield. Res. Rep. DB-245-A, RAND Co. Santa Monica, USA (1998)
3. Benton, J.R., Iyengar, S.S., Deng, W., Brener, N., Subrahmanian, V.S.: Tactical route planning: new algorithms for decomposing the map. In: Proceeding of IEEE International Conference on Tools for AI, Herndon, pp. 268–277 (1995)
4. Campbell, C., Hull, R., Root, E., Jackson, L.: Route planning in CCTT. In: Proceedings of the 5th Conference on Computer Generated Force and Behavioral Representation, pp. 233–244. Institute for Simulation and Training (1995)
5. Chen, H., Du, X., Gu, W.-K.: Genetic algorithm based global path planning in a static environment. Journal of Zhejiang University 32(1), 49–53 (2005)
6. Jiang, M.: Research on mobile robot path planning method based on genetic algorithm, pp. 28–59. Shenyang University of Technology, Shenyang (2007)

Nelder-Mead Method with Local Selection Using Neighborhood and Memory for Optimization via Simulation

Noocharin Tippayawanakorn and Juta Pichitlamken

Department of Industrial Engineering, Faculty of Engineering,
Kasetsart University, Bangkok 10900, Thailand
noocharin@yahoo.com, juta.p@ku.ac.th

Abstract. We consider the Nelder-Mead (NM) simplex algorithm for optimization of discrete-event stochastic simulation models. We propose new modifications of NM to reduce computational time and to improve the optimal solutions. Our means include utilizing past information of already seen solutions, expanding search space to their neighborhood, and using adaptive sample sizes. We compare performance of these extensions on two test functions with 3 levels of random variation. We find that using past information leads to reduction of computational effort. Comparing with the non-adaptive modification, the adaptive one needs more resources but gives better solution.

Keywords: Nelder-Mead Simplex, adaptive Nelder-Mead Simplex, continuous stochastic optimization, neighborhood search, local selection.

1 Introduction

Optimization via simulation (OvS) is the problem of finding possible set of input variables or decision variables that gives an optimal solution. The objective is to minimize the resources spent while maximizing the information obtained in a simulation experiment [6]. Most OvS problems in practice involve stochastic objective functions. We are interested the OvS problems with a single objective function and continuous decision variables (see, [1], [2], [18] and [24] for OvS survey).

Optimization algorithms are essential for effective use of simulation. For example, in ore processing, ore that flows from production faces, ore-bodies, or seams may have different qualitative features [16]. The objective is to determine an appropriate blending rate of each material so as to minimize the total material cost of the finished product while satisfying the blending requirement.

Many OvS tools are developed for unconstrained continuous problems. Most of them are based on the random search method that takes objective function values from a set of sample points and uses that information to continue sampling. Various techniques differ in the choice of sampling strategies [2]. A point-based strategy involves sampling points in a neighborhood of the current solution, e.g., the Stochastic Ruler [27] and the Simulated Annealing [15]. A set-based strategy

J.-H. Kim et al. (Eds.): AsiaSim2011, PICT 4, pp. 134–143, 2012.

generates a set of candidate solutions from a subset of the feasible region, e.g., the Nested Partitions Method [22], Hook-Jeeves pattern search [12], the Nelder-Mead Simplex [17] and the COMPASS [10]-[11]. In contrast, a population-based strategy creates a collection of candidate solutions using some properties of the previously visited solutions, for example, the Genetic Algorithm [9], the Evolutionary Strategies [21] and the Ordinal Optimization [8].

We focus on the Nelder-Mead (NM) simplex algorithm [17], which is originally developed for unconstrained optimization of deterministic models, but which demonstrates wide versatility and ease of use for solving different optimization problems. At first, the NM is designed for deterministic optimization, but the method is still robust with respect to small perturbations in the function's values; and therefore, this method has been used for optimizing stochastic functions as well [26], e.g., a food refluent treatment model for plant design [4], a microsimulation model for breast cancer screening [5], and a quality assurance of manufactured parts [14]. However, in the case that the random perturbations in the function's values are large enough, the method may terminate before reaching the optimum of the expected function. In this work, we create parsimonious variants of the NM by utilizing past information and/or proximate points.

In this paper, we formulate, implement, and evaluate some simulation optimization algorithms that adapt from the NM simplex algorithm. We present the NM and its variants, RS9 [3], by using the idea of incorporating

- information collected since the search begins, and
- search neighborhood.

With numerical experiments, we show that our algorithms provide better solutions while requiring less computational effort than the original NM.

Generally, optimization problem can be defined as follows: The objective is to determine an optimal solution, $\mathbf{x}^*$, that minimizes the unknown objective function, $\mu : \Theta \to \mathbb{R}$ over a continuous feasible region, $\Theta \in \mathbb{R}^d$; that is

$$\mathbf{x}^* = \arg\min_{\mathbf{x}\in\boldsymbol{\Theta}} \mu(\mathbf{x}).$$

The objective function $\mu(\mathbf{x})$ is estimated with a stochastic simulation output, $G(\mathbf{x}, \xi_{\mathbf{x}})$, i.e.,

$$\mu(\mathbf{x}) = \mathrm{E}[G(\mathbf{x}, \xi_{\mathbf{x}})],$$

where $G(\mathbf{x}, \xi_{\mathbf{x}})$ is a simulation output evaluated at $\mathbf{x}$, which can be an implicit or explicit function. An example of implicit function is a total material cost of finished product from ore-processing which depends on $\mathbf{x}$, a blending rate of each material.

In this paper, we consider an explicit function where

$$G(\mathbf{x}, \xi_{\mathbf{x}}) = g(\mathbf{x}) + \xi_{\mathbf{x}}, \tag{1}$$

where $G(\mathbf{x}, \xi_{\mathbf{x}})$ is an unbiased white noise with $\mathrm{E}[\xi_{\mathbf{x}}] = 0$ and $\mathrm{Var}[\xi_{\mathbf{x}}] = \sigma^2$ and $g(\mathbf{x})$ is deterministic test problems. The estimate of $\mu(\mathbf{x})$, $\hat{\mu}(\mathbf{x})$, is a sample mean of independent simulation outputs,

$$\hat{\mu}(\mathbf{x}) = \frac{1}{m}\sum_{i=1}^{m} G(\mathbf{x}, \xi_i). \tag{2}$$

2 The Nelder-Mead Simplex Algorithm

This section describes the original NM algorithm and its variants.

2.1 The Original NM

The first of the simplex methods is due to [23] for deterministic problems. They assume that any point in the domain of search can be constructed by taking linear combination of the edges adjacent to any given vertices. The original simplex consists of the reflection of one vertex through the centroid of the opposite face and the resulting reflection simplex. Sometimes, the sequence of reflections brings the search back to where it starts. Nelder and Mead [17] add an expansion and contraction moves to accelerate the search and add a shrink step to reduce the lengths of edges adjacent to the current best vertex by half, in case that none of the steps brings acceptable improvement to solve the problems of the original simplex. The NM is also implemented in MATLAB as a function `fminsearch` which can be applied to stochastic functions. By default, `fminsearch` takes one simulation output for estimating an objective function. The NM algorithm can be outlined as follow:

1. **Create an initial simplex:**
   (a) Select a starting point $\mathbf{x}_0$, a vector in d dimension.
   (b) Form an initial simplex of $d+1$ points, by defining

   $$\mathbf{x}_i = [x_1, x_2, \ldots, x_i + s_i, \ldots, x_d], i = 1, \ldots, d,$$

   where d_i are user-defined step sizes.
   (c) Estimate the objective function at each of the $d+1$ simplex points by running m simulation runs at each point and using (2) to get $\hat{\mu}(\mathbf{x}_0), \hat{\mu}(\mathbf{x}_1), \ldots, \hat{\mu}(\mathbf{x}_d)$.
   (d) Re-order these points in an ascending order so that $\hat{\mu}(\mathbf{x}_0) \leq \hat{\mu}(\mathbf{x}_1) \leq \ldots \leq \hat{\mu}(\mathbf{x}_d)$.
2. **Calculate the Reflection Point:** The worst point on the simplex (recall that we consider a minimization problem) $\mathbf{x}_d$ is replaced with another point which has a lower objective function. Let $\mathbf{x}_d^R$ be the reflection of the worst point, and $\mathbf{x}_d$ passes through the centroid C of the d-best points. The centroid is computed as

   $$\mathbf{C} = \frac{1}{d} \sum_{j=0}^{d-1} \mathbf{x}_j.$$

   The reflection point is determined as $\mathbf{x}_d^R = \mathbf{C} + \alpha(\mathbf{C} - \mathbf{x}_d)$, where α is a reflection parameter, typically 1. Then their objective function values $\hat{\mu}(\mathbf{x}_d^R)$ is estimated via m simulation runs.
3. **Update the simplex.** Figure 1 shows an initial simplex with dash-lines and updated simplex as solid-lines. The updated simplex depends on the relationship between $\hat{\mu}(\mathbf{x}_d^R)$ and $\hat{\mu}(\mathbf{x}_0), \hat{\mu}(\mathbf{x}_1), \ldots, \hat{\mu}(\mathbf{x}_d)$:

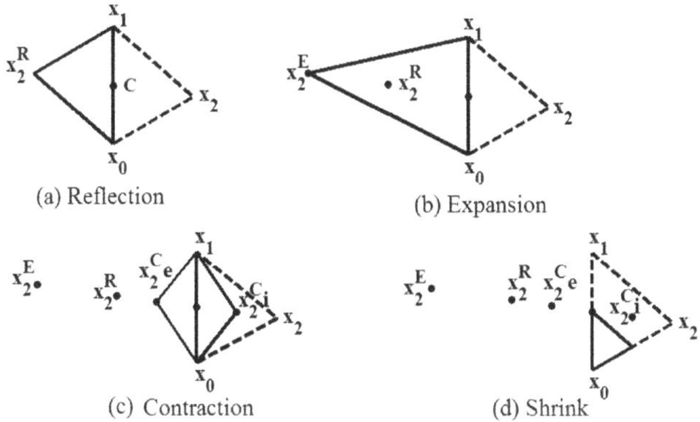

Fig. 1. Nelder-Mead Operations

(a) If $\hat{\mu}(\mathbf{x}_0) \leq \hat{\mu}(\mathbf{x}_d^R) < \hat{\mu}(\mathbf{x}_{d-1})$, set $\mathbf{x}_d \leftarrow \mathbf{x}_d^R$ and $\hat{\mu}(\mathbf{x}_d) \leftarrow \hat{\mu}(\mathbf{x}_d^R)$ as shown in Figure 1(a). Then go to Step 5.
(b) If $\hat{\mu}(\mathbf{x}_d^R) < \hat{\mu}(\mathbf{x}_0)$, the search continues in the same direction by calculating the expansion point, $\mathbf{x}_d^E = \mathbf{C} + \gamma(\mathbf{x}_d^R - \mathbf{C})$, where $\gamma > 0$ is an expansion parameter, typically 2. Then $\hat{\mu}(\mathbf{x}_d^E)$ is estimated from m simulation runs. The expansion point is accepted when it improves over the best point in the simplex as shown in Figure 1(b); otherwise, the reflective point is accepted. Go to Step 5.
(c) If $\hat{\mu}(\mathbf{x}_{d-1}) \leq \hat{\mu}(\mathbf{x}_d^R)$, the search reduces simplex size by calculating the contraction point, $\mathbf{x}_d^C = \mathbf{C} + \beta(\bar{\mathbf{x}}_d - \mathbf{C})$, where $\beta > 0$ is a contraction parameter (generally 0.5) and $\bar{\mathbf{x}}_d$ is $\mathbf{x}_d^R$ if $\hat{\mu}(\mathbf{x}_d^R) < \hat{\mu}(\mathbf{x}_d)$ and $\mathbf{x}_d$ otherwise. The objective function $\hat{\mu}(\mathbf{x}_d^C)$ is estimated. If $\hat{\mu}(\mathbf{x}_d^C) \geq \hat{\mu}(\mathbf{x}_d)$, go to Step 4; otherwise, the contraction point is accepted, i.e., $\mathbf{x}_d \leftarrow \mathbf{x}_d^C$. The updated simplex can be one of two dash-simplex in Figure 1(c) depending on which contraction point ($\mathbf{x}_2^{C_i}$ or $\mathbf{x}_2^{C_e}$) is used. Go to Step 5.

4. **Shrink the simplex.** If the reflection point and contraction point provide no improvement, then the entire simplex is shrunk toward the best point $\mathbf{x}_0$ as shown in Figure 1(d). Compute the new simplex:

$$[\mathbf{x}_0, \tau\mathbf{x}_1 + (1-\tau)\mathbf{x}_0, \ldots, \tau\mathbf{x}_d + (1-\tau)\mathbf{x}_0],$$

where τ is a shrink parameter, typically 0.5. The objective function values of these new points are estimated. Then go to Step 5.
5. **Re-order the simplex points** in ascending orders. Go to Step 2.

The search iterates until one of the following conditions are satisfied: i) the maximum coordinate difference between the current best point and the other points in the simplex is less than or equal to $\epsilon_{\mathbf{x}}$; ii) the corresponding difference in estimated objective function values is less than or equal to $\epsilon_{\hat{\mu}}$; or iii) the maximum number of iterations reaches $N_{\hat{\mu}}$ or function evaluations reaches N_{search}. When the search terminates, the current solution and its corresponding estimated objective function are used as the optimum $\hat{\mathbf{x}}^*$ and $\hat{\mu}(\mathbf{x}^*)$, respectively.

2.2 Barton and Ivey Stochastic Modification of NM

Barton and Ivey [3] adapt the NM algorithm to optimize the expected response of stochastic systems. From the empirical results, they see that the NM algorithm relies on the ranks of the objective function values at the simplex vertices. Thus, NM can make progress in presence of relatively small randomness, since small random perturbations do not change the rank of the function values at the simplex points. However, if the perturbations are large enough to affect the relative ranks of the simplex vertices, it can mislead the algorithm.

Barton and Ivey recommend the shrinkage coefficient (τ) of 0.9 instead of a standard 0.5 to increase the extent of reduction after shrink. This change improves the performance effectively for the case where the original NM fails. Resampling the best point after shrink reduces the frequency of contraction which is not effective in improving algorithm performance.

Moreover, Barton and Ivey apply an appropriate stopping criterion for stochastic problems as suggested by Dennis and Woods [7]. The criterion based on simplex size as follows:

$$\frac{\sum_{i=1}^{d} \|\mathbf{x}_j - \mathbf{x}_0\|}{\max(1, \|\mathbf{x}_0\|)} \leq \epsilon_{\mathbf{x}},$$

where $\hat{\mu}(\mathbf{x}_0) < \hat{\mu}(\mathbf{x}_1) < \cdots < \hat{\mu}(\mathbf{x}_d)$, and $\|\cdot\|$ is the Euclidean norm. Through the numerical experiments, Barton and Ivey show that random noise leads to premature simplex size reduction during the contraction and shrink steps because search algorithm reaches rapidly.

Their so-called RS9 is NM with the following modifications: the objective function is estimated with 6 replication runs ($m = 6$), every solution in the simplex is resampled in every search iteration, and the shrinkage coefficient τ is 0.9.

3 Our NM Variants

We are motivated by several general-purpose optimization algorithms for deterministic problems that are based on a neighborhood search. At each iteration, the search may iteratively move from the current solution to one of its neighbors which is better than itself and any other solutions in the neighborhood. Neighborhood search strategy and statistical selection of the best are used in OvS in [19], followed by a framework for OvS in [20].

In this paper, we express the neighborhood distance and the revisited distance as follows: The neighborhood of solution $\mathbf{x} = [x_1, x_2, \ldots, x_d]$ are $[x_1 - e, x_2, \ldots, x_d]$, $[x_1 + e, x_2, \ldots, x_d]$, $[x_1, x_2 - e, \ldots, x_d]$, $[x_1, x_2 + e, \ldots, x_d]$,...,$[x_1, x_2, \ldots, x_d + e]$ where $e \in [0, \epsilon]$, and ϵ is the maximum neighborhood distance. We exclude any neighbor that lies outside the feasible space Θ.

Similarly, the already-seen solution is defined as $\mathbf{x}$ that is not "too far" from the one already seen $\mathbf{v}$:

$$\|\mathbf{v}-\mathbf{x}\|_\infty \equiv \max_{1\leq i\leq d} |\mathbf{v_i}-\mathbf{x_i}| = \mathbf{e_v}$$

where $e \in [0, \epsilon_\mathbf{x}]$ and $\epsilon_\mathbf{x} = 10^{-4}$, similar to the terminating criterion in `fminsearch`. We select the neighbor $\mathbf{v}$ which provides the minimum $e_\mathbf{v}$ as not distinguishable from $\mathbf{x}$.

The advantage of using past information is in avoiding resampling at every encounter. We propose two NM-based algorithms with memory as follows:

1. **The Nelder-Mead algorithm with memory (NMSM).** Memory of revisited solutions and their candidate solutions replace new sampling. Nevertheless, for already seen solutions, NMSM adds one simulation output at a time for them so as to protect the search from unusually good or bad observations.
2. **The Nelder-Mead selection with memory and using neighborhood (NMSMN).** The NMSMN constructs neighborhood for every vertex of the simplex and estimate their objective function values. The best solution in the neighborhood replaces the original vertex.
3. **The adaptive Nelder-Mead Selection with memory (ANMS).** Tomick [26] proposes an adaptive sample size of the next iteration m^{j+1} based on the smallest sample size of every solution on iteration j m^j. (See [25] for details).
4. **The adaptive Nelder-Mead Selection with memory and using neighborhood (ANMSMN).**

4 Numerical Experiments

We describe our test functions in Section 4.1 and our search performance measures in Section 4.2. We explain our experimental setups and results in Sections 4.3–4.4.

4.1 Test Functions

We test our algorithms on two deterministic test functions ($g(\mathbf{x})$ in (1)) in 2 dimensions ($d = 2$) with added white noise whose mean is zero and standard deviations of 0.75, 1.00 and 1.25 times of the optimal objective function values. Common random numbers are used. These test functions were also used in [3] and [13] for optimization of noisy responses. Our test functions are of the form

$$g(\mathbf{x}) = \sum_{i=1}^{d} [f_i(\mathbf{x})]^2 + 1,$$

where $f_i(\mathbf{x})$ is defined as follows:

1. *Trigonometric function:*

$$f_i(\mathbf{x}) = d - \left[\sum_{j=1}^{d} \cos(x_j - 1)\right] + i[1 - \cos(x_i - 1)] - \sin(x_i - 1),$$

   where the optimal solution is $[1 + 2\pi k_1 \dots, 1 + 2\pi k_d]$, where $k_j = 0, \pm 1, \pm 2, \dots$ for $j = 1, \dots, d$ and $g(\mathbf{x}^*) = 1$.
2. *Extended Rosenbrock function:*

$$f_{2i-1}(\mathbf{x}) = 10(x_{2i} - x_{2i-1}^2) \text{ and } f_{2i}(\mathbf{x}) = (1 - x_{2i-1}),$$

   where the optimal solution is $[1, 1, \dots, 1]$ and $g(\mathbf{x}^*) = 1$.

4.2 Search Performance Measures

Following [13], we evaluate the search performance via the average of the following performance measures over many macroreplications:

1. Logarithm of the number of function evaluations, L: to assess the computational efforts used
2. Relative deviant of final estimated optimal function value from the true optimal value:

$$D = \left| \frac{\mu(\hat{\mathbf{x}}^*) - \mu(\mathbf{x}^*)}{\mu(\mathbf{x}^*)} \right|. \tag{3}$$

4.3 Experimental Setup

The NM coefficients are as follows: $\alpha = 1, \gamma = 2, \beta = 0.9$ and $\tau = 0.5$. Minimum deviance $\epsilon_{\mathbf{x}}$ and $\epsilon_{\hat{\mu}}$ are 10^{-4}. Maximum budget consumption $N_{\hat{\mu}}$ and N_{search} are 10^5. To estimate the objective function, the sample size is 6. Maximum neighborhood distance ϵ is 0.01. Three levels of standard deviation of random noise $\epsilon_{\mathbf{x}}$ is $\{0.75\mu(\mathbf{x}^*), 1\mu(\mathbf{x}^*), 1.25\mu(\mathbf{x}^*)\}$. The factor for increasing simulation size b in [26] is 1.25. We perform 20 macroreplications (i.e., experiments) for each search algorithms and test problem.

4.4 Experimental Results

Table 1 shows budget consumption of each algorithm until computational budget is exhausted or until the search is unable to get any improvement, and Table 2 contains the relative error in estimating the objective function values (3). As the level of random noise increase, a given test problem becomes harder; the simulation efforts required increase, while the relative errors increase. Using memory appears to help save the simulation efforts for both test problems and across all noise levels, yet it does not cause much harm to the performance of the search. Surprisingly, using adaptive sample sizes is not significantly helpful; perhaps because using memory provides greater benefits. Four other test functions and other search performance measures are presented in [25].

Table 1. Average of logarithm of number of function evaluations

Test Function	Levels of variance	NMSM	ANSMN	ANMS	NMSMN	NM	RS9
Trigonometric	0.75	5.900	6.716	7.086	7.378	11.513	11.511
	1.0	5.859	6.685	7.025	7.433	11.513	11.513
	1.25	5.950	6.624	7.047	7.385	11.513	11.513
Rosenbrock	0.75	6.166	6.751	6.576	7.799	11.513	11.513
	1.0	6.188	6.780	7.489	7.587	11.513	11.513
	1.25	6.052	6.623	7.127	7.554	11.513	11.513

Table 2. Relative deviants in objective function values from the true optima

Test Function	Levels of variance	NMSM	ANSMN	ANMS	NMSMN	NM	RS9
Trigonometric	0.75	0.555	0.317	0.288	0.181	0.280	1.048
	1.0	0.805	0.376	0.428	0.285	1.059	1.067
	1.25	0.804	0.504	0.524	0.320	0.483	0.901
Rosenbrock	0.75	3.166	3.338	3.182	3.007	3.083	3.853
	1.0	3.189	3.353	3.197	3.141	3.206	3.645
	1.25	3.148	3.637	3.407	3.140	3.137	3.827

5 Conclusion

We show that the Nelder-Mead algorithm which was designed for deterministic optimization can be modified to accommodate stochastic outputs. Using past information and exploiting neighborhood generally decreases computational efforts. Currently, we are in the process of incorporating Barton and Ivey's stopping criterion [3] into the search, and extend our algorithm to discrete search space.

Acknowledgment. This work was funded by the Industrial Engineering Department, Faculty of Engineering, and the Graduate School of Kasetsart University, Thailand. The authors would like to thank Dr. Walailuck Chavanasporn, Mathematics Department, King Mongkut University of Technology North Bangkok, for her editorial comments.

References

[1] Allon, G., Kroese, D., Raviv, T., Rubinstein, R.: Application of the Cross Entropy Method to the Buffer Allocation Problem in Simulation Based Environment. Annals of Operations Research 134, 137–151 (2005)

[2] Andradóttir, S.: An Overview of Simulation Optimization Via Random Search. In: Handbooks in Operations Research and Management Science: Simulation, pp. 617–631. Elsevier, North-Holland (2006)

[3] Barton, R.R., Ivey, J.S.: Nelder-Mead Simplex Modifications for Simulation Optimization. Management Science 42, 954–973 (1996)

[4] Bonello, A.F., Mosca, R., Cotto, S.P.: Multi-Objective Optimisation of Food Refluent Treatment Plant Design Parameters through Designed Simulation with Significant Experimental Error. In: Proceedings of the 14th European Simulation Symposium, Dresden, Germany (2002)
[5] Branco, D.M.: Nonlinear Optimization of a Stochastic Function in a Cell Migration Model. M.S. Thesis, Worcester Polytechnic Institute, MA, USA (2006)
[6] Carson, Y., Maria, A.: Simulation Optimization: Methods and Applications. In: Proceedings of the 1997 Winter Simulation Conference, Atlanta, GA, USA, pp. 118–126 (1997)
[7] Dennis, J.E., Woods, D.J.: Optimization on Microcomputers the Nelder-Mead Simplex Algorithm. In: New Computing Environments: Microcomputers in Large Scale Computing, pp. 116–122. The Society for Industrial and Applied Mathematics, Philadelphia (1987)
[8] Ho, Y.C., Sreenivas, R.S., Vakili, P.: Ordinal Optimization of DEDS Discrete Event Dynamic Systems, vol. 2, pp. 61–88 (1992)
[9] Holland, J.H.: Genetic Algorithms. Scientific American 267, 66–72 (1992)
[10] Hong, L.J., Nelson, B.L.: An Indifference-Zone Selection Procedure with Minimum Switching and Sequential Sampling. In: Proceedings of the 2003 Winter Simulation Conference, New Orleans, LA, USA, pp. 474–480 (2003)
[11] Hong, L.J., Nelson, B.L.: A Framework for Locally Convergent Random-Search Algorithms for Discrete Optimization Via Simulation. Association for Computing Machinery Transctions on Modeling and Computer Simulation 17, 1–22 (2007)
[12] Hooke, R., Jeeves, T.A.: Direct Search Solution of Numerical and Statistical Problems. Association for Computing Machinery 8, 212–229 (1961)
[13] Humphrey, D.G., Wilson, J.R.: A Revised Simplex Search Procedure for Stochastic Simulation Response Surface Optimization. Institute for Operations Research and the Management Sciences: Journal on Computing 12, 272–283 (2000)
[14] Gosavi, A., Phatakwala, S.: A Finite-Differences Derivative-Descent Approach for Estimating Form Error in Precision-Manufactured Parts Manufacturing Science and Engineering, vol. 128, pp. 355–359 (2006)
[15] Kirkpatrick, S., Gelatt, C.D., Vecchi, M.P.: Optimization by Simulated Annealing. Science 220, 671–680 (1983)
[16] Kumral, M.: Solution of Ore Blending Problem by Stochastic Approach. In: Proceedings of the 10th International Mining Congress and Exhibition of Turkey, Antalya, Turkey, pp. 485–490 (2003)
[17] Nelder, J.A., Mead, R.: A Simplex Method for Function Minimization. Computer Journal, 308–313 (1965)
[18] Ólafsson, S., Kim, J.: Simulation Optimization: Simulation Optimization. In: Proceedings of the 2002 Winter Simulation Conference, Atlanta, GA, USA, pp. 79–84 (2002)
[19] Pichitlamken, J., Nelson, B.L.: A Combined Procedure for Optimization Via Simulation. Association for Computing Machinery Transactions on Modeling and Computer Simulation 13, 155–179 (2003)
[20] Pichitlamken, J., Nelson, B.L., Hong, J.L.: A Sequential Procedure for Neighborhood Selection-of-the-Best in Optimization Via Simulation. European Journal of Operational Research 173, 283–298 (2006)
[21] Schwefel, H.-P.: Evolution and Optimum Seeking: The Sixth Generation. John Wiley & Sons, New York (1993)
[22] Shi, L., Ólafsson, S.: An Integrated Framework for Deterministic and Stochastic Optimization. In: Proceedings of the 1997 Winter Simulation Conference, Atlanta, GA, USA, pp. 358–365 (1997)

[23] Spendley, W., Hext, G.R., Himsworth, F.R.: Sequential Application of Simplex Designs in Optimisation and Evolutionary Operation. Technometrics 4, 441–461 (1962)

[24] Swisher, J.R., Hyden, P.D., Jacobson, S.H., Schruben, L.W.: A Survey of Recent Advances in Discrete Input Parameter Discrete-Event Simulation Optimization. Institute of Industrial Engineers Transactions 36, 591–600 (2004)

[25] Tippayawannakorn, N., Pichitlamken, J.: Global Search with Local Selection for Stochastic Problems with Finite Discrete Feasible Space. To appear, Ph.D. Thesis, Kasetsart, Bangkok, Thailand (2012)

[26] Tomick, J.J.: On Convergence of the Nelder-Mead Simplex Algorithm for Unconstrained Stochastic Optimization. Ph.D. Thesis, The Pennsylvania State University, PA, USA (1995)

[27] Yan, D., Mukai, H.: Stochastic Discrete Optimization. Society for Industrial and Applied Mathematics Journal on Control and Optimization 30, 594–612 (1992)

New Advances of the Research on Cloud Simulation

Bo Hu Li[1,2], Xudong Chai[2], Lin Zhang[1], Baocun Hou[2], Ting Yu Lin[1], Chen Yang[1], Yingying Xiao[1], Chi Xing[1], Zhihui Zhang[1], Yabin Zhang[1], and Tan Li[1]

[1] School of Automatic Science and Electrical Engineering, Beijing University of Aeronautics and Astronautics, 100191 Beijing China
bohuli@moon.bjnet.edu.cn, lintingyu2003@sina.com

[2] Beijing Simulation Center, Second Academy of Aerospace Science & Industry Co., 100854 Beijing China
Xdchai@263.net, houbc2002@yahoo.com.cn

Abstract. Based on the research fruits of Cloud Simulation Platform [1], this paper expounds the latest research results of our team in cloud simulation, including the further research on the technology content and features of cloud simulation, architecture and service patterns of the cloud simulation system, technology system and several improved key technologies (individuation virtual desktop technology, multi-users oriented dynamic building technology of virtual simulation environment, fault-tolerant migration technology for simulation resources, high performance cloud simulation supported co-simulation platform technology) of cloud simulation and typical application demonstration system. The primary research and practice show that the proposed latest research results can better support "cloud simulation" pattern, which users can access services of simulation resource and capability on demand anytime and anywhere through network and cloud simulation platform, to accomplish varied activities in the whole simulation life-circle. Finally, this paper gives the prospect of the future work of cloud simulation.

1 Introduction

In June 2005, the U.S. President's Information Technology Advisory Committee pointed out in a report recommended to the President named "Computational Science: Ensuring America's Competitiveness" that Computational Science, which consisted of three distinct elements including the algorithm and the modeling and simulation software, computer and information science and computing facilities, had gradually become the third pillar of the scientific enterprise, a peer alongside theory and physical experiment [2]. Currently, the "cloud computing" is becoming a hot research field of information technology [3], and has made great progress for the users to access computing service at anytime and anywhere by the network and cloud computing platform. The authors have proposed a new networked modeling and simulation platform named "Cloud Simulation Platform" [1]. It is based on the concept of Cloud Computing and integrates complex systems modeling and simulation technology with the emerged information technologies such as virtualization technology, pervasive computing technology, artificial intelligence technology and high performance computing

J.-H. Kim et al. (Eds.): AsiaSim2011, PICT 4, pp. 144–163, 2012.

technology etc."Cloud Simulation Platform"can support users to access the services of M&S resource and capability on demand through the network.

Based on the research mentioned above, this paper introduce the latest research results of our team in the cloud simulation, including the further studies on the technology content and features of cloud simulation, architecture and service patterns of the cloud simulation system, technology system and several improved key technologies of cloud simulation and typical application demonstration systems. Finally, we give the prospect of the future work of cloud simulation.

2 Technology Content and Features of the Cloud Simulation

Cloud simulation is a new network-based (including Internet, internet of things, telecommunication network, broadcasting network, mobile network, etc.) and service-oriented simulation mode. It has combined the current networked M&S technology with the newly emerging information technologies including cloud computing, SOA, virtualization, high performance computing, internet of things and intelligent science, etc, and then developed them. It encapsulates various simulation resource and capability into virtual resource or service-oriented resource, which constitutes the cloud pool of simulation resource and capability. The unified and centralized management and operation of the cloud pool enable users to acquire service of the simulation resources and capacity anytime on demand through the network and cloud simulation platform, to complete various activities in the whole lifecycle of simulation. Cloud simulation technology has accelerated the speed of promoting simulation technology developing towards the "service-oriented, agile, green and intelligent" direction.

Cloud simulation is an extension and development of cloud computing, which provides IaaS (Infrastructure as a Service), PaaS (Platform as a Service), SaaS (Software as a Service). It enriches and expands the sharing content of resource, service pattern and supporting technology of cloud computing, as shown in Fig. 1.

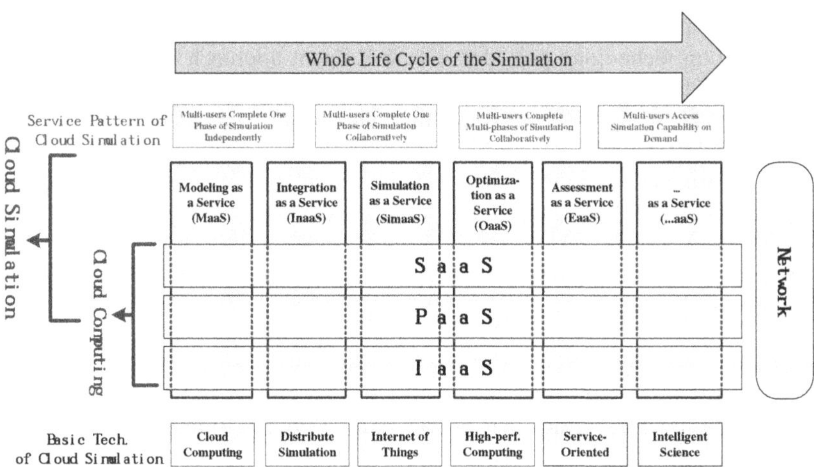

Fig. 1. The expansion of the cloud simulation to the cloud computing

Firstly, in terms of resources and capabilities sharing, Cloud simulation can firstly share soft simulation resources (different kinds of model, data, software, information, knowledge in the Simulation process), hard simulation resources (various types of computing devices, simulation equipment, test equipment, etc.), and M&S capability (Supporting the capacity of modeling, simulation, results-analyzing, evaluation, application etc in virtual simulation, live simulation and constructive simulation) .

Secondly, in terms of service pattern expanding, Cloud simulation can provide users to submit task online, and Interactive, collaborative and life-cycle simulation services. The supported patterns include: action of one single user completing one phase simulation (such as modeling), action of multi-user collaboratively completing one phase simulation (such as collaborative modeling), actions of multi-user collaboratively completing multi-phase simulation (such as collaborative modeling, running and analyzing simulation results), and multi-user access a variety of simulation capabilities on demand.

Thirdly, in terms of supporting technology, it is to expand and combine the existing networking M&S technology and the emerging information technologies such as cloud computing, service-oriented architecture, virtualization, high performance computing, the internet of things, intelligent science and etc. Among them,

- The cloud computing technology provides a series of enabling technologies and new simulation patterns for simulation resources or capabilities in cloud simulation;
- The internet of things technology provides a series of enabling technologies for interconnection between things in simulation domain and realization of wisdom simulation;
- Service-oriented technology provides a series of enabling technologies for constructing and executing service-oriented virtual simulation environment;
- Intelligent science provides the enabling technology for making simulation resources or capabilities intelligent;
- Virtualization technology provides a new efficient approach for cloud simulation system in the aspects of building the simulation application system, efficient and transparent use of resources in simulation, universal collaboration of running environment and so on.
- High performance computing technology provides the enabling technology to solve complex problems and achieve large-scale collaborative simulations in cloud simulation;
- Networked M&S technology is the basic technology of the cloud simulation.

Cloud simulation has the following characteristics,

- On-demand dynamic structure: Providing simulation services anywhere at any time, according to users' needs.
- Interoperability: Supporting the interoperation between simulation resources and simulation capabilities.
- Collaboration: Supporting the collaboration in large-scale complex simulation task with multi-users.

- Heterogeneous integration: Supporting the integration of distributed heterogeneous simulation resources and capabilities.
- Strong, quick and infinite abilities: Supporting fast and flexible composition of various types of simulation services (infinite) to respond a demand.
- Wisdom simulation in whole life-cycle: Providing services in the life-cycle of the simulation, and using the intelligent information technologies to realize wisdom simulation in multi-phases of the life-cycle.

It is worth to point out that the cloud manufacturing is becoming a new research issue [4] in the manufacturing field, and cloud simulation exactly is an important basis and technique of the cloud manufacturing.

2 Architecture of the Cloud Simulation

The architecture of Cloud Simulation System includes resource/capability layer, Cloud Simulation Platform layer (including perception/connection layer, virtual resource/capability layer, core function layer and user interface layer) and service application layer, as shown in Fig. 2.

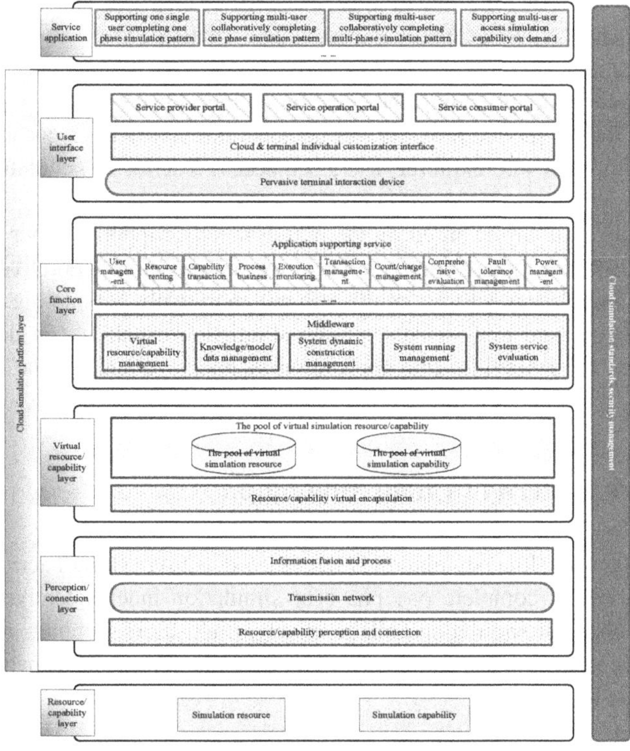

Fig. 2. Architecture of the cloud simulation

Resource/capability layer: where simulation resource consists of soft resource (simulation model, simulation data, simulation software, simulation knowledge, etc), and hard resource (computing device, simulation hardware, testing equipment, etc); simulation capability consists of the capability to support different stages of activities in Virtual, Constructive, Live simulation, such as the capability of modeling, simulation execution, results analysis, evaluation and application, and so on.

Perception/connection layer: supporting the perception, connection, networking interconnection, information fusion and processing of all kinds of simulation resources/capabilities.

Virtual resource/capability layer: encapsulating simulation resource/capability using virtualization technology and giving unified description of them, mapping physical resource/capability to logic resource/capability to form one virtualized pool of simulation resource and capability.

Core function layer: providing the middleware supporting for simulation platform, including the middleware for virtualized resource/capability management, knowledge/model/data management, system construction management, system running management, system service evaluation, etc. with the support of middleware, providing supporting services for simulation applications, which include services such as user management, resource/capability transaction, runtime observation and control, charge management, etc.

User interface layer: providing the function of individual interface customization and the portal for the service provider, service operator, and service consumer. Entering the portal through web explorer, users can access a series of simulation resource and capability.

Service application layer: based on the portal, mainly supporting four modes: single user completing one phase simulation activity, multi-user collaboratively completing one phase simulation activity, multi-user collaboratively completing multi-phase simulation activity, and multi-user acquiring simulation capability on demand. Service application layer provides two application manners: “cloud service” and “cloud plus terminal”.

3 Service Pattern of Cloud Simulation

As mentioned above, cloud simulation is represented as four types of service patterns, including multi-users complete one phase of simulation independently, multi-users complete one phase of simulation collaboratively, multi-users complete multi-phases of simulation collaboratively, and multi-users access simulation capability on demand, shown as follow.

3.1 Support One Single User to Complete One Phase of Simulation Independency

On a phase of simulation life-cycle, such as the phase of the M&S for component, the user selects the interactive software for design or job processing software for analysis, and submit the requirements of the computing resources (such as the number of CPU cores and memory size) through the application portal. Cloud simulation system will automatically allocate an exclusive simulation environment, until the end of the application, shown in Fig. 3.

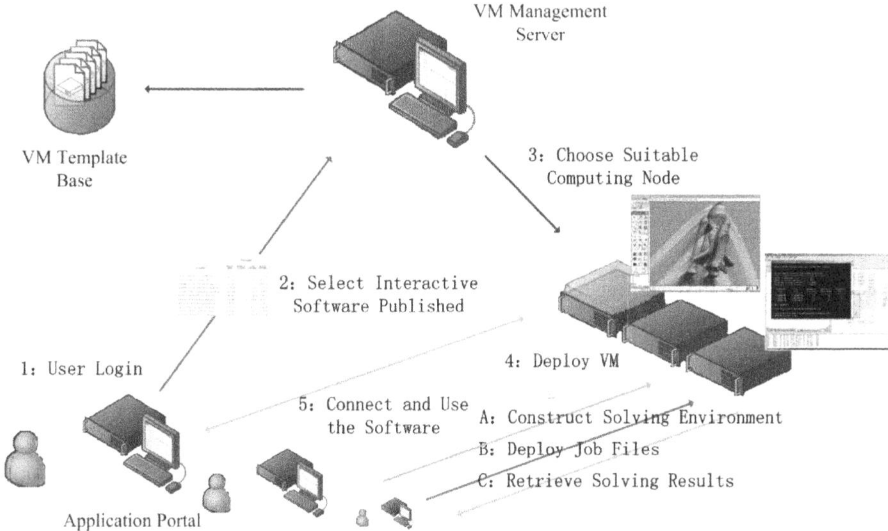

Fig. 3. Service pattern supporting one single user to complete one phase of simulation independently

3.2 Support Multi-users to Complete One Phase of Simulation Collaboratively

On a phase of simulation life-cycle, such as the phase of the M&S for multi-disciplinary virtual prototype, the overall system designer (user) and the sub-system designers (users) complete the top-level system modeling and sub-systems' modeling respectively in their own virtual desktops, which are the custom virtual research and development (R&D) environments, and then upload the corresponding model files through the application portal. Cloud Simulation System then constructs the running environment of the collaborative simulation dynamically, and builds the system-level virtual prototype automatically, so that achieving the aggregation and collaboration of the simulation resources, shown in Fig. 4.

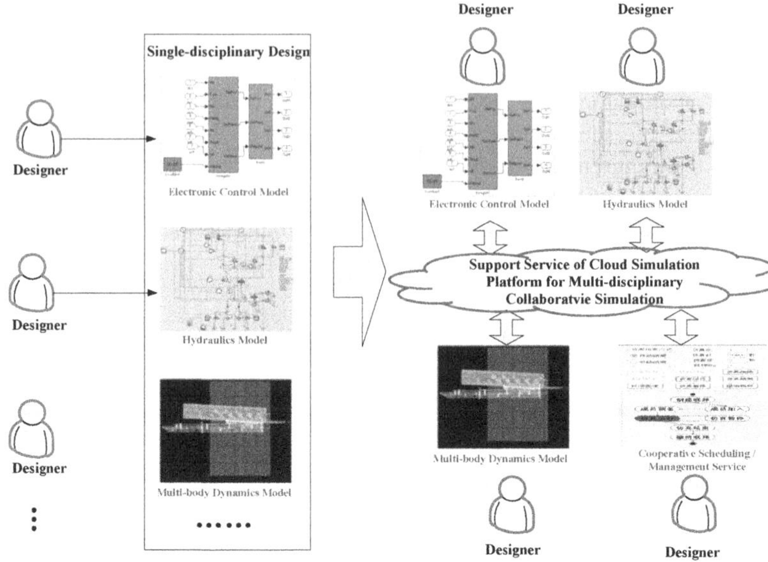

Fig. 4. Service pattern supporting multi-users to complete one phase of simulation collaboratively

3.3 Support Multi-user to complete Multi-phases of Simulation Collaboratively

On the multi-phase of simulation life-cycle, such as from the M&S of the system component phase, to the M&S of multi-disciplinary virtual prototype phase and the analysis and assessment on large-scale system level phase, the project manager

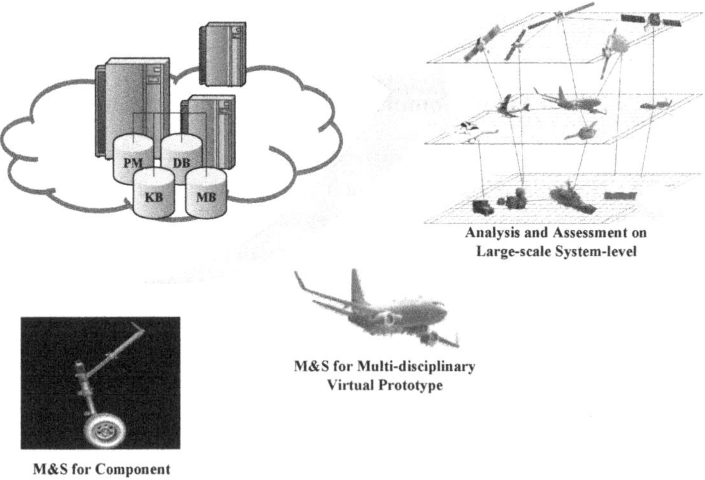

Fig. 5. Service pattern supporting multi-users to complete multi-phase of simulation collaboratively

defines the process of the M&S on multi-phases, and members of project team select the related resource services through application portals. Cloud Simulation System then real-time perceives the status of the related resource services and the progress of the process, so as to dynamically drive the process, and maintain the information chain between the resource services, which support fault tolerance and optimization of the process, shown in Fig. 5.

3.4 Support Multi-users to Access Simulation Capability on Demand

In this service pattern, enterprises, such as the M&S institute, publish the M&S capabilities which include system-of-systems combat M&S capability, hardware-in-the-loop M&S capability, virtual reality M&S capability and multi-disciplinary virtual prototype M&S capability. In addition, the enterprises of the product R&D, which are the users of the capabilities, publish the requirement of the capabilities. Based on the domains knowledge base, Cloud Simulation System then achieves the aggregation and scheduling of the capabilities and controls the process of the transaction, shown in Fig. 6.

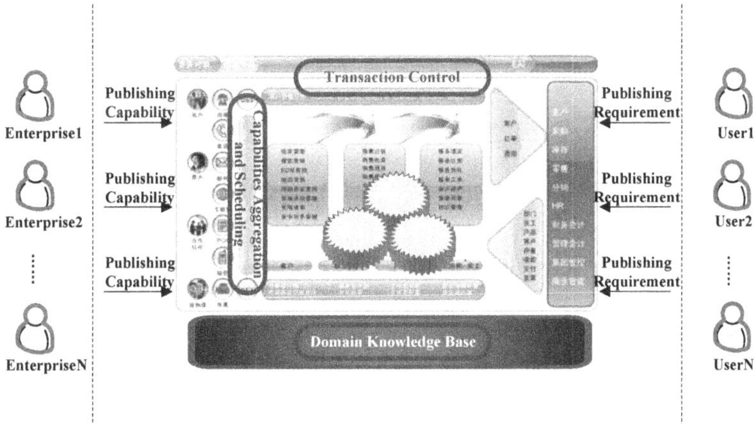

Fig. 6. Service pattern supporting multi-users to access simulation capability on demand

4 Technology System of Cloud Simulation

The technology system of cloud simulation includes overall technology, sensing technology for simulation resource and capability, virtualization and service technology for simulation resource and capability, construction and management technology for virtual simulation environment, operation technology for virtual simulation environment, evaluation technology for virtual simulation environment, trusted virtual simulation technology, pervasive human-computer interaction technology, application technology of service platform, business model for system operation, shown in Fig. 7.

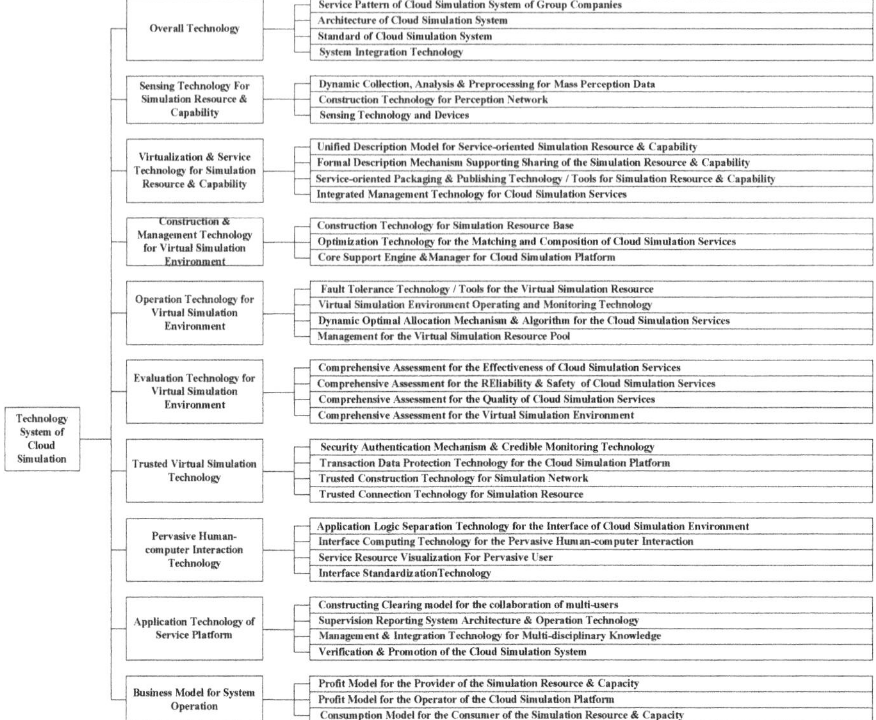

Fig. 7. Technology system of cloud simulation

5 New research Results on Key Technologies of Cloud Simulation

The author in paper [1] has proposed some relevant key technologies in 2009. This article focuses on the new research results of our team, including:

- Individuation virtual desktop technology;
- Multi-users oriented dynamic building technology of virtual simulation environment;
- Fault tolerance technology for resources in cloud simulation;
- Efficient cloud simulation-supported Co-simulation platform technology and etc.

In the following paper, the research results of these key technologies will be introduced.

5.1 Individuation Virtual Desktop Technology [5]

Individuation virtual desktop technology is on the basis of virtual desktop technology, and to build an exclusive and security desktop environment dynamically on the demand of users. In this virtual environment, the ordered simulation software has been installed previously, and users can carry out modeling and designing tasks in the virtual desktop based on those software.

Actually, the environment is running on the high-performance computer cluster. So, users could obtain the computing capability of high-performance computers easily, and this virtual desktop pattern will improve the utilization of high-performance computers.

This individuation virtual desktop pattern is shown in Fig. 8 and its techniques are corresponded to the universal human-computer interaction technology and VM technology in Fig. 7. The specific process to implement this pattern is shown as follows:

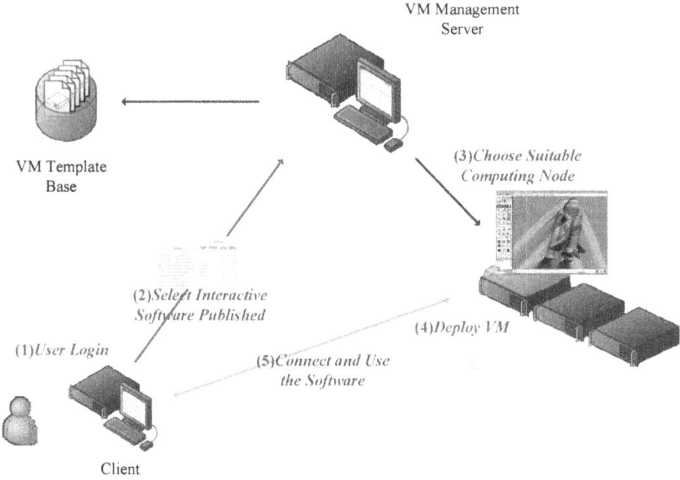

Fig. 8. Individuation virtual desktop pattern

- The user login the portal.
- The user selects the interactive software published, and submits the requirements of the computing resources (such as the number of CPU cores and memory size) through the application portal.
- Cloud simulation system will automatically choose suitable computing node.
- Cloud simulation system then deploys VM on the allocated node, and creates an exclusive simulation environment,
- The user connects and uses the software through the network.

5.2 Multi-users Oriented Dynamic Building Technology of Virtual Simulation Environment [6]

In the building and managing technology of virtual simulation environment, shown in Fig. 9, multi-users oriented dynamic building technology of virtual simulation environment automatically builds running environment for multi-users and multi-tasks application through analysis requirements of the simulation tasks. Specifically, according to requirements of simulation tasks, computing resources, software resources, platform resources, and model resources are dynamically composited to provide an efficient collaborative virtual simulation environment. Thus users can get simulation services fast, efficiently and flexibly to meet their individual needs.

Multi-users oriented dynamic building model of virtual simulation running environment proposed in this paper is showed in Fig. 9. This model is based on virtualization technology, and its core components are "two bases and two centers", which include the simulation resource base, the simulation knowledge base, simulation service center and virtual machine management center.

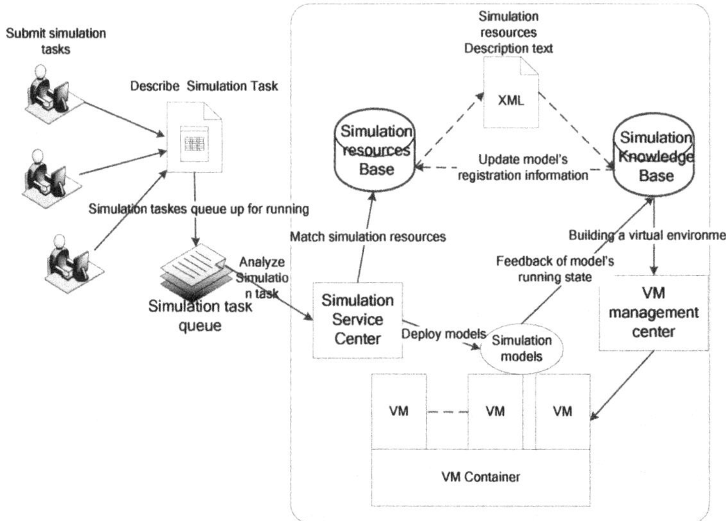

Fig. 9. Multi-user oriented dynamic building model of virtual simulation running environment

Fig. 10 illustrates the process of virtualization-based simulation environment dynamically building.

- Users submit simulation tasks through the application portals, and then the simulation service center parses simulation tasks and obtains the requirements of model computing environment. It is supported by the simulation task cases in the simulation knowledge base.
- Virtual machine management center revises the information of simulation resource requirements with the help of evaluation information of simulation resources in simulation knowledge base. According to the feedback information of physical nodes, virtual machine management center selects the appropriate physical nodes and deploys virtual machines on it, and also deploys software and models into virtual machines. All of virtual machines, software and models are selected from simulation resource base.

– When the simulation system is running, the status information of computing node is collected and put into simulation knowledge base, which is considered as the evaluation foundation of the simulation resource requirements. The simulation results solved in the virtual node are returned to the portal.

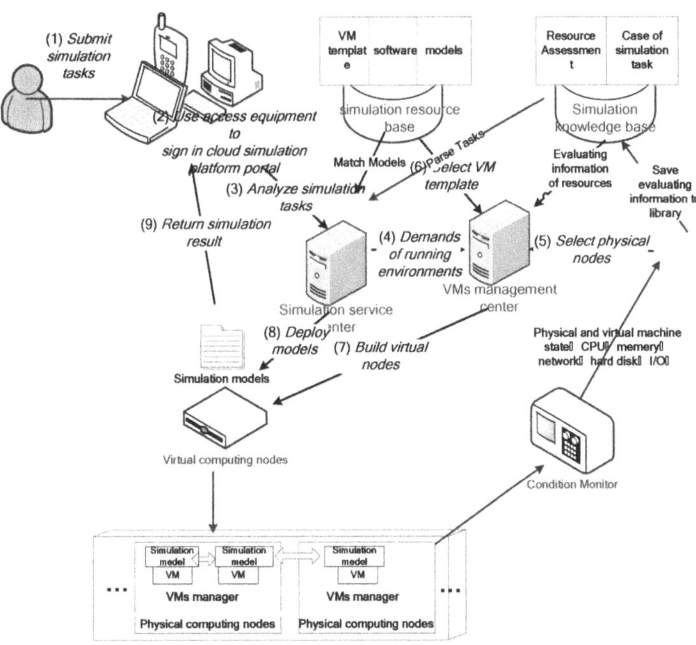

Fig. 10. Dynamic building process for virtual simulation running environment

5.3 Fault-Tolerant Migration Technology for Simulation Resources [7]

For the characteristics, such as more apparent distribution, complexity, dynamics, heterogeneity, etc. of simulation resources in Cloud Simulation Platform, which bring strong dependence on network bandwidth, unbalance of resource load, and instability of network connection, can easily lead to the problem that the whole cloud simulation platform could not achieve effective, credible, and robust running. The research on fault-tolerant migration technology for simulation resources can support users to transparently acquire simulation services, that are consistent, high-quality, and suitable for individual need, anytime and anywhere through cloud simulation platform.

In the technology of virtualization-based simulation running environment, the fault-tolerant migration technology for simulation resources which is on the basis of virtualization technology, has realized load-balance oriented and high-availability oriented fault tolerant migration algorithms respectively, for the two kinds of errors: the performance decease and the crash of simulation computing nodes. The fault-tolerant migration model for simulation resources is shown in Fig. 11.

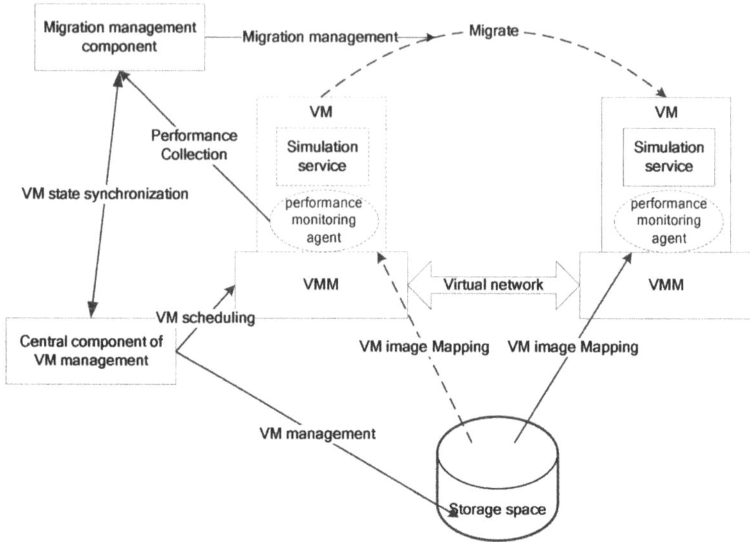

Fig. 11. Model of virtualization-based fault-tolerant migration for simulation resources

It can be seen from the model of fault-tolerant migration that the core component of migration process is migration management component, which can determine the error type of the simulation independently, work out a strategy of migration intelligently, and control the migration process. Migration management component which is an important supporting module for simulation fault-tolerant migration service in the core services of cloud simulation platform, provides transparent and efficient mechanism to use resources in cloud simulation system for upper applications, as shown in Fig. 12.

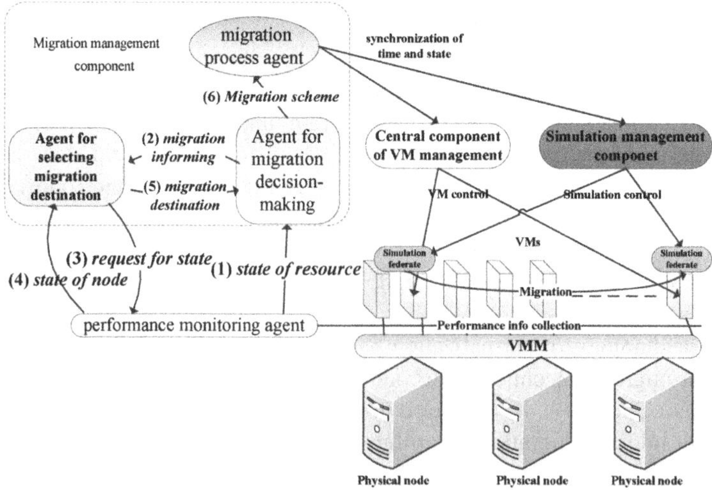

Fig. 12. The instruction of migration management component and the migration process

The specific migration process performed by Migration management component is shown as follows:

- The agent for migration decision-making collects and makes statistics of the running state information of VM and the state information of physical nodes provided by performance monitoring agent and VMM. Based on the information, it gives out the necessity of migration and decision of migration types, and sends the decision to the agent for selecting migration destination.
- After receiving the decision, using the current state of the physical computing environment, the agent for selecting migration destination selects the right physical nodes, and notified to agent for migration decision-making.
- After receiving the destination from the agent for selecting migration destination, the agent for migration decision-making will notify the migration process agent with the destination and scheme.
- After receiving the scheme from the agent for migration decision-making, the migration process agent starts the whole process and achieves time and state synchronization of simulation resources via VM management component and simulation management component.

5.4 High Performance Cloud Simulation Supported Co-simulation Platform Technology [8]

With the increased complexity of research questions, cloud simulation platform faces the need for high-end computing power, high-performance communication capabilities and high scalability from the M&S practitioners in the area of engineering, national security and research [9]. Research on high-performance collaborative simulation technology based on Run Time Infrastructure (RTI) architecture of HLA [10] lays solid foundations for providing services of high-performance simulation engine for users.

In [11, 12] they researched HP-RTI, and we further researched HP-RTI on multi-core environment. In order to improve the performance such as the running speed of simulation, the throughput of interactive data, the scalability of simulation system, Research on high-performance collaborative simulation technology should face the new high-performance computing environment and combine with Multi-core computing technology, etc. So, in this article, our team have studied and realized one kind of multi-core and high-performance computing environment-oriented, high-performance Collaborative simulation platform. As shown in Fig. 13, the core ideas are as follows:

- For the problem that centralized processing in CRC (Core RTI Component) has become the bottleneck of system performance, we employ a parallel processing approach to balance the load and distribute the processing work originally done by CRC to the LRC (Local RTI Component) as much as possible.
- For the problem that there are none connections between CRC and all LRC or the system performance are limited by low communication capability of network, it established full connection communication structure based on shared memory to achieve high speed, high throughput communication between CRC and LRC.
- In order to maintain the consistency of Global data, combining with the demand of HLA/RTI core function services, such as federation management services, time management services, data exchange services (consisting of declaration management

services, object management services and data distribution services, etc.), it sets up the dedicated shared memory area for core processing between CRC and all LRC, in order to maintain data consistency between CRC and LRC.

- Based on the above research, it optimizes time management and data exchange technology of collaborative simulation.

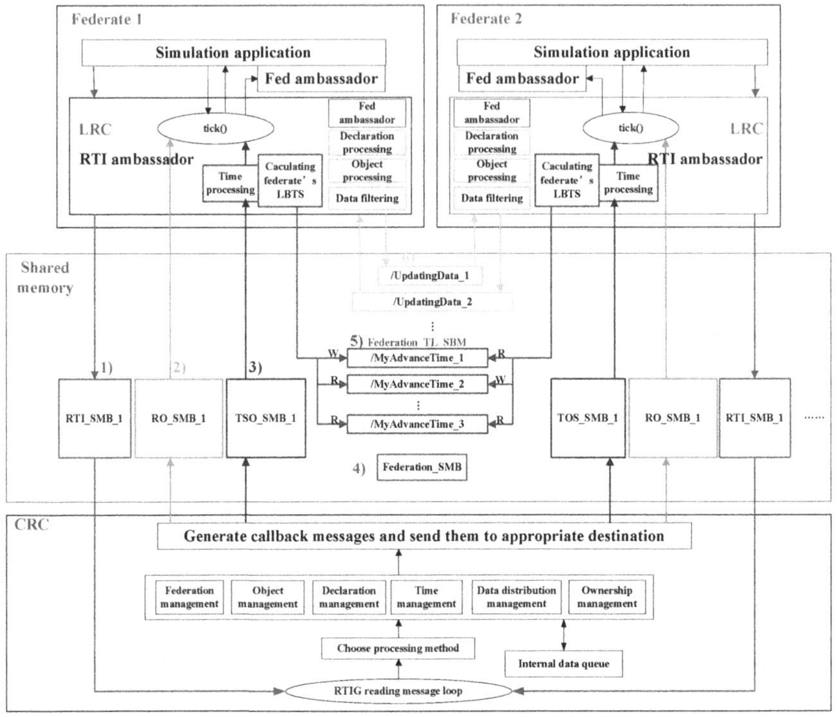

Fig. 13. The architecture of HP-RTI-SHM [9]

5.5 Monitoring and Evaluation Technology for Cloud Simulation

In order to optimize the performance of virtual simulation system, the performance and topology evaluation information is needed to guide: the static scheduling of virtual machine(VM)-based simulation tasks - schedule VMs in which simulation tasks will run to physical computing nodes that are optimally selected; the dynamical scheduling of VM-based simulation tasks - migrate some virtual machines on high-load physical machines to low-load physical machines to achieve high efficient execution of simulation task. To address the aforementioned requirement, one evaluation technology of virtual simulation system is proposed, and it has realized:

- The test for running speed of virtual simulation system;
- The test and evaluation of performance of physical computing resource allocated to virtual simulation system, the topology of virtual simulation system, and their rationality.

The evaluation technology consists of two main parts:

- Performance monitoring: the structure - one single server and several distributed monitoring agents – is adopted, and the state information of the simulation system and its running environment (e.g. simulation time, execution time, CPU utility, memory usage, network I/O traffic, and network bandwidth, etc.) is accessed.
- Performance evaluation: based on the collected information, the evaluator will give out different evaluation items according to the type of simulation system (computation intensive, communication intensive and mixed ones) with certain principles. The principles include: computation intensive simulation models should not be scheduled on one computing node; the group of models which communicate with each other intensively should be allocated on one computing node.

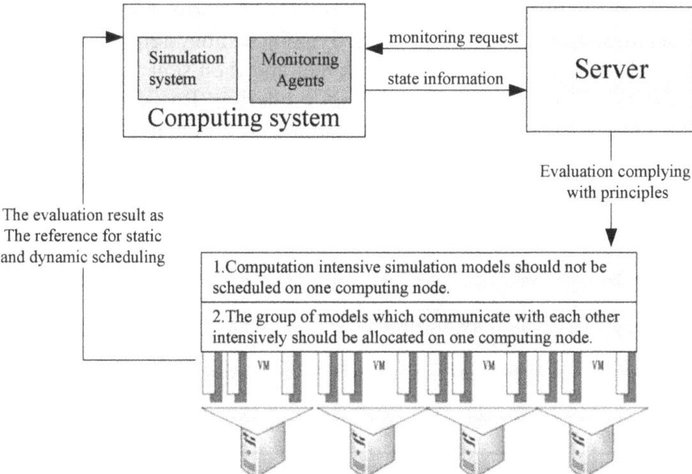

Fig. 14. Virtualization-based simulation system monitoring and evaluation

The specific monitoring and evaluation process is shown as follows:

- Monitoring agents collect state information of simulation system, VMs and their host physical machines in real-time to realize the monitoring of simulation system.
- The single server receives and stores the data sent by monitoring agent, and it can display the state information graphically.
- The evaluator tests the running speed of virtual simulation system (real-time, ultra-real-time, due to real-time) based on the data collected.
- The evaluator gives out different evaluation content automatically corresponding to the type of simulation system, and it can evaluate the performance of simulation system while complying with certain principles.
- The evaluation results guide the static or dynamical scheduling of simulation system to get better performance.

6 Typical Application of COSIM-CSP

Currently, based on the research fruits of cloud simulation, the author has developed one cloud simulation platform prototype "COSIM-CSP", which has been applied in areas of multidisciplinary collaborative simulation of virtual prototyping and large-scale system-level co-simulation. This paper presented an example of aircraft virtual prototype collaborative simulation design system based on COSIM-CSP.

The collaborative simulation design system of an aircraft virtual prototype is mainly made up of six subsystems: control subsystem, engine subsystem, gas subsystem, multi-body dynamics subsystem, pneumatic subsystem, and visualization subsystem, which involve many disciplines, such as mechanical discipline, control discipline, etc. Many domain modeling and simulation tools are used, such as MATLAB (for control subsystem, engine subsystem, and hydraulics subsystem), ADAMS (for multi-body dynamics subsystem), CFX (pneumatic subsystem). And visualization subsystem employs three-dimensional animation simulating to display the simulation running statement. In the course of simulation advance, the control subsystem started the control system solving with gesture information from platform, and then sent the result (control force) to platform. The multi-body dynamics subsystem started the computation of aircraft drive using data of control force, and then sent gesture information to visualization subsystem for three-dimensional rendering. Based on Cloud Simulation Platform, the internal logic of the simulation application system is shown in Fig. 15.

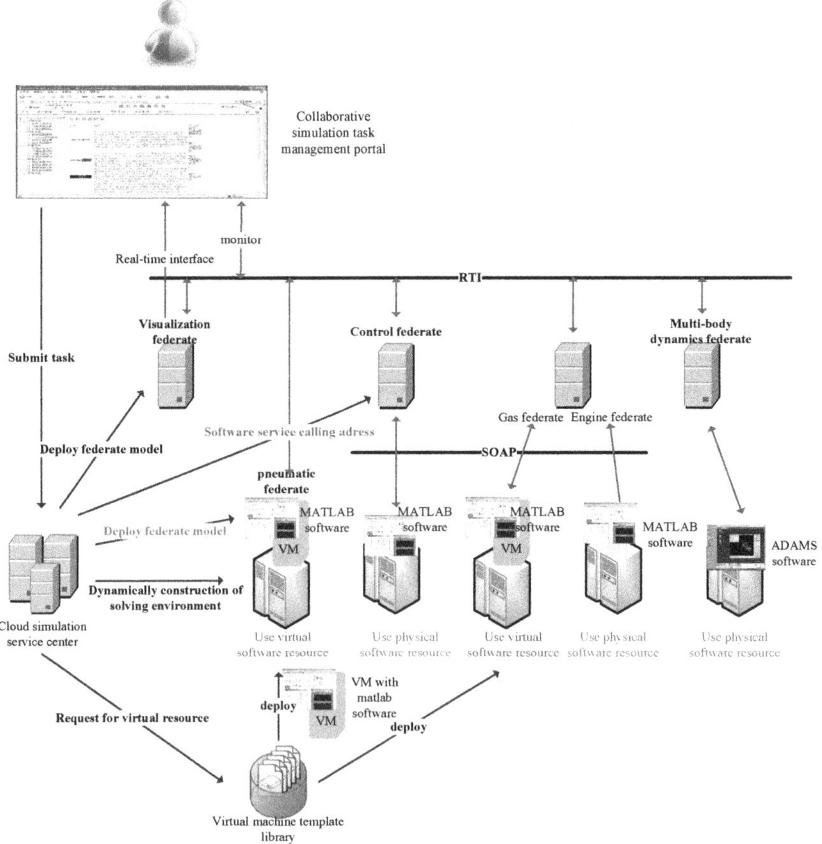

Fig. 15. The internal logic of the collaborative simulation application system of an aircraft virtual prototype

Based on COSIM-CSP, the high performance collaborative simulation application system of complex product virtual prototype provides high-performance cluster and distributed interactive simulation to meet different simulation demand for multi-users. As shown in Fig. 16, users can aggregate modeling and simulation service resource on demand to construct simulation cloud in high-performance cluster and distributed environment. Then users can run their own collaborative simulation system of complex product virtual prototype. Specifically, the powerful computing capability and high-speed internal communication network in high-performance cluster can improve speed of the solving and interaction, so as to meet users' demand of high performance. In addition, based on COSIM-CSP, the high performance collaborative simulation application system of complex product virtual prototype can support the usage mode of "cloud plus client". With cloud simulation platform, users can not only add their computing resource to cloud base, but also obtain good simulation resource with the help of only one networking terminal device installed platform client.

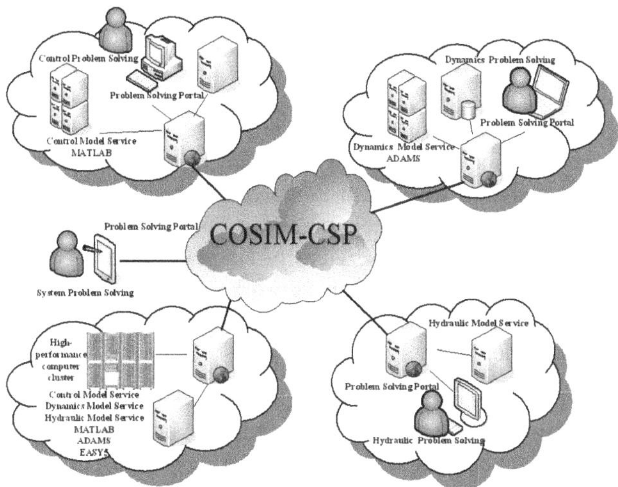

Fig. 16. The structure of the collaborative simulation application system of an aircraft virtual prototype

Take one cloud simulation user for an example. Before the setting up of simulation system, the simulation model services and software tools are encapsulated based on resource virtualization technology and deployed dynamically in the distributed computing environment (including distributed computing environment and high-performance computer cluster). The user can give the description of simulation problem requirement through individual virtual user environment, and dynamically construct the simulation system on demand by discovering, scheduling, and combining various services of simulation models and software tools, to carry out various simulation problem solving tasks. The user can execute simulation system via portal and high-performance RTI, and the platform will perform fault-tolerant migration automatically. Finally, the simulation result can be returned through visualization portal. The screenshots of the aircraft simulation system are shown in Fig. 17.

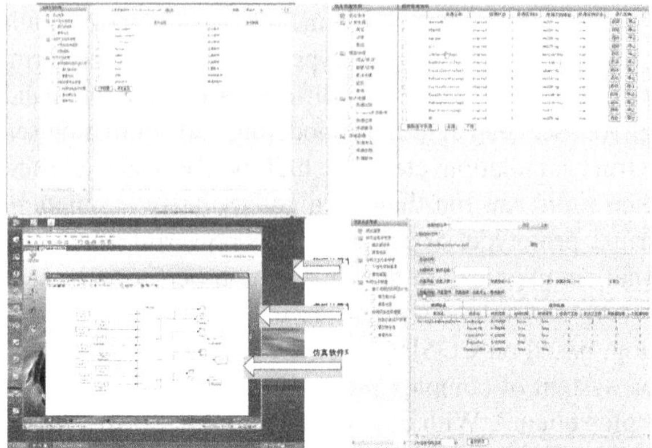

Fig. 17. The screenshots of the collaborative simulation application system of an aircraft virtual prototype

7 Conclusion and Future Work

The primary practice shows that: the aforementioned research results can upgrade the ability of cloud simulation system in several aspects, such as on-demand sharing of multi-granularities resource with multi-users, collaboration, fault-tolerant migration, high performance co-simulation execution, etc, so as to better support "cloud simulation" pattern that users can access services of simulation resource and capability on demand anytime and anywhere through network and cloud simulation platform, to accomplish varied activities in the whole simulation life-circle.

Further works include:

- Further research on semantic-based unified description for varied services of simulation resource (especially for simulation hardware) and capability, in order to support unified management and high efficient sharing of simulation resource in cloud simulation.
- Further research on high efficient collaboration technology of virtual simulation resource, in order to support high efficient and collaborative execution of virtual simulation resource in cloud simulation.
- Further research on the simulation resource and capability share on demand for mass users, supporting mass users to access M&S services on demand based on cloud simulation platform.
- Further focus on and enhance the application research of cloud simulation system.

References

1. Li, B.H., et al.: A network modeling and simulation platform based on the concept of cloud computing — "Cloud Simulation Platform". Journal of System Simulation 12, 5292–5299 (2009)

2. President's Information Technology Advisory Committee: Computational Science: Ensuring America's Competitiveness. Report To the President, pp. 10–13 (2005)
3. Armbrust, M., Fox, A., Griffith, R., et al.: Above the Clouds: A Berkeley View of Cloud Computing (2009), http://www.eecs.berkeley.edu/Pubs/TechRpts/2009/EECS-2009-28.pdf
4. Li, B.H., Zhang, L., Wang, S., et al.: Cloud manufacturing—a new service-oriented networking manufacturing mode. Computer Integration and Manufacture system 1, 1–7, 16 (2010)
5. Zhang, Y., Chai, X., Hou, B., et al.: Research on the Application Pattern of Cloud Simulation based on Virtualization-Based Simulation Technology. In: 2010 International Conference on Future Information Technology, Changsha (2010)
6. Zhang, Y., Li, B.H., Chai, X., Hou, B., et al.: Research on Virtualization-based Simulation Environment Dynamically Building Technology for Cloud Simulation. In: 2010 International Conference on Information Security and Artificial Intelligence, Chengdu (2009)
7. Zhang, Y., Li, B.H., Chai, X.: Research on virtualization-based cloud simulation resource migration technology. Journal of System Simulation 6, 1268–1272 (2011)
8. Zhang, Z.H., Li, B.H., Chai, X.D., et al.: Research on High Performance RTI Architecture Based on Shared Memory. In: ICCEE 2010, Chengdu (2010)
9. Perumalla, K.: Switching to High Gear: Opportunities for Grand-scale Real-time Parallel Simulations. In: 13th IEEE International Workshop on Distributed Simulation and Real-Time Applications, Singapore (2009)
10. Adelantado, M., Bussenot, J.L., Rousselot, J.Y., Siron, P., Betoule, M.: HP-CERTI: Towards a high Performance, high Availability Open Source RTI for Composable Simulations. In: 2004 Fall Simulation Interoperability Workshop (2004)
11. Fujimoto, R.M., McLean, T., Perumalla, K.S., Tacic, I.: Design of High Performance RTI Software. In: Fourth IEEE International Workshop on Distributed Simulation and Real-Time Applications, DS-RT 2000, pp. 89–96 (2000)
12. Steinman, J.: WarpIV Kernel: Real Time HPC-RTI Prototype (2005), http://www.warpiv.com/Documents/Papers/HPC-RTI.pdf

Development of Simulation Service Platform Based on Cloud Computing for Manufacturing Industry

Sang-Hyun Cho

e-Design Center, #2007, Korea Institute of Industrial Technology, Songdo-Dong, Yeonsu-Ku, Incheon, South Korea
chosh@kitech.re.kr

Abstract. Computer Aided Engineering (CAE) is very helpful field for every manufacturing industry including foundry. It covers CAD, CAM, and simulation technology also, and becomes as common sense in developing new products and processes. In South Korea, more than 600 foundries exist, and their average employee number is less than 40. Moreover, average age of them becomes higher. To break out these situations of foundry, software tools can be effective, and many commercial software tools had already been introduced. But their high costs and risks of investment act as difficulties in introducing the software tools to SMEs (Small and Medium size Enterprise). So we had developed cloud computing platform to propagate the CAE technologies to foundries. It includes HPC (High Performance Computing), platforms and software. So that users can try, enjoy, and utilize CAE software at cyber space without any investment. In addition, we also developed platform APIs (Application Programming Interface) to import not only our own CAE codes but also 3rd-party's packages to our cloud-computing platforms. As a result, CAE developers can upload their products on cloud platforms and distribute them through internet.

1 Introduction

We had developed and propagated CAE software package to foundries over last 10 years, and it has many functions based on CFD, and heat transfer analysis to predict and analyze physical phenomena of casting processes. But up to date, the CAE market of foundry does not recognize necessities of CAE technologies enough, moreover price of software packages does not drop down. These situations make CAE market smaller than ever. In South Korea, we have more than 600 foundries, and average sale of them is less than 1 million US dollars per year. These environmental elements delay the introduction of scientific tools to workplaces, and concerns about industry weakening are on increasing.

From the viewpoint of CAE developer, development of new CAE technology is marking time, so that investments for CAE are being focused mainly not on solvers, but on user interfaces including pre and post processors.

To resolve situational difficulties, we have constructed cloud computing platforms for CAE, and started to provide services to SMEs (Small and Medium sized Enterprises).

J.-H. Kim et al. (Eds.): AsiaSim2011, PICT 4, pp. 164–169, 2012.

2 Development of CAE Services on Cloud Computing Concept

In this time, Cloud Computing is most important keyword in the world, and many services are already being provided by huge IT providers. Amazon Web service [1] and Google Apps[2] are providing computing resources through internet, and Microsoft is supporting his Windows Azure platforms [3] and providing development environments from Visual Studio 2010. These cloud computing services basically are hidden by "cloud", so that users don't know how services act and from where services come, but they can access to same services and data at anywhere in the world.

Cloud computing platform consists of three core elements, and those are 1) IaaS, 2) PaaS and 3) SaaS as shown in Fig. 1.

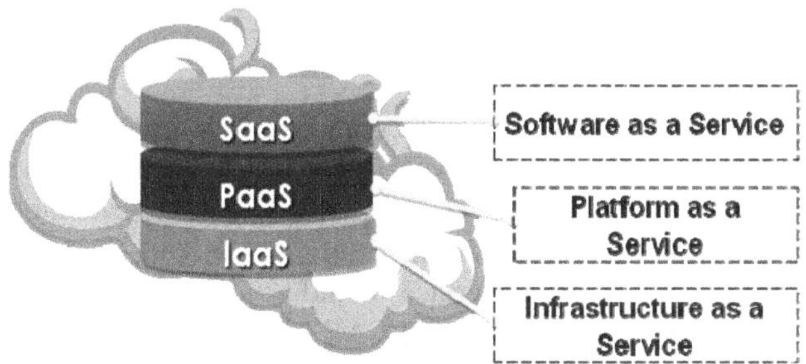

Fig. 1. Components of Cloud Computing

As shown in Fig. 1, cloud computing works on infrastructures of computing resource, and systems are operated by platform structures including operating system, networks, database and security-related platforms. Web services are also important parts of PaaS. Finally, software for users will be loaded on cloud computing platforms and distributed, so that both hardware and software resources become as service.

2.1 ISC(Internet Simulation Center)

ISC is the name of our CAE services based on cloud computing, and it includes IaaS, PaaS and SaaS. If users under clouds try to access to CAE cloud through internet, necessary plug-in applications will be installed automatically on local desktop or laptop computers running Windows OS. Then it becomes possible to log in to CAE service portal page through account validation processes. This every step plays on Microsoft Windows platforms to guarantee user friendly environments.

To provide PaaS to users, we built up cluster computer center which consists of more than 100 computing nodes for simulation jobs with additional peripherals. Each node has multi connection lines to WAN and inter-nodes to guarantee non-stop service. Computational resources are composed by cluster nodes, control servers, network equipments and massive storages. Between cluster nodes, infiniband connection lines were installed with DDR (Double Data Rate) switches, and HCA cards were also installed to all clusters and servers, so that MPI (Message Passing Interface) parallelism can obtain maximum performances when data communication occurs to exchange boundary data of each domain for every time step.

To guarantee data safety from unexpected data loss, massive storage system was introduced. The storage system was built up by RAID 6 technology to obtain fast recovering time, and was divided into several virtual volumes for users' data space, data backup and SAN booting disks for important control servers. Users' data, especially results of CAE jobs, would be stored in distributed file system to increase I/O speed under multi-user environment, and then will be also back-up by secondary and tertiary storage equipments. As a result, totally, past two days' data will be maintained.

This hardware system is sheltered by multi-layered security shells and connected to WAN by exclusive internet line with 500 Mbps speed.

2.2 Distribution of ISC Accessing Methods

Aims of ISC services are to build up cloud computing platform and to propagate CAE technologies to SMEs. It is different with desktop virtualization which runs under server-client concept, so that all CAE contents of ISC should have uniform structures to be loaded into ISC and to be distributed through internet. Namely, solvers and pre/post processors including user interface should be separated completely, run independently. In Fig. 2 (a), structure of normal CAE software for off-line usage is shown and the structure of on-line driven CAE software is shown in Fig 2(b).

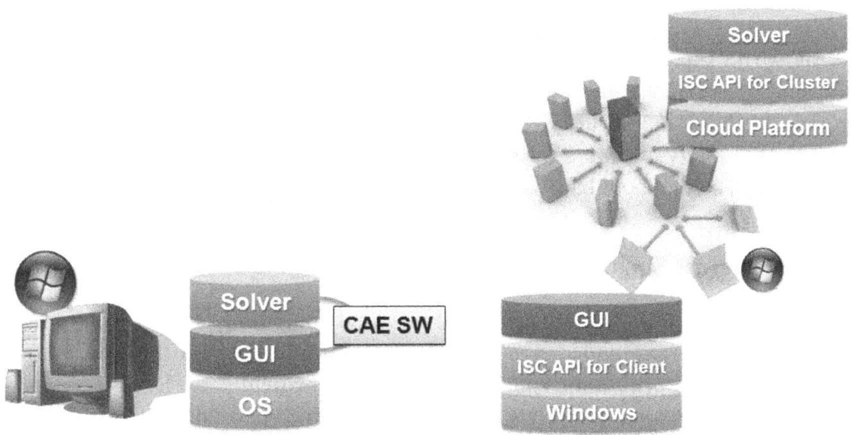

(a) Structure of off-line CAE software

(b) Structure of CAE software on cloud platform

Fig. 2. Structures of off-line and cloud CAE software

To help 3rd party developers in transferring their own codes to cloud computing environment, we developed API (Application Programming Interface) for ISC platform and its functions are shown in table 1.

Table 1. Functions of ISC API

	Solver API	Client API
Function	• Interface for file I/O on distributed storage by virtual storage skill • Information achiever of each simulation job • Job management functions for solving, pause, completion, cancellation and unexpected interruption • Monitoring functions such as job progress reporting and obtaining • Functions for networking and message passing among clusters • High efficient functions for data compression & decompression acts	• Firewall accessing methods • Simplified network related methods • Job-related information achiever • Job managing request functions such as launching and cancellation • Data transferring functions like upload, download with request function of data deletion • Data decompression method • Job listing function
Supported Binaries	x64 platform	x86, x64 platform (Windows XP, Windows Vista, Windows 7)
GUI	Not included	Included

In developing ISC contents such as solver and UI, testing bed is necessary for developers. Although ISC APIs are provided with simplicity, but they have complex processes running in background. They all are inter-connected with network, database and even WEB, so that solver or UI developers can't test their codes with ISC API at isolated local computer. Moreover, most of ISC API runs under service-client environment.

Most simple way to provide the testing beds is to open ISC clouds to developers. But, this way is not free from data hacking acts. Furthermore service structure of ISC can be exposed to developers, as same reason, contents developers hate to expose their own source codes to ISC service provider simply because they use testing bed of ISC.

ISC built the testing bed system for developers. It is not ISC service zone, but located in front of 1st firewall, and services for these testing zones run in between 1st and 2nd firewall. The testing bed consists of not real servers but virtual machines, so that if developer requests some testing bed (machines), the virtualized machine will be furnished with ISC environments and static IP address. Developers can upload their developed binaries to supplied virtual machines, and can access to that machines by using ISC API for clients. Under these environments, developers can test solvers and clients including ISC API.

3 ISC Services

Up to now, ISC provides three CAE fields, and will increase more contents. The supported fields by CAE are casting analysis, heat treatment analysis and injection molding of plastic. Until the end of this year, analysis of forging process will be added by 3rd party developer.

To use CAE contents registered at ISC, users should make user accounts twice, because CAE contents based on cloud computing transfer important data like as CAD files, ISC administrator checks out user's account form, and issues new access password for ISC services. Once users obtain permission for ISC services, they can move to ISC service page, and then select the field to use. If users click the "run" button, ISC checks user's local computer if ISC application was installed or now. If none of ISC applications were found, ISC application will be automatically installed onto user's computer. With automatic installation functions, ISC provides automatic version maintenance function also. User can choose "run" or "upgrade" button at ISC service page. If "upgrade" button is clicked, ISC WEB server transfer the upgrade information of xml format to user's computer and the pre-installed active-X based parser checks inconsistencies among pre-installed ISC application files and upgrade-needed files. After this process, WEB server sends files and pre-installed active-X application receives them. The ISC administrator can maintain version of ISC application which was installed onto users' local computer by using WEB.

Now, we have three contents on services listed in Table 2 and Fig. 3.

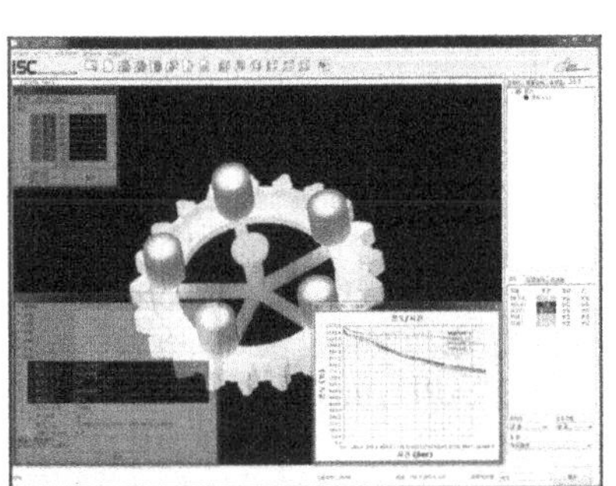

(a) ISCCast (casting)

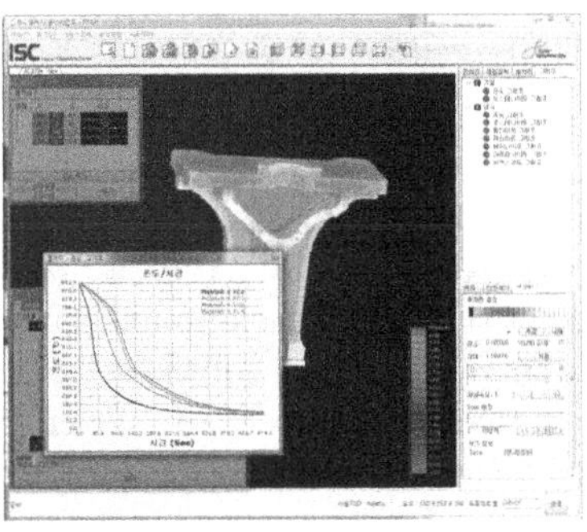

(b) ISCHeat (Heat treatment)

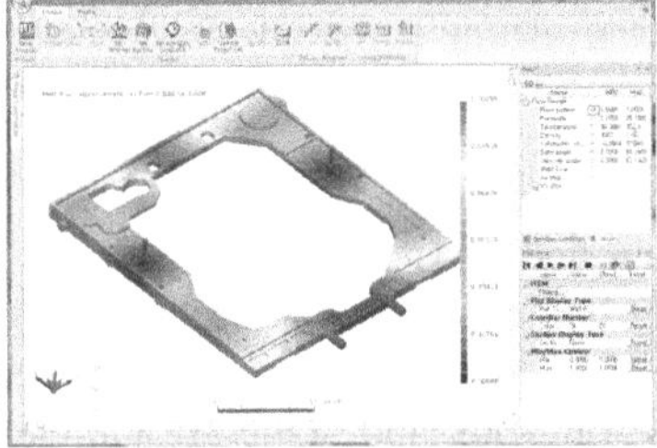

(c) ISCMold (Injection Molding for Plastic)

Fig. 3. ISC Contents on services

Table 2. ISC Contents

Field	Service Launching Date	Analysis Method
Casting	2009. 12	FDM
Heat treatment	2010. 5	FDM
Injection Molding	2011. 1	FEM

4 Conclusion

In South Korea, we have lots of SMEs, and they all are supporting for basement of all industries. But they are encountering with many problems in sustaining the industry. To help and encourage them to improve their technical competitiveness, we introduced cloud computing concept to CAE fields. Furthermore, to expand the CAE clouds, we had developed APIs to help migration processes of 3rd party developers. With this endeavor, ISC started to provide CAE contents in three industrial fields, and will increase the variety and volumes of contents gradually.

References

1. http://aws.amazon.com
2. http://www.google.com/apps/
3. http://www.microsoft.com/windowsazure/

Integrated Real-Time Acoustic Communication System for Controlling of Bio-inspired Underwater Robot

Yung-Pyo Kim[1], Nam-Yeol Yun[1], Soo-Young Shin[1], Ji-Eon Kim[1], Soo-Hyun Park[1], Jun-Ho Jeon[2], Sung-Joon Park[2], Chang-Hwa Kim[2], and Young-Sun Ryuh[3]

[1] Ubiquitous System Lab., Graduate School of BIT, Kookmin University, Korea
{kuniv003,anuice,sy-shin,un1730,shpark21}@kookmin.ac.kr
[2] Ocean Sensor Network System Technology Research Center, Gangneung-Wonju National University, Korea
musician1222@nate.com, psj@ieee.org, kch@gwnu.ac.kr
[3] Underwater Robot Center, Korea Institute of Industrial Technology, Korea
ysryuh@kitech.re.kr

Abstract. Robot control system construction that uses an underwater robot is essential in the oceanic environment. In this paper, we designed a robot system optimized for the underwater environment and proposed a hybrid MAC protocol considering both real time operation and energy efficiency, which is especially important for underwater communication system. Also, we examined the designed hardware of underwater robot control system in detail, and carried out simulations to evaluate the performance of the proposed MAC protocol using Matlab. The simulation results showed that the proposed model achieves better performance than conventional models.

1 Introduction

Underwater communication technology can be applied to many fields, such as oceanic data collection, undersea exploration and development, disaster prevention, underwater environmental monitoring, tactical surveillance, etc [1] [2]. Many areas mentioned above can be revitalized like AUV or ROV by underwater robots, and underwater robot control system construction is essential using underwater wireless communications networks for undersea exploration and development system operations. Currently, research exists on robot control systems on the ground, and many robot control systems are operational. However, research on underwater robot control remains insufficient due to various restrictions of the underwater environment. Restrictions exist in achieving underwater communications of underwater robot control due to the acoustic signal properties such as electromagnetic waves and light wave that are weakened rapidly unlike RF communication used on the ground [3]. In acoustic communication, a small amount of information can be sent at a time as we transmit in limited bandwidth in the underwater environment. Further, underwater channels can be seriously damaged due to the multi-path and fading phenomena [4]. Also, acoustic communication has an issue of battery depletion due to the long propagation

J.-H. Kim et al. (Eds.): AsiaSim2011, PICT 4, pp. 170–179, 2012.

delay, packet collisions, etc [5]. Therefore, the underwater robot control system construction should be designed considering various environments.

In this paper, we develop the hardware according to each layer for underwater robot control system construction, and we design the integrated underwater robot control system. Also, we propose a MAC protocol that guarantees real-time and energy that are essential in the robot control system. In section 2, we examine related works to underwater robot control system. We explain the design process of the underwater robot control system in section 3. We propose a MAC protocol suitable for an underwater robot control system in a section 4, and we evaluate the performance using Matlab in section 5. Finally, we present the conclusion and future work direction in section 6.

2 Related Works

Many studies have been conducted in the area of ground-based robot system with the assistance of various sensor technologies and high speed communication technology. The environment of underwater robots is very different from that of ground-based robots. In addition, there is limited room for using necessary devices due to the lack of sensors and devices that are suitable for the underwater environment. In [6], the underwater robot that is in the process of development was introduced, and the characteristic of underwater acoustic signal detection system was explained with respect to its feature and system configuration. The underwater robot that was in the process of development in reference [6] has 5 degrees of freedom control except Right and left movement Using 4 horizontal propellers and 2 vertical propellers. In addition, the robot system has two cameras, two single lights, 8-channel sonar for obstacle distance measurement, sonar detection, and 2 high draw phones for underwater sound signal acquisition. The underwater acoustic system that was installed in the underwater robot system measures signals from 2 high draw phones, amplifies the measured signal, and digitizes the amplified signal using a signal processor detached on AD board. Based on the obtained information, the signal characteristics and direction are analyzed. The analyzed information is provided to the operator as well as to the underwater robot for conducting a hypothetical mission (avoidance, tracking, ignored, etc.). In [7], Hyun-Taek Choi et al. proposed the design of control and sensor system for the intelligent implementation of underwater robot. They designed separate systems of control, and internal sensor and external sensor for intelligent implementation, respectively. The internal sensor system consists of a temperature sensor, humidity sensor, and battery measuring sensor. The value of internal sensor system has higher priority than any sensor's input value; the internal sensor system will proceed with the situational process after stopping the other operation when some errors occur. The purpose of external sensor systems is to get the surrounding information of an underwater robot. Its environmental information includes the information required to control underwater robots and to simply present the status of an environment. The above research is mainly organized sensor for the measurement of information that proposed controlling the underwater robot. Furthermore, the development of MAC protocol that guarantees real-time is important for constructing a robot control system using underwater acoustics. Presently, research on underwater MAC protocol considering the distinct characteristics of underwater environment is in progress [8] [9]. The UWAN-MAC

[10] proposed an energy efficient MAC protocol in underwater acoustic wireless networks. This MAC protocol has a structure that transmits data in given channels after securing channels in advance. UWAN-MAC transmits the control message to secure the channels in an initialization period. This protocol then sets up the transmission channel of each node by notifying its own channel using time to neighbor nodes. However, it causes unnecessary waste of channels with no data transmission due to setting channels in advance. It also has a disadvantage in that energy consumption by channel collision and propagation delay increase the number of channels.

To avoid conflicts in the conventional RTS/CTS mechanism, the RTS length should be greater than the maximum transmission delay and CTS length should be greater than the sum of RTS lengths, 2 times maximum transmission delay and hardware transmit-to-receive transition time. Slotted FAMA [11] overcomes this problem by introducing the slot concept to the conventional FAMA [12] protocol. All packets are sent only at the beginning of the slot, and the length of slot is set to have the value of the sum of CTS packet transmission time and maximum transmission delay to avoid the collision of data packets. In this way, RTS or CTS packets transmitted at the beginning of a slot are received by all nodes within transmission range in the slot section. In Slotted FAMA, the nodes are always in the active state for exchanging RTS/CTS packets and, consequently, energy waste increases because of the continuous idle state and overhearing due to the exchange of RTS/CTS packet. In addition, since the slot duration size reflects the maximum transmission delay time, there is a limit on providing real-time service because of the increased slot duration due to the long transmission delay time of sound waves.

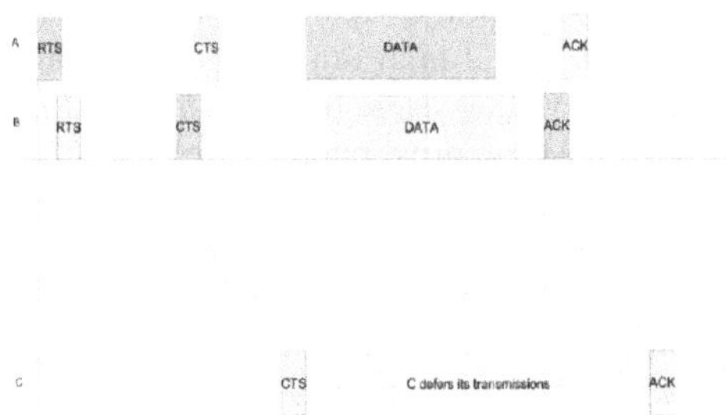

Fig. 1. Successful handshake between A and B in slotted FAMA

3 The Design of Underwater Robot Control System

The Operation flow of robot control system is as follows. The gateway receives a control message from the base stations. Then, the gateway delivers a command until robot nodes through acoustic modems, and gateway controls a robot. The following figure a robot fish picture and a formation table during the current development. The proposed underwater robot fish is equipped with a buoyancy pack and a motor for smooth movement underwater. Also, all kinds of sensors that can detect the environment are loaded, and an underwater communication module that can transmit detected data is added. A camera, GPS, and gyroscope sensor are also added.

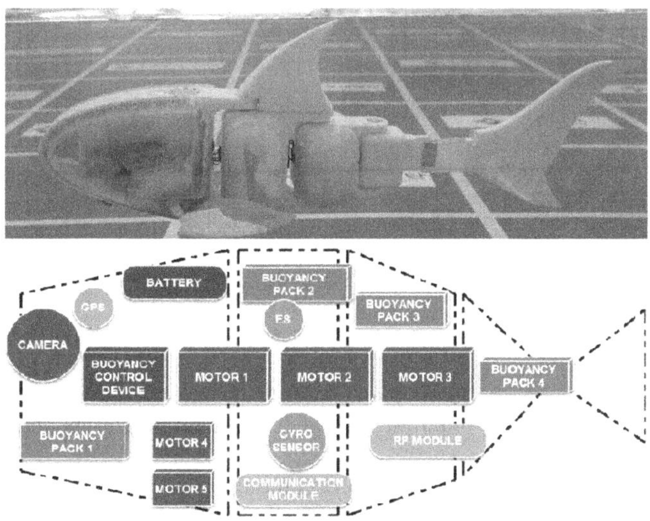

Fig. 2. Underwater robot fish and system configuration

In this study, we separately developed hardware for the MAC and PHY layers in order to transmit the data acquired underwater; SPI (Serial Peripheral Interface) is used for the interface between boards. The hardware developed for the MAC layer consists of a MCU based on Cortex-M3 to achieve low-power operation, and it uses the MAC protocol for optimizing the underwater control system. Also, we made miniaturization of a robot fish possible by minimizing the size of hardware. The following figure and table show the form of hardware between each levels and detailed specification.

Fig. 3. Underwater communication base-board

Table 1. Base-board specifications

BASEBOARD SYSTEM	
Integral Parts	**Description**
MCU	Cortex-M3 (STM32F103CB)
Operating System	Firmware
MAC Protocol	Proposed MAC
MAC/PHY Interface	SPI
Size	30mm × 30mm
Power	3.3V or 5V

Fig. 4. Underwater acoustic modem

Fig. 5. Underwater acoustic transducer

Figures 4 and 5 show the implemented underwater acoustic modem that is equipped with a commercial omni-directional acoustic transducer having the resonant frequency of 70 kHz.

The transmitting acoustic modem conducts the following functions. First, it makes the physical layer transmit a frame suitable for underwater communication via digital signals received from the MAC-layer. Then, it converts the frame into an analog signal and amplifies it via the analog circuits. At the transducer, the electric signal is converted to an acoustic wave that is sent through an underwater channel.

At the receiving acoustic modem, the transducer transforms the received acoustic wave into an electric signal via an underwater channel. After amplifying the transformed signal and then filtering out the tiny and noisy signal, the receiving modem converts the analog signal into digital signal and obtains an estimate of the transmitted physical-layer frame. Finally, the MAC-layer signal is extracted via estimating the physical-layer frame and then transferred to MAC module by the protocol of SPI.

The detailed modem specification is given in Table 2.

Table 2. Underwater acoustic modem specifications

UNDERWATER ACOUSTIC MODEM	
Integral Parts	**Description**
MCU	ATmega128
Operating frequency	70 kHz
Battery	±14.8 V, 8.8 AH Li-ion
Max. data rate	200 bps
Max. operating range	100 m
Dimensions	70 × 40 mm (ø × H)

4 Adjusted MAC Protocol of Underwater Robot Control System

The purpose of this chapter is to design a MAC protocol optimized for the underwater robot control system for effectively controlling underwater robots (AUV, ROV). The following figure shows the architecture of the robot control system based on underwater acoustic communication.

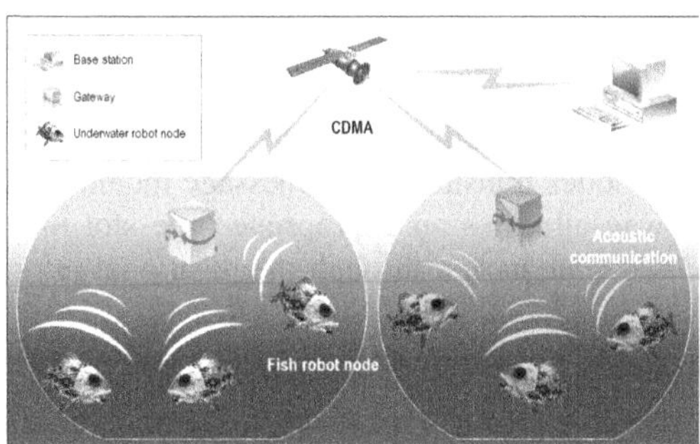

Fig. 6. Architecture of the robot control system

Until recently, the MAC protocol generally has had a structure that deploys nodes to collect surrounding information, and the buoy, which is cluster head, then sends the gathered data to the base-station. It also sends the data by reserving a slot in advance prior to transmitting the data, like the RTS/CTS mechanism. However, research on underwater MAC protocols, that can control aperiodic acoustic signals such as those of robot control systems, is still in its early stage. The MAC protocol proposed in this paper effectively transmits periodic robot control messages through acoustic signal and receives information from robots through a time slot that is allocated in advance. The proposed MAC also transmits integrated messages at regular intervals by buffering in order to efficiently send data to many robots. Figure 7 shows the transmission cycle of the proposed MAC protocol, and Figure 8 shows the operating example of the proposed MAC protocol. The transmission cycle of our proposed MAC protocol consists of *Initialization period* and *Superframe period, as* is shown in Figure 7. The *Initialization period* consists of an *Advertisement (**Ad**) period* and *Join (**Jn**) period;* the *Superframe period* consists of a *Construction (**Ct**) period*, *Data transmission (**Dt**) period* and *Maintenance & Sleep (**M&S**) period.*

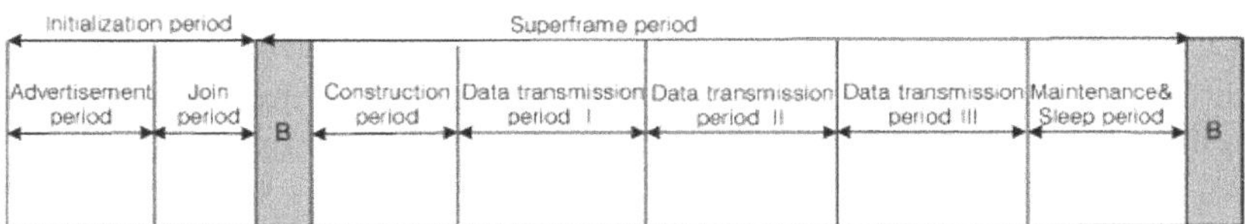

Fig. 7. The cycle of the proposed MAC protocol

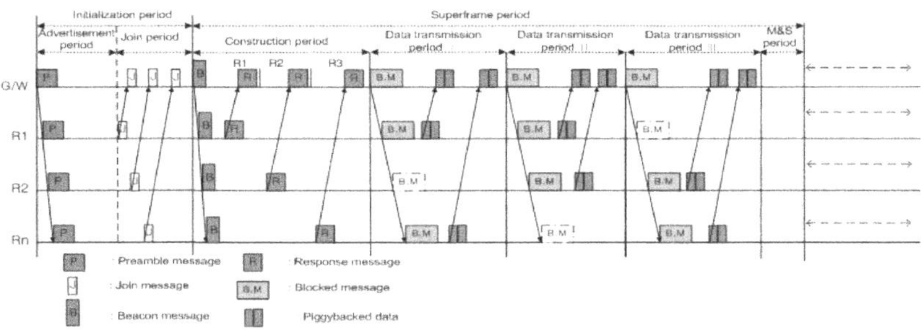

Fig. 8. The proposed MAC protocol

A. Initialization period

Ad period: The gateway broadcasts a preamble message to all robot nodes in a cluster. The preamble message requests the positional information of each robot node.

Jn period: The robot nodes transmit a join message including their own positional information. The gateway allocates an appropriate time slot to each node in the ***Ct*** *period* of the next *Superframe period* through the positional information included on the join message.

B. Superframe period

Ct period: In this period, the gateway allocates an appropriate time slot through the positional information of robot nodes acquired in *Initialization period* and then broadcasts a beacon message. Here, the number of time slots that can be allocated is variable, and we will set the maximum number of nodes according to a rule. The number of nodes set up in advance because real time cannot guarantee since many time allocate to reception of data in ***Ct*** *period* if it allocates an unlimited number of time. The robot nodes transmit the current positional information to the gateway during this period.

Dt period: The gateway broadcasts a message to the robot nodes via an encapsulation processing when the gateway receives an aperiodic robot control message from the base station. In this process, the gateway merges into one frame with each robot message via the buffering process. Further, the robot nodes transmit the gathered data and positional information to the gateway via the piggyback scheme. Thus, this method is very efficient in terms of data processing and energy consumption because the gateway transmits the merged message without separately generating the frame according to each robot control command. Considering the real-time guarantees of a control message and the data error rate in underwater environment, the maximum buffering time is set to one second. And the ***Dt*** *period* is repeated three times in one *superframe period.*

M&S period: In the ***M&S*** *period*, a new robot node is assigned a new time slot in the next ***Ct*** *period* through transmitting a join message to the gateway by using the CSMA/CA method when a new robot node enters the cluster. The existing robot nodes, which do not need to be allocated a new time slot, operate in sleep mode, and these nodes then remain in an inactive state to prevent unnecessary energy consumption.

5 Simulation Results

To prove the superiority of the proposed MAC protocol in underwater robot control system against conventional protocol, performance evaluations were conducted using Matlab. For the comparison of their performances, a series of simulations of end-to-end delay and power consumption were conducted. In the simulations, we set up the range of underwater network size at 500m x 500m, and we randomly deploy the underwater robot nodes of 1 to 20. We applied a slot size of one second considering the sizes of propagation delay and data packet underwater. In addition, a commercialized WHOI micro-modem was used for the experiment. The maximum WHOI micro-modem data rate is 5 kbps, and power consumption along the state of each robot node is specified in Table 3 in detail.

Table 3. Simulation parameters

Parameter	**Values**	**Parameter**	**Values**
Network Area	500m × 500m	Number of Nodes	1~20 ea
Data Rate	5 kbps	Transmit Power	< 50 W
Max. Frame Size	1024 bits	Idle Power	~158 mW
Data Packet Size	160 bits	Receive Power	~158 mW
Acoustic Speed	1500 m/s	Sleep Power	~5.8 mW
Time Slot Size	1 sec		

The equation used in the simulation is as follows:

<Power consumption model>

Proposed MAC model:
$$A(Rx\_p * Ts) + N(Id\_p * Ts) + (Tx\_p * Ts) \tag{1}$$

TDMA model:
$$N(Rx\_p * Ts) + (N-1)(Id\_p * Ts) + N(N-1)(Sl\_p * Ts) + (Tx\_p * Ts) \tag{2}$$

RTS/CTS model:
$$N(Rx\_p*Ts') + (Rx\_p*Ts) + (N-1)*(Sl\_p*Ts') + 2(N-1)(Sl\_p*Ts) + (Tx\_p*Ts) + (Tx\_p*Ts') \tag{3}$$

<Total End-to-End delay model>

Proposed MAC model:
$$2 * Max\_pdelay + (L\_bm + L\_bc)/k + \sum_{i=1}^{N}(pdelay\_Ri + L\_pdata/k) \tag{4}$$

TDMA model:
$$Max\_pdelay + (L\_bc)/k + \sum_{i=1}^{N}(2 * pdelay\_Ri + L\_cm/k + L\_pdata/k) \tag{5}$$

RTS/CTS model:
$$Max\_pdelay + (L\_bc)/k + \sum_{i=1}^{N}(3 * pdelay\_Ri + L\_cm/k + L\_pdata/k + L\_ack/k) \tag{6}$$

The notations used in the numerical model are as follows:

Table 4. Notations of the numerical model

Notations	**Descriptions**	**Notations**	**Descriptions**
Rx_p	Power consumption of Receive period	K	Data rate
Tx_p	Power consumption of Transmission period	pdelay	Propagation delay
Id_p	Power consumption of Idle period	Max_pdelay	Maximum Propagation delay
Sl_p	Power consumption of Sleep period	L_bm	Length of Blocked message
N	Number of Nodes	L_bc	Length of Beacon message
A	Number of Blocked message	L_pdata	Length of Piggyback data
Ts	Time slot size	L_ack	Length of Acknowledgment message
Ts'	Time slot size (RTS/CTS)		

In the simulation, we assumed that the message transmission to each of the robot nodes occurs at all time slots, and we performed the evaluation based on the data transmission cycle. Figure 9 shows the amount of energy consumption during the total transmission cycle of one robot node. The energy consumed by one robot node is the sum of the energy required to receive the control message from the gateway and the energy consumed to transmit the piggybacked data to the gateway. Also, energy consumption in the interval of inactivity is included in the total energy consumption. The following figure shows a node on a network increases, total energy consumption. The proposed MAC protocol can achieve low energy consumption compared to that of the TDMA and RTS / CTS methods.

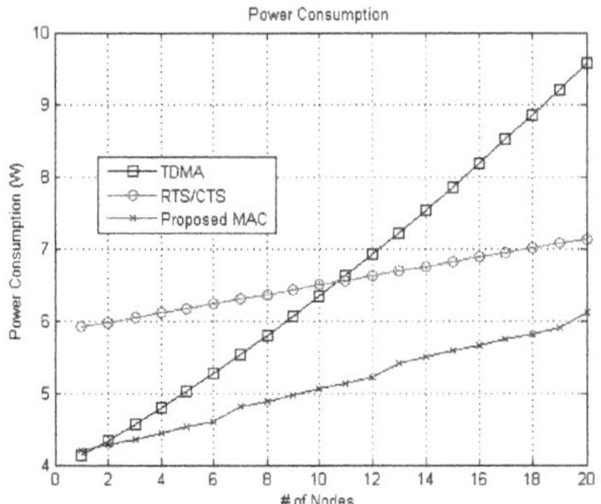

Fig. 9. Power consumption

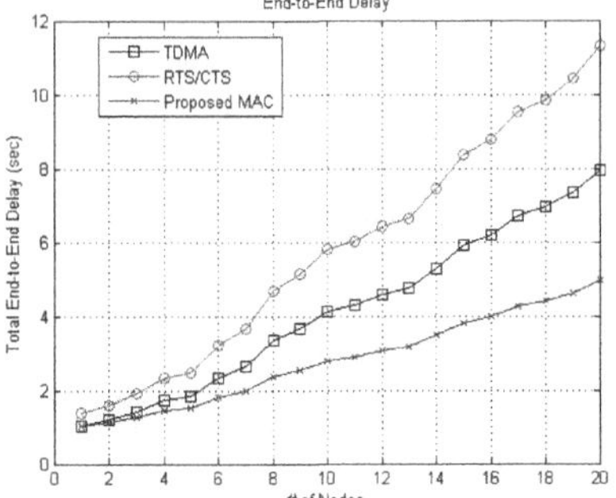

Fig. 10. Total end-to-end delay

Figure 10 shows the total end-to-end delay according to the change of node number. In underwater network of range which is fixed, randomly deploy the 1 to 20 nodes and total end-to-end delay changes measured according to data transmission. Increasing the number of nodes will increase the messages being sent. So, the total end-to end delay will increase. The simulation results prove the best performance of the proposed MAC protocol, and the RTS / CTS method produced the longest total end-to-end delay. According to the RTS / CTS methods, the end-to-end delay increases in order to transmit data after exchanging the RTS/CTS packets, and this increases the total end-to-end delay of all cycles.

6 Conclusion and Future Work

Until now, research on underwater robot system has been in its early stages compared ground robot systems mainly due to the special characteristics of underwater environment. As interest in the huge resources of ocean increases, the need for underwater robot control system research has increased also. In this paper, we designed a robot control system suitable for underwater environment and also proposed a MAC protocol that guarantees both real time operation and energy efficiency. In the process of developing the underwater robot control system, independent hardware between the MAC and PHY layers was developed, and, by using a MAC protocol that is optimized to underwater environment, a compact and low-power operational system was implemented.

For comparative analysis of low-power and real time operation performance of the proposed and conventional MAC protocols, total end-to-end delay and power consumption rate were calculated using Matlab and compared with each other. The comparative analysis result showed that the proposed protocol achieves excellent performance over the conventional one. With the proposed protocol, however, there are possibilities of collision in case of simultaneous joining multiple nodes, so further research to solve the problem of collision is needed. Based on the designed underwater robot control system, a robot fish has been implemented.

Acknowledgment. This research was supported by the MKE (The Ministry of Knowledge Economy), Korea, under the ITRC (Information Technology Research Center) support program supervised by the NIPA (National IT Industry Promotion Agency) (NIPA-2011-C1090-1121-0001).

References

1. Akyildiz, I.F., Pompili, D., Melodia, T.: Underwater acoustic sensor networks: research challenges. Ad Hoc Networks 3(4), 257–279 (2005)
2. Pompili, D., Melodia, T., Akyilidiz, I.F.: A CDMA-Based Medium Access Control for Underwater Acoustic Sensor Networks. IEEE Transactions on Wireless Communications 8(4) (April 2009)
3. Heidemann, J., Ye, W., Wills, J., Syed, A., Li, Y.: Research challenges and applications for underwater sensor networking. In: IEEE WCNC (2006)
4. Sozer, E., Stojanovic, M., Proakis, J.: Underwater acoustic networks. IEEE J. Oceanic Engineering 25(1), 72–83 (2000)
5. Proakis, J., Sozer, E., Rice, J., Stojanovic, M.: Shallow water acoustic networks. IEEE Commun. Mag., pp. 114–119 (November 2001)
6. Choi, H.-T., et al.: Development of Underwater Robot & Underwater Acoustic Signal Recognition System. In: Proceedings of CICS (October 2010)
7. Choi, H.-T., et al.: Design of a Control and Sensor System of Underwater Robot for Implementing Robot Intelligence. Proceedings of the IEEK 33(1) (June 2010)
8. Shin, S.Y., Namgung, J.I., Park, S.H.: SBMAC: Smart Blocking MAC mechanism for Variable UW-ASN (Underwater Acoustic Sensor Network) Environment. Sensors 10(1) (January 2010)
9. Kim, Y.-P., Namgung, J.-I., Yun, N.-Y., Cho, H.-J., Park, S.-H.: Design and Implementation of the Test-bed for Underwater Acoustic Sensor Network based on ARM9 Processor. IEEE Computer Society (2010)
10. Rodoplu, V., Park, M.K.: An energy-efficient MAC protocol for underwater wireless acoustic networks. In: Proceedings of the IEEE OCEANS 2005 Conference (September 2005)
11. Molins, M., Stojanovic, M.: Slotted FAMA: a MAC protocol for underwater acoustic network. In: Proceedings of the IEEE OCEANS 2006 Asia Conference (May 2006)
12. Fullmer, C.L., Garcia-Luna-Aceves, J.J.: Floor acquisition multiple access (FAMA) for packetradio networks, June 07 (1995)
13. WHOI Acoustic Communications: Micro-Modem, `http://acomms.whoi.edu/umodem`

Application of Pipelining Technique in Concatenated Tomlinson Harashima Precoder for Downlink MIMO Systems

Thanakorn Bamrungkitjaroen[1], Prapun Suksompong[1], Chalie Charoenlarpnopparut[1], Keattisak Sripimanwat[2], and Kazuhiko Fukawa[3]

[1] School of Information, Computer, and Communication Technology (ICT) Sirindhorn International Institute of Technology,
Thammasat University, Thailand
tanakorn95@gmail.com, {prapun,chalie}@siit.tu.ac.th
[2] National Electronics and Computer Technology Center
keattisak.sripimanwat@nectec.or.th
[3] Tokyo Institute of Technology
fukawa@radio.ss.titech.ac.jp

Abstract. We investigate the use of pipelining Tomlinson Harashima precoding (PIPTHP) for downlink multiple-input multiple-output (MIMO) systems. PIPTHP were previously analyzed in SISO systems but to our knowledge, implementation and analysis of PIPTHP in MIMO systems do not exist in the literature. Earlier studies have shown that PIPTHP technique can improve the BER performance and speed of the SISO systems. Here, the technique is extended to and analyzed in implementation for MIMO system. Furthermore, we develop and explore the idea of concatenated PIPTHP (C-PIPTHP) to improve the BER performance even more. The improved speed and bit error rate (BER) performance of C-PIPTHP are described in our simulation result. Our fast precoder can be implemented in 802.11 WLAN systems where MIMO has become the standard. We are also investigating an extension into multi-user (MU-) MIMO which is the technique of choice for the next generation of 802.11.

Keywords: Multiple-Input Multiple-Output (MIMO), High speed, Pipelining, Multiuser MIMO, and Tomlinson Harashima Precoding.

1 Introduction

Multiple-input multiple-output (MIMO) techniques were proposed for improving performance of wireless communication systems in terms of bit error rate (BER), throughput and range. Recently, MIMO schemes have become the most attractive choice for modern wireless systems [1]. In general, MIMO systems use several antennas at the base station and several antennas at each receiver. This paper focuses on downlink MIMO communications.

Tomlinson Harashima precoding (THP or TH precoding) is related to dirty paper coding (DPC) in terms of desired throughput and THP is the suboptimal technique, which can achieve the near sum rate of channels with much simpler computational

J.-H. Kim et al. (Eds.): AsiaSim2011, PICT 4, pp. 180–189, 2012.

complexity as compared to DPC approach [1,10]. THP, which is a non-linear precoding technique, was initially used for combating intersymbol interferences (ISI) for single-input single-output (SISO) channels [3,4] and it was later proposed for pre equalizing the interferences between users [10]. This scheme was further extended to be used in MIMO system (MIMO THP) in which the interferences can be suppressed with the help of QR decomposition of the channel matrix [1, 5].

The processing speed for the original THP is slow because the TH precoder uses a modulo device and a feedback filter to implement a pre-equalizer. Consequently, it is highly challenging to implement the TH precoders which can operate in high-speed application. In such scenario, pipelining Tomlinson Harashima precoding (PIPTHP) has been proposed. There were two pipelining schemes (hereafter, referred to as PIPTHP1 and PIPTHP2, respectively) that were originally developed this schemes to have low complexity to achieve faster speed and mitigate the hardware overhead [6–8]. However, these two schemes have been studied only in SISO systems and there has been no known implementation in MIMO systems. In this paper, one of the goals is to apply pipelining technique to MIMO approach.

We also explore the idea of concatenation of two THPs. We first studied to develop concatenated THP (C-THP), which is a modified version of THP, where two pre-equalizers are concatenated to improve the overall performance. Of course, with two pre-equalizers, the speed is poor. Therefore, we propose C-PIPTHP, which concatenates two pre-equalizers from low complexity PIPTHP and MIMO THP at transmitter while the BER performance is still similar. THP schemes for downlink MIMO and our proposed system is shown in Table 1. In summary, C-THP gives good BER performance; on the contrary it is very slow processing, because IIR filter and feedback filter are used as two pre-equalizers. Our design attempts to improve communication performance and operates in MIMO scheme by applying C-PIPTHP and compare with the MIMO THP, and C-THP.

2 Downlink MIMO Tomlinson Harashima Precoding

We consider downlink MIMO where Nt transmitting antennas are located at the base station and Nr receiving antennas where Nt = Nr = K. The channel fading is assumed to be the Rayleigh: the channel coefficients between the transmitting antennas and receiving antennas are assumed to be zero-mean circularly symmetric complex Gaussian random variables with unit variance. These coefficients can be represented in terms of an Nt×Nr matrix H. Under such H, the received signal is given by:

$$y = Hx+n$$

$$\begin{bmatrix} y_1 \\ y_2 \\ \vdots \\ y_k \end{bmatrix} = \begin{bmatrix} H_{11} & \cdots & H_{1k} \\ H_{21} & \cdots & H_{2k} \\ \vdots & \ddots & \vdots \\ H_{k1} & \cdots & H_{kk} \end{bmatrix} \begin{bmatrix} x_1 \\ x_2 \\ \vdots \\ x_k \end{bmatrix} + \begin{bmatrix} n_1 \\ n_2 \\ \vdots \\ n_k \end{bmatrix} \tag{1}$$

where x and n are input signal and noise respectively.

Fig. 1 shows the block diagram for downlink MIMO THP system .The input is the complex-valued vector a = $[a_1 \ldots a_{Nr}]^T$ of QAM symbols from the signal constellation defined by $\{\tilde{a}+jb|\tilde{a},b \in \{\pm 1, \pm 3, \ldots \pm(\sqrt{M}-1)\}\}$ [5], where M is the modulation order. The feedback filter in the THP is capable of pre-eliminating the inter-stream interference at transmitting side, while a feedforward filter can be applied to decrease the vector detection at receiving side to a scalar detection [10,11].

Table 1. Comparison between THP based systems, which is discussed in this paper. Orders of THP schemes in table below are arranged by capability to operate in MIMO technqiue.

Non-linear Precoding in downlink MIMO				
Schemes		**Pipeline**	**MIMO**	**Concatenated THP**
(1)	Traditional THP[3,4]	×	×	×
(2)	PIPTHP [6,7,8]	√	×	×
(3)	MIMO THP [5,9]	×	√	×
(4)	MIMO C-THP	×	√	√
(5)	Proposed MIMO C-PIPTHP	√	√	√

Due to the triangular nature of a feedback filter, the transmitted signal can be determined by equation below:

$$X_K = a_K - \sum_{m=1}^{k-1} b_{km} X_m \qquad m=1,..,K-1 \qquad (2)$$

If a modulo device is not applied at transmitting side, the magnitude can increase by recursive operation at a feedback filter. Therefore, modulo devices are capable of confining the amplitude of the signal to be within [-$\sqrt{M}, \sqrt{M}$). The modulo operation is generated precoding signal at the transmitter, so its output can be given by:

$$X_K = a_K + d_K - \sum_{m=1}^{k-1} b_{km} X_m \qquad m=1,..,K-1 \qquad (3)$$

where d is the precoding signal and it can be determined as an equation below:

$$d_k \in \{2\sqrt{M}\,.\,d_R + id_I |\, d_R , d_I \in Z\} \qquad (4)$$

The transmitted signal from the modulo device is separated systematically in real and imaginary part by integer multiples of $2\sqrt{M}$ represent the same data [5,11], and their input output relationship is shown in Fig.2. The output signal of modulo function can be used to eliminate the precoding signal at receiving side, so the amplitude without precoding signal can be recovered.

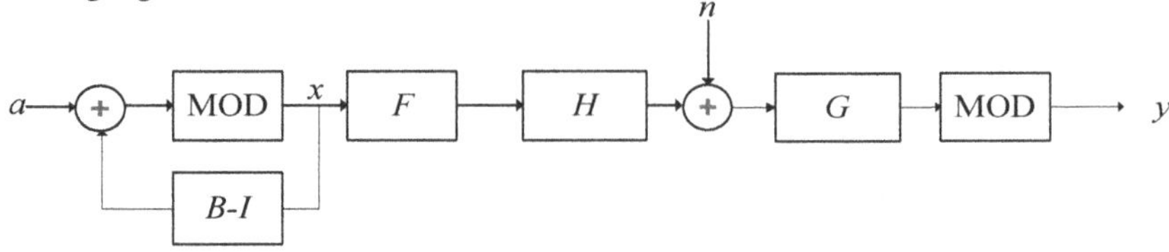

Fig. 1. Downlink Multiuser MIMO TH precoding describe in [5], [9]. (3)

Performing QR decomposition to H^H generates the channel matrix (H)

$$H^H = FR \tag{5.1}$$

$$H = F^H S \tag{5.2}$$

where *F*, *R*, and *S* are the unitary feedforward matrix , a upper triangular matrix, and a lower triangular matrix respectively. *S* is represented by:

$$S_k = \begin{bmatrix} s_{11} & \cdots & 0 \\ s_{21} & \cdots & 0 \\ \vdots & \ddots & \vdots \\ s_{k1} & \cdots & s_{kk} \end{bmatrix} \tag{6}$$

The feedback filter (*B*) =*GS* where *G* is a scaling matrix, which is exploited for unit-gain transmission and the matrix (*G*), are given by:

$$G = \operatorname{diag}\left(\begin{bmatrix} s_{11} & \cdots & 0 \\ s_{21} & \cdots & 0 \\ \vdots & \ddots & \vdots \\ s_{k1} & \cdots & s_{kk} \end{bmatrix}\right)^{-1} \tag{7}$$

The matrix *B* is the unit lower triangular matrix, which is lower left triangular matrix with one on the main diagonal [5, 9]. The received signal can be defined as:

$$y = \text{mod}\,(GFH(x+d) + Gn) \tag{8}$$

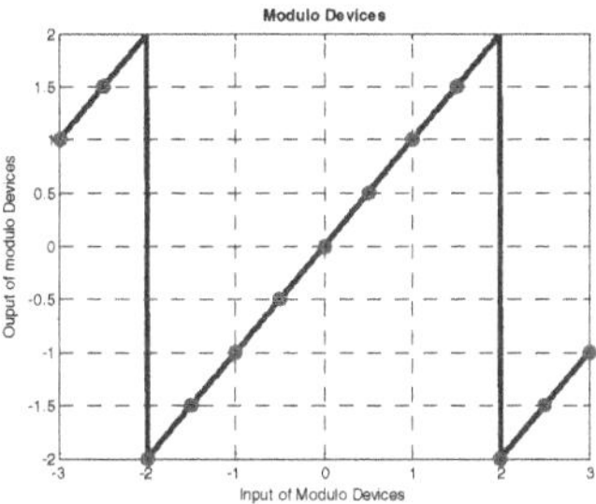

Fig. 2. Output of modulo devices

3 Low Complexity Tomlinson Harashima Precoding

This section explains the process of a low complexity pipelining THP. The theory of low complexity was improved from PIPTHP 1 and PIPTHP 2 and their structures are shown in Fig. 3, where v, t are compensate signal and transmitted signal respectively. The transfer function in Z-transform for finite impulse response filter (FIR) *N*(*z*), *Ne*(*z*) and IIR Filter (*D*(*z*)) [6,7] are given as following:

$$N(z) = \left(1 + \sum_{i=1}^{LN} N_i Z^{-i}\right) \tag{9.1}$$

$$Ne(z) = \left(\sum_{i=1}^{LN} N_i Z^{-i+1}\right) \tag{9.2}$$

$$D(z) = (1 + \sum_{i=1}^{LH} d_i Z^{-iK}) \tag{9.3}$$

where *LN*,*LH*,*K* denotes as the pipelining parameter, channel memory length and pipelining level respectively. Moreover, N_i and d_i are filter coefficients. In MIMO, the channel length is an Nt×Nr matrix, and pipeline parameter is the nth order of pipelining filter *N*(*z*).

In order to design the system that can eliminate the drawback of these two pipelining technique and also can operate in the high speed communication design, the low-complexity PIPTHP, has been designed. This high speed design has been developed based on the method for computation of compensate signal in PIPTHP2 and also reduce the hardware overhead. The speed of PIPTHP 2 is highly slow because it is limited by the a non-linear non-pipelined loop (*Ne*(*z*)). The architecture of low complexity PIPTHP is shown in Fig. 4 and Z- transform equations of FIR filters $N_1(z)$ and $N_2(z)$ are given as:

$$N_1(z) = (\sum_{i=1}^{L1} N_i Z^{-i-1}) \tag{10.1}$$

$$N_2(z) = (\sum_{i=L1+1}^{LN} N_i Z^{-i-L1-1}) \tag{10.2}$$

where L_1 is the tap FIR filter $N_1(z)$.

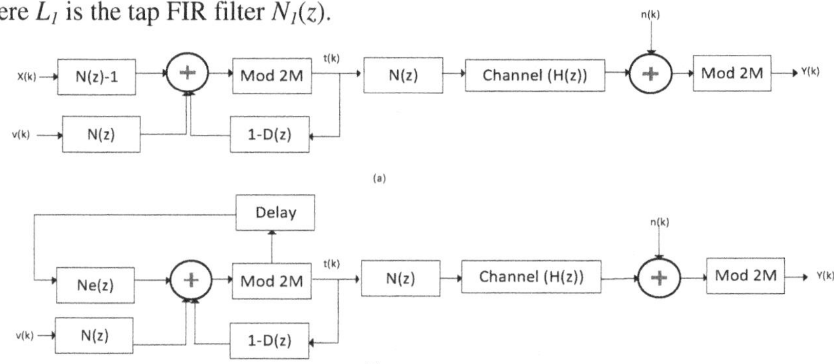

Fig. 3. Block diagram for PIPTHP proposed in [6], [7] (a). PIPTHP 1. (b) PIPTHP 2 (2).

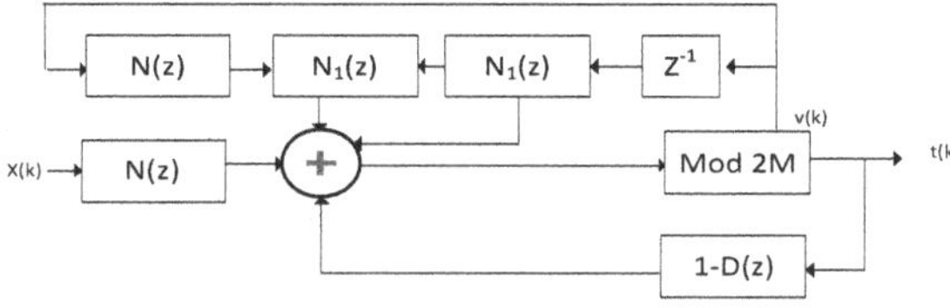

Fig. 4. Block diagram for low complexity PIPTHP proposed in [6, 7] (2)

Performances of THP system for both traditional design and low complexity design are similar, which are determined in [6,7] and they are shown below:

$$\text{PIPTHP, Traditional THP: } Y(z) = X(z) + V(z) + n(z) \tag{11}$$

Y(*z*), *X*(*z*), *V*(*z*) and *n*(*z*) are denoted as received signal, input, compensate signal and noise respectively.

4 Implementation of Low Complexity PIPTHP and MIMO THP

In this section, we explain the implementation of concatenated PIPTHP for downlink MIMO at transmitting and receiving side. A PIPTHP, a feedback filters and modulo device are concatenated to improve the speed of our proposed system and further pre-eliminate interstreams interferences, which are described in this section. This section is also presented technical structures of our proposed MIMO system as well.

In Fig. 5, we first design the communication system for pipelining THP at transmitter. The input signal is generated as a random signal. The modulation scheme of our proposed system is QPSK and input signal is mapped to$\pm\frac{1}{\sqrt{2}} \pm \frac{1}{\sqrt{2}}j$. A mapper is applied to change the dimension of symbol vector in order to operate from one stream to Nt streams. A demultiplexer is applied to separate many streams to be one stream due to the one-dimensional filter so that the PIPTHP system can be performed. In Fig. 6, this structure shows the diagram for low-complexity PIPTHP block. The Nt streams are operated in PIPTHP block, and is capable of manipulating one streams per one process. The transfer functions of filters in discrete time are shown below:

$$N(z): y(k) = t(k) + n_1 t(k-1) + n_2 t(k-2) + \cdots + n_{LN} t(k-LN) \quad (12.1)$$

$$D(z): y(k) = -d_1 t(k-K) - d_2 t(k-2K) - \cdots - d_{LH} t(k-LHK) \quad (12.2)$$

$$N_1(z): y(k) = n_1 t(k-2) + n_2 t(k-3) + \cdots + n_{L1} t(k-L1-1) \quad (12.3)$$

$$N_2(z): y(k) = n_{L1+1} t(k-2L1-2) + n_2 t(k-2L1-3) + \ldots + n_{L1} t(k-LN-L1-1) \quad (12.4)$$

where *t(.)* and *y(k)* are denoted as the input and output of the filters and these equations are transformed from Z-domain transfer functions, which are proposed in [6, 7]. After operating the PIPTHP process, a multiplexer is added in our proposed system to combine many streams to one stream. The symbols are transmitted to modulo device and a feedback filter (*B-I*) to decrease the transmitting symbol to [-$\sqrt{M}, \sqrt{M}$) and pre-subtract the interferences from previous users simultaneously [5]. The signals from the feedback filter are given as:

$$X_K = a_K + p_K - \sum_{m=1}^{k-1} b_{km} X_m \qquad m=1,..,K-1 \quad (13)$$

where a_K, p_K, b_{km} are the pipelining symbol, precoding symbol and coefficients of the feedback filter. This is the operation of the transmitting scheme for our proposed C-PIPTHP. The speed of our proposed system is increased because of PIPTHP and also gets rid of interstreams interferences by operating the feedback filter.

Fig. 7 shows the proposed receiver. The output of channel is transmitted to the scaling matrix (*G*) after combining with noise and a function of scaling matrix is to adjust the transmitted power. After that, a modulo device performs to recover the amplitude for the received symbol. The output of modulo (*y*) is shown below in (8), and then it is rechanged the dimension by exploiting de-mapper.

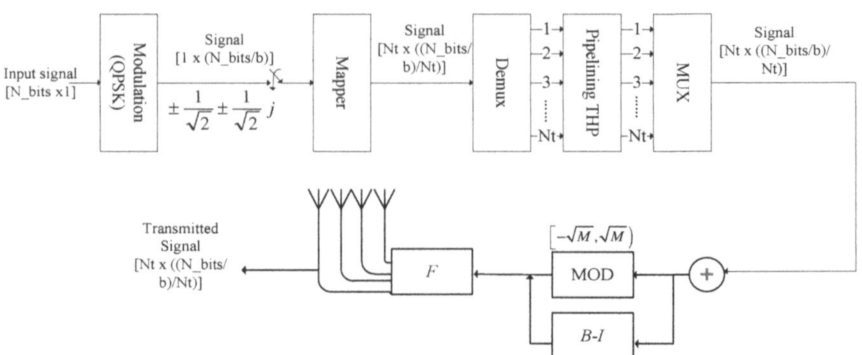

Fig. 5. Proposed system for downlink MIMO at transmitter (5)

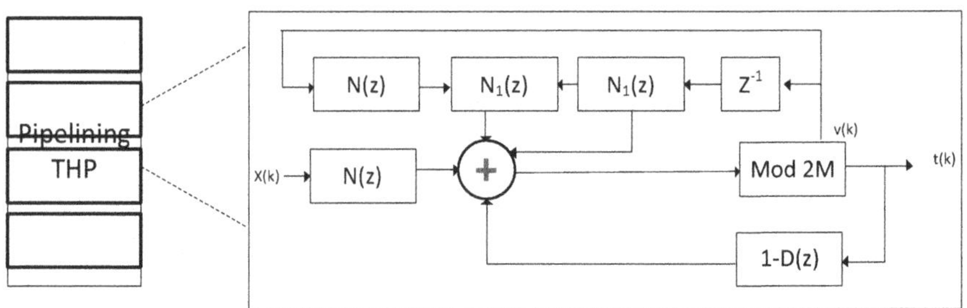

Fig. 6. Inside PIPTHP Block diagram of low complexity PIPTHP proposed in [6, 7] (5)

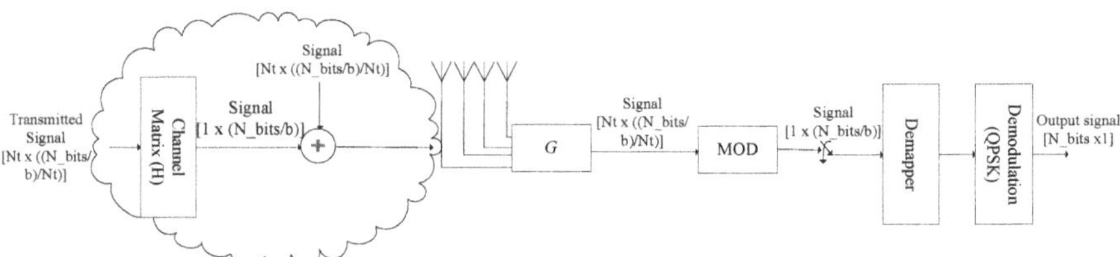

Fig. 7. Proposed system for downlink MIMO at channel and receiver (5)

$$y = \text{mod}\,(GFH\,(X_K + d_K) + Gn_K) \tag{14}$$

The last step for our proposed scheme is to transform symbols to binary signal by demodulation in order to measure our proposed performance.

There are several main advantages for our proposed C-PIPTHP system over THP based schemes. The first one is that our proposed system comprises more pre-equalizer at the transmitter, so THP schemes, which have only a pre-equalizer part, are given the worse error rate performance than our proposed system due to more channel distortion and interferences. The second one is the speed of C-PIPTHP to be operated much faster than C-THP and we develop C-THP before C-PIPTHP and get the good BER performance. Nonetheless, it takes a lot of time to process in order to reduce the channel distortion because of the non-pipelining pre-equalizer and extra pre-equalizer (i.e. a modulo device and a feedback filter) in the C-THP; therefore, C-PIPTHP is developed.

Originally, the PIPTHP, which was developed by [6, 7], is implemented in SISO only; consequently, the third advantage is that our proposed scheme to be capable of operating in MIMO system, which is a main benefit for improving for the precoding in MIMO scheme. The last advantage for our proposed MIMO design is to procure better BER performance than the SISO PIPTHP technique.

Nevertheless, in our proposed architecture, there are two main disadvantages. First one is to operate one dimensional filter of pipelining THP, which means the PIPTHP is capable of performing one streams per one time only. The speed can be improved faster if the filters are multi-dimensional filter, which can perform all Nt streams simultaneously, The second disadvantage is the cost of the components. In spite of adding more FIR at the transmitter in the communication schemes and the hardware overhead not being arduous, the cost of the components will be the most expensive of other THP schemes.

5 Simulations Results

We analyze the proposed system architecture by the simulation on MATLAB. The proposed scheme is simulated for downlink MIMO by applying C-PIPTHP for high speed application and improves the BER performance.

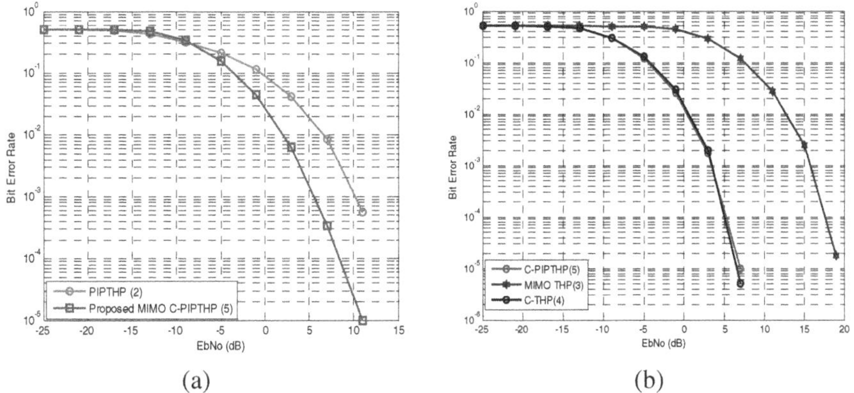

Fig. 8. (a) BER performance of PIPTHP for SISO and MIMO Schemes, (b) BER performance for Downlink MIMO TH precoding Based Schemes

Fig. 8(a) is shown error rate comparison between SISO and MIMO communication schemes based on THP. Multi-antennas scheme gives the better performance than single user system, because the multi signals can reduce the probability of error for transmitting the signal by increasing accurate data from base stations to mobile stations and also provides the high data rates. For our scheme, the bit rate of MIMO method is four times greater than that of SISO method by the use of four streams.

The base station and the receiver, that numbers of transmitting and receiving antennas are equal, are assumed to know the channel state information perfectly. For the C-PIPTH precoding consists of 4 single antenna receiving antennas and 4 transmitting antennas at base station. Fig. 8(b) illustrates the comparison of BER performance between proposed system and other TH precoding schemes. The BER performance of proposed system is significantly better than MIMO THP and highly similar as C-THP.

The better performance between C-PIPTHP, C-THP and MIMO THP is to add more equalizers at the transmitter.The proposed MIMO system, which is composed of only one pre-equalizer of PIPTHP at transmitter, acquire worse error rate performance than two pre-equalizer MIMO THP system, because there is interstreams interference occurs while signal from a modulo device is dispatched to transmitted antennas. Fig. 9 elucidates the effect of the number of equalizers in terms of BER performance.BER performance of our proposed THP system outperforms one pre-equalizer MIMO PIPTHP.

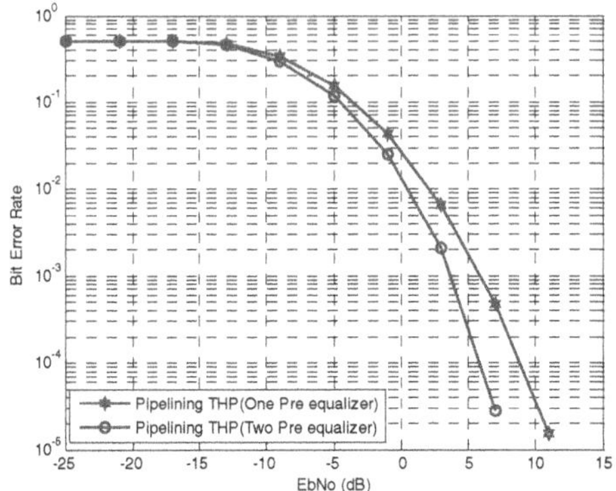

Fig. 9. BER Performance of downlink MIMO THP based schemes for one pre-equalizer and two pre-equalizer schemes

Table 2. Processing time for downlink precoding schemes [4×4]

Scheme (5)	Scheme (4)	Scheme (3)
2.53	506.19	0.25

Table 2 illustrates the processing time of precoding scheme for downlink MIMO. The speed of the proposed C- PIPTHP is much faster than the C-THP and it obtains the same processing time as the others precoding techniques. For proposed system, the processing time is only 2.53 seconds. However, the operating time for concatenated THP, which is the precoding system before improving C-PIPTHP, requires 506.19 seconds, because the two pre-equalizers take a lot of time to be operated due to not pipelining design. The time of C-PIPTHP,C-THP, and MIMO THP have done the same across all cases in Fig. 9.

6 Conclusion

MIMO is implemented to improve the spectrum efficiency in the wireless communication systems. Before we develop concatenated pipelining Tomlinson Harashima precoding(C-PIPTHP), we first improve the concatenated THP (C-THP) for improving the BER performance, but the speed of C-THP is very slow. After that, we implement the pipelining technique in C-THP; therefore C-PIPTHP has been design to increase the speed of THP architecture. In this paper, we have implemented C-PIPTHP by developing based on the low complexity PIPTHP and MIMO THP in our proposed system. Our proposed method is mainly the combination of PIPTHP and an extra pre-equalizer, which consists of a modulo and a feedback filter. Our method can achieve the high speed performance and also is capable of pre-eliminating interferences from others streams. Moreover, the performance of our proposed scheme is much better than MIMO THP due to the two pre-equalizers at the transmitter. In the simulations, we have shown the BER performance and processing time. Our proposed system is capable of obtain not only better system performance, but also it can operate in high speed design. We are currently investigating the alternative methods that the PIPTHP can be implemented such as Generalized Triangular Decomposition (GTD) for downlink MU-MIMO system, which is applied to separate the user's stream and also we will develop the PIPTHP to be able operate for parallel processing in MATLAB simulation.

Acknowledgment. This work was supported by the NTC Telecommunication Research lab, the National Research University Project of Thailand, Office of the Higher Education Commission, TRIDI Advanced Research Grant, and TAIST-Tokyo Tech.

References

1. Khalid, F., Speidel, J.: Advances in MIMO technique for Mobile Communications – A Survey. Int. J. Communications, Network and System Sciences, 213–252 (March 2010)
2. Spencer, Q.H., Peel, C.B., Swindlehurst, A.L., Haardt, M.: An Introduction to the Multi-User MIMO Downlink. IEEE Communications Magazine, 60-67 (October 2004)
3. Tomlinson, M.: New Automatic Equaliser Employing Modulo Arithmetic. IEEE Lectr. Lett. 7(5,6), 138–139 (1971)
4. Harashima, H., Miyakawa, H.: Matched-Transmission Technique for Channels with Intersymbol Interference. IEEE Trans. Commun. COM-20, 774–780 (1972)
5. Windpassinger, C., Fischer, R.F.H., Vencel, T., Huber, J.B.: Precoding in multiantenna and multiuser communications. IEEE Trans. Wireless Commun. 3(4), 1305–1316 (2004)
6. Gu, Y., Parhi, K.K.: Pipelining Tomlinson-Harashima precoders. In: Proc. IEEE Int. Symp. Circuits Syst. (ISCAS), Kobe, Japan, pp. 408–411 (May 2005)
7. Gu, Y., Parhi, K.K.: High-speed architecture design of Tomlinson-Harashima precoders. IEEE Trans. Circuits Syst. I, Fundam. Theory Appl. 54(10), 1929–1937 (2007)
8. Yen-Liang, Wu, A.-y.: Generalized Pipelined Tomlisson Harashima Precoder Desugn Methodology With Build-In Arbitrary Speed-Up Factors. IEEE Transactions on Signal Processing 58(4) (April 2010)
9. Fischer, R.F.H., Windpassinger, C., Lampe, A., Huber, J.B.: MIMO precoding for decentralized receivers. In: Proc. Int Symp. Inform. Theory, Lausanne, Switzerland, p. 496 (July 2002)
10. Khaleghi Bizaki, H., Falahati, A.: Tomlinson-Harashima precoding with imperfect Channel state information. IET Commun. (1), 151–158 (2008)
11. Liu, H., Hao, L., Fan, P.: A low-complexity scheduling scheme for multiuser MIMO system based on Tomlinson-Harashima Precoding. In: Information Theory and Information Security (ICITIS), pp. 981–984 (2010)

A Reliable Error Detection Mechanism in Underwater Acoustic Sensor Networks

Imtiaz Ahmed Khan, Nam-Yeol Yun, Sardorbek Muminov, and Soo-Hyun Park

Ubiquitous System Lab, Graduate School of Business IT, Kookmin University
{fish786,anuice,smuminov,shpark21}@kookmin.ac.kr

Abstract. Underwater Acoustic Sensor Networks (UWASNs) are not secure due to various characteristics of underwater environments. The error rate is very high in underwater environment due to factors such as long propagation delay, fading, multi-path and noises. Moreover, malicious attacks to the network add one more threat to reliable data transport problems. With respect to terrestrial sensor network, UWASNs are very different in nature. Reliable and faster data communication is our main issue. As the signal has to travel inside a noisy wireless acoustic channel so the processing of the signal is also take to measure. Moreover, we know the error rate is always high in underwater, so detecting the error and correcting them is the main key to secure the communication among nodes. The energy efficiency as well as the performance of any mechanism depends on collusion free data transmission and reception among sensors. Data link layer plays a crucial part in the Media Access Control (MAC) operation by detecting errors. For secure communication it is necessary to detect the errors during the transmission and reception. Especially in data link layer, the error detection and correction used some techniques like Cyclic Redundancy Check (CRC), checksum, parity bit, Forward Error Correction (FEC) and different error-correcting codes like Humming code, Reed Solomon code, binary convolution code and others tailored made program. In UWASNs data throughput depends on the packet size and the MAC protocols, where different protocols perform differently. A valid packet size has a great effect on the network. The ideal packet sizes depend on the specific application and the pattern of message it generates. We have to use small size packets which do help faster communication and higher data throughput in UWASNs. In this paper, depends on our practical experience, instead of using any kind of FEC techniques, we used CRC-8 because the process of data transmission usually contains the risk of unwanted modification as the channel is open. Signals arrive to the physical layer at different power and delay due to acoustic effects. So packets need to recheck for reliable data communication in UWASNs.

1 Introduction

Three decades ago building a scalable UWASNs was a great challenge. But last decade it was answered by inter-disciplinary efforts of acoustic communications, signal processing and ground acoustic network protocol design. Not only that, most of the research projects organized by the different countries are also successful. According to the experienced research groups UWASNs are being used to monitor the underwater world.

J.-H. Kim et al. (Eds.): AsiaSim2011, PICT 4, pp. 190–199, 2012.

The UWASNs can help us to monitor and research the underwater micro organisms for biological matters, protection against environmental disasters, like tsunamis, that could reshapes the portion of earth causing massive disaster and treats to valuable human lives. The study can help us find out the underwater titanic plate and monitor sea beds; protections for technical surveillances by neighboring countries, UWASNs are the easiest and most effective solution for all these applications.

In this paper, we first try to find out the characteristics of UWASNs. Then we discuss about how many ways errors can occur in underwater data communication, and how can we detect errors. After that we talk about the CRC mechanism. Then we describe related works about CRC in UWASNs technology. After that we proposed our 1 byte CRC for UWASNs. Then we shall discuss about simulation and our experiment. Finally we conclude our paper.

2 How to Detect Communication Errors in UWASNs

During the data communication between two underwater wireless sensor nodes there are possibilities to occur a data error due to the sending device accidental fault or due to noisy burst channel. UWASNs used synchronous data transmission which is also called bit–oriented protocol. The data link layer provides single hope data transmission capability only. Sender node and the distention node must be accessible directly. Due to the secure data communication reliability among sensors some degree of error detection capability like checksum, CRC and FEC is used for shared medium.

Moreover, due to channel noise, many types of random errors may occur. Sometimes it was also found that error or packet losses caused by traffic jam in networks, beside these packets altered or created intentionally by malicious nodes and create errors in the header of packets [1]. So, checking packets in the receiver side is logical for reliable data communication, for detecting error, among others like parity bit, checksum, Longitudinal Redundancy Check (LRC), Vertical Redundancy Check (VRC), CRC and FEC. By the virtue of being codes of fixed length, the encoding and decoding becomes much simpler when compared to convolution turbo codes that encode streaming data. Now we discuss about three different error detecting way they are linear systematic block codes.

2.1 Checksum Mechanism

Checksum is closely related to group of parity bits. A group of parity bits is one example of a checksum. There are other stronger checksums based on running sum of the data bits of the message. It is usually placed at the end of the message. This is probably one of the oldest methods of ensuring that data is correct and has not been corrupted during transmission or encryption.

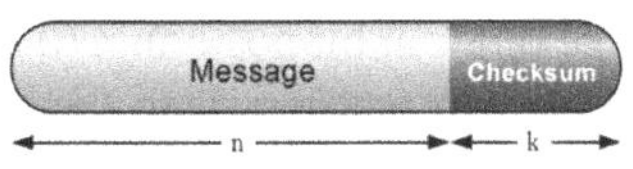

k = Number of binary digits in the message before encoding
n = Number of binary digits in the encoded message
(n - k) = number of check bits

Fig. 1. Working principal of checksum mechanism

Checksums also provide a form of authentication because an invalid checksum suggests that the data has been compromised in some fashion. A checksum is determined in one of two ways. Let's say the checksum of a packet is one byte long. A byte is made up of 8 bits and each bit can be one or two states that leading to a total of 256 (2^8) possible combinations. Since the first combination is zero so 1 byte can have a maximum value of 255. The logical point is if the sum of the other bytes in the packet is 255 or less then checksum contains that exact value. If the sum of the other bytes is more than 255, the checksum is the remainder of the total value after it has been divided by 256.

Byte-1	Byte-2	Byte-3	Byte-4	Byte-5	Byte-6	Byte-7	Byte-8	Total	Checksum
201	244	150	154	160	75	85	125	1,194	170

Fig. 2. Checksum calculation for receiver

The checksum variables is usually appended to the end of the data field and called the checksum field. It can be one or more bytes. The ability to detect errors in a multi byte sequence is affected by the number of bytes in the sequence and size of the checksum field. Four different types of checksums are: a) *Single-Precision Checksum*: each byte of the data is summed into a single byte. b) *Double-Precision Checksum*: each location of the data of n-bits is summed into a location of 2n-bits. c) *Honeywell Checksum*: it is an alternative to the double precision checksum. d) *Residue Checksum*: it is a variation of the single precision checksum.

The checksum technique just described is remarkably weak. For example, if we were to transpose two of the characters of the transmission, the result would be the same, so although the wrong packet is received, a correct checksum is delivered. Certain kinds of noise injection on the line can also introduce undetectable errors because the noises that mangle one byte cancelled by the noise that mangles another byte.

2.2 Cyclic Redundancy Check (CRC)

CRC is the similar concept to checksums but used "Modulo -2 binary division" polynomial to determine the value of the CRC. The CRC general process is a short simple message block and a small CRC generator. The CRC performs its arithmetic calculation on binary data which is presented by a polynomial. Each bit position of a message block is represented as a coefficient of polynomial [2]. The general message of a polynomial is like:

$$M(x) = b_nX^n + b_{n-1}X^{n-1} + b_{n-2}X^{n-2} + \ldots\ldots\ldots\ldots b_2X^2 + b_1X + b_o1$$

Where, b_n = the value of message block at bit position n (0 or 1)

The degrees of the polynomial will one less than the total number of bits in the message block. The message block "1101 10011" has 9 bits message, so k will be 9. The polynomial representation for this message block will have a degree of 8. With respect to LSB to the right, we can represent the message like:

$$M(x) = 1\cdot X^8 + 1\cdot X^7 + 0\cdot X^6 + 1\cdot X^5 + 1\cdot X^4 + 0\cdot X^3 + 0\cdot X^2 + 1\cdot X^1 + 1\cdot 1$$

After simplification, the polynomial for the message block is:

$$M(x) = X^8 + X^7 + X^5 + X^4 + X + 1$$

For generator polynomial, we will use:

$$G(x) = X^8 + X^7 + X^4 + X^3 + X^2 + X + 1, \text{ binary value} = 11001111$$

Since the degree of the generator polynomial is 8, the number of bits in the generated CRC will be 8. The total number of bits transmitted, therefore, will be 17. The calculation for sender side CRC is multiply the message polynomial by $X^{n-k,}$

The value will be $X^{17-9} = X^8$.

$$X^{n-k} \cdot [M(x)] = X^8 \cdot [X^8 + X^7 + X^4 + X^3 + X^2 + X + 1]$$

$$= X^{16} + X^{15} + X^{12} + X^{11} + X^{10} + X^9 + X^8 = 1\ 1001\ 1111\ 0000\ 0000$$

Now, divide $X^{n-k} \cdot [M(x)]$ by the generator polynomial G(x) and discard the quotient. The remainder will be the CRC or Block Check Character (BCC) which will be called B(x).

Append the CRC value, B(x) in the message block, M(x), to create a transmitted message T(x).

$$T(x) = X^{n-k} \cdot [M(x)] + B(x)$$

1 1001 1111 0000 0000	: X^{n-k} [M(x)]
+ 1001 1110	: B(x)
1 1001 1111 1001 1110	: T(x)

Receiver will check the transmitted message block T(x) 1 1001 1111 1001 1110 with same polynomial G(x) 1100 1111 and discard the quotient. A reminder zero indicates that the block was received without error.

If the CRC value not matches then the message treated as errors. So retransmission action starts by the protocol. The idea behind CRC calculation is to look at the data as one large binary number. This number is divided by a certain value and the remainder of the calculation is called the CRC value. Receiver nodes do the same calculation and check the CRC value. CRCs are popular because they are simple to implement in binary hardware, also easy to analyze mathematically, and these are particularly good at detecting common errors creates by noise in transmission channels. Cyclic codes have favorable properties in that they are well suited for detecting burst errors. So CRC commonly used in digital networks. RF based CRC is usually 4 to 64 bit length; the good thing is that CRC works very accurately. If a single bit is incorrect the CRC value will not match up. CRC very efficient to detect errors and CRC codes are shortened cyclic codes that are widely used for error detection on digital communications links.

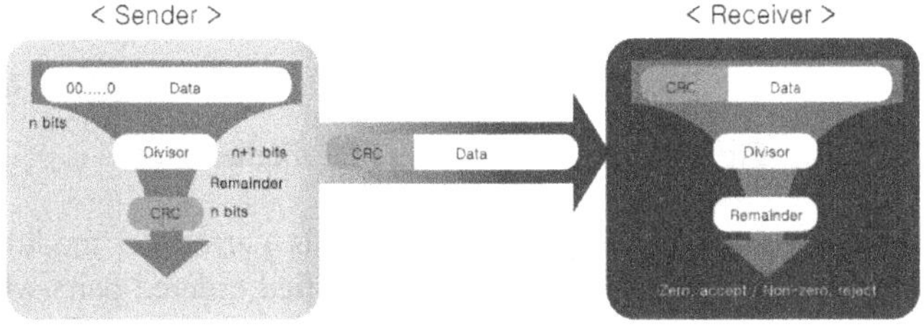

Fig. 3. Working principal of CRC mechanism

They owe their popularity to their extremely simple and fast encoder and decoder implementations and the very good error detection performance. That is why a number of CRC codes have been adopted as standards: ATM, IEC TC 57, IEEE WG 77.1, CCITT X.25, ANSI, IBM-SDLC, and IEEE-802. It is interesting to test the error detection performance of these standards, and compare them to other CRC codes, and to find the codes of the best performance [3]. CRC-CCITT -16, CRC-32, CRC-64 and the ISO standard CRC is CRC-16-IBM, CRC-16-DECT and CRC-32 IEEE 802.3. Most of the cases 32 bit CRC is used for internet applications. But underwater networks there is no more standard CRC as it is a developing technology and very different form RF based networks.

2.3 Forward Error Checking (FEC)

One of the best uses for the redundancy in a multi-level code is to implement error checking and correcting at the same time. So the check bits are transmitted along with the message. The ability to detect error and perform correction error after receiving data without any retransmission is called FEC. That is one of the best methods when the communication is simplex and broadcast to many sensors. Two popular systems to check and correct a message using block code, combination of parity check coding and hamming coding. It is so called "memory less" as the output of the block code only depends on the present state of its inputs. To measure the performance of the different type of FEC codes one parameter is used. It is called hamming distance. The common 4 error correcting codes are a) *Hamming codes* b) *Binary Convolution Codes* c) *Reed-Solomon Codes* d) *Low-Density Parity Check Codes.*

The main concept is a frame consists of m data (i.e., message) bits and c (i.e., check) bits in its output. All hamming codes are classified as linear block code is an (m + c) codeword. The extra bits are appended to the original data by a device which implements the generator matrix. The codeword is transmitted to the receiver and checked by another device which implements the parity check matrix. The output of the parity check device is a string of data bits with the same number of bits as the check bits. The output is called hamming syndrome and a single bit fault can be identified and corrected.

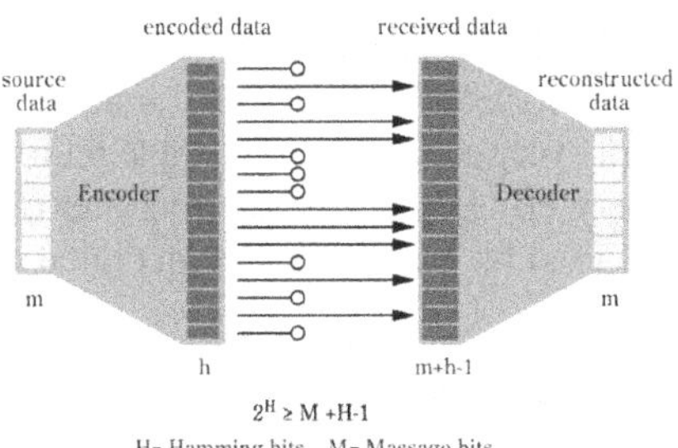

Fig. 4. Working principal of FEC mechanism

The calculation based on matrix multiplication is performed modulo arithmetic operation. Hamming code works with a set of identified ordered pairs which indicate there number of message bits and number of bits in the codeword. For a (n/k) Hamming code, k is the number of input bits and n is the number of output bits. The numbers of hamming bits are determined by the number of data bits. We can know from the equation, which determine the number of Hamming bits we required during coding.

$2n\text{-}k \geq n+1$, where, k is number of data bits, n is total number of bits transmitted is a message stream. Most of the FEC code has the minimum hamming distance 3 regardless of the number of check bits. The greater the code rate value, the grater the efficiency. Error bits, called hamming bits, are inserted into the message at random location as because these hamming bits themselves may be face errors and both the sender and the receiver should know the actual position of the Hamming bit inside the message stream.

From the equation, $2^H \geq M + H - 1$, where, the first step in the process is to determine how many Hamming bits (H) are to be inserted between the message (M) bits. Then their actual placement is selected. The number of bits in the message (M) are counted and used to solve the mentioned equation to determine the number of hamming (H) bits, which will use later in the simulation part.

3 Consideration Factors for Choosing Error Detection Mechanism for UWASNs

Checksum, CRC and FEC all three error detection tools are well suited in RF based wireless sensor networks and digital data communication system. UWASNs are totally different from RE based networks not only in view of physical effects like multipath, fading, noise and propagation delay beside this, we have very small size of hardware infrastructure like Cortex-M3 MCU, memory is limited and fixed amount of power is total resource of our underwater sensor nodes.

Most of the CRC mechanisms' error detection rate is 99.9978% also specially burst error detection capabilities but checksum mechanism error detection rate 95% not strongly detected the burst errors due to its construction. FEC mechanism is typically used when retransmission is not possible or unrealistic. Depending on the nature of the error-generating mechanism in the transmission medium, error can be random (isolated) or (clustered) and correspondingly influence after detection [3]. Different kind of FEC Codes like Reed Solomon Codes, Binary Block Codes, used in RF based sensor networks but depends to network overhead as 30% (depends on design) extra bits are injected in FEC based mechanism, so the data packet size increases which is not efficient due to transmission efficiency as our bandwidth is also limited. But it is good for multicast broadcasting protocol only. FEC based protocols message encoding and decoding time is longer than CRC and checksum measure time. As our energy is limited so FEC is not suitable for most of general sensor networks. So, based on underwater factors, we like to discuss here among the suitable error detection mechanism for our underwater protocol suite. FEC mechanism is good for the application where multimedia signal based communication, and signal Retransmission not possible and bit error rate is very high. Error detection capability both CRC-8 byte and CRC-16 is almost same.

3.1 Burst Signal Errors

During data communication among the sensors, instead of bit 1or 2 bit errors we find burst error in the signal. Due to the multipath and fading and unexpected noise badly affect some part of our signals. These types of burst signals are detected by the CRC mechanism very easily as CRC use modulo-2 binary arithmetic operation. And it was proved by the mathematicians earlier [4].

3.2 Data Packet Size

According to the design challenge of underwater sensor network very large bulk data transmission is not suitable in underwater sensor networks. In underwater sensor network most nodes are mobile. Therefore, there is no long-lasting fixed neighborhood, i.e., the communication time between any pair of sender and receiver is limited. Moreover, the bandwidth of communication channels is relatively low, and the propagation delay is extremely high. All these factors hamper the sender from sending very large bulk data at one time [5]. So, large size FEC or checksum cannot be good for data throughput. So, standard error detection capability 8 bit CRC might be the best error detection tool in underwater sensor networks.

3.3 Energy Efficiency

Energy is a great factor in underwater because sensor energy is equal to sensor lifetime. FEC consumed more energy by encoding message and decoding message during transmitting and receiving signals. It consume more power during signal processing and as the extra bit injected inside it, modem are staying more busy than CRC and checksum, CRC-8 and checksum consume less energy than FEC [6].

3.4 Network Throughput

Packet size is a great factor in underwater sensor network, it's directly related to network bandwidth and protocol design. Same packet size of data transmit in different protocol perform differently due to network protocol design. CRC or checksum based packet perform faster than FEC based data packets. There related time for the complete one cycle is also less than FEC based data packets [7]. So, CRC-8 is good for our networks.

3.5 Distance, Depth and Noise Factor

In minimum distance and depth like 100m meters it was found normal but when we change the distance to 500m then we found more CRC errors. That means the strength of the signal, the bit energy rate and the transmitted power to noise per bit. E_b/N_o are commonly used with modulation and coding noise-limited rather than interference limited communication. In the data link layer this FEC codeword sending to the physical layer for modulation and send the signal to the noise channel. So it is why the FEC works well in the long distance and deep ocean communication [8], [9].

Some authors study and prove that the checksum is not strong enough like CRC. The concept of a checksum is less well defined. In the field of data communication, the goal of a checksum algorithm is to balance the effectiveness at detecting errors. Furthermore, it also founds that a checksum works in conjunction with other, stronger, data checks such as a CRC. MAC layer used a CRC to check that data was not corrupted during transmission on the local media, and checksums are used by higher layers to ensure that data was not corrupted in intermediate routers or by the sending or receiving host [10] in the RF based network.

Checksum is not suitable for underwater environment as so many factors as involved in UWASNs. FEC mechanism with 3~6 bits humming distance error correction capable mechanism is good for underwater but this is not easy to fix in UWASNs. And extra error check bit is expensive for nodes energy. Some authors also proved that the CRC has a good capability to detected burst errors and it was mathematically proves and approved by the IEEE and ISO standards.

4 Performance Analysis

We analyze the performance of each error detection mechanism, and present the result in this section.

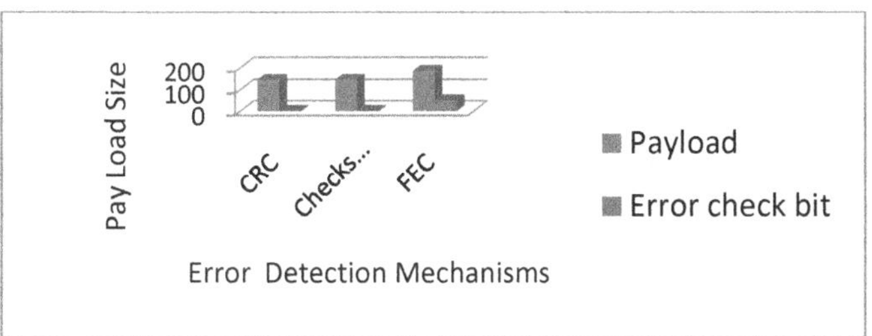

Fig. 5. Comparison different error detection mechanism

Figure 5 shows that FEC mechanism used 40% extra bit or error check for correction which enlarge our packet size.

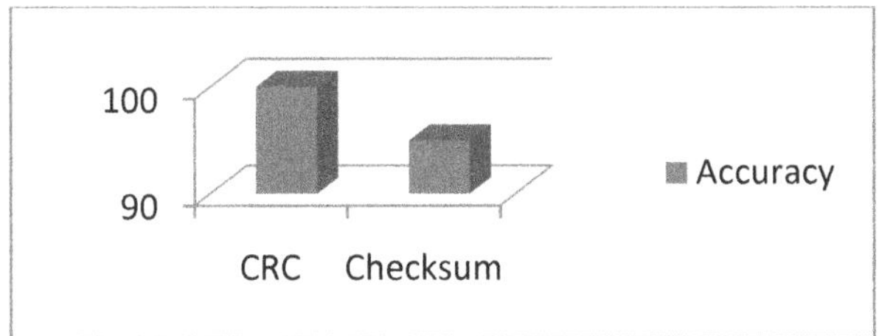

Fig. 6. Comparison between CRC and checksum error detection rate

Figure 6 shows that though CRC and the Checksum uses same amount of extra bit for error detection but the accuracy rate of CRC is much better than checksum.

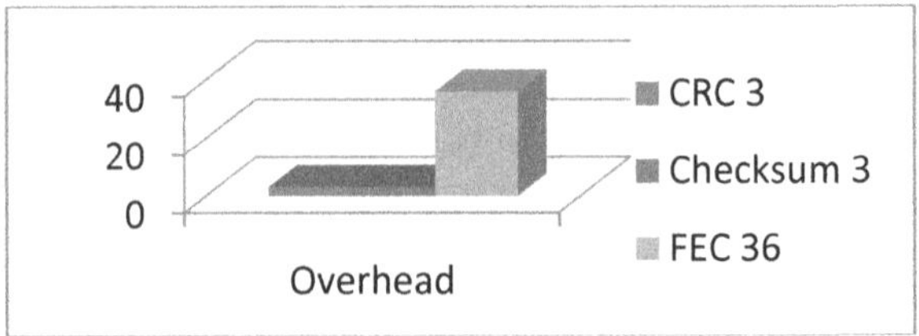

Fig. 7. Energy consumption of CRC, checksum and FEC mechanism

Figure 7 shows based on energy consumption FEC mechanism use more energy during data encoding and decoding time for the circuit and keep the modem busy.

Fig. 8. CRC implementation in real field

Figure 8 is a photo of our performance analysis of acoustic communication among 4 nodes; 3 sensor nodes and 1 sink (gateway) node. We used Cortex-M3 MCU processor in our main board with our own firmware. For secure communication and for error detection we used CRC-8 in our protocol during our experiments.

We used a pseudo code for underwater CRC.

```
WHILE
   IF byte_length not EQUAL 0 THEN
      total_length = total _length + 1
      crc_value = crc_value XOR cycle_value
      FOR i=0 to 4 BEGIN i = i + 1
         IF one_bit_shifted_cv = 1 THEN
            Shift_bit_right (crc_value)
            crc_value = crc_vallue XOR 0xFF
            ENDIF    shift1bit_rightcrc_value
       END FOR
   END IF
END WHILE
```

It was found that when we change packet size 6 to 17 byte, we didn't find any abnormal error symptom when we changed our location and distance in real field which is located in Han River, Seoul, Korea.

5 Conclusions

This paper is one of the efforts to try CRC mechanism for building UWASNs. We were able to characterize and localize some of the problems emerging in the implementation experiments when compared to other CRC mechanism results. Our implementations show that a more adaptive error detection model must be provided to improve the performance of UWASNs. We restricted our attention in simulating our proposed CRC mechanism and we showed how this mechanism can actually improve the performance of network making them a better and faster at error detecting. In the future we will make researches on how to build universal model for error detection and correcting for underwater acoustic communication.

Though so many limitations exist in underwater environment, UWASNs are developing research area. More and more experiments are carried out around the world to develop strong UWASNs with integrated application field like terrestrial network.

Acknowledgements. This research was supported by the MKE (The Ministry of Knowledge Economy), Korea, under the ITRC (Information Technology Research Center) support program supervised by the NIPA (National IT Industry Promotion Agency) (NIPA-2011-C1090-1121-0001).

References

1. Dong, Y., Liu, P.: Security Considerations of Underwater Acoustic Networks. In: Proceedings of 20th International Congress on Acoustics, ICA 2010 (August 2010)
2. Tanenbaum, A.S., Wetherall, D.J.: Computer Networks, 5th edn.
3. Koopman, P., Chakaravarty, T.: Cyclic Redundancy Code (CRC) Polynomial Selection for Embedded Networks. In: The International Conference on Dependable System and Networks (July 2004)
4. Ramabadran, T.V., Gaitonde, S.S.: A tutorial on CRC Computations. IEEE Micro. 8(4) (August 1988)
5. Akyildiz, I.F., Pompili, D., Melodia, Y.: Underwater acoustic sensor networks: research challenges. Ad Hoc Networks 3(3), 257–279 (2005)
6. Ovaliadis, K., Savage, N., Kanakaris, V.: Energy efficiency in underwater sensor networks: A research review. Journal of Engineering Science and Technology Review 3(1) (2010)
7. Chitre, M., Shahabudeen, S., Potter, J., Freitag, L., Sozer, E., Stojanovic, M.: An Architecture of Underwater networks. In: OCEANS 2006 – Asia Pacific (2006)
8. Zhang, Z.: Some recent progresses in network error correction coding theory. In: The Fourth Workshop on Network Coding, Theory and Applications 2008 (January 2008)
9. Labrador, Y., Karimi, M., Pan, D., Mille, J.: Modulation and Error Correction in the Underwater Acoustic Communication Channel. IJCSNS International Journal of Computer Science and Network Security 9(7) (July 2009)
10. Stone, J., Greenwald, M., Partridge, C., Hughes, J.: Performance of Checksums and CRC's over Real Data. IEEE/ACM Transactions on Networking 6(5) (October 1998)

Distributed Consensus and Mitigating Risk Propagation in Evolutionary Optimized Networks

Takanori Komatsu and Akira Namatame

Department of Computer Science, National Defense Academy, Yokosuka Japan
{ed10004,nama}@nda.ac.jp

Abstract. The consensus model has been extensively applied to solve coordination problems of distributed systems. A consensus algorithm specifies the information exchange rule between a system (node) and all of its neighbors on the network. The design issue of sensor networks capable of reaching a consensus quickly has received considerable attentions. The conditions for achieving a consensus depend on the properties of the graph of the nodes. The network is also subjected to external shocks both with respect to the size of the shock and the spatial impact of the shock. There is empirical evidence that as the connectivity of a network increases, we observe an increase in the performance, but at the same time, an increase in the chance of risk propagation which is extremely large. In this paper, we address the issue of designing desirable networks for any fixed number of nodes and links by employing an evolutionary optimization procedure. The resulting networks are optimal not only for fast consensus but also for minimizing risk propagation.

1 Introduction

It is broadly recognized that most complex systems in nature are organized as intricate network patterns [1,2]. Network topologies are often not static. The application on networks changes and evolves network topology [3]. This observation has triggered an intense research effort aimed at understanding the organizing principles of these networks, and the interplay between topology and network dynamics [4, 5].

Despite the fundamental importance of the network topologies in many social and engineering settings, there is still a lack of analysis of how desirable networks for specified purposes should be designed. Although most theories concerning the origin of these topologies use growing graphs [4, 6], some optimization methodologies can account for the observed regularities displayed by most complex networks. We can observe some aspects of network optimization in transport network, power grid network, the Internet and so on. In such complex networks, there are some statistical quantities to optimize. For example, average hop distance between nodes and betweeness of each node are some important for communication networks [7]. Many problems in engineering and related areas require the simultaneous optimization of many objectives that may be conflicting.

J.-H. Kim et al. (Eds.): AsiaSim2011, PICT 4, pp. 200–209, 2012.

Topological design of communication networks, for instance, is a typical multi-objective problem involving simultaneous optimization of cost of the network and various performance criteria such as efficiency, average delay, and robustness.

Distributed computation over networks has a tradition in systems and control theory on agreement problem for distributed decision-making systems [8]. Consensus problems have a long history in computer science and control theory [9]. In network systems, consensus means reaching an agreement regarding a certain item of interest that depends on the state of all nodes. The consensus model has also been extensively applied to solve coordination problems of distributed systems. For instance, the design of sensor networks capable of reaching a consensus on a globally optimal decision without the need for a fusion center is a problem that has received considerable attentions [10, 11]. The conditions for achieving a consensus ultimately depend on the properties of the graph modeling the interaction among the nodes. Coordination among independent elements is also affected by the network topology that describes how they are connected. The level of stability of a coordinated state is a measure of the systemfs ability to yield a coherent response and to distribute information efficiently among its elements [11].

Real-world networks increase interdependencies and this creates challenges for handling network risks like cascading failure. Important processes studied within this framework include contact processes such as information and epidemic spreading. Studies like these have clarified that certain topological properties have strong impacts on the dynamics of networks [1, 2, 5].

The paper studies the problem of designing the topology of a network. The topology of the network has a major impact on the convergence of distributed consensus algorithms, namely, that the distributed consensus algorithm converge much faster for certain connectivity patterns than for others, thus requiring much less inter sensor communication and power expenditure [12, 13].

In this paper, we address the issue of designing optimal networks for fast consensus as well as minimizing risk propagation. In general there is usually a trade-off between risk contagion and risk sharing. Optimizing a network requires the selection of a proper fitness function. The maximum eigenvalue can be used as the index of robustness against cascading failure. The network with a smaller maximum eigenvalue of the adjacency matrix has a better property to mitigate risk contagion. We use an evolutionary algorithm involving minimization of the weighted fitness function of the eigenvalues of the Laplacian matrix and the inverse of the average degree for consensus problems. We compare the convergence speed of a consensus problem in an evolutionary optimized network with random networks, regular network and scale-free networks.

2 Distributed Consensus on Networks

Consensus problems have a long history in computer science and form the foundation of the field of distributed computing. Formal study of consensus problems in groups of experts originated statistics. The original idea was in aggregation of information with uncertainty obtained from multiple sensors or experts [8]. In networked systems consensus means to reach an agreement regarding a certain quantity of interest that depends on the initial values of each system (node). A consensus algorithm is an interaction rule that specifies the information exchange between nodes and all of its neighbors on the network. The theoretical framework for posing and solving consensus problems for networked systems is well surveyed by Olfati-Saber and Murray in [9].

The synchronization of mutually coupled oscillators also has been a typical field of application of consensus problems [4]. Synchronization often occurs unexpectedly and little is known what is the best network topology is the best for synchronization. In recent studies, the reason for the occurrence of synchronized networks becomes clear and the underlying network topology is important [6,14].

The consensus problems rely heavily on matrix theory and spectral graph theory. The interaction topology of a network is represented using a directed graph G with the set of nodes and edges, in which neighbors of node i are denoted as N_i. Consider a network of nodes with the following dynamics:

$$\dot{x}_i = \sum_{j \in N_i} \alpha_{ij}(x_j(t) - x_i(t)) \tag{1}$$

where α_{ij} is the weight of node i on node j.

Reaching a consensus means asymptotically converging to the same state values by an agreement characterized by the following equation:

$$x_1 = x_2 = \cdots = x_n = \alpha \tag{2}$$

Assuming that the underlying graph G is undirected ($\alpha_{ij} = \alpha_{ji}$ for all ij), the dynamics in Eq.1 converges to the average of the initial states of all nodes,

$$\alpha = \frac{1}{n} \sum_{i=1}^{n} x_i(0) \tag{3}$$

The dynamics of the system in Eq.1 can be expressed as

$$\dot{x} = -Lx(t) \tag{4}$$

where L is the Laplacian matrix of G which is defined as

$$L = D - A \tag{5}$$

where D is the diagonal matrix with elements of the degree and A is the adjacency matrix ($n \times n$matrix) with elements $\alpha_{ij} = \alpha_{ji}$ for all i, j where α_{ij} is 1 if node i and node j is connected or 0 if they are disconnected. Because in our model networks are undirected, L is a symmetric matrix with all real entries. The Laplacian L is a positive semi-definite matrix, and all eingenvalues are nonnegative, and we order the Laplacian eigenvalues as

$$0 = \lambda_1(L) \leq \lambda_2(L) \leq \cdots \leq \lambda_n(L) \tag{6}$$

The question is how consensus performance depends on network topology. An answer to this question is given by the ratio($\lambda_n(L)/\lambda_2(L)$) of the second smallest eigenvalue $\lambda_2(L)$(λ_2 which is known as algebraic connectivity), and the largest eigenvalue $\lambda_n(L)$.

3 Risk Propagation in Networks

Many real-world processes are diffusive in nature, giving rise to optimization problems where the goal is to maximize or minimize the spread of some entity through a network. For example, in epidemiology, the spread of infectious diseases should be minimized. In marketing activity, one individual's adoption of a certain marketed product may trigger his or her friends or fans to adopt that product as well. In this case diffusion should be maximized in the social network.

Cascading failure is one of big problems of complex network [4]. Many types of real networks increase interdependencies and this creates challenges for managing risks. This is especially apparent in areas such as financial institutions and enterprise risk management, where the actions of a single actor in an interconnected network can impact all the other actors in the network. While more links between banks might be expected to increase the risk of contagion. A densely connected network may be less susceptible to contagion than those with a sparse network structure with few connections. Incomplete networks are more prone to contagion than complete structures. For instance, better-connected networks are more resilient, since the proportion of losses in one node is transferred to more nodes through a network. On the other hand, in the case of an incomplete network, the failure of one node may trigger the failure of the entire network [15].

There is empirical evidence from a range of disciplines that as the connectivity of a network increases, we observe an increase in the average fitness of the system. But at the same time, there is an increase in the proportion of failure events which are extremely large. The probability of observing an extreme event remains very low, but it is markedly higher than in the system with lower degrees of connectivity [16].

The network is also subjected to external negative shocks both with respect to the size of the shock and the spatial impact of the shock. In this context, there are entire classes of problems, ranging from the control of cascade failures to information diffusion, which are naturally defined as optimization problems. More interconnections means higher levels of insurance, but also higher risk of contagion, and small shocks can wipe out large parts of the whole system.

Our basic question is how the network structure impacts macroscopic diffusion patterns, especially the ratio of infected nodes. This question can be investigated by obtaining epidemic thresholds of various network topologies. Diffusion only starts when there is a high propagation rate β in a network that has a sizable largest eigenvalue.

The principal eigenvalue of the adjacency matrix A ($\lambda_1(A)$) provides useful information for determining how risk such as an epidemic disease diffuses. Wang suggests the largest eigenvalue of the adjacency matrix which also characterizes the network topology is closely related to the ease of risk contagion [17].The risk diffusion process can be characterized as a threshold phenomenon, and the threshold is characterized by the inverse of the largest eigenvalue, i.e.,

$$\beta_{cri} \simeq 1/\lambda_1(A) \tag{7}$$

This threshold provides the conditions under which risk does not diffuse and it dies out ($\beta_{cri} < 1$) or it diffuses in a large-scale ($\beta_{cri} > 1$).

4 An Evolutionary Design of Optimal Networks

The optimization has a long history and it is most often connected to a function that the system performs. It comprises the variation principles or the survival-of-the-fittest principles that pervade biology and engineering, the funding hypotheses of numerous computer algorithms, and the frameworks for addressing the improvement of efficiency in various contexts. However, most traditional optimization problems have a strict mathematical definition, which refers to obtaining the solutions that strictly optimize a well-defined objective function. Here we adopt a looser definition of the word by extending it to include a tendency of the system to improve its behavior as a result of a selection pressure based on artificially imposed fitness function.

<Case 1> The Fitness Function for Fast Consensus

We study the design of the network topology that optimizes the rate of convergence of the iterative consensus algorithm. We reformulate the topology design problem as a spectral graph design problem, namely, maximizing the ratio $\gamma = \lambda_n(L)/\lambda_2(L)$ of two eigenvalues of the graph Laplacian L. It is obvious that the network with a minimum eigenvalue ratio is a fully connected complete network. Therefore we consider the optimization under constraint of the link density. Then the fitness function to be minimized is

$$E(\omega) = \omega(\frac{\lambda_n}{\lambda_2}) + (1-\omega)\langle k\rangle \tag{8}$$

where $\omega(0 \leq \omega \leq 1)$ is a parameter controlling the two objects. The average degree $\langle k\rangle$ of a network with the adjacency matrix $A = (a_{ij})$ is defined as

$$\langle k\rangle = \frac{1}{n}\sum_{i=1}^{n}\sum_{j=1}^{n} a_{ij} \tag{9}$$

We use the MGG model for the change of generations [18]. Especially we select two best networks and their adjacency matrices are crossover as shown in Fig.1 to generate two better networks. We use crossover rate at 0.7, and mutation rate is set at $2/{}_nC_2$, i.e. reverses of two links per one generation. We use the multi-point crossover. After crossover, each element in the matrix switches to a reverse state with a specific probability. There is the possibility that an isolated network appears after crossover and mutation. In this simulation, when an isolated node appears in a new network that the node has zero distance to another node, we dump the network. Therefore, we can use non-isolated matrices. After long generations have passed, we can obtain an optimal network which minimizes the fitness function defined in Eq.10.

The obtained optimal network is characterized as a homogeneous network where each node has almost the same degree. In Table 1 we show the eigenvalue ratio λ_n/λ_2 of a regular network with the degree $k = 4$, a small-world network generated from a regular network with rewiring probability $p = 0.3$, a random network with the average degree $k = 4$, a random regular network with the degree $k = 4$, and an evolutionary optimized network. All networks have the number of nodes, $n = 500$. An evolutionary optimized network has the smallest the eigenvalue ratio and it is optimal in the sense to maximize the speed of convergence in consensus problem.

Table 1. The algebraic connectivity λ_n/λ_2 of various networks: all networks have the same network size $n = 500$, and the degree (average) is $k = 4$

Network	Regular	Small World	Random	Random Regular	Optimized
λ_n/λ_2	7812.5	46.7	19.3	13.2	12.3

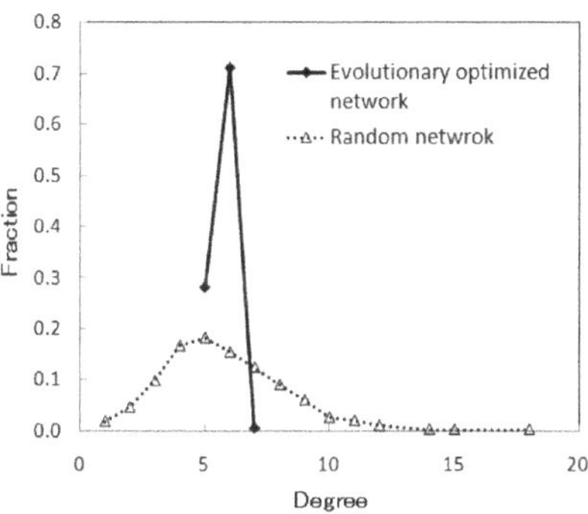

Fig. 1. Degree distribution of the networks for $\omega = 0.97$: In the dashed graph - the optimized network; dotted graph - the random network that has the same number of links

<Case 2> The Fitness Function for Minimizing Risk Propagation

In Section 3, we discuss that the diffusion in a network is a threshold phenomena and the threshold is determined by the inverse of the largest eigenvalue($\lambda_1(A)$) of the adjacency matrix. Another important factor to determine diffusion is the number of link. In the many cases not only risk managements, the network with a lot of links has a good performance. For example, in the case of the airway network, if every airport is directly connected by an airline, travel time will be reduced drastically. And in the case of the computer network, if hosts are connected directly by a communication link, the congestion may not happen.

The network with a larger maximum eigenvalue helps faster risk contagion on the network. On the other hand, average number of links on each node (the average degree) $\langle k \rangle$ of the network is related to risk sharing. When nodes have more links, nodes becomes more stable against external failures because they still have alive nodes after some neighboring nodes become failure.

In our GA process, we evaluate a given network by maximum eigenvalue $\lambda_n(A)$ and average degree $\langle k \rangle$, which is associated with average degree $\langle k \rangle$, Eq.11 defines the fitness function to minimize in our GA.

$$E = \omega\lambda_1(A)/(N-1) + (1-\omega)/\langle k \rangle \tag{10}$$

where $\omega(0 \leq \omega \leq 1)$ is a parameter controlling the linear combination of the $\lambda_n(A)/(N-1)$ and $1/\langle k \rangle$.

In typical cases of ω, we can easily imagine optimal network topology. For instances, if we set $\omega = 1$, GA minimizes only maximum eigenvalue then we have a tree type network which has $\lambda_n(A)$. And if we set $\omega = 0$, GA minimizes only $1/\langle k \rangle$ then we have complete graph which has $\lambda_n(A) = N - 1$ and $\langle k \rangle = N - 1$. By varying parameter $\omega(0 < \omega < 1)$, we have the network which has the smallest maximum eigenvalue.

5 Simulation Results

We show that the convergence speed is much faster in evolutionary optimized networks than Ramanujan graph or random regular networks which are currently known as the best networks in the literatures [7, 11].

<Case 1> Comparison of Convergence Speeds in Consensus Problems

The speed of convergence are compared with the initial value of each node

$$x_i(0) = i \quad i = 1, 2, \ldots, n \tag{11}$$

We consider the network topologies, the evolutionary optimized and the random network with the same degree $\langle k \rangle = 4$. In Fig. 3 we show a comparison of the convergence to the agreement (the average of initial values) under the optimized network for $\omega = 0.97$, and random network. We can observe that the convergence to agreement is much faster under the optimized networks.

For an evolutionary optimized network topology, the convergence is much faster to the global consensus than for structured graphs, such as random graphs, or graphs exhibiting small-world characteristics. Finally, we compare with random regular graphs that are easy to construct, can be designed with arbitrary number of sensors, and whose spectral and convergence properties make them practically equivalent to evolutionary optimized network topology.

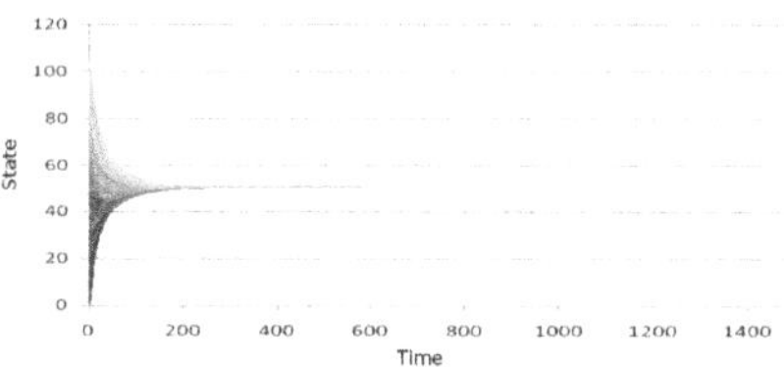

(a) Convergence on the optimized network

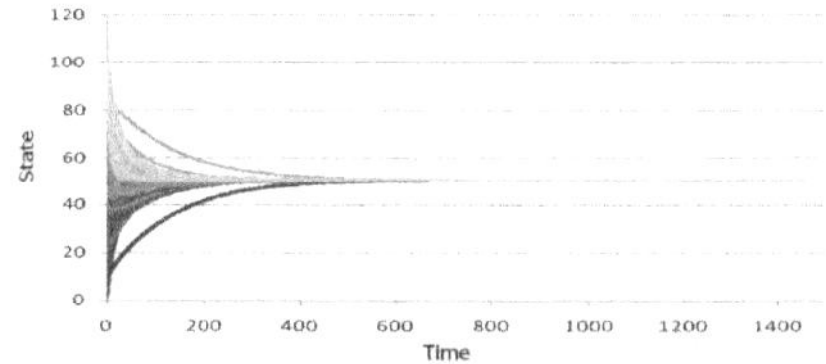

(b) Convergence on the random network

Fig. 2. Comparison of the convergence speeds

<Case 2> Comparison of Risk Propagation

Different networks exhibit different features with different largest eigenvalues. We can optimize a network topology using an evolutionary approach, where we select a proper fitness function. To maximize diffusion, for instance, we should have a network with the largest possible eigenvalue of the associated adjacency matrix. It is obvious that a fully connected complete network fulfills that requirement. Since this is rare, we should consider optimization where the network is constrained by the average degree.

Our basic question is to investigate how the network structure impacts on macroscopic diffusion patterns, especially the ratio of infected nodes. This question can be investigated through by obtaining epidemic thresholds of various network topologies. The diffusion only starts with high the relative propagation rate β in a network with a higher largest eigenvalue. We compare the largest eigenvalues of a random network of a scale-free network and of the evolutionary optimized network with the same average degree. We define the score s of an epidemic on a graph

$$s = \beta / \beta_{KN} \tag{12}$$

Figure 3 shows the diffusion rates over time for various values of the score s in log-log scales. The dotted lines shows the case for $s = 1$. We observe a clear trend and the below the threshold ($s = 1$), the diffusion dies out while the diffusion starts above the threshold ($s > 1$) on the network with the largest eigenvalue. However the diffusion only starts with a higher score in a scale free-network and it requires a much higher score (high propagation probability) to diffuse in a random network. Therefore we can observe that diffusion is easier to start with a lower propagation probability β in a network with a higher largest eigenvalue (KN network) [19–21].

We simulate risk contagion by threshold risk contagion model on three type networks: an evolutionary optimized network by our proposed method, an Erdös and Rény (ER) random network and a scale free network.

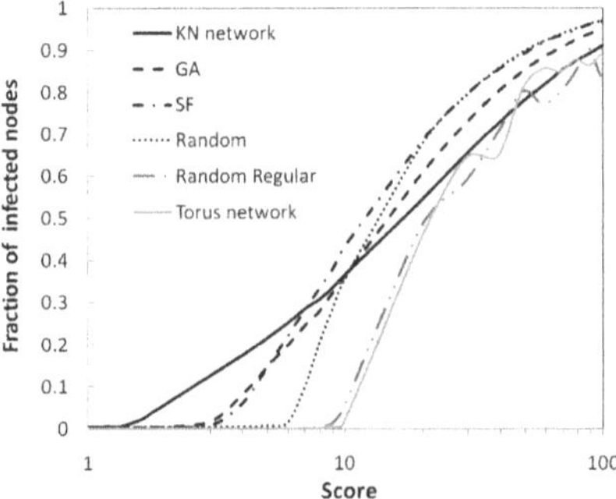

Fig. 3. The ratios of infected nodes ρ as a function of propagation probability $\beta = Score \times \beta_{KN}$ in different network topologies: random network, scale-free network (SF) and the network with the maximum eigenvalue (KN network)

6 Conclusion

Designing desirable networks is complex and it may pose a multi-constraint and multi-criterion optimization problem. We presented a genetic optimization approach to designing an optimal network for best consensus while simultaneously minimizing risk propagation. We investigated and compared convergence speeds of evolutionary optimized networks. Results obtained are compared with those obtained by traditional approaches.

Optimized networks are homogeneous networks having almost the same degree and node-to-node distance. However the optimized network is not Ramanujan graph (random regular network) in which all nodes have the same degrees. Most of the nodes have the same degrees but some nodes have more nodes and some other nodes have less nodes than the majority of nodes. Solving a network design problem of additional complexity by including a larger number of objectives and constraints is an area of further investigation.

References

1. Albert-László, B.: Linked: The New Science of Networks. Perseus Books Group, New York (2002)
2. Strogatz, S.H.: Exploring complex networks. Nature 410(6825), 268–276 (2001)
3. Durrett, R.: Random Graph Dynamics. Cambridge University Press, Cambridge U.K. (2006)
4. Albert, R., Barabási, A.-L.: Statistical mechanics of complex networks. Rev. Mod. Phys. 74(1), 47–97 (2002)
5. Krapivsky, P.L., Redner, S.: A statistical physics perspective on web growth. Computer Networks 39, 261–276 (2002)
6. Watts, D.J., Strogatz, S.H.: Collective dynamics of small-world networks. Nature 393(6684), 440–442 (1998)

7. Donetti, L., Neri, F., Muñoz, M.A.: Optimal network topologies: expanders, cages, ramanujan graphs, entangled networks and all that. J. Stat. Mech. 26, 134–144 (2006)
8. Degroot, M.H.: Reaching a consensus. Journal of the American Statistical Association 69, 118–121 (1974)
9. Olfati-Saber, R., Fax, J.A., Murray, R.M.: Consensus and cooperation in networked multi-agent systems. Proceedings of the IEEE 95(1), 215–233 (2007)
10. Boyd, S., Diaconis, F., Xiao, L.: Fastest mixing markov chain on graph. SIAM Review, 121–155 (2004)
11. Kar, S., Aldosari, S., Moura, J.M.F.: Topology for distributed inference on graphs. IEEE Trans. on Signal Processing 56, 2609–2613 (2008)
12. Kar, S., Aldosari, S., Moura, J.M.F.: Consensus based detection in sensor networks. In: International Workshop on Information Theory in Sensor Networks (June 2007)
13. Xiao, L., Boyd, S.: Fast linear iterations for distributed averaging. Systems and Control Letters 53, 65–78 (2003)
14. Wang, X.F., Chen, G.: Synchronization in scale-free dynamical networks: robustness and fragility. IEEE Transactions on Circuits and Systems I: Fundamental Theory and Applications 49, 54–62 (2002)
15. Sornette, D.: Critical Phenomena in Natural Sciences: Chaos, Fractals, Selforganization and Disorder: Concepts and Tools. Springer, Heidelberg (2006)
16. Allen, F., Gale, D.: Financial contagion. Journal of Political Economy 108(1), 1–33 (2000)
17. Chakrabarti, D., Wang, Y., Wang, C., Leskovec, J., Faloutsos, C.: Epidemic threshold in real networks. ACM Trans. on Information and System Security 10(4), 131–156 (2006)
18. Sato, H., Isao, O., Shigenobu, K.: A new generation alternation model of genetic algorithms and its assessment. Journal of Japanese Society for Artificial Intellligence 12(5), 734–744 (1997)
19. Komatsu, T., Namatame, A.: Dynamic diffusion process in evolutionary optimized networks. International Journal of Bio-Inspired Computation 3(6), 384–392 (2011)
20. Komatsu, T., Namatame, A.: An evolutionary optimal network design to mitigate risk contagion. In: 2011 Seventh International Conference on Natunal Computation (ICNC), pp. 1954–1959. IEEE (2011)
21. Komatsu, T., Namatame, A.: Optimal diffusion in evolutionary designed networks. Proceedings of Econophysics Colloquium 2010 (November 2010)

A Study on Mobile Cloud Anomaly Detection Model

Hyung-Jong Kim[1], Ji-Yeon Kim[1], JinHyung Kim[1], Dal-Nim Choi[1], Jeong Wook Kim[2], and HyunCheol Jeong[2]

[1] Department of Information Security, Seoul Women's University, Seoul, Korea
{hkim,jykim07,jinny,dnchoi}@swu.ac.kr
[2] Security R&D Team, Korea Internet & Security Agency, Seoul, Korea
{kjw,hcjung}@kisa.or.kr

Abstract. The mobile cloud service is inevitable alternative which provides mobile device users with high computing power and large storage. The cloud servers provide various services such as IaaS, PaaS and SaaS to mobile device users seamlessly and the situation means that the device's apps or mobile browser can interact with the VMs in clouds at anytime and anyplace. In this mobile cloud environment, our concern is the abnormal access of cloud resources. Since the users of cloud services pay based on the usage of the cloud resources, if there are a certain kind of malicious access to the cloud resources, it can cause financial damages to mobile cloud users. Especially, the mobile devices are always turned on, even if we are moving around the street. Thus the malicious behavior of mobile device can do more harm to users compared to PC-based cloud services. This research suggests an anomaly detection model in mobile cloud services. In our suggested model, the users' behavior and resource usages are considered and the model contains the usual behaviors and resource usages as the users' profile which is supposed to be compared with the current users' behavior for detecting the abnormalities. The research result gives insight to security managers about monitoring factors and recognizing the anomaly in the mobile cloud service.

1 Introduction

The mobile cloud service users make use of mobile devices to interact with the cloud services such as IaaS, PaaS and SaaS. The mobile device enables users to access those cloud services in anytime and any places. Through using the mobile devices, the physical coverage of cloud services has been getting larger than before. The situation of prevailing mobile cloud services means that the attack surfaces for cloud computing environment is also getting larger. Especially, as mobile cloud users can access the cloud resource, the malicious codes in terminals also can communicate with cloud servers whenever they want to. On the other hand, in the mobile cloud environment, the mobile device is provided with the computing power, memory and storages by cloud servers [1].

This research is motivated by the ubiquitous characteristic of the mobile device. This characteristic enables the malicious codes of mobile devices to access the cloud services also. To handle this issue we suggest an anomaly detection models which can

J.-H. Kim et al. (Eds.): AsiaSim2011, PICT 4, pp. 210–216, 2012.

be used to recognize the malicious behaviors of mobile cloud service users. Our anomaly detection model can be used as a user's profile of abnormal behavior detection. The model has 6 monitoring factors which can be observed in cloud servers and 4 factors of mobile devices. This paper shows what factors should be monitored to detect the abnormal behaviors and how those factors can be aggregated using rule-based knowledge representation.

The remaining part of this paper consists of background knowledge of this research, enumeration of essential monitoring factors, rule-based anomaly detection method and research summary.

2 Background

There are two kinds of mobile cloud services such as mobility services and seamless services. The mobility services are making use of mobile devices' facilities such as GPS location, SMS/MMS and voice call. The seamless services are for providing cloud service users with service continuity. In the seamless type mobile cloud services, users can see the movies which were watched on TVs or PCs with their mobile devices.

The mobile cloud service has three main components that we need to regard to elicit the factors for anomaly detection. One is mobile devices, second is the cloud computing environment and the third is the wireless access networks. Fig. 1 shows the mobile cloud service architecture. The mobile device can access the cloud server farms through WIFI, Wibro and 3G access networks. Since the access networks and the IP backbone network just provide routes between the mobile devices and cloud servers, they don't have meaningful information for the anomaly detection. For this reason, in our work we choose the data gathering points as mobile devices and cloud servers.

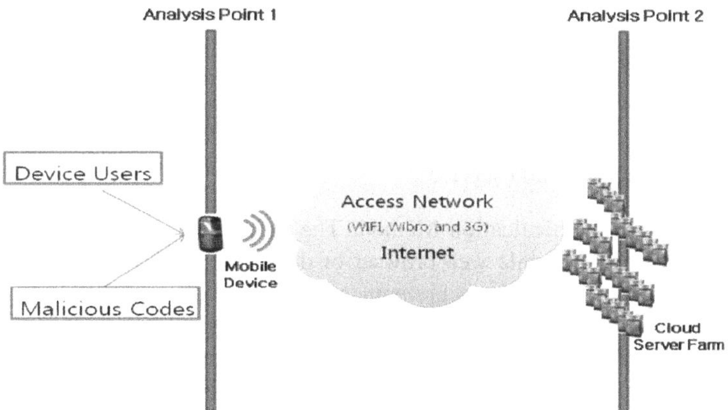

Fig. 1. Analysis Points for Mobile Cloud Services

3 Monitoring Factors

In the previous section, two analysis points are mentioned. The cloud server consists of set of virtual machines and the servers can also be recognized as services such as IaaS, PaaS and SaaS. In this work, we have elicited 6 monitoring elements for detecting malicious activities originated from the mobile device.

o Access Time: The access time of the mobile devices is logged and the log information can be turned into statistics. Whenever service requests are arrived, servers can estimate the how the event is exceptional using below expression (1). The 24-hour is divided using the variable *n* and *24/n* time-windows are derived. In the log file, each time-window has access count that is used for estimating abnormality.

$$Factor_{AccessTime} = \frac{NumAccess_{current}}{\sum_{i=1}^{n} NumAccess_i}, \quad (1)$$

where

$0 \leq Factor_{AccessTime} \leq 1$

$Factor_{AccessTime}$: Factor of mobile device's access time

$NumAccess_{current}$: The service request number of the current time window

$NumAccess_i$: The service request number of i_{th}'s time window

n : The number of time window in a day

o Average Service Time: The service time means the time of the users' remaining in the servers to get the service from the cloud servers. It usually depends on the services types. For instance, if it is a movie steaming service, it may last until the content is ended. If it is a social network service, the service time would be very short period. In this factor, we see if there is exceptional service time, and for this purpose we can use central limit theorem.

o Resource Usage: Cloud servers are monitoring the usage of their resource using tools such as XenCenter, vCenter and CloudStack. For each virtual machine, CPU and memory usage, amount of HDD read and write and inbound/outbound traffic for each NIC are monitored. If there are unusual resource usages, it can be considered an abnormal status of mobile cloud services.

o Accessing and Authentication Method: The mobile cloud service users can access the services using mobile web browser or dedicated apps. Both access methods are acceptable but there can be a preference. In addition, to authenticate the users, ID/password, digital certificate, one time password and biometrics can be used. In the authentication method, we can classify them based on their authentication strength. In our decision making for abnormality, if the accessing method is unusual and authentication method is weak, we considered the situation is abnormal.

o Physical Location: If the mobile devices' physical location is unusual, it should be considered that the user may not be the authentic user. The devices' physical location is identified by the GPS information and IP address. We can see the characteristics of IP address assignment in 3G network in [2].

o Vulnerability Exploitation Trial: According to [3], [4] and [5], there are vulnerabilities in VMM and virtual machine management tools. In addition, the hacking tools of virtual machine environment are emerging. VASTO is a representative hacking tool and it contains 18 hacking script codes exploiting vulnerabilities in Xen, VMware and OracleVM [5]. In our research, if there are vulnerability exploitation trials, we consider that the abnormality of the user is very high.

For the situation that the mobile device is used for malicious purpose, we choose below 4 factors to identify the abnormality of users.

o Access Network and Authentication Method: When we use the mobile devices to access the cloud service, we need to choose the access network to let the device go through the Internet. In addition, the mobile device should be authenticated and authorized to use the access network. For instance, mobile devices can use WEP key, WPA and WPA2 in WIFI, and PKMv2 in Wibro. The access network choice and authentication method can be a clue to recognize the abnormality of mobile cloud service usages. Especially, in our work, we are trying to see if there is unusual choice of access network and authentication method.

o Malicious Code: There are a lot of reports about malicious codes in mobile devices. Sometimes the malicious code let mobile devices to generate call, send SMS message while users do not know that situation. If this kind of trial is monitored or malicious codes are detected, we can assume the abnormality of mobile devices. The detection of malicious codes is conducted using dedicated tools.

o Network Traffic: To cause a bad effect to the cloud service, the mobile device can generate traffic too much. Sometimes, it can cause cloud server's bandwidth consumption and cause HDD's saturation by sending garbage data. To identify this kind of situation, we need to estimate the mobile devices' inbound and outbound traffic every minute or every hour. In addition, we can also monitor each application's average inbound and outbound traffic to find the unusual status.

o Battery Usages: The mobile device is very sensitive to battery usage. If the battery drains too much in a short time period, we can assume that the device is in unusual state. We can consider this situation as an evidence of the abnormality of mobile cloud service usage.

Table 1 shows the summary of the monitoring factors which are mentioned in this section.

Table 1. Mobile Cloud Service's Anomaly Detection Monitoring Factors

<table>
<tr><td rowspan="6">Cloud Servers</td><td>Access Time</td><td>o Uses User's Service Log Statistics and Current Access Time
o Evaluate the New Access Using the Statistics</td></tr>
<tr><td>Average Service Time</td><td>o Duration of Service Usage
o Evaluate the Unusual Service Time</td></tr>
<tr><td>Resource Usage</td><td>o Resource Usage for Each VM
o Resource Usage for Each User and Service Request</td></tr>
<tr><td>Accessing and Authentication Method</td><td>o Service Access Method (Apps, Mobile Web)
o Authentication Method(Non-Authentication, ID/PW, Digital Certificate, OTP, Biometrics)</td></tr>
<tr><td>Physical Location</td><td>o GPS Location: Profiling Primary User's Location
o IP Address: Registering Blocking IP Address</td></tr>
<tr><td>Vulnerability Exploitation</td><td>o Dedicated Detecting Tools</td></tr>
<tr><td rowspan="4">Mobile Devices</td><td>Access Network and Authentication Method</td><td>o Profiling as a User's Preference
- WIFI (Non-Authentication/Authorization, WEP Key, WPA, WPA2)
- Wibro (Non-Authentication /Authorization, PKMv2)
- 3G Access Network</td></tr>
<tr><td>Malicious Code</td><td>o Dedicated Detecting Tools</td></tr>
<tr><td>Network Traffic</td><td>o Application's Traffic Statistics
o Traffic Statistics over Time</td></tr>
<tr><td>Battery Usage</td><td>o Monitors Rapid Battery Drain</td></tr>
</table>

4 Anomaly Detection System Architectures

In this section, the aggregation method of the anomaly detection factors which are enumerated in previous section is presented. As an aggregation method, we make use of the rule-based knowledge and all factors of the server and mobile devices are used as facts of the knowledge. Fig 2 shows the aggregation rules for anomaly detection. There are three impact values HIGH, MID and LOW which mean the strength of the evidence. As the more evidence is gathered, the value of user's abnormality is getting higher. In addition, there are threshold for abnormality decision, NormalSTD. If the value of the abnormality is bigger than the threshold, system makes decision that the mobile cloud user is abnormal.

```
// Mobile Cloud Anomaly Detection Rules
// Impact Definition
#define LOW 1
#define MID 2
#define HIGH 3
#define NormalSTD 15

// Rules for Cloud Servers
Rule C01: If AccessTime < SVAccTimeThre
          Then UserAnormality += LOW
Rule C02: If UsageDurationDeviation > SVDurDevThre
          Then UserAnormality += MID
Rule C03: If ResourceUsage > SVResUseThre
          Then UserAnormality += MID
Rule C04: If (AccessMethod is Unusual) and
                       (Without AA or ID/PW)
          Then UserAnormality += HIGH
Rule C05: If ServerVulExploit is Occurred
          Then UserAnormality += HIGH

// Rules for Mobile Devices
Rule M01: If AccessNetwork is deviated
          Then UserAnormality += MID
Rule M02: If ClientMaliciousCode = Detected
          Then UserAnormality += HIGH
Rule M03: If ClientDataVolumeTime > CLDataVolTimeThre
          Then UserAnormality += MID
Rule M04: If ClientAppDataUsage > AppDataUsageThre
          Then UserAnormality += MID
Rule M05: If ClientBatteryUsage > CLDataVolUseThre
          Then UserAnormality += LOW

// Final Decision Rule
Rule F01: If UserAnormality > NomralSTD
       Then Alert "The user's behavior is unusual."
```

Fig. 2. Mobile Cloud Anomaly Detection Rules

Using the facts mentioned in previous section and the rules of Fig 2, we can design the system architecture of our suggesting mobile cloud anomaly detection system as shown in Fig 3. The system gathers values from mobile devices and cloud servers. The system's rules are used to conduct inference with the facts and make a decision whether the behavior is abnormal or not.

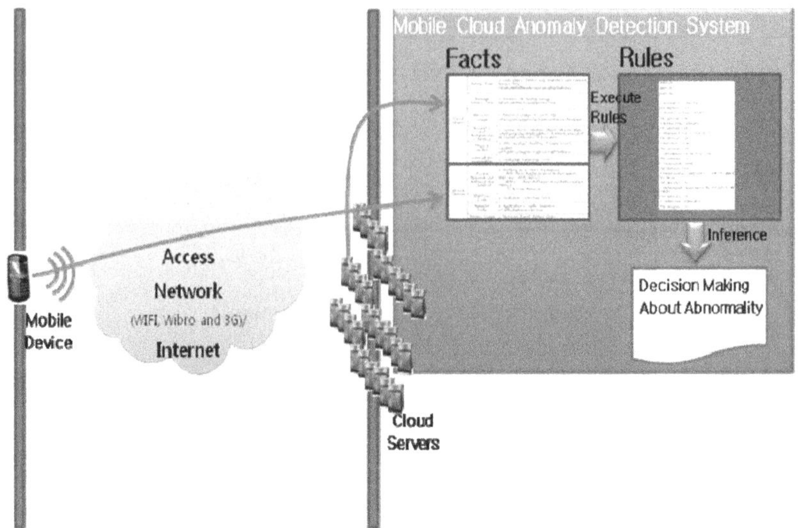

Fig. 3. Mobile Cloud Anomaly Detection System Architecture

5 Conclusions

This research presents an anomaly detection model which can be used to recognize the malicious behaviors of mobile cloud service user. The model has 6 monitoring factors which can be observed in cloud servers and 4 factors from mobile devices. Those factors are aggregated using rule-base knowledge and the inference is preceded by gathering evidence of abnormal behaviors. In this research, the mobile cloud anomaly detection system architecture is proposed.

As future works, we consider development of the mobile cloud attack scenarios and our anomaly detection model can be validated based on the scenarios. In addition, the rules should be specified and tuned to be applicable to real anomaly detection.

Acknowledgement. This research was supported by the KCC (Korea Communications Commission), Korea, under the R&D program supervised by the KCA (Korea Communications Agency) (KCA-2011-11914-06002).

References

1. Chun, B.-G., Maniatis, P.: Augmented Smartphone Applications Through Clone Cloud Execution. In: Proceedings of the 8th Workshop on Hot Topics in Operating Systems, HotOS (2009)
2. Balakrishnan, M., Mohomed, I., Ramasubramanian, V.: Where's that Phone?: GeolocatingIP Addresses on 3G Networks. In: Proceedings of IMC (2009)
3. Wojtczuk, R.: Adventures with a certain Xen vulnerability. Technical Report, Invisible Things Lab (2008)
4. Rutkowska, J.: Security challenges in virtualized environments (April 2008), `http://www.invisiblethingslab.com`
5. Criscione, C.: Virtually Pwned Pentesting Virtualization. In: BlackHat Conference, USA (2010)

Study on the Indoor SLAM Using Kinect

Yoshiaki Takeda[1], Norifumi Aoyama[1], Takahiro Tanaami[1], Syoto Mizumi[1], and Hiroyuki Kamata[2]

[1] Graduates School of Science and Technology, Meiji University
{ce11053,ce11001,ce11055,ce01095}@meiji.ac.jp
[2] School of Science and Technology, Meiji University
kamata@isc.meiji.ac.jp

Abstract. The simultaneous localization and mapping (SLAM) using kinect is examined. Building map is very useful to avoid obstacles and probed object. In order to solve the rapidly SLAM problem, we focus on the vision data and depth data(:RGB-Ddata) from Kinect that is a 3D sensor provided by Microsoft. The proposed method detects some landmarks extracted from images. Depth data of landmark points is used for estimation of consecutive frame status (rotation matrix, translation vector) by ICP algorithm. Putting detected landmark points into Smirnov-Grubbs test to remove outliers, the accuracy is improved. In this paper, we report the results obtained by actual environment.

Keywords: SLMA, SURF, ICP, Kinect.

1 Introduction

In recent years, robotic technology has evolved significantly. However, the automatic movement of the robot technology has not yet been completed. In particular, traveling in an unknown environment is difficult because it is necessary for the robot itself to rapidly calculate the self-localization and surrounding information.

Making of 3D map is very important to realize the automatic driving because the maps help robots avoiding obstacles, recognizing the wall, and probing unknown environment.

In order to build the maps, the simultaneous localization and mapping(SLMA) problem is needed to solve. The various methods to solve the SLAM problem have been studied and proposed. For example, GPS(Global Position System) method[1], wheel odometry method[2] and visual odometry method[3] are the typical outcomes. GPS method performs to solve SLAM accurately. However, GPS cannot be used indoor.

Besides, wheel odometry method troubles errors data due to wheel slip. Therefore, we are focused on Visual odometry because this is stable in indoor where brightness is not changed rarely.

Visual odometry solves the SLAM problem using a combination of image feature and depth data from laser sensors. Image feature is a process to extract the image where the color change is strong. Examples of the image features are SIFT(Scale

J.-H. Kim et al. (Eds.): AsiaSim2011, PICT 4, pp. 217–225, 2012.

Invariant Feature Transform)[4] and SURF(Speed-Up Robust Feature)[5]. The two approaches are resilient to lighting and scale changes. Calculation time of SURF is faster than that of SIFT. So, we select SURF to extract image feature.

In this paper, We propose Visual odometry method using kinect shown in Fig. 1. Kinect is equipped with one camera and one laser sensor was developed by Microsoft in 2010.

Moreover, it is reasonable. Therefore, our method which is costly and non-special equipment can be expected to be used a lot of automatic movement robot.

Our proposed method consists of three steps as follows:

- First, corresponding points are found by SURF among two different images.
- Second, false corresponding image points are removed by Smirnoff-Grubbs.
- Third, movement parameters are calculated by ICP algorithm.

By repeating the three steps, the surrounding information and self-localization are updated.

This paper describes the results of the proposed measurement method.

Fig. 1. Kinect sensor

2 SURF(Speed-Up Robust Feature)

This chapter describes SURF that is robust to image scale and brightness changes[5].

It is based on two steps as follows:

Step1: Keypoint detection.
Step2: Extracting descriptor.

We would like to stress that the class/style files and the template should not be manipulated and that the guidelines regarding font sizes and format should be adhered to. This is to ensure that the end product is as homogeneous as possible.

2.1 Integral Image

Rectangular features can be computed rapidly using Integral Image[6]. In a object detection, a detection window is scanned by the whole image and detects the target property. So, many addition of pixel value is required. The integral image can be created by Eq.(1) in advance, this process is faster.

$$ii(x,y) = \sum_{x' \leq x} \sum_{y' \leq y} i(x', y') \tag{1}$$

Where i(x,y) is the brightness value of the pixel in original image and ii(x,y) is the integral value. Using the integral image, any rectangular sum be calculated in four array references(Fig. 2).

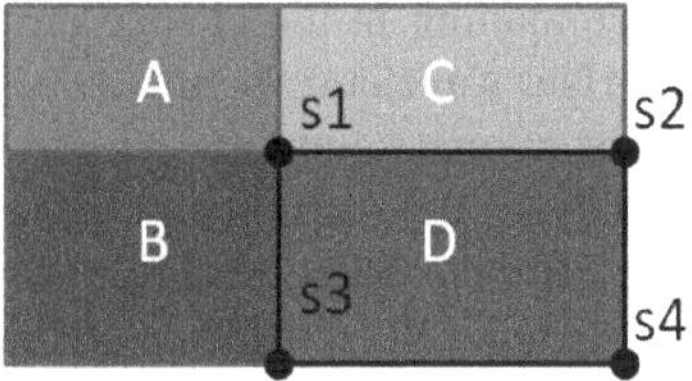

Fig. 2. Integral image

s1 is the sum of the pixel in rectangle A. Similarly, s2, s3 and s4 show A+B, A+C and A+B+C+D, respectively. For example, sum of brightness in area D is calculated as s4+s1−(s2+s3).

2.2 Keypoint Detection

Keypoint detection is based on Hessian matrix. A point of x=[x,y] in original image I, the Hessian matrix $H(\mathbf{x},\sigma)$ in x at scale σ is defined as follows:

$$H(x,\sigma)=\begin{bmatrix} L_{xx}(\mathbf{x},\sigma) & L_{xy}(\mathbf{x},\sigma) \\ L_{xy}(\mathbf{x},\sigma) & L_{yy}(\mathbf{x},\sigma) \end{bmatrix}$$

where $L_{xx}(\mathbf{x},\sigma)$ is the convolution of second order Gaussian derivative with the image I in point x, and similarly for $L_{xy}(\mathbf{x},\sigma)$ and $L_{yy}(\mathbf{x},\sigma)$. SURF is to substitute box filters instead of second order Gaussian derivatives. $D_{xx}(\mathbf{x},\sigma)$ is an approximation to $L_{xx}(\mathbf{x},\sigma)$. Likewise $D_{xy}(\mathbf{x},\sigma)$ is $L_{xy}(\mathbf{x},\sigma)$.

$D_{yy}(\mathbf{x},\sigma)$ is $L_{yy}(\mathbf{x},\sigma)$. Fig.3 shows an example of the box filter. Image convolution of second order Gaussian derivatives by using box filter can be obtained quickly by using the integral image.

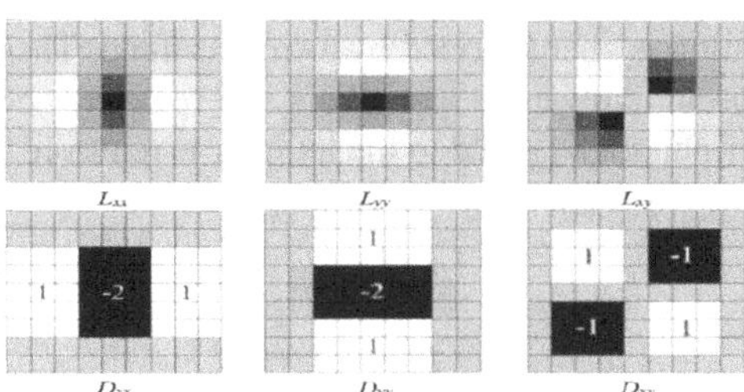

Fig. 3. Box filter

Keypoint of localization and scale are determined by the determinant of the Hessian. Keypoints are localized in scale and image space by applying a non-maximum suppression in a 3×3×3 neighborhoods. If the determinant of the H(x,σ) over the threshold, x is keypoint of scale factor σ.

2.3 Extracting Descriptor

This section explains setting the orientation and extracting descriptor of the keypoints are found in Step1. First, to create a circle from the position of the keypoints. The orientation is determined by computing in the created circle area of the image. The orientation is calculated using Haar wavelet responses.

Next, creating square regions around keypoints. At that time, the square is rotated in the orientation. This process bring the rotation invariance to SURF. The square are split up in 4×4 sub-regions to compute descriptor. Each block is determined the v=[dx,dy,|Σdx|,| Σdy|]. 1 block contains 4 elements and there are 16 blocks in one keypoint. So, SURF is composed 64(4×16) factors, which is called SURF-64. In addition to SURF-64, there are different variations such as SURF-32, SURF-128, SURF-256 according to Bay, H. and Tuytelaars, T. and Van Gool, L [5], SURF-64 is a good balance of computation time and the recognition performance. Thus our method uses SURF-64.

2.4 SURF Matching

SURF matching is done by Sign of the Laplacian and the euclidean distance. Sign of the Laplacian is landmark that the keypoint brighter or darker than the surrounding image. If sign of the Laplacian is corresponding between each SURF feature, euclidean distance of each feature is calculated.

This process saves matching-time without decreasing matching precision. Finally, the euclid distance is below the threshold, the two SURF feature are considered the same feature.

3 Smirnov-Grubbs Test

There are little miss pairs in making SURF feature pairs session among two images because of strong brightness changes, the shadows and new object. It is necessary to remove the wrong pairs due to adversely affects the measurement. Most of the pair is correct and that transition is characterized by a similar trend. Therefore, the pair have a very different transition is regarded as outliers and removed. For example, most of the pairs move to the right, then moved to the left pair is removed.

To remove them, we adapt the Smirnov-Grubbs test.

$$\tau = \frac{(N-1)t}{\sqrt{N(N-2)+Nt^2}} \tag{4}$$

Our method sets α 0.05. For example, when N is 100, significance points is 0.321. So, the data computed data's T is treated as outliner.

4 ICP(Iterative Closest Point)

ICP algorithm calculates iteratively one local minimum of equation[7]. ICP algorithm estimates transformation (R,t) for corresponding points in different BGR or Depth data. Many variances of ICP algorithm have been proposed, we used SVD(singular value decomposition) method.

Because SVD solution of ICP is stable when SVD calculation is easy[8]. This section states ICP algorithm used by our method.

ICP is composed of two steps.

Step1: Searching for duplicate data.
Step2: Movement parameter estimation by a local minimum of equation.

4.1 Searching for Duplicate Data

Finding the intersection of two independently acquired data is a first process. Duplication data are searched across over the entire depth data by computing euclid distance. In general, most smallest euclid distance data pair are linked. This process has been said in the ICP algorithm calculates the most time-consuming. So, we decided to use the depth data obtained by matching pairs of SURF feature points as described in Chapter 2. Due to the fast processing of SURF, this method can reduce the processing time for Searching for duplicate data.

4.2 Movement Parameter Estimation

The amount of movement parameter is calculated by the least squares method. Movement parameter is consist of $\boldsymbol{R}$(rotation:3×3) and $\boldsymbol{t}$(transport:1×3). $\boldsymbol{R}$, $\boldsymbol{t}$ are obtained by Eq. (5)

$$E(R,t)min = \sum_{i=1}^{k} \left\| f_i - Rs_i - t \right\|^2 \tag{5}$$

Each of the two independent 3D data is called first data as f and second data as s. k is the number of pairs which is passed SURF mathcing and Smirnov-Grubbs test. $\boldsymbol{R}$, $\boldsymbol{t}$ is acquired ,s'is computed. s' is transformed s with the $\boldsymbol{R}$, $\boldsymbol{t}$.

For calculating the parameter, determine the covariance matrix(Σfs). Be given to the singular value decomposition as follows.

$$\sum fs = U_{fs} S_{fs} V_{fs}{}^{t} \tag{6}$$

Optimum rotation angle is given as follows.

$$R = U_{fs} \begin{bmatrix} 1 & 0 & 0 \\ 0 & 1 & 0 \\ 0 & 0 & det(U_{fs} V_{fs}{}^{t}) \end{bmatrix} V_{fs}{}^{t} \tag{7}$$

There are several ways to calculate the parameters. This solution has been rated as stable and easy[8].

If the distance between f and s' is below the threshold , complete the calculation. Otherwise back to the Eq.(5).

That way, we can determine the amount of movement between the two D-data. It is found by moving distance from the starting point to continue this process sequentially without using rotation of the tire and the GPS.

Next chapter says our proposed method.

5 Proposed Method

In 2～4 chapter, SURF ,Smirnov-Grubbs test and ICP algorithm are described. This chapter says the proposed method combined them to solve the SLAM problem. The proposed method consists of three steps.

Following are instructions for proposed method.

5.1 Key Point Detection

First, getting two images and the corresponding depth data.

Data is defined as the image and depth information. Acquisition time of **Data:1** and **Data:2** are different. **Data:1** is gotten prior to **Data:2.**

The key points from Image1 and Image2 are extracted by SURF. Then, associated the pair with the euclid distance matching.

5.2 Delete of Outlier

Second, the Smirnov-Grubbs test removes the pair data were regarded as outliers found a significant difference from the overall trend of a pair of feature points.

5.3 Calculation of Movement Parameters

ICP algorithm calculates the parameter or how the sensor moves in **Data:1** and **Data:2** from the remaining pair of depth data.

Finally, **Data:1** is set to **Data:2**. New **Data:2** is acquired now image and depth data.

By the repeat this process, it is possible to self-localization.

Fig.4 is flowchart of the proposed method.

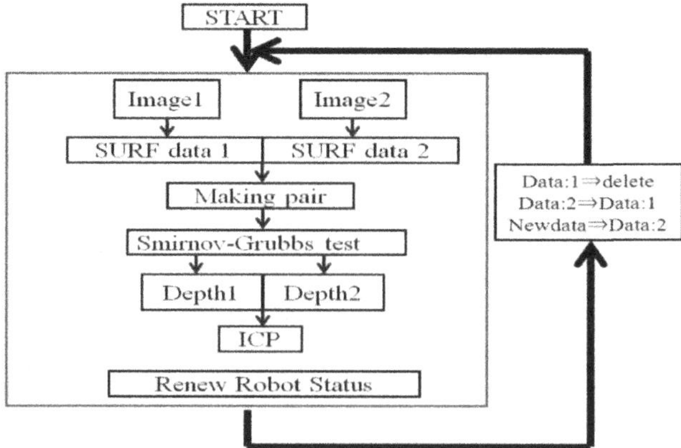

Fig. 4. Flowchart of the proposed method

6 Result

We moved the kinect under various situations to evaluate proposed method.

- Horizontal movement
- Straight movement
- Free movement in our laboratory.

In these three situations, the accuracy of our method is tested. Our method was implementd by C++, OpenCV, OpenNI and OpenGL.

6.1 Horizontal Movement

The kinect was moved to Horizontal movement.

Fig. 5 shows the start point image. Fig. 6 is the kinect status in Fig. 5. Moreover, Fig. 7 shows the after movement. Fig. 8 depicts as kinect status in Fig.7.

Fig. 5. First image

Fig. 7. Kinect has moved to left-side

Fig. 6. Kinect status at Fig. 1.
Red: Start position
Blue:kinect rotation

Fig. 8. Kinect status at Fig. 3.
Red: Start position
Blue:kinect rotation
Green: localization

6.2 Straight Movement

This experiment verifies computing kinect's status during moving straight.

Fig.9 shows the start position. Fig.10 is kinect status in Fig.10. Fig.11 shows the after movement. Fig.12 depicts as kinect status in Fig.11.

Fig. 9. Start image

Fig. 11. Kinect has gone straight

Fig. 10. Kinect status at Fig.10.
Red: Start position
Blue:kinect rotation.

Fig. 12. Kinect status at Fig.12.
Red: Start position
Blue:kinect rotation.
Green:localization.

6.3 Capturing Our Laboratory

This section is described the verification of free movement.

Fig. 13. Building Map in our laboratory

6.4 Evaluation

These results using our method can solve to SLAM problem.

In Fig.5～Fig.8, these four pictures are depicted as the success of calculation to Horizontal movement.

In Fig.9～Fig.12, these four pictures obtained by the proposed method can solve to compute precisely during straight movement.

Fig.13 shows the building map that are combined multiple images and depth datas which are revised by our method. This map includes 3-D data. This data is useful to capture the object and avoid obstacles.

However, after the experiment shown in **6.3**, we put the kinect at start poison, Green square and Red square were not overlap.

This result shows our calculation contain false data. If our method is all-right, Green and Red squares is overlap. We want to remove false data with SURF-matching and Smirnov-Grubbs test. However, removal of these two methods is not yet perfect.

To improve the measurement accuracy, we suppose that the other processing is required.

7 Conclusion

The purpose of this study is to obtain the building map and estimate the self-location using kinect which is equipped with one camera and one laser sensor. By this way, the cost becomes low. So, the implementation of this approach is expected to be widely used in autonomous mobile robot.

The proposed method consists of three steps.

1. Key point detection among two images by SURF.
2. Outliers deletion using the Smirnov-Grubbs test
3. Calculation of movement parameters by ICP.

By repeating these processing, a robot can be making the modeling map and performing the self-localization. The experimental results have shown by this way is successful to solve SLAM. However, it is found accuracy fall as time goes by.

After this, we would like to make further improvements of the proposed method in three points.

First, enhancing accuracy of computed parameter is realized. Proposed method is concerned about the accuracy in the long run. Probably, there are some false pairs after step 2 which deletes outliers. The remaining false pairs may negative impact on accuracy. We consider the other image feature which is high performance of corresponding image point , not SURF.

Second, we would like to reduce computation time in order to fast automatic movement.

Third, we consider combination our method and other information. For instance, robot's wheel data aids answer computed by our method.

Finally, our method is simple and easy to use due to reasonable and non-special equipment. If whoever wants to develop a convenient and affordable automatic driving machine, this method would be useful.

References

1. Talbot, N.C.: Centimeters in the Field, a Users Perspective of Real-Time Kinematic Positioning in a Production Environment. In: Proc. ION GPS 1993, Slat Lake City U.S.A., pp. 1049–1057 (September 1993)
2. Muir, P.F., et al.: Kinematic Modeling for Feed-back Controul of an Omnidirectional Wheel Mobile Robot. In: Cox, I.J., Wilfong, G.T. (eds.) Autonomous Robot Vehicles, pp. 25–31. Springer (1990)
3. Frintrop, S., Jensfelt, P., Christensen, H.I.: Intelligent Robots and System. In: 2006 IEEE/RSJ International Conference, pp. 2582–2587 (2006)
4. Lowe, D.G.: Lowe, "Distinctive image features fromscale-invariant keypoints". Journal of Computer Vision 60(2), 91–110 (2004)
5. Bay, H., Tuytelaars, T., Van Gool, L.: SURF: Speeded Up Robust Features. In: Leonardis, A., Bischof, H., Pinz, A. (eds.) ECCV 2006. LNCS, vol. 3951, pp. 404–417. Springer, Heidelberg (2006)
6. Viola, P., Jones, M.: Rapid Object Detection using a Boosted Cascade of Simple Feature. In: IEEE Computer Society Conference on Computer Vision and Pattern Recognition, vol. 1, pp. 511–518 (2001)
7. Paul, J., McKay, N.D.: A method for registration od 3-D shapes. IEEE Transaction on Pattern Analysis and Machine Intelligence 14(2), 33–80 (1986)
8. Eggert, D.W., Lorusso, A., Fisher, R.B.: Estimating 3-D rigid body transfor-mations:a comparison of four major algorithms. Machine Vison and Applications 9, 272–290 (1997)
9. Kanatani, K.: Geometric Computation for Machine Vision, Oxford (1993)

Simulation of Gas Purge Process in City Gas Pipe

Chongho Youn[1], Masahiro Arakawa[2], Akira Kogo[3], and Toshiharu Kagawa[1]

[1] Precision and Intelligence Laboratory, Tokyo Institute of Technology,
4259 Nagatsuta, Midori-ku, Yokohama 226-8503, Japan
youn.c.aa@m.titech.ac.jp
[2] Tokyo Gas Engineering Co., Ltd., Nissay Aroma Square, 5-37-1,
Kamata, Ota-ku, Tokyo 144-8721, Japan
[3] Sony Corporation, Atsugi Tec.No.2
4-16-1 Okata Atsugi-shi, Kanagawa, 243-0021, Japan

Abstract. In this study, the simulation method of gas purge process in city gas pipe is proposed. In the concentration simulation, more detailed mesh was used. In addition, the CIP method was used to reduce the numeric diffusion. In experiment, the gas mixture that consists of air and carbon dioxide is used. The purge process of gas is measured by using a venturi flowmeter and a laminar flowmeter, and the simulation results were compared with experimental results. As a result, the simulation results have a good agreement with experimental results.

1 Introduction

In process industries, such as city gas industry, many pipeline and tank exist. These pipelines need reconstruction and maintenance service for safety. In that case, the work called a "purge process" which replaces the city gas in the piping with nitrogen gas or air is done. It is important to do this purge process by safety, cheapness, and a method with little influence on environment. For this reason, the analysis of a flow and diffusion of the gas in a pipeline is required.

As an example, the pipeline network of Tokyo Gas Co., Ltd. is shown in Fig. 1. The total extension of a pipeline is 52,310 km and large-scale purge process is done 15 -20 times a year in this pipeline network. In this purge process, the nitrogen gas of 1,000-2,000 m^3 is used. The expense is millions of yen including facilities expense, nitrogen gas expense, personnel expenses. Furthermore, using biogas for a gas pipe passage is assumed in future. The importance of the analysis of the gas distribution in a pipeline is increasing.

On the diffusion phenomenon by the flow in a circle pipe, the analysis and the experiment are conducted by Taylor [1], Flint [2] , Shih [3], Ekambara[4]. However, as for these researches, the diameter of the pipeline was performed by tens of mm, and pipeline length was mainly performed by the direct pipe of tens of m or less. Fluid is also using the liquid.

J.-H. Kim et al. (Eds.): AsiaSim2011, PICT 4, pp. 226–234, 2012.

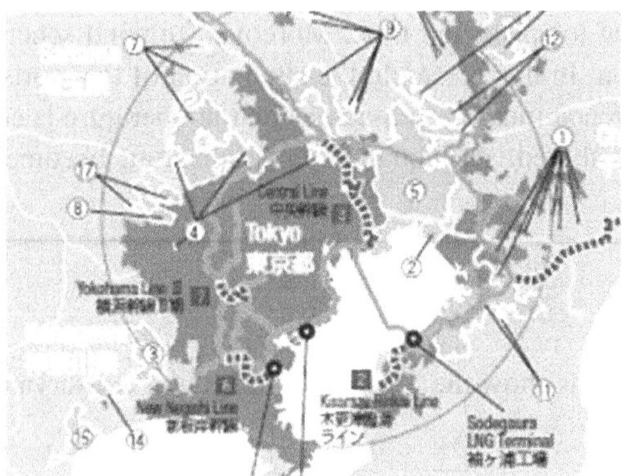

Fig. 1. The pipeline network of Tokyo Gas Co., Ltd.

In this research, the numerical computation technique which solves the flow and diffusion of gas is proposed, and confirm the effectiveness by an experiment. Chapter 2 explains a theoretical formula and its composition about the flow in a circle pipe. Chapter 3 and Chapter 4 explain the numerical analysis and the experimental device. In Chapter 5, a simulation result is shown and comparison with an experimental result is performed.

Nomenclature

C : concentration	[-]
D : inner diameter of a pipeline	[m]
K : diffusion factor	[m^2/s]
L : length of a pipeline	[m]
M : molecular weight	[kg/mol]
P : pressure	[Pa]
R : Gas constant	[J/(kg·K)]
R_e: Reynolds number	[-]
t: time	[s]
u: velocity	[m/s]
X: molecular density	[mol/ m^3]
ρ: density of gas	[kg/ m^3]
θ: temperature	[K]
λ: coefficient of friction	[-]

2 Theoretical Formula

In the simulation in a pipeline of single gas, a finite difference method is common. A finite difference method divides a pipeline into some short sections, and calculates a differentiation value to space and time. Moreover, upwind scheme is used in the numerical computation. In this research, the simulation of two sorts of gas is performed using a finite difference method. It assumes that temperature is constant and the energy equation is not solved. An equation of momentum becomes like the following equation.

$$\frac{\partial u}{\partial t} + u\frac{\partial u}{\partial x} = -\frac{1}{\rho}\frac{\partial P}{\partial x} - \frac{\lambda}{2D}u^2 \tag{1}$$

The following formulas show the formula of molecular conservation [6].

$$\frac{\partial X}{\partial t} + \frac{\partial (Xu)}{\partial x} = 0 \tag{2}$$

A one-dimensional diffusion phenomenon is denoted by the following equation.

$$\frac{\partial C}{\partial t} + u\frac{\partial C}{\partial x} = K\left(\frac{\partial^2 C}{\partial x^2}\right) \tag{3}$$

The diffusion factor k is a value from experiment, and is calculated from the following approximate expressions [7].

$$C(x,t) = \frac{C_0 + C_1}{2} + \frac{C_0 - C_1}{2} erf\left\{\frac{x - \bar{u}t}{2\sqrt{K(t - t_0)}}\right\} \tag{4}$$

The equation of state of an ideal gas is expressed as follows.

$$P = \rho R \theta \tag{5}$$

A gas constant is given by the following formulas.

$$R = \frac{R_0}{M} \tag{6}$$

Molecular weight changes with change of concentration and the change is expressed below.

$$M = M_0 + (M_1 - M_0)\cdot C \tag{8}$$

Density is shown to the following formulas by molecular density and molecular weight.

$$\rho = M \cdot X \tag{9}$$

3 Numerical Simulation

In numerical computation, as shown in Fig. 2, a Staggered grid is used. Since the diffusion velocity of concentration was slower than the flow velocity, the mesh of calculation of concentration was made finer than other meshes. When calculating the molecular weight of the equation (8), the average value of the concentration calculation result was used. The difference equations of the momentum, the molecular conservation, and the diffusion are shown below.

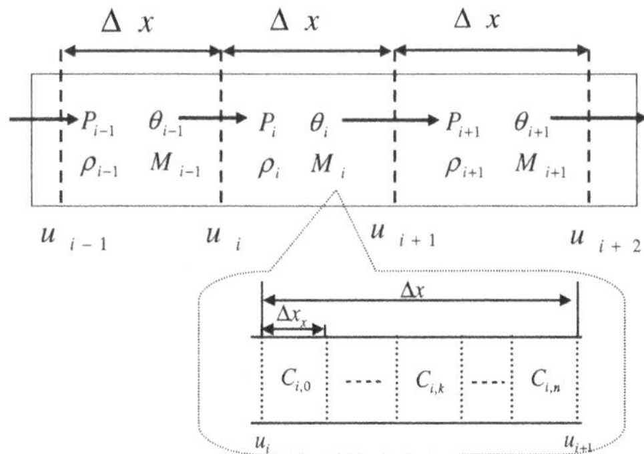

Fig. 2. Mesh of the pipeline by a staggered grid

1) Momentum equation

$$u_{i,j+1} = u_{i,j} - \frac{\Delta t}{\Delta x} u_{conv} - \frac{1}{\hat{\rho}_{i,j}} \frac{\Delta t}{\Delta x} (P_{i,j} - P_{i-1,j}) - \frac{\lambda \Delta t}{2D} |u_{i,j}| u_{i,j} \tag{10}$$

By upwind scheme, a convection term becomes below.

$$u_{conv} = \frac{u_{i,j} + |u_{i,j}|}{2} (u_{i,j} - u_{i-1,j}) + \frac{u_{i,j} - |u_{i,j}|}{2} (u_{i+1,j} - u_{i,j}) \tag{11}$$

$\hat{\rho}_{i,j}$ is the average of the density of a front and a back cell. A coefficient of friction changes with Reynolds numbers. In the case of a laminar flow, it is as follows.

$$\lambda = \frac{64}{\text{Re}} \tag{13}$$

In the case of a turbulent flow, it is as follows from the formula of Blasius.

$$\lambda = 0.3164 \, \text{Re}^{-\frac{1}{4}} \tag{14}$$

2) Molecular conservation equation

$$X_{i,j+1} = X_{i,j} - \frac{\Delta t}{\Delta x} X_{i,j} (u_{i+1,j} - u_{i,j}) - \frac{\Delta t}{\Delta x} X_{conv} \tag{15}$$

By upwind scheme, a convection term becomes below.

$$X_{conv} = \frac{\hat{u}_{i,j} + |\hat{u}_{i,j}|}{2}(X_{i,j} - X_{i-1,j}) + \frac{\hat{u}_{i,j} - |\hat{u}_{i,j}|}{2}(X_{i+1,j} - X_{i,j}) \quad (16)$$

$\hat{u}_{i,j}$ is the average of the density of a front and a back cell.

3) Diffusion equation

In the convective diffusion equation of concentration, there is a problem of the numerical diffusion by a finite difference method. For this reason, concentration is calculating using the fine calculation mesh. Furthermore, in order to make numerical diffusion small, not a finite difference method but the CIP (Constrained Interpolation Profile Scheme) method was used [8].

4 Experimental Device

Generating of mixed gas and the composition of a measuring device are shown in Fig. 3 [9]. This device consists of a gaseous mixture generating part and a test section part. The gaseous mixture generating part consists of a pressure regulator, a buffer tank, a spool type servo valve, and a laminar type flow meter. In this research, the gaseous mixture which consists of air and carbon dioxide is set as the object of an experiment. Let air concentration 0.1 be the gas A, and let air concentration 0.9 be the gas B. There is gas A in a pipe with the initial state and performs an experiment to replace gas A in gas B afterwards.

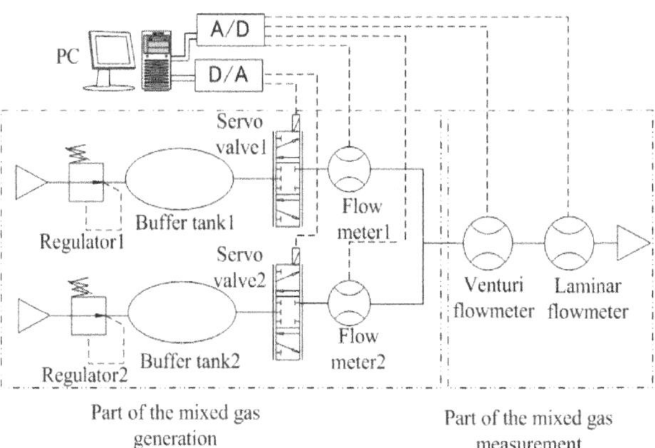

Fig. 3. Experimental device

Two kinds of the pressure regulator are used. The regulator of the air supply line is a product of SMC Corp. and the regulator of the carbon dioxide supply line is a product of Yutaka Eng. Corp.. The capacity of a buffer tank is 5.0 m^3. The pressure of buffer tank was set to 0.5 MPa. Servo valves are a product of Festo Corp. (type: MPYE-010B). Moreover, the thermometer is attached to the flow meter to in order to rectify a gaseous temperature. The flow rate of each gas is loaded to PC through an A/D board, and controls a servo valve through an A/D board. The long tube is attached between the gaseous mixture generating part and the test part. The inner diameter of a tube is 13 mm.

The experiment method is the following. At first, pipe is filled with the gas A by passing the gas A. Then, a flow is stopped by closing a servo valve for 2.5 seconds. After then the servo valve was opened, the gas B was passed and gas was replaced.

5 Evaluation of Simulation

5.1 Simulation Conditions

A simulation model is shown in Fig. 4. A simulation parameter is shown in Table 1. The flowrate of a gaseous mixture is set to 5.0×10^{-4} m^3/s. The number of meshes of the flow velocity, pressure and molecular density were set to 40 (Δx =2.5 m). The mesh of concentration divided the flow velocity mesh into 50 meshes (Δx_x =0.05 m). Time interval is 1.0×10^{-4} s.

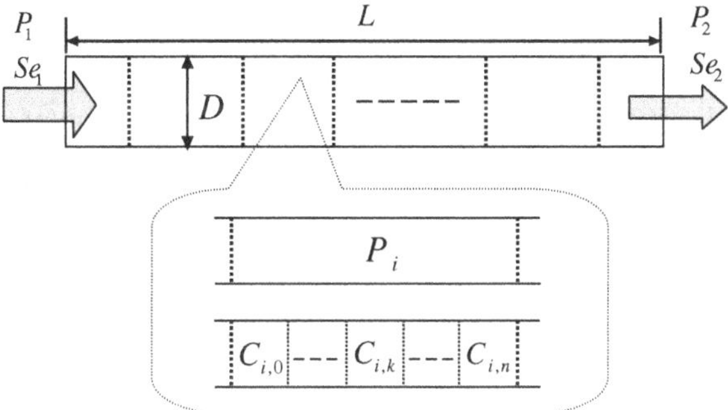

Fig. 4. Simulation model

Table 1. Simulation parameter

Symbol	Value
L	100.3[m]
D	13[mm]
P_1	500 [kPa(gage)]
P_2	0 [kPa(gage)]
Se_1	0.43[mm^2]
Se_2	63.6 [mm^2]

5.2 An Experimental Result and a Simulation Result

The experimental results and simulation results are compared, and the result is shown in Fig. 5. A vertical axis is air concentration and a horizontal axis is time. This result shows that the simulation results have a good agreement with experimental results. From this result, the effectiveness of the simulation method was confirmed.

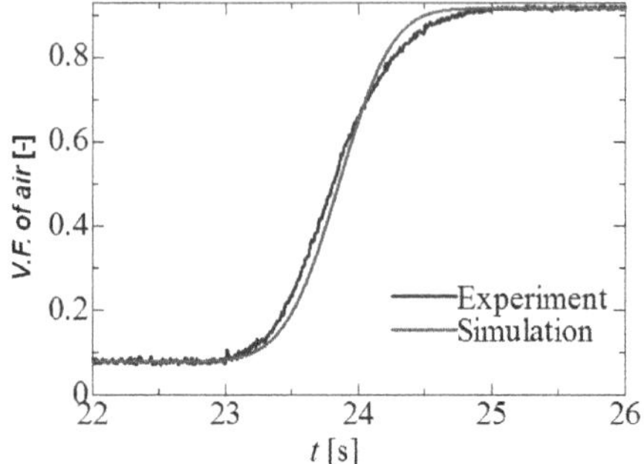

Fig. 5. Comparison of the experimental results and simulation results

A simulation result is shown below. The simulation results of the density of 0 seconds, 10 seconds, 20 seconds, and 30 seconds are shown in Fig. 6. A vertical axis is density and a horizontal axis is the distance from the starting point of piping. It shows that density changes by being replaced with gas B by gas A.

The simulation results in each time of the velocity are shown in Fig. 7. The upper figure is the figure which enlarged the vertical axis of the lower figure. The simulation results in each time of the pressure are shown in Fig. 8. Since density becomes light in being replaced by the gas B from the gas A, the pressure loss in piping after 30 seconds is small.

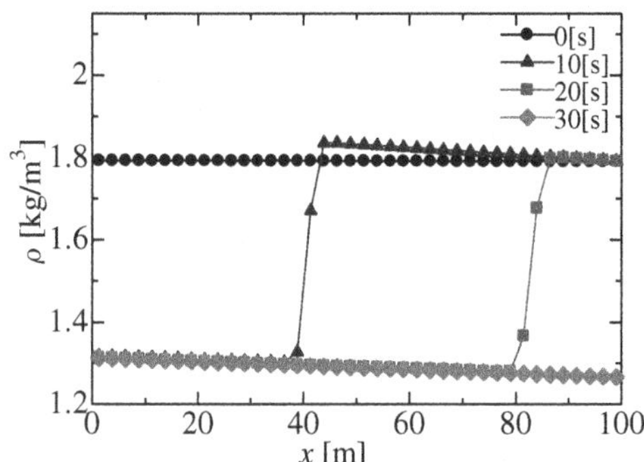

Fig. 6. Simulation results (density)

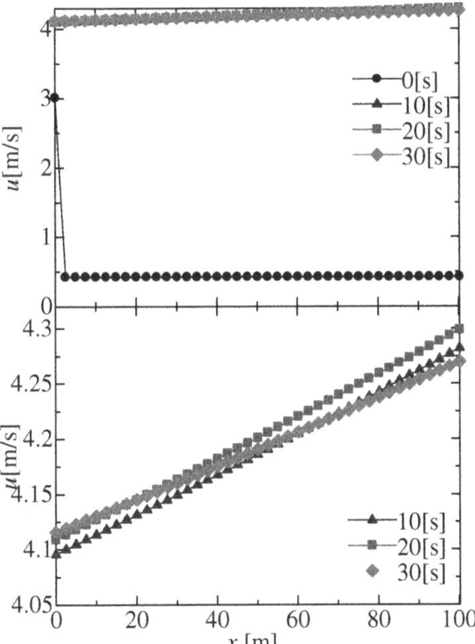

Fig. 7. Simulation results (velocity)

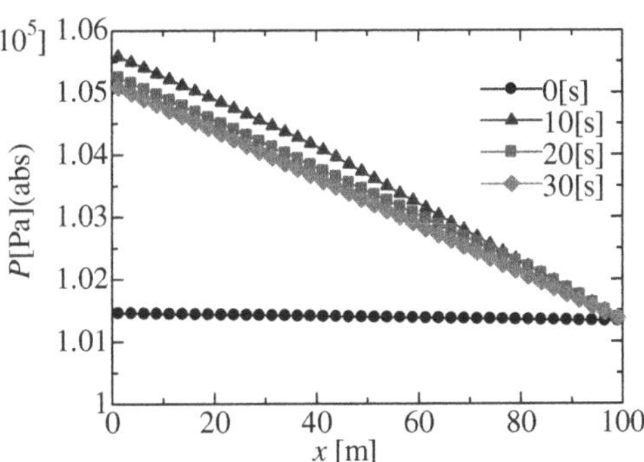

Fig. 8. Simulation results (pressure)

6 Conclusions

In this research, the simulation method of the purge process of city gas was proposed. Moreover, the experiment of purge process was conducted by 100m tube and the experimental results and simulation results were compared. The obtained conclusions become below.

1) The simulation of the gaseous mixture in piping on purge process was proposed by using CIP method.
2) It was confirmed that the simulation results have a good agreement with experimental results.

References

1. Taylor, G.I.: Dispersion of soluble matter in solvent flowing slowly through a tube. Proceedings of the Royal Society of London 219, 186–203 (1953)
2. Flint, L.F., Eisenklam, P.: Longitudinal gas dispersion in transitional and turbulent flow through a straight tube. The Canadian Journal of Chemical Engineering 47(2), 101–106 (1969)
3. Shih, S.-D., Tung, Y.-K.: Internal Layers of a Transient Convection-Diffusion Problem by Perturbation Methods. Advances in Water Resources 18(22), 111–120 (1995)
4. Ekambara, K., Joshi, J.B.: Axial mixing in pipe flows: turbulent and trasition regions. Chemical Engineering Science 58, 2715–2724 (2003)
5. Zhang, X., Yoshida, M., Kagawa, T., Iio, S.: Characteristics of Compressible Fluid System including Branch and Junction. Japan Fluid Power System Society 40(3), 50–55 (2009)
6. Takeda, T., Hishida, M.: Study on Diffusion and Natural Circulation of Two Component Gases. The Japan Society of Mechanical Engineers 60(569), 208–214 (1994)
7. Arakawa, M., Kagawa, T., Takeuchi, M., Rokuka, T., Someya, T.: Consideration of mixing effect of different gas in pipes 26(4), 10–16 (2009)
8. Yabe, T., et al.: Constrained Interpolation Profile Scheme. Morikita Publishing Co., Ltd. (2003)
9. Yamazaki, S., Funaki, T., Kawashima, K., Kagawa, T.: A concentration measurement system for binary gas mixtures using two flowmeters. Measurement Science and Technology 18, 2762–2768 (2007)

A Simulation Study for Micro Assembly Cell in a Micro Lens Module Industry

Dug Hee Moon[1], Geun Hyun Lee[1], Jun Yeob Song[2], and Choun Ho Kim[3]

[1] Department of Industrial and Systems Engineering, Changwon National University, Changwon 641-773, Korea
dhmoon@changwon.ac.kr, kkulyut@naver.com
[2] Intelligent Machine Systems Research Center
Korea Institute of Machinery and Materials, Daejeon 305-343, Korea
sjy658@kimm.re.kr
[3] Division of Automation, Yoosung Precision Co. Ltd, Changwon 641-847, Korea
chounho.kim@yoo-sung.com

Abstract. This paper introduces a case study of implementing simulation for developing a micro assembly cell in a Korean optical factory. This factory produces a micro lens module that is the part of a phone-camera. The newly designed assembly cell pursues the concept of micro-machine. From the simulation experiments, we can estimate performances of the system.

1 Introduction

A camera-phone is a phone having the combined function of a mobile phone and a camera. After the camera-phone is developed, it rapidly replaced the traditional mobile phone. In the report of Gartner Dataquest [1], the global demand of mobile phone as of 2008 was 1.2 billion, and it would be increased up to 1.4 billion in the year 2013. Recently, the 3G technology has been adapted to the mobile phone and dual cameras are equipped on a phone, where the first camera is for taking a picture and the second is for taking a movie. Lens module is a core part of camera module as shown in Figure 1, and thus, the demand of the lens module increases twice of the increased demand of mobile phone. Furthermore, the demand of camera module increases explosively in an automotive industries, and thus, optical manufacturing factories are trying to increase their production capacity and enhance their competitiveness in terms of quality and cost [6].

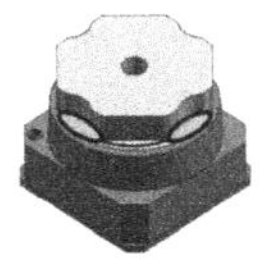

Lens Module Camera Module

Fig. 1. Examples of camera module and lens module

J.-H. Kim et al. (Eds.): AsiaSim2011, PICT 4, pp. 235–240, 2012.

As shown in Figure 2, a lens module is composed of multiple unit lenses, spacers, and shield. In the case of a three (or five) mega-pixel lens module, four lenses, three spacers, and one shield are assembled together in the lens module assembly process. Then the lens module is assembled with the PCB(Printed Circuit Board) in the packaging process and then become a camera module.

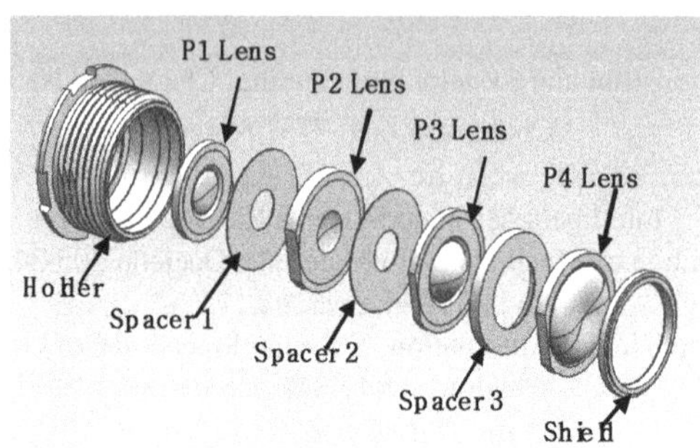

Fig. 2. Assembly chart of three mega-pixel lens module

The factory considered in this paper produces various phone-camera modules, from VGA to eight mega-pixel lens module. Although the global demands of phone-camera are increasing explosively, the assembly and inspection processes of the lens modules are conducted manually in most of the factories, because high-precision technologies are required for the assembly operations. Recently, the factory became interested in changing its operations from manual to automatic because labor and material costs are increasing. Furthermore, in spite of the substantial increase in market demand, the selling price is continuously going down. Thus they have developed various kinds of automatic assembly machines (See Moon et al. [2], [3], [4], [5] and [6]).

This paper introduces a case study of applying 3D simulation technology to the development of an automatic assembly line in a Korean optical factory. 3D simulation technology is adopted from the early stage of the development. In the conceptual design and the initial design phases for individual equipment, 3D mechanical simulation using CATIA® and DPM-Assembly® was applied. For the design of assembly line, system simulation using QUEST® was conducted for the evaluating the performance of assembly line.

2 Lens Module Assembly Line

2.1 Lens Module Assembly Processes

Figure 3 shows assembly processes of lens module which are composed with nine parts. At first, a lens is inserted to the lens holder (barrel), and then three additional lenses, three spacers, and a shield are stacked into the holder one by one as in Figure 3.

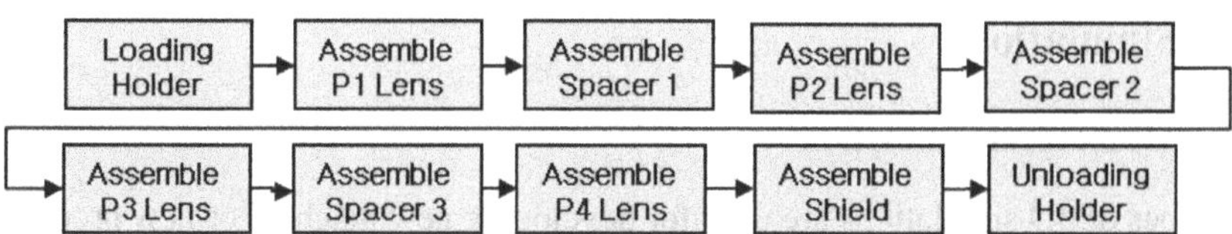

Fig. 3. Assembly processes for lens module

2.2 Concepts of Lens Module Assembly Lines

There are two layout concepts for the design of assembly line layout. The first concept is that only two parts are assembled in a machine-cell, and therefore four machine-cells are connected as a serial line to finish all assembly processes (see Figure 4(a)). The second concept is that all parts are assembled in a machine-cell, and thus only one machine-cell is enough for assembly. In this case four machines are considered as a parallel line as shown in Figure 4(b). The merits of first layout concept are the short cycle time in a machine and the low investment costs caused by the simple structure. However, the merit of second layout concept is the flexibility because four different types of lens modules can be assembled simultaneously. Figure 5 shows the 3D models of the two types of machine-cells introduced in this paper.

MC 1	MC 2	MC 3	MC 4
Loading, Assy. P1, Assy. S1	Assy. P2, Assy. S2	Assy. P3, Assy. S3	Assy. P4, Assy. S4, Unloading

(a) Concept 1 (serial line)

MC 1	MC 2	MC 3	MC 4
Loading ~ Unloading	Loading ~ Unloading	Loading ~ Unloading	Loading ~ Unloading

(b) Concept 2 (parallel line)

Fig. 4. Two layout concepts of assembly line

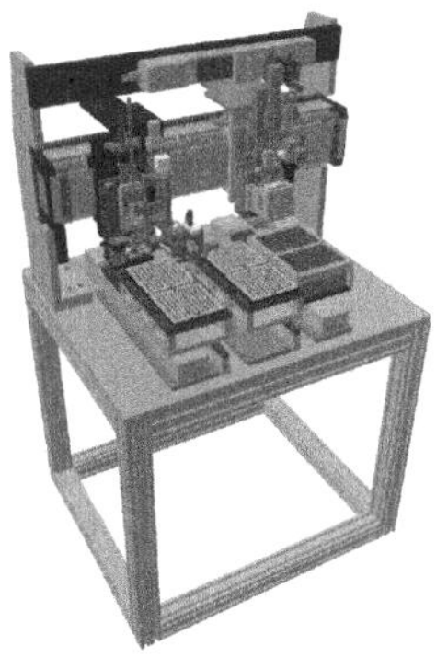

(a) Concept 1 (serial line)

(b) Concept 2 (parallel line)

Fig. 5. 3D CAD models of assembly machine-cells

3 Simulation

3.1 Mechanical Simulation

Two types of 3D simulations are used for developing new machine or new production line. These included the mechanical simulation for determining the mechanism of the equipment, and the system simulation for evaluating the performance of the line.

Mechanical simulation was conducted using DPM-Assembly® which has been developed by the Dassault Systemes Co. Ltd. With the mechanical simulation, many design errors were detected and corrected in the basic design and detailed design phases. After finishing the 3D modeling and defining kinematics, the moving speed and acceleration/deceleration should be defined to estimate the cycle time. In this phase, we can reduce the cycle time using the precedence network as shown in Figures 6 and 7. To reduce the cycle time, various motion sequences are tested and simulated to evade the collision among devices. As a result, the cycle time of first machine (concept 1) is reduced from 2.46 seconds to 2.36 seconds and that of second machine (concept 2) is reduced from 11.5 seconds to 8.9 seconds respectively.

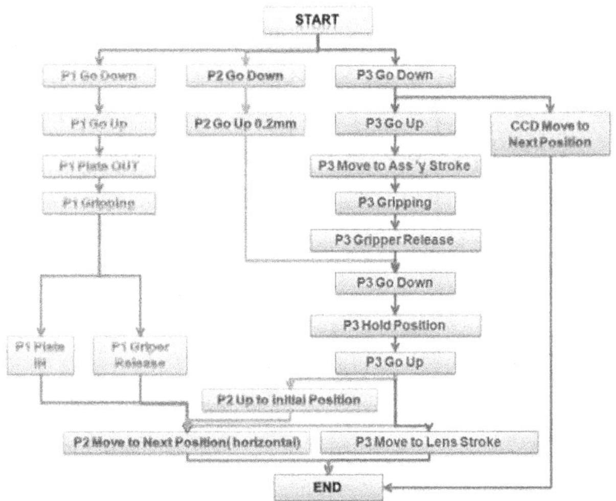

Fig. 6. Precedence network of operations (Concept 1)

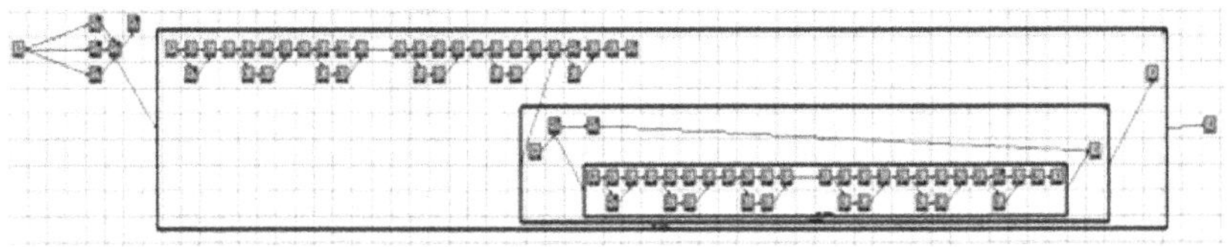

Fig. 7. Precedence network of operations (Concept 2)

3.2 System Simulation

After finishing the 3D mechanical simulation, system simulation model was developed with QUEST® as shown in Figure 7. The simulation model considered in this paper is composed of four identical assembly machine-cells to compare the productivity with the first type of assembly line that was explained in section 2.2. The number of operator is one person, and his major job is the material handling of trays which contains 100 parts in a tray.

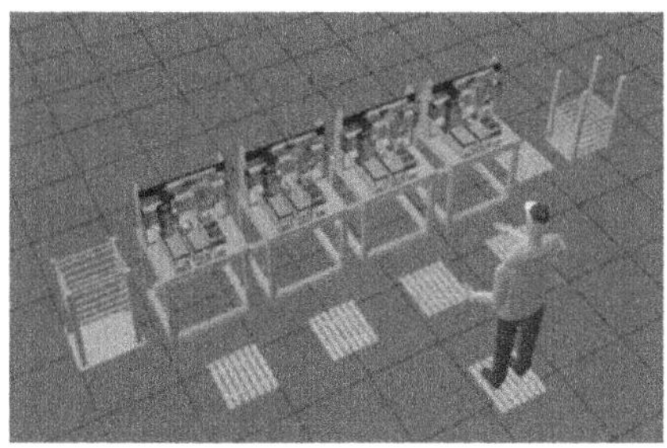

(a) Concept 1 (serial line)

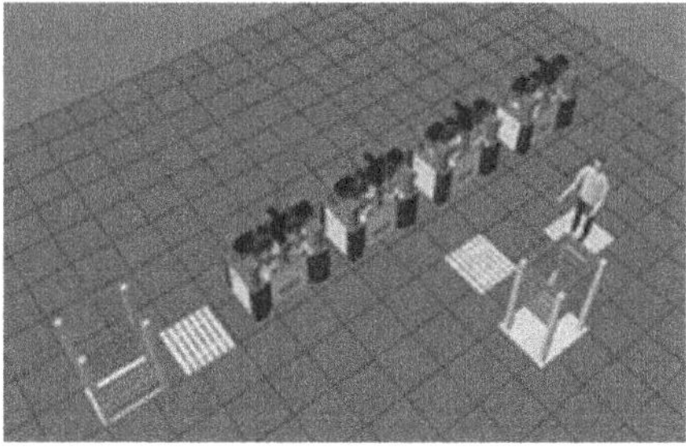

(b) Concept 2 (parallel line)

Fig. 8. System simulation models

The simulation run length was set to 250 days (24 hours per day) excluding the warm up period. The distribution functions of changing tray time are assumed with the normal distributions, NORM(40,5) and NORM(20,2.5) respectively. The simulation result indicates that the number of lens modules assembled was 8,816,373 and the takt time was estimated to 9.80 seconds. It means that the percentage of downtime due to the changing trays is about 10%.

Table 1. Results of system simulations

	Utilization		Throughput		Takt Time
	Machine	Laborer	Trays	Lens Modules	
Concept 1	89.7%	21.2%	82,086	8,208,600	2.63 sec
Concept 2	90.8%	10.4%	88,163	8,816,373	2.45 sec

4 Conclusion

3D simulation is a powerful methodology for developing a new product, equipment, and manufacturing system. It enables the verification of errors in the design and estimation of system performance before production. This paper is a case study of 3D simulation in an optical manufacturing factory, in which a phone-camera lens-module is assembled. 3D model of individual equipment was built with CATIA®, and the CATIA® model was transformed into DPM Assembly® to conduct the mechanical simulation (verifying the collision in a dynamic state) and the estimation of the cycle time. Finally, simulation models were developed with QUEST®, and we compared various performance measures. The company developed both of two types of assembly lines. The first concept has been used in mass production models and the second concept has been used in the models with relatively small quantities.

References

1. Gartner Inc., Gartner Dataquest Market Insight (2009)
2. Moon, D.H., Baek, S.G., Zhang, B.L., Jung, J.Y., Kim, Y.K., Song, J.Y.: Application of Virtual Manufacturing in Developing Manufacturing System for the Lens Module of Phone-Camera. Preprints of 12th International Symposium on Information Control Problems in Manufacturing (INCOM 2006) (Industrial Supplement and Discussion Paper), Saint Etienne, France, pp. 93–98 (May 2006)
3. Moon, D.H., Baek, S.G., Lee, J.S., Zhang, B.L., Shin, Y.W., Kim, Y.G.: A Case Study of 3D Simulation for Developing Automatic Assembly Line in the Phone-Camera Industry. In: Proceedings of Spring Simulation Multiconference 2007, Norfolk, U.S.A., vol. 2, pp. 90–95 (March 2007)
4. Moon, D.H., Lee, J.S., Baek, S.G., Zhang, B.L., Kim, Y.G.: Developing Automatic Lens Module Assembly System Using 3D Simulation. Journal of the Korea Society for Simulation 16(2), 65–74 (2007)
5. Moon, D.H., Xu, T., Lee, J.S., Song, J.Y., Kim, Y.G., Song, C.: Implementation of 3D Simulation for Developing Lens Module Production Line. In: Proceedings of 6th International Conference of Information and Management Sciences, Lhasa, Tibet, China, pp. 311–318 (July 2007)
6. Moon, D.H., Zhang, B.L., Song, J.Y., Lee, C.W.: A Simulation Study of the Micro Assembly Machine for the Lens Module of Phone Camera. In: Proceedings of 4th International Workshop on Microfactory and Technology (IWMT 2008), Jeju-do, Korea, pp. 79–84 (July 2008)

Simulation-Based Study on the Effect of Scan Pitch for Ultrasonic Non-destructive Inspection in Cast Billet with Computerized Tomography

Yoko Norose, Koichi Mizutani, and Naoto Wakatsuki

Graduate School of Systems and Information Engineering, University of Tsukuba, 1-1-1 Tsukuba, Ibaraki 305–8573, Japan
norose@aclab.esys.tsukuba.ac.jp

Abstract. This paper deals with ultrasonic non-destructive inspection of cast billet using computerized tomographic method, which has been proposed by the authors. This method reconstructs pseudo sound velocity distributions, which reflect defect existence in cast billet as the decrease of the velocity. Although this method enabled to detect the defects whose size was 2.0 mm in our previous work, transducer scanning with small pitch, which resulted in long testing time, was taken. In this paper, we analyze the effect of scan pitch for ultrasonic non-destructive inspection in cast billet with computerized tomography, with aid of numerical simulation of wave propagation, to find the necessary and sufficient scan pitch to detect defects.

1 Introduction

Square billets are manufactured by continuous casting and processed into many metal products such as plates and pipes. In casting process, defects are produced by remaining inclusions during cooling and extrusion in the billets [1, 2]. Gas porosity is one of the most common types of defects. Gas porosity usually arises because gas, such as oxygen and hydrogen, forms bubbles in the billets in cooling process. Defects with diameters of about 2 to 3 mm are called pinholes, and larger defects are called blowholes, those debase the billets [3]. We treat the holes those diameters are 0.5 – 2.0 mm as defects because they are generated the most in high-temperature casting in a steel plant. Defect detection is important for quality improvement of the billets and metal products [4]. Nondestructive testing with ultrasound is efficient for the detection [5–8].

We have proposed a method to detect defects using the transmission method [9, 10]. Time-of-flight (TOF) of longitudinal waves is measured by this method. In addition, pseudo sound velocity distribution, which is different from actual sound velocity of the billet, is reconstructed from the TOF using ultrasonic computerized tomography (CT) [11]. The defects are visualized as a decrease in pseudo sound velocity because the TOF increases owing to diffraction of ultrasonic waves at the defects. We have succeeded in detecting a defect, whose diameter is 2 mm [10]. On the other hand, it was clear that the measurement took too long

J.-H. Kim et al. (Eds.): AsiaSim2011, PICT 4, pp. 241–247, 2012.

time. In the paper [10], the scan pitch was set to 2 mm, and then the number of all the measurement points are 15,000.

In this paper, to solve the above problem and to find the optimum number of measurement points, we evaluated a relationship between the scan pitch and the imaging ability of defects by numerical simulation. In this method of defect detection,the longitudinal wave is the most dominant, and only this wave is used to measure. Therefore the TOF is calculated based on scalar wave propagation model using Transmission-Line Matrix (TLM) method in the numerical verification.

2 Principles of Imaging

2.1 TOF Determination

The transmitter and receiver are set on the billet surface and scanned along the surface. On each paths, the ultrasonic longitudinal waves are measured by the transmission method. A direct wave is usually received when there are no defects on the propagation path. However, when a defect is located on the path, the received wave is not the direct wave but a wave diffracted at the defect [12]. TOFs are derived by the cross correlation of the transmitted and received waves. The TOF of diffracted wave at the defect is larger than that of the direct wave. We detect the defect to utilize the TOF difference between measured data and calculated data by ray theory, which means increase in TOF because it contains information relating to defect existence [9]. TOF difference data measured by above way are used in ultrasonic CT.

2.2 Imaging of Defects Using Ultrasonic CT

The pseudo sound velocity distribution is reconstructed by ultrasonic CT. The pseudo sound velocities at the defect areas are lower than those of defect-free areas because the TOF increases owing to extension of propagation path by the diffraction at the defects. Therefore, the decrease in the pseudo sound velocity indicates the existence of defects at the location [10].

Here, we explain about ultrasonic CT method for the square billet. Figure 1 shows a schematic view of measurement plane in the billet, transducers, and projection data used in reconstruction. The size of square billet is defined as D. The projection data, $\Delta\tau(X, \theta)$, denotes the TOF difference of the propagation path in the Y–direction passing through point (X, θ). The distribution of pseudo sound velocity, $c_{\mathrm{a}}(x, y)$, is reconstructed from the projection data using filtered back-projection [13]. To use this method, the projection data at equal intervals is required. However, intervals of measured projection data are irregular because the billet is square and the transducers are only linearly scanned. Therefore, projection data are interpolated to get equal interval data [14]. A width between one measured projection data and the other increases with increase in measured

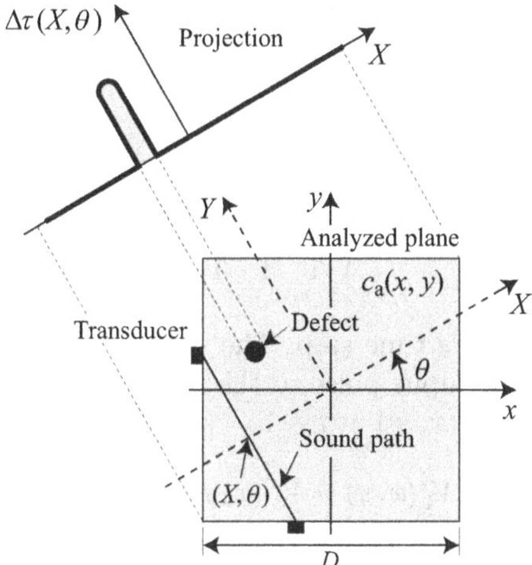

Fig. 1. Schematic view of measurement plane, transducers and projection data used in CT

points, and then the imaging ability becomes worse. Therefore, we intend to find out required scan pitch for the imaging.

3 Numerical Verification

3.1 TLM Method

TLM method is way to calculate the wave propagation sequentially based on Huygens' model discretized spatially and temporally. This study treats potential wave in view of only longitudinal wave. The shear wave is neglected. Therefore, TLM method is used in this study. In this study, the analyzed plane is considered as two-dimensional plane. The analyzed plane is gridironed by TLM elements. A TLM element has four branches as shown in Fig. 2(a). The pulse, V, incidents

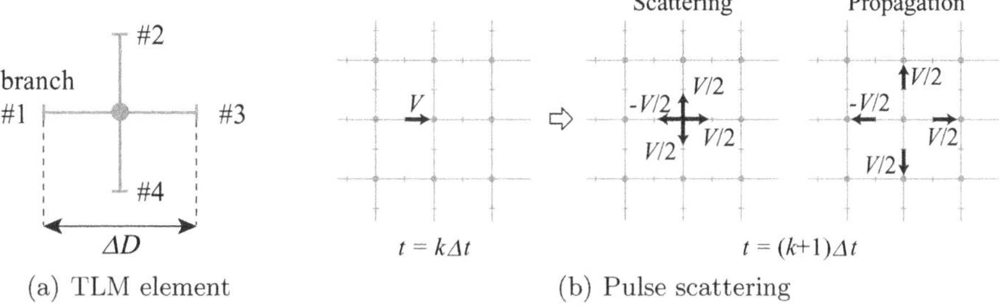

Fig. 2. Scheme of TLM

into one TLM element as shown in Fig. 2(b). The pulse scatters int four directions at the intersection. The scattering pulses on the four branches, V_1, V_2, V_3, V_4 are given by

$$\begin{pmatrix} V_1 \\ V_2 \\ V_3 \\ V_4 \end{pmatrix}_{k+1}^{\mathrm{r}} = \frac{1}{2}\begin{pmatrix} -1 & 1 & 1 & 1 \\ 1 & -1 & 1 & 1 \\ 1 & 1 & -1 & 1 \\ 1 & 1 & 1 & -1 \end{pmatrix}\begin{pmatrix} V_1 \\ V_2 \\ V_3 \\ V_4 \end{pmatrix}_{k}^{\mathrm{i}}. \tag{1}$$

where, $k = \Delta t/t$, and Δt is time step. The scattering pulse propagates on the branch, and becomes the input pulse to the next TLM element. Therefore, the input pulse on coordinate $(x,\ y)$ at $t+1(=(k+1)\Delta t)$ is given by

$$\begin{aligned} V_1^i(x,y) &= V_3^r(x, y - \Delta D) \\ V_2^i(x,y) &= V_4^r(x + \Delta D, y) \\ V_3^i(x,y) &= V_1^r(x, y + \Delta D) \\ V_4^i(x,y) &= V_2^r(x - \Delta D, y) \end{aligned} \tag{2}$$

where, ΔD is mesh size. Δt and mesh size ΔD are set in accordance with the relation given by

$$\Delta t = \frac{1}{\sqrt{2}}\frac{\Delta D}{c}, \tag{3}$$

where, c is sound velocity in the measurement piece. The pulse propagation and scattering are represented by sequential calculation of Eq. (1) and (2). In this calculation, dispersion error and time discretization should be considered. To minimize that error, ΔD should implement the following equation.

$$\frac{\Delta D}{\lambda} \leq \frac{1}{10}. \tag{4}$$

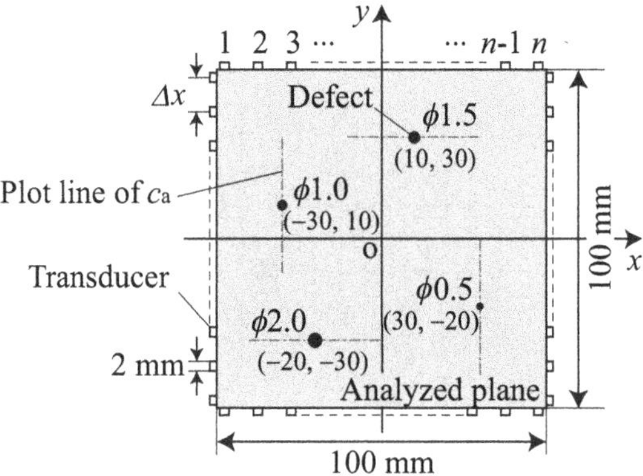

Fig. 3. Analyzed plane

In this study, it is assumed that the billet is made of duralumin. ΔD, Δt, and c are about 0.1 mm, 0.011 μs, and 6320 m/s, respectively. Therefore, the ratio of ΔD and λ is 1/20, satisfies Eq (4). In this study, V is sound pressure.

3.2 Simulation Condition

Figure 3 shows the analyzed plane. The shape of cross section of the billet is square whose size is 100 × 100 (mm^2). Four defects, whose diameters are 2.0, 1.5, 1.0 and 0.5 (mm) respectively, are located in the analyzed plane of billet as shown in Fig. 3. Measurement signal is up-chirp signal swept from 1 to 3 (MHz) in 5 μs. The wavelength of input signal, λ, is about 2 to 6 (mm). Aperture of the transducer is 2 mm. Scan pitch, Δx, varied from 1 mm to 7 mm every 2 mm. The measurement points on one side, n, are 100, 33, 20, and 14, respectively.

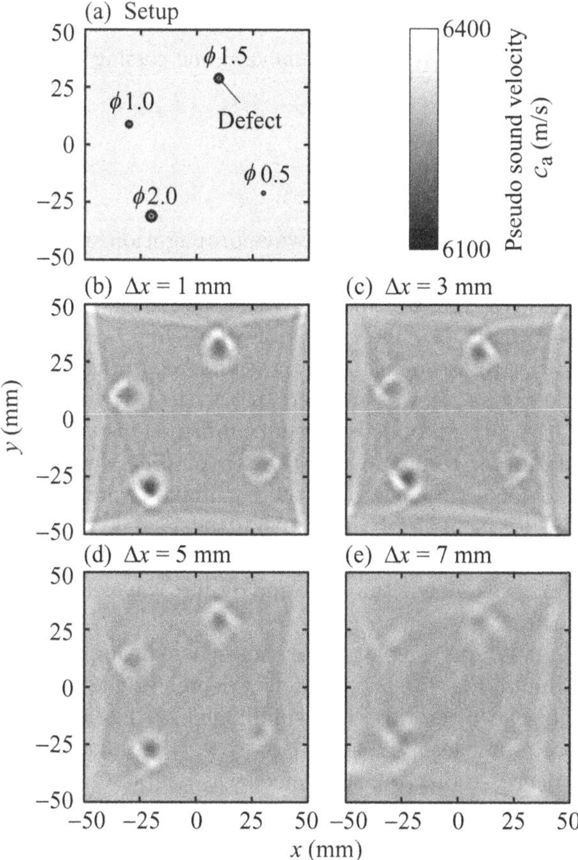

Fig. 4. Pseudo sound velocity distributions obtained by calculation; (a): calculation setup, (b)–(e): images for Δx = 1, 3, 5, and 7, respectively

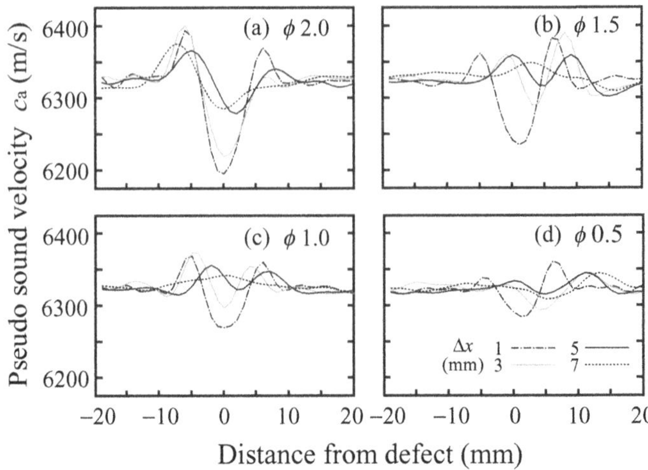

Fig. 5. Pseudo sound velocity on analyzed line passing through defects

4 Simulation Result

TOFs of sound paths were obtained in wave propagation simulation using above conditions and then the distribution of pseudo sound velocity, c_a, in the billet was reconstructed by ultrasonic CT method. The distribution of the pseudo sound velocity is shown in Fig. 4. The circles of white line on the defects in setup, shown in Fig. 4(a), are the defect size to scale. According to Fig. 4, the defect was visualized clearly as Δx was smaller. In Fig. 4(e), it was difficult to find the defects. In Fig. 4(b)–(d), all of defects were confirmed visually. However, the size of the defects are larger than those of the setup. This was caused by the wave nature, and the defects size in reconstructed images were approximately equal to the maximum wavelength of the transmitted signal, i.e. about 6 mm. The high velocity areas surrounding the defects were also equal to the wavelength. In addition, when Δx was larger than the maximum wavelength, detection of the defects were difficult.

Next, Fig. 5 shows c_a on the plot lines, those pass through the defects as shown in Fig. 3. According to Fig. 5, c_a at the defect decreased along with the size of the defect. When the scan pitch, Δx, was small, such as 1 or 3 (mm), the decrease of c_a at the defect was clearly visible. However, when Δx was large, such as 5 or 7 (mm), the decrease of c_a at the defect was vague. On the other hand, the increment of c_a was found around the defect. This increment was caused by the measurement error of TOF, which means the difference from TOF calculated by ray theory. TOF was derived from the cross-correlation between the transmitted and received signal. The received signal contained the effect of the interference between scattered and diffracted wave by the defects. Therefore, the phase of the received signal was led and the peak of the cross-correlation was shifted to the earlier. Hence, c_a around the defect increased. Although the decrease of c_a at the defect was vague in Fig. 5(d) $\Delta x = 1$ mm, the decrease at the defect could be found out owing to the increase of c_a around the defect.

As a result, the scan pitch should be smaller than the wavelength of the signal to make detection of defects possible. With respect to the defect size, although we could not confirm the size visually in the reconstructed image, we could estimate by the decrease of the pseudo sound velocity at the defect. It is required that the defect whose size is 0.5 mm in diameter can be detected for actual use. In this study, the defect which was 0.5 mm in diameter could be detected by the chirp signal swept from 1 to 3 (MHz).

5 Conclusion

We evaluated the relationship between the scan pitch and ability of defect imaging in billet by calculation. TOF was calculated by the wave propagation simulation using TLM method as the scan pitch was varied from 1 to 7 (mm) every 2 mm. Then the pseudo sound velocity distribution was reconstructed from the obtained TOF data. In consequence of the simulation, the defect whose diameter is 0.5 mm could be detected under the condition of $\Delta x = 5$ mm. In our previous paper, the scan pitch had been set to 2 mm, and then the number of measurement points are 15,000. In this study, it was found out that the scan pitch could be set to 5 mm, and this corresponds to 2,400 points of measurement. On the other hand, it was indicated that the wavelength of the transmitted signal might affect the scan pitch and the size of defect detected. For the future, we intend to investigate the defect size which can be detected by varying the frequency of the transmitted signal.

References

1. Shimura, M.: Trans. Jpn. Inst. Met. 19, 589 (1978)
2. Gupta, R.K., Ramkumar, P., Ghosh, B.R.: Eng. Failure Anal. 13, 1 (2006)
3. Kumar, S., Kumar, V., Nandi, R.K., Suresh, T.S., Datta, R.: J. Fail. Anal. and Preven. 8, 492 (2008)
4. Karamis, M.B., Nair, F.: Wear 265, 1741 (2008)
5. Ohtani, T., Kawashima, K., Drew, M., Guagliard, P.: Jpn. J. Appl. Phys. 46, 4577 (2007)
6. Salim, M.N., Hayashi, T., Murase, M., Kamiya, S.: Jpn. J. Appl. Phys. 48, 07GD06 (2009)
7. Nishino, H., Ogura, K., Komatsu, T., Yoshida, K.: Jpn. J. Appl. Phys. 48, 094501 (2009)
8. Nishino, H., Masuda, S., Mizobuchi, Y., Asano, T., Yoshida, K.: Jpn. J. Appl. Phys. 49, 116602 (2010)
9. Mitsui, H., Mizutani, K., Wakatsuki, N.: Jpn. J. Appl. Phys. 48, 07GD05 (2009)
10. Mitsui, H., Mizutani, K., Wakatsuki, N.: Jpn. J. Appl. Phys. 49, 07HC13 (2010)
11. Minamide, A., Mizutani, K., Wakatsuki, N.: Jpn. J. Appl. Phys. 48, 07GC02 (2009)
12. Kuo, M.K., Lin, T.R., Liu, P.L., Wu, T.T.: Ultrasonics 36, 803 (1998)
13. Zhu, N., Jiang, Y., Kato, S.: Energy 30, 509 (2005)
14. Minamide, A., Wakatsuki, N., Mizutani, K.: Jpn. J. Appl. Phys. 49, 07HC07 (2010)

The Study on the Validity of the Conceptual Cost Estimation Model by Using Monte Carlo Simulation

Doo-Suk Kim[1], Young-Joo Lee[2], and Seong-Yong Jang[3]

[1] Dept. of Industrial and Information Systems Engineering,
Seoul National University of Science & Technology,
Seoul City in South Korea
doosukkim@hanmail.net

[2] Electrical Engineering & Construction Headquarters,
Signaling and Communications Division, Korea Rail Network Authority,
TaeJeon City in South Korea
ktx21c@hanmail.net

[3] Dept. of Industrial and Information Systems Engineering,
Seoul National University of Science & Technology,
Seoul City in South Korea
syjang@snut.ac.kr

Abstract. Normally, the Railroad Constructions need the huge budgets due to the large-scale works including civil, tunnel and bridge. Thus, it is so difficult to estimate the budgets out of data in the previous works. This research presents two models; (1) the *conceptual estimation model* by using the parametric method and regression analysis (2) the *range estimation model* for estimation of accuracy. Also, the simulation has been used for verifying the validity of the results from the conceptual estimation model to find the expected cost of High-Speed Railroad construction. In addition, the range estimation model for estimation of accuracy is generated by using Monte Carlo Simulation. Crystal Ball Program as a tool of the simulation is used in the research. The performance measures are used Percentage Error and the criterion according to the ACCE [1]. It is expected to improve the accuracy of the model and to contribute to effective total cost management.

1 Introduction

1.1 Research Background

The investment to the large-scale national projects, such as Railroad, Highway, etc., should need to check the validity of the budget firstly. In the preliminary planning at the beginning stage, the estimation of the budget is critical factor to determine the bidding price, budgeting and financing of the project as well as take an important role as a standard for analysis of the profitability of the project. However, the reasonable budget cannot be made and may be hard works in the beginning stage of the project. Because it can't determine the project scope due to the deficiency of detailed project

J.-H. Kim et al. (Eds.): AsiaSim2011, PICT 4, pp. 248–257, 2012.

planning information. Therefore, budgeting the national project must have been allocated from the conceptual estimation. Accordingly, the accuracy of the budget estimation can always provoke a controversy in the budget-decision stage. Since the conceptual estimation model generally should use the historical data of similar projects, it must be limited to improve the accuracy, even if the model is very useful. The limit can lead an excess or a shortage of national project budget in the practical operating stage. Therefore, the analysis for the reliability to the process and result of conceptual estimate has better do with the confidence level. Moreover, the project risk can be minimized in the project budgeting determination stage.

In spite of this important fact, research has not been developed to combine a budgeting estimation method and the contingency estimation method considering the project risk. Thus, this research focuses on the analysis of the accuracy and the validity of the results from the conceptual estimation model by using the parametric method [2]. In addition, the range estimation model by using the Monte Carle Simulation is proposed followed by the conceptual estimation model. This model is useful as a tool to help to make a decision of the budgeting for the national project.

Therefore, by using the conceptual cost estimation and the range estimation model, the accuracy of the project cost estimation should be improved as high as possible.

1.2 Research Scope

This research is following as;

(1) The stage of the conceptual estimation for the cost of construction
  - Do the statistical analysis of the accuracy and the risk of a project
  - Research the methodology and the steps to calculate the cost of construction in the literature review

(2) Analysis of the characteristics of a project for High-Speed Railroad Construction
  - Research for the technical parameters considering the costs of the civil, tunnel and bridge works
  - Conversion of the present price into the price in the previous works

(3) Development of the conceptual estimation model with parametric method
  - Development of the model including the regression analysis and statistical verification
  - Computing the difference between actual costs and estimated costs in the model to find proper values of the model parameters
  - Verification of the value in the model through the Percentage Error

(4) Calculation of the range estimation in the project
  - Find the range estimation in the Monte Carlo simulation

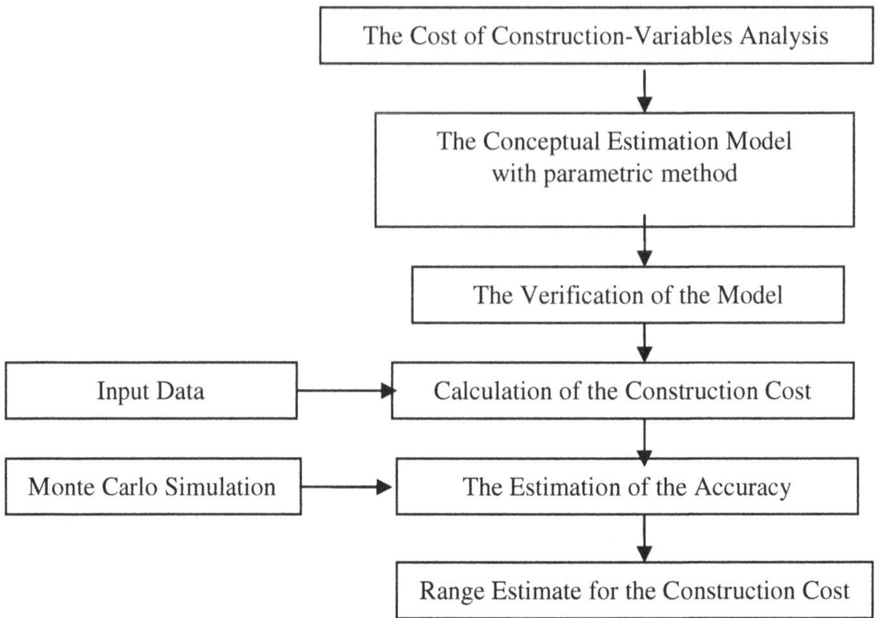

Fig. 1. The Flow and Method of the Research

2 Estimate Classification System

The cost estimation of project is defined the price of resources used to build the designed construction. Generally, the good criterion is the Estimate Classification System (Estimate Type) [1]:

Table 1. Estimate Classification System

<table>
<tr><th>ANSI standard</th><th>AACE</th><th>Expected Accuracy Range</th></tr>
<tr><td>Order of Magnitude Estimate (Expected Accuracy Range -30% ~ +50%)</td><td>Class 5</td><td>L: -20% to -50%
H: +30% to +100%</td></tr>
<tr><td rowspan="2">Budget Estimate (Expected Accuracy Range -15% ~ +30%)</td><td>Class 4</td><td>L: -15% to -30%
H: +20% to +50%</td></tr>
<tr><td>Class 3</td><td>L: -10% to -20%
H: +10% to +30%</td></tr>
<tr><td rowspan="2">Definitive Estimate (Expected Accuracy Range -5% ~ +15%)</td><td>Class 2</td><td>L: -5% to -15%
H: +5% to +20%</td></tr>
<tr><td>Class 1</td><td>L: -3% to -10%
H: +3% to +15%</td></tr>
</table>

3 The Conceptual Cost Estimation Model

The model is developed by two methods; Parametric method and regression analysis. The Parametric method is defined as the computing the estimated values about cost, budget and period with the previous data or other variables such as the construction location or area. The reliability can be high as following cases [2];

- The previous data is right and same when the model is developed
- The exchange of variables in the model into real date easily
- Scalable model

MRA(Multiple Regression Analysis) is used to build the model with the dependent values and independent variables. The basic model in MRA is [3]

$$\mathrm{Y} = \mathrm{a} + \beta_{1 \ldots\ldots\ldots} \beta_n X_n + \varepsilon$$

Coefficient of Determination is

$$R^2 = \frac{SSR}{SST} = \frac{SST - SSE}{SST} = 1 - \frac{SSE}{SST}$$

- SST= Sum of Squares of Total
- SSR= Sum of Squares of Regression
- SSE= Sum of Squares of Error

Significance Level and Standardized Coefficient are also used[4].

4 MCS and the Criterion

4.1 Monte Carlo Simulation

MCS are a class of computational algorithms that rely on repeated random sampling to compute their results [5].

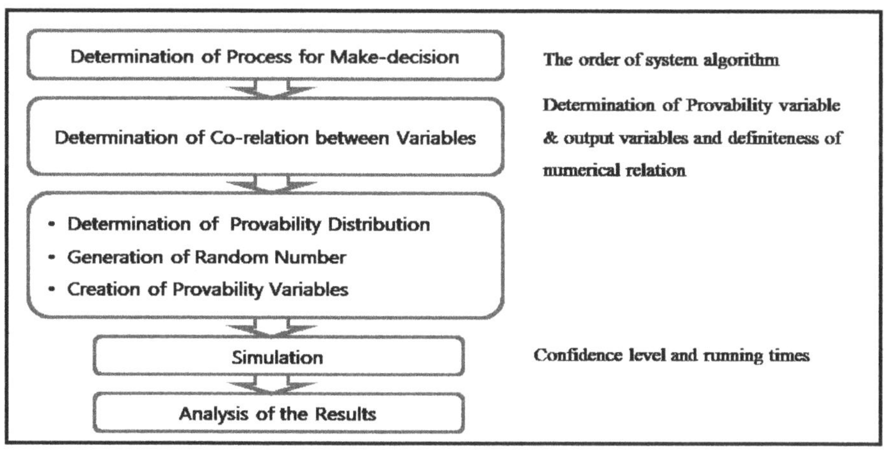

Fig. 2. Flow Chart of MCS

Monte Carlo methods are often used in simulating physical and mathematical systems. These methods are most suited to calculation by a computer and tend to be used when it is infeasible to compute an exact result with a deterministic algorithm. MCS is used to model phenomena with significant uncertainty in inputs, such as the calculation of risk in business. When MCS has been applied in space exploration and oil exploration, their predictions of failures, cost overruns and schedule overruns are routinely better than human intuition or alternative "soft" methods. The MCS is a tool for take an experiment in virtual environments. The process of MCS[6] show as following in fig. 2

4.2 The Criterion of Measurement

This paper performs the verification of the accuracy as the measurements as follows;

- Input data for validity: actual data obtained from high-speed railroad zone (Table 2)
- PE (percentage error) = [(output data from the model – actual data)/actual data] X 100
- MPE (mean percentage error) = (sum of PE)/number of data
- MAPE (mean absolute percentage error) = |Sum of PE|/number of data

Also, the criterion for judge in the AACE international and ANSI is used in the research.

Table 2. Input date with parameters in the civil work

Zone	Length (m)	Cost (Vm)	Geological Structure			
			Earth	Weathered rock	Soft-rock	Hard-rock
11-2	1,235	5,352,001	22.1%	6.8%	19.7%	51.3%
8-2	5,595	5,417,105	32.9%	38.3%	27.2%	1.6%
6-1	1,903	3,275,597	42.9%	40.3%	10.1%	6.7%
5-1	2,305	9,438,221	45.0%	36.4%	11.4%	7.1%
7-2	3,842	5,882,110	56.2%	15.1%	16.4%	12.2%
12-5	3,250	6,601,624	21.5%	13.4%	30.8%	34.3%

5 Simulation

5.1 Modeling

The MCS is performed by using the simulation software with running time determined.

- o Simulation Software: Crystal Ball 11.1
- o Running Time: 5000
- o Confidence Level: 95%

In analysis of variation and estimation of accuracy, a risk neutral approach is used to set as 50%:50%, which means that the increased and the decreased costs of railroad construction are equal [2].

The results from MCS indicate the variation range with probability. So the expected cost from the model includes the assumption that the risk of estimation happens in 50%. The input data for implement the MCS considering the Triangle distribution show in table 3. The values are obtained from the conceptual estimation model with parametric method.

Table 3. Input data for MCS

Zone	Min	Max	Most Likely	Actual Average	Average Cost
Civil	1.7	8.5	5.5	5.7	5.2
Tunnel	10.6	33.5	16.9	18.5	15.3
PC Bridge	23.4	39.8	28.6	30.1	27.1
PSC Bridge	16.9	24.0	21.9	23.1	20.7

5.2 The Results of Simulation

The output results from the simulation are obtained from the steps in Figure 3.

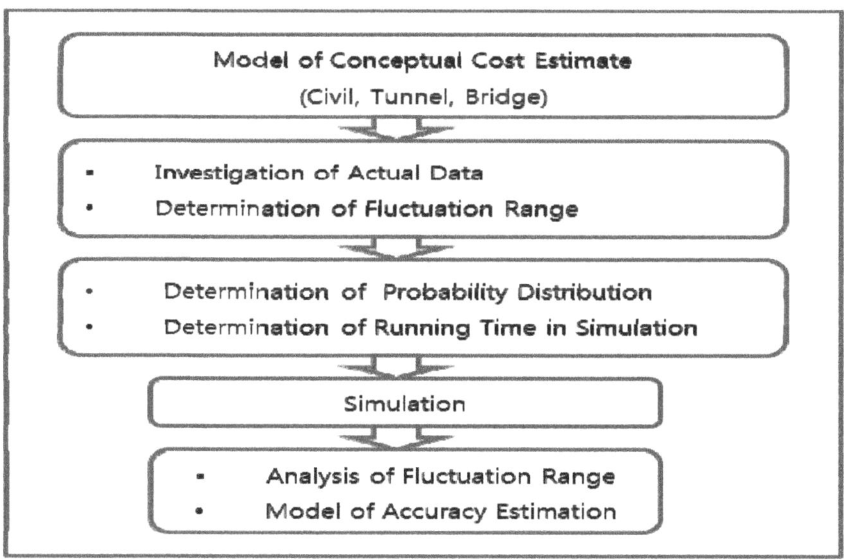

Fig. 3. Flow Chart for Development of the Model

First, the variables for the multiple regression analysis for implement the conceptual estimation model are set as follow in table 4. The Table 4 shows variable for the parametric method in only civil works. The results from the conceptual estimation model are applied to verify the validity with the criterion in AACE. 3 measures, PE, MPE and MAPE are used and showed in the table 5.

Table 4. Data for construction cost & percentage error

Zone	The cost of geological structure (Vm)				Soil Tw (Weight)	Results (Vm)	PE
	f(earth)	f (Weathered rock)	f(soft-rock)	f(Har-rock)			
11-2	6,840,715	3,372,713	8,426,572	2,805,844	4,740,842	4,371,056	-18.3%
8-2	7,610,330	4,365,313	15,740,092	1,553,903	5,184,422	4,780,037	-11.8%
6-1	7,330,594	2,868,494	2,250,533	5,288,181	5,392,717	4,972,085	51.8%
5-1	7,176,546	5,520,891	2,853,535	5,532,250	5,966,954	5,501,531	-41.7%
7-2	6,060,746	6,725,426	5,872,942	7,523,595	6,734,371	6,209,090	5.6%
12-5	6,752,517	6,147,023	20,068,348	5,243,599	6,886,545	6,349,394	-3.8%

Table 5. The results of the criterion

Works	Average value(million won/km)			Range of Error	
	Real value	Model value	PE	MPE	MAPE
CIVIL	57.1	52.0	-9.5%	-3.1%	27.1%
TUNNEL	185.1	152.5	-17.6%	-17.3%	28.2%
PC BRIDGE	301.9	270.6	-10.4%	-2.7%	22.0%
PSC BRIDGE	230.2	207.0	-10.1%	+1.9%	22.8%

The plots present the percentage error in civil, tunnel and bridge in the High-Speed Railroad Project in figure 4.

5.3 The Estimation of the Accuracy

The distribution chart of the input data in Table 3 and the results in MCS is showed in figure 5. In the civil work, the probability which the actual cost is 54.7 million won/meter is 35%. If the risk of cost estimation is set 50%, the results of most likely value in MCS are the estimation range between -9.9% and +8.4%, 40.93~50.93 million won/m according to above distribution. So, the average value (50.43 million won/m) must be between them.

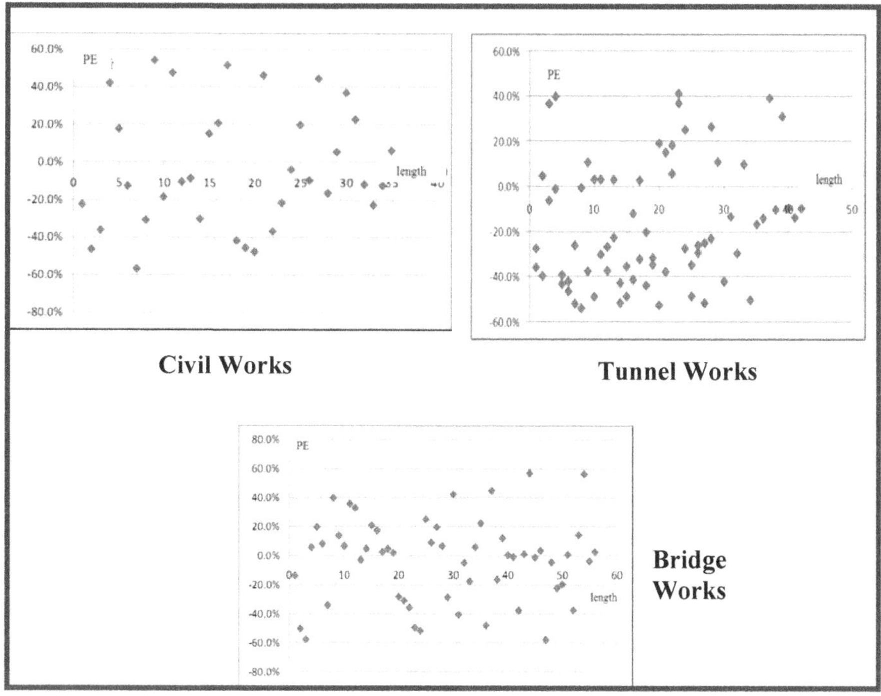

Fig. 4. The plot of PE

6 Results and Analysis of the Model

6.1 Results of Hybrid Model

The final results of the conceptual estimation model and range estimation model are presented in table 6.

Table 6. Hybrid Model in Railroad Works

Zone	50% (Distribution)			Cost in models		Actual cost
	Low	Mean	Max	Most	Cost	
Civil	4.9	5.4	5.9	5.5	5.2	5.7
Tunnel	16.9	20.4	24.3	16.9	15.3	18.5
PC Bridge	28.6	30.6	34.0	28.6	27.1	30.1
PSC Bridge	19.9	21.0	22.1	21.9	20.7	23.1

- Mean: Average value from MCS (5000 times)
- Most: (Average value + Actual value)/2

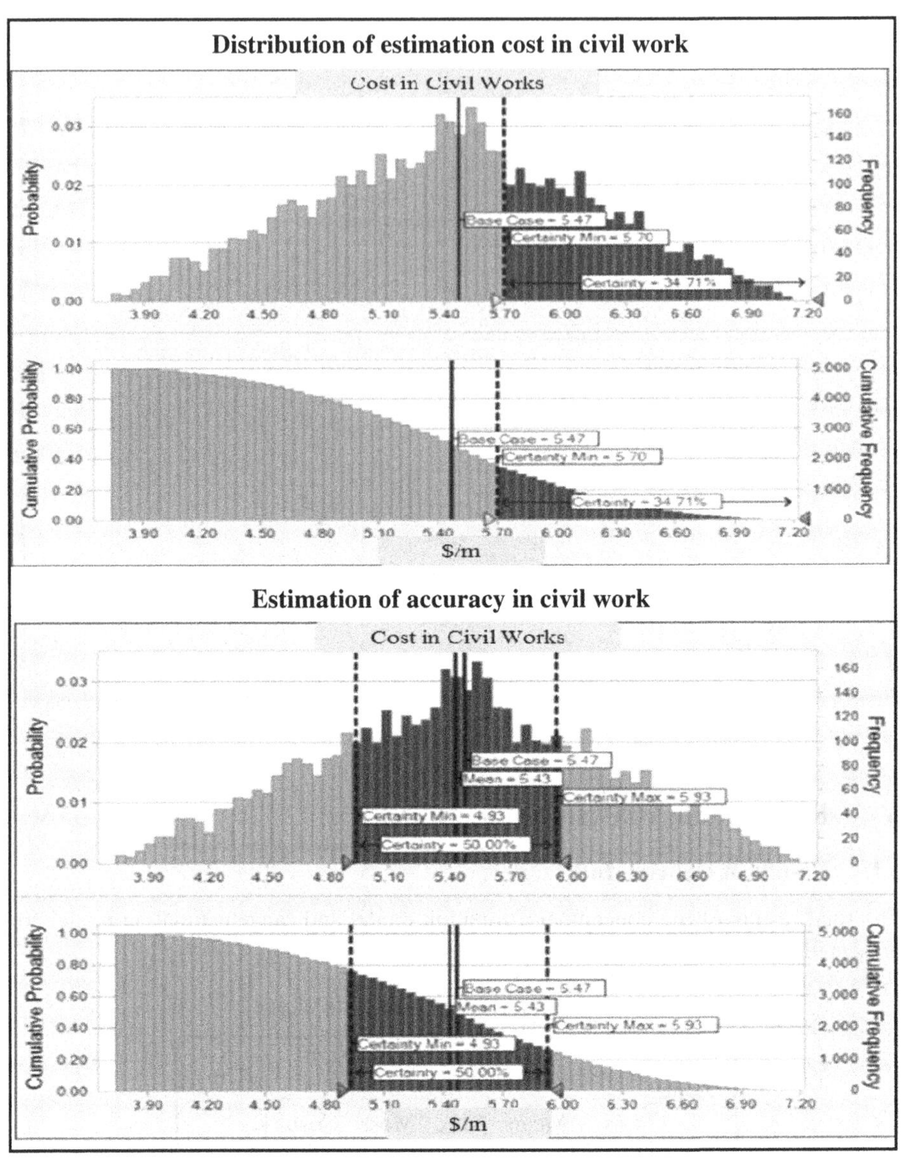

Fig. 5. Distribution and estimation of accuracy

6.2 Analysis of Hybrid Model

The first model is applied to MRA (multiple regression analysis) in high-speed railroad projects including civil, tunnel and bridge works. Also, the probability model is showed with actual cost in the works and technical particulars. The independent variables such as length and geological materials in the civil works and dependent variable such as cost-per unit-length are considered in the experiment. The performance measures are percentage error, mean percentage error, and mean absolute percentage error, which means the difference between the unit costs estimated using the developed parametric cost model and the actual costs date. The results show that PE is presented in the range between -9.5% and -17.6%, MPE is presented in the range between -17.3% and +1.9% and the range of MAPE is between 22% and 28%. This range is acceptable compared the typical range -30% to +50% during the beginning stage of the large-scale project.

Second model uses the MCS(Monte Carlo Simulation) as a tool to verify the estimation of accuracy about the results from the conceptual estimation model. Crystal Ball program is utilized to develop the range estimating model and the Triangle distribution is adopted as input data. Simulation is performed 5000 times under 50% of probability. It is believed that the accuracy of expectation value of the civil work KRW5.47million/meter ranges between -9.9% and +8.4% and the Tunnel KRW 16.9 million/meter ranges between -0.0% and +43.8%. Likewise, the accuracy of expectation value of PC Bridge KRW28.6 million/meter ranges between 0.0% and +18.9% and of PSC bridge KRW21.9mil./meter ranges between -9.1% and +0.9%.

7 Conclusion

This paper presents the hybrid model combining the conceptual cost estimation with parametric method and the range estimation model for estimation of accuracy.

This Hybrid Model will improve the accuracy of existing method and be expected to contribute to effective total cost management and the economic aspects, and reduce the financial expenditure. Also, by using the accuracy estimation model in the beginning stage of the project, decision-makers can expect to determine reasonably the budgets of large-scale national projects.

References

1. AACE, International, Skill & Knowledge of Cost Engineering. AACE, International, 4th edn., ch. 2 (2002)
2. Project Management Institute (PMI), Project Management Body of Knowledge (PMBOK), PMI, Inc., 4th edn., Pennsylvania, USA (2009)
3. Seek, P.J., Seon, Y.Y.: The Probability and Statistics, Dasan, vol. 4 (2004)
4. Sup, K.D., Joon, K.N.: Regression Analysis (Basic and Applicaiotn), Nanam (2000)
5. `http://www.wikipedia.org`, Wikimedia Foundation, Inc. (2011)
6. Werckman, C., Hardy, T., Wainwright, E., Harris, C., Gentry, B. (eds.): Crystal Ball® User Manual. Colorado, Decisioneering, Inc. (2004)
7. AACE, International, Risk analysis and Contingency determination using Range Estimating. AACE, International (2008)

Improving Electric Vehicle Conversion's Ride and Handling Performance Using Active Suspension System

Saiful Anuar Abu Bakar[1], Ryosuke Masuda[1], Hiromu Hashimoto[2], Takeshi Inaba[1], Hishamuddin Jamaluddin[3], Roslan Abdul Rahman[3], Pakharuddin Mohd Samin[4]

[1] Department of Applied Computer Engineering, Tokai University, Japan
saiful@fkm.utm.my, masuda@keyaki.cc.u-tokai.ac.jp, inaba@tokai.ac.jp

[2] Department of Mechanical Engineering, Tokai University, Japan
hashimoto@tsc.u-tokai.ac.jp

[3] Department of Applied Engineering, Universiti Teknologi Malaysia, Malaysia
{hishamj,roslan}@fkm.utm.my

[4] Department of Automotive Engineering, Universiti Teknologi Malaysia, Malaysia
pakhar@fkm.utm.my

Abstract. This paper present an evaluation on passenger vehicle's ride and handling performance if being converted into an electric vehicle. The mathematical modelling of vehicle's ride and handling model as well as its validations are described. The simulation model was used to evaluate the vehicle's ride and handling performance when converted into an electric vehicle. The validated simulation model was also integrated with an active suspension system in order to improve the EV conversion's ride and handling performance. It was found that the modifications towards an EV conversion only affect the vehicle's handling performance significantly. The modifications which change the weight distribution, which biased to the rear of the vehicle, will cause the vehicle to travel off from its original traveling path. The application of active suspension is possibly being used in correcting this condition as well as further improving the EV conversion's ride comfort performance.

Keywords: electric vehicle conversion, ride and handling, active suspension system.

1 Introduction

The development of electric vehicle (EV) from the commercially available vehicle model is becoming a trend nowadays due to the global concerns in reducing the green house effect which one of its contributing factors came from the pollution of vehicle. An electric vehicle that is being converted from a normal production vehicle model usually known as a electric vehicle conversion or EV conversion.

Many research works were found to focus on the electric vehicle's electric and electronic systems but not many were found to focus in improving EV conversion's

J.-H. Kim et al. (Eds.): AsiaSim2011, PICT 4, pp. 258–267, 2012.

stability while manoeuvrings. The researches on EV conversion's stability are mainly related to the yaw stability control and traction system.

In yaw stability control on electric vehicle [1] it focuses in controlling yaw motion of the electric vehicle by controlling the operation of the drive motor. The drive motor, either two or four are basically controlled in terms of its torque generation. While in traction control system [2] the generation of electric motor torque is controlled to ensure the wheel does not skid while accelerating ensuring full control over the vehicle. This is done by controlling the slip ratio of the wheel.

It is not clear how the modifications towards an EV conversion affect the vehicle's ride comfort and handling performance; the level of isolations of passenger compartment from being affected by harsh road profile and vehicle's steerability after the conversions. Typically, any conversion of internal combustion engine vehicle to electric vehicle involves some weight addition (or weight reduction), due to the installations of the electric vehicle systems i.e. battery system. Any weight addition or weight reduction on the chassis will cause the vehicle's weight distribution to change and this can compromise vehicle's ride comfort and handling performance, as current suspension system tuning was not being designed specifically for the new weight and load distribution at front and rear axles. This paper will investigate the affects of weight distribution changes on EV conversion's ride comfort and handling performance and the improvements that can possibly been done by only using an active suspension system.

2 14-DOF of Ride and Handling Model

The development of fourteen-degrees (14DOF) of vehicle's ride and handling model in this study are based on [3]. The first model is the vehicle's ride model, consists of seven degrees of freedom namely roll, pitch, bounce and vertical motion of each four wheels. Figure 1 shows the vehicle' ride model.

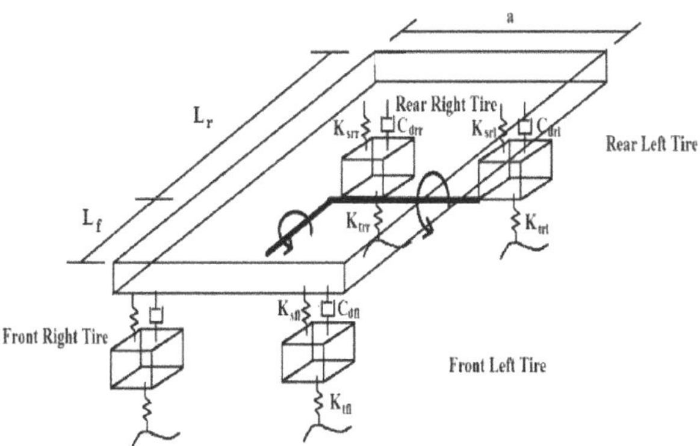

Fig. 1. Seven degree of freedom of vehicle ride model

Based on the Figure 1, the displacements of the sprung masses at each corner of the vehicle model is given by

$$Z_{sij} = Z_b + \frac{a_{car}}{2}\theta_{car} - L_i\alpha_{car} \tag{1}$$

with Z_{sij} is the total sprung mass displacement (*i* =*f* for front, *r* for rear and *j*=*l* for *le*ft, *r* for right), Z_b is the sprung mass vertical displacement at the center of gravity, θ_{car} is the roll angle and α_{car} is the pitch angle. The distance of centre of gravity to the front axle and rear axle are given by L_f and L_r respectively. The forces acting at each of the suspension (F_{ij}) is the sum of the spring force (F_{sij}) and damper force (F_{dij}). The suspension forces are given by

$$F_{ij} = F_{sij} + F_{dij} \tag{2}$$

The spring forces, F_{sij} in each of the suspension system is given by;

$$F_{sij} = K_{sij}(Z_{uij} - Z_{sij}) \tag{3}$$

with K_{sij} is the spring stiffness of the spring, Z_{uij} and Z_{sij} are the unsprung mass vertical displacement and the sprung mass vertical displacement respectively at each side of the vehicle. The damper forces, F_{dij} is given by;

$$F_{dij} = C_{sij}(\dot{Z}_{uij} - \dot{Z}_{sij}) \tag{4}$$

with C_{sij} are the damping coefficient of the dampers, $\dot{Z}_{uij}$ and $\dot{Z}_{sij}$ are the unsprung mass vertical velocity and the sprung mass vertical velocity respectively. For the vehicle tires, it is modelled as a spring and the force acting at tires is usually known as dynamic tire loads, F_{tij}. For each tires, their dynamic tire loads is given by;

$$F_{tij} = K_{tij}(Z_{rij} - Z_{uij}) \tag{5}$$

where K_{tij}, Z_{rij}, and Z_{uij}, are the tire stiffness, road input displacement and unsprung mass displacement respectively.

Using Newton's Second Law at the vehicle's sprung mass, the body vertical acceleration, $\ddot{Z}_b$ can be determined by

$$F_{fl} + F_{fr} + F_{rl} + F_{rr} = M_b\ddot{Z}_b \tag{6}$$

where M_b is the total mass of the vehicle. Angular acceleration during the roll effect, $\ddot{\theta}_{car}$ is given by;

$$(F_{fl} + F_{rl})\frac{a_{car}}{2} - (F_{fr} + F_{rr})\frac{a_{car}}{2} = I_{xx}\ddot{\theta}_{car} \tag{7}$$

where a_{car} is the vehicle's track width and I_{xx} is the moment of inertia about x-axis. The angular acceleration while the vehicle is in pitch effect, $\ddot{\alpha}_{car}$ it is given by;

$$(F_{rl}+F_{rr})L_r-(F_{fl}+F_{fr})L_f=I_{yy}\ddot{\alpha}_{car} \tag{8}$$

with I_{yy} is moment of inertia about y-axis respectively. Acceleration of each wheel can be calculated using

$$F_{tij}-F_{sij}-F_{dij}=M_{uij}\ddot{Z}_{uij} \tag{9}$$

As for the handling model (Figure 2) used in this study it consists seven degrees of freedom, as it takes into account three degrees of freedom for the vehicle body (roll, pitch and yaw motions) and one degree of freedom due to the rotational motion of each tire.

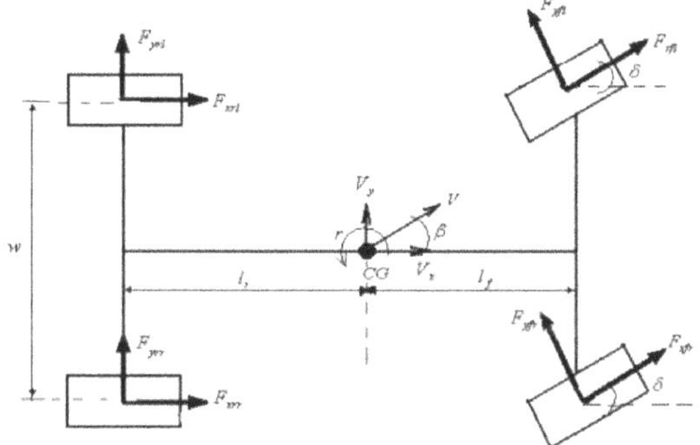

Fig. 2. Seven degree of freedom of vehicle handling model

The vehicle experiences motion along the longitudinal x-axis and the lateral y-axis, and the angular motions of yaw around the vertical z-axis. The motion in the horizontal plane can be characterized by the longitudinal and lateral accelerations, denoted by a_x and a_y, and the velocities in longitudinal and lateral direction, denoted by v_x and v_y:

For longitudinal x-axis:

$$\dot{v}_x=a_x+v_y\dot{\psi} \tag{10}$$

Longitudinal acceleration, a_x:

$$a_x=\frac{F_{xfl}\cos\delta+F_{yfl}\sin\delta+F_{xfr}\cos\delta+F_{yfr}\sin\delta+F_{xrl}+F_{xrr}}{m} \tag{11}$$

For lateral y-axis:

$$\dot{v}_y=a_y-v_x\dot{\psi} \tag{12}$$

Lateral acceleration, a_y:

$$a_y = \frac{F_{yfl}\cos\delta - F_{xfl}\sin\delta + F_{yfr}\cos\delta - F_{xfr}\sin\delta + F_{yrl} + F_{yrr}}{m} \tag{13}$$

where F_{xij} and F_{yij} denote the tire forces in the longitudinal and lateral directions, respectively, with the index (i) indicating front, (f) or rear I tires and (j) indicating left (l) or right I tires. The steering angle is denoted by δ, the yaw rate by $\dot{\psi}$ and m denotes the vehicle mass. The longitudinal and lateral vehicle velocities v_x and v_y can be obtained by the integration of $\dot{v}_y$ and $\dot{v}_x$. It can be used to obtain the side slip angle, denoted by α. Thus, the slip angle of front and rear tires is found as

For the side slip angle of front tires is,

$$\alpha_f = \tan^{-1}\left(\frac{v_y + L_f r}{v_x}\right) - \delta \tag{14}$$

For the slip angle of rear tires is,

$$\alpha_r = \tan^{-1}\left(\frac{v_y - L_f r}{v_x}\right) \tag{15}$$

To calculate the longitudinal slip, longitudinal component of the tire velocity should be derived. The front longitudinal velocity component is found by,

$$v_{wxf} = V_{tf}\cos\alpha_f \tag{16}$$

where, the speed of the front tire is,

$$V_{tf} = \sqrt{(v_y + L_f r)^2 + v_x^2} \tag{17}$$

The rear longitudinal velocity component is,

$$v_{wxr} = V_{tr}\cos\alpha_r \tag{18}$$

where, the speed of the rear tire is,

$$V_{tr} = \sqrt{(v_y + L_r r)^2 + v_x^2} \tag{19}$$

Then, the longitudinal slip ratio of front tire,

$$S_{af} = \frac{v_{wxf} - \omega_f R_w}{v_{wxf}} \quad \text{(under braking conditions)} \tag{20}$$

The longitudinal slip ratio of rear tire is,

$$S_{ar} = \frac{v_{wxr} - \omega_r R_w}{v_{wxr}} \quad \text{(under braking conditions)} \tag{21}$$

where, ω_r and ω_f is angular velocity of rear and front tire, respectively. R_w, is the wheel radius. The yaw motion is also dependent on the tire forces F_{xij} and F_{yij} as well as on the self-aligning moments, denoted by M_{zij} acting on each tire:

$$\ddot{\psi} = \frac{1}{J_z}\begin{pmatrix} \frac{t_f}{2}F_{xfl}\cos\delta - \frac{t_f}{2}F_{xfr}\cos\delta + \frac{t_r}{2}F_{xrl} - \frac{t_r}{2}F_{xrr} + \\ \frac{t_f}{2}F_{yfl}\sin\delta - \frac{t_f}{2}F_{yfr}\sin\delta - cF_{yrl} - cF_{yrr} + \\ bF_{yfl}\cos\delta + bF_{yfr}\cos\delta - bF_{xfl}\sin\delta - bF_{xfr}\sin\delta + \\ M_{zfl} + M_{zfr} + M_{zrl} + M_{zrr} \end{pmatrix} \tag{22}$$

where L_f and L_r denote the distance between the centre of gravity and the front and the rear axle, respectively, T_f and T_r are the track width front and rear respectively and J_z is the moment of inertia around the z-axis.

The roll and pitch motion depend very much on the longitudinal and lateral accelerations. Since only the vehicle body undergoes roll and pitch, the sprung mass, denoted by M_b has to be regarded:

$$\ddot{\phi} = \frac{-M_b h_s a_y + \phi(M_b g h_s - k_\phi) + \dot{\phi}(-\beta_\phi)}{J_{sx}} \tag{23}$$

$$\ddot{\theta} = \frac{-M_b h_s a_y + \theta(M_b g h_s - k_\theta) + \dot{\theta}(-\beta_\theta)}{J_{sy}} \tag{24}$$

h_s, the height of the sprung mass center of gravity, g the gravitational acceleration and k_ϕ, β_ϕ, k_θ and β_θ are the damping and stiffness constant for roll and pitch, respectively. The moments of inertia of the sprung mass around x-axes and y-axes are denoted by J_{sx} and J_{sy} respectively. As for the tire dynamic (Figure 3) it can be modelled as

$$T_i - (F_{xij}R_w) = I_w \times \dot{\omega}_{ij} \tag{25}$$

where I_w is the moment of inertia of each wheel about its axis of rotation, $\dot{\omega}_{ij}$ is its angular speed, R_w is the effective tyre radius and T_i is the braking(negative) or traction(positive) moment. F_{xij} denote the tire forces in the longitudinal direction. The index *i* indicate front, *f* or rear, *r* tires and *j* indicating left, *l* or right, *r* tires.

The forces acting on the tyres, which are required in the above equation, can be obtained by use of tyre models. The full mathematical equation of Pacejka tyre model was used in this study. This model calculates the longitudinal and lateral tyre forces as well as the self-aligning moments based on the vertical forces denoted by F_z. The Pacejka tyre model calculates lateral force and aligning torque based on slip angle and longitudinal force based on percent longitudinal slip. The model parameters are depends on the normal force, F_z or the vehicle's weight acting on each tires. The Pacejka tire modelling can be summed as follow

The lateral force is:

$$F_y = D\sin\left(C\tan^{-1}(B\phi)\right) + S_y \tag{26}$$

Aligning moment is:

$$M_z = D\sin\left(C\tan^{-1}(B\phi)\right) + S_v \tag{27}$$

Longitudinal force is:

$$F_x = D\sin\left(C\tan^{-1}(B\phi)\right) \tag{28}$$

3 EV Conversion with Active Suspension

An active suspension system is considered in this study to improve EV conversion's ride comfort and handling performance. A slow-active suspension system type is used where it consist a conventional spring combined with hydraulic actuator.

3.1 Modelling of Hydraulic Actuator in Active Suspension System

The modelling of hydraulic actuator for active suspension system is based on [4]. The dynamics of hydraulic actuator is given by

$$\dot{F}_A = A_p\,\alpha\left[\begin{array}{l} C_{d1}\,w u_1\sqrt{\dfrac{P_s - \mathrm{sgn}(u_1)\,P_L}{\rho}} - \\ - C_{tm}\,P_L - A_p\,(\dot{x}_s - \dot{x}_u) \end{array}\right] \tag{29}$$

where α, C_{d1}, C_{d2} and C_{tm} are constants while A_p $w, \dot{x}_s - \dot{x}_u$, P_s and P_L are spool length, spool relative velocity, supply pressure and pressure different respectively. Spool valve positions u_1 and u_2 are controlled by a current-position feedback loop.

$$\tau\dot{u} + u = kv \tag{30}$$

The parameters of used for hydraulic actuator model were taken from [9] as the followings: Ap= 0.0044 m2 , α = 2.273e9 N/m5, $Cd1$=0.7, $Cd2$= 0.7, w = 0.008 m, Ps = 20684 kN/m2, ρ =3500, Ctm = 15e-12, τ = 0.001 sec.

3.2 Stability Augmentation System Algorithm

The stability augmentation system (SAS) [5] was used to control the operation of the actuator in active suspension system. The SAS operates by considering the motion of sprung and unsprung mass as well as the vertical roll and pitch motions. The SAS algorithm which considers handling motions is given by

$$\begin{bmatrix} F_{afl} \\ F_{afr} \\ F_{arl} \\ F_{arr} \end{bmatrix} = \begin{bmatrix} \frac{L_r}{2(L_f + L_r)} & -\frac{1}{2(L + L_r)} & \frac{1}{2T} \\ \frac{L_r}{2(L_f + L_r)} & -\frac{1}{2(L + L_r)} & -\frac{1}{2T} \\ \frac{L_f}{2(L_f + L_r)} & \frac{1}{2(L + L_r)} & \frac{1}{2T} \\ \frac{L_f}{2(L_f + L_r)} & \frac{1}{2(L_f + L_r)} & -\frac{1}{2T} \end{bmatrix} \begin{bmatrix} K_z(Z_{ref} - Z_s) - B_z \dot{Z}_s \\ K_\theta(\theta_{ref} - \theta) - B_\theta \dot{\theta} \\ K_\alpha(\alpha_{ref} - \alpha) - B_\alpha \dot{\alpha} \end{bmatrix} + C_d \begin{bmatrix} (\dot{Z}_{ufl} - \dot{Z}_{sfl}) \\ (\dot{Z}_{ufr} - \dot{Z}_{sfr}) \\ (\dot{Z}_{url} - \dot{Z}_{srl}) \\ (\dot{Z}_{urr} - \dot{Z}_{srr}) \end{bmatrix} + Y|\mathrm{sgn}(\delta)| \begin{bmatrix} (0.3W_T - 0.2W_T) \\ (0.3W_T - 0.2W_T) \\ (0.2W_T - 0.3W_T) \\ (0.2W_T - 0.2W_T) \end{bmatrix} \tag{31}$$

where $K_z, K_\theta, K_\alpha, B_z, B\theta$, B_α, Y, C_d and W_T are constants, damping coefficient and total vehicle weight respectively, The first two terms in eq.(14) is used improve the ride comfort of the vehicle by attenuating vibrations from the harsh road profile while the last term is used to improve the handling of the vehicle model by inserting external forces to the vehicle. In this study the value for the constants in eq. (31) were manually tuned and the tuning started by assuming

$$[K_z \;\; K_\theta \;\; K_\alpha] = K \tag{32}$$

$$[B_z \;\; B_\theta \;\; B_\alpha] = B \tag{33}$$

Based on the tuning work done, the value for these constants are 1e6, 1e5 and 100 Ns/m respectively, while the value used for Y is 13.

4 Simulation Results

The ride and handling performance of the EV conversion, before and after conversions (with and without active suspension system) were evaluated using the validated ride and handling model.

In ride comfort evaluations, the vehicle model was simulated to move at a constant speed of 20 km/h before only one side of the vehicle hop on the bump. The bump dimension is 7.5 cm, 40 cm, and 244 cm in height, width and length respectively. As in the handling evaluation test, the vehicle model was simulated to move at the speed of 50 km/h.

Figure 3 show the ride comfort performance on the vehicle before the conversion, after the conversion and the EV conversion's responses with the active suspension system. It is assumed that before the modifications the vehicle is having 60:40 of

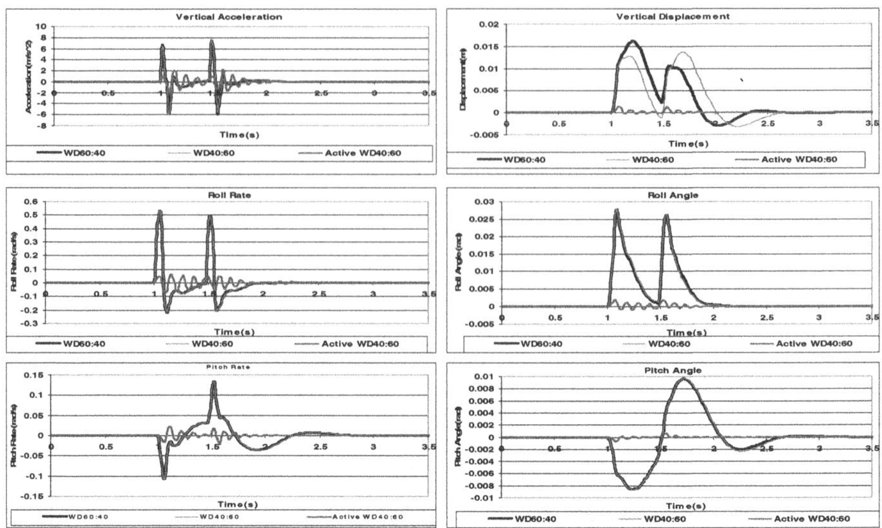

Fig. 3. Ride comfort results

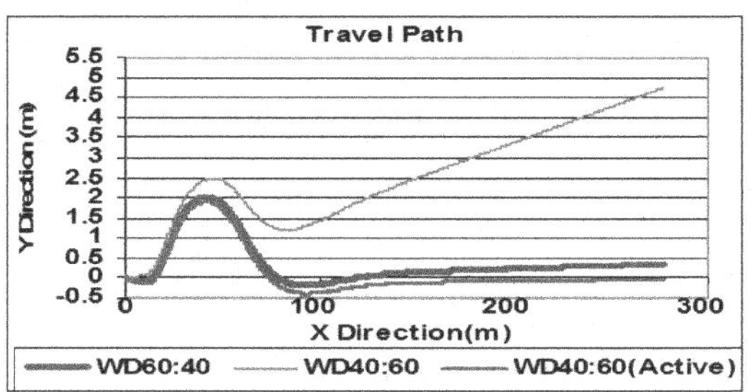

Fig. 4. EV conversion's traveling path

weight distribution while after the modifications, the weight ratios is 40:60. Based on these figure, it can be seen that the changes of weight distributions due to the modifications did not affects the EV conversion's ride comfort significantly, instead the implementation of active suspension system in the EV conversion improves significantly the EV conversion ride comfort by almost 81 percents where the most significant improvements can be seen were made in terms of EV conversion's vertical displacement, roll angle and pitch angle.

As for the handling evaluation (Figure 4), it can be seen that the changes of the weight distribution in EV conversion, which biased to the rear of the vehicle had caused the vehicle to travel off from its original path, using the same steering input used by the vehicle model with 60:40 weight distribution ratio. In practical, this would not be happening as driver will put more steering effort to ensure the vehicle is travelling on the desired travelling path. The same steering inputs as before the modifications were used in EV conversion to simulate what will actually happen if the vehicle is steered, as if no modifications had been done.

The reduction of the weight at the front axle of the vehicle had caused less normal load to act upon the tire, causing less tire lateral force to be generated. This is however is possibly being improvised by the application of active suspension system, as the system is able to 'replace' back the loss normal load at the front axle by introducing external forces at the front axle activated by the steering angle. The increased normal forces will increased the tire lateral force; based on Pacejka tire model, the lateral force are as a function of normal load, lateral slip and longitudinal slip. As the results, the EV conversion's desired travel path can be achieved without extra efforts done on steering the steering wheel.

5 Conclusions

As for the conclusions, the studies on vehicle modifications towards an electric vehicle were found to affects significantly vehicle's handling stability. The EV conversion's handling instability is due to the changes of the of weight distribution in EV conversion which typically biased to the vehicle's rear. In terms of ride comfort performance, the modification towards an EV conversion does not give any significant effects on the ride comfort performance. The instability of EV conversion handling while maneuvering, as well as further ride comfort improvement could possibly being improved by an active suspension system. The actuators in the active suspension system are used to subject external forces on the wheels, increasing the tire normal loads which results an increases in tire lateral forces. In terms of ride comfort improvement, the active suspension system is able to improve the studied ride comfort responses significantly. The improvement made by the active suspension system in both ride comfort and handling performance has made it as a good alternative to be considered to be applied as a primary suspension system in an EV conversion.

References

1. Niasar, A.H., Moghbeli, H., Kazemi, R.: Yaw Moment Control Via Emotional Adaptive Neuro-Fuzzy Controller for Independent Rear Wheel Drives of an Electric Vehicle. In: Proceedings of 2003 IEEE Conference on Control Applications (2003)
2. Shino, M., Miyamoto, N., Wang, Y.Q., Nagai, M.: Traction Control of Electric Vehicles Considering Vehicle Stability Advanced Motion Control. In: Proceedings of the 6th International Workshop on Advanced Motion Control (2000)
3. Hudha, K., Jamaluddin, H., Samin, P.M., Rahman, R.A.: Vehicle Modelling and Validations: Experience with Proton Car. In: International Association of Vehicle System Dynamics (IAVSD), Kanagawa, Japan, August 24-30 (2003)
4. Donahue, M.D.: Implementation of an Active Suspension and Preview Controller for Improved Ride Comfort. MSc. Theses, University of California at Berkeley (2001)
5. Samin, P.M., Jamaluddin, H., Rahman, R.A., Anuar, S., Hudha, K.: Semi-Active Suspension System For Handling Quality and Longitudinal Stability Improvements Using Hybrid Stability Augmentation System-Force Control Algorithm. In: 2nd Regional Conference on Vehicle Engineering and Technology, Kuala Lumpur, Malaysia, July 5-16 (2008)

Object-Oriented Simulation for Service Supply Chain

Jong-Hyuk Moon and Young Hae Lee*

Department of Industrial and Management Engineering
Hanyang University, S. Korea
moonjhie@gmail.com, yhlee@hanyang.ac.kr

Abstract. Most existing supply chain research focuses exclusively on the manufacturing sector. However, there is a lack of research on the service sector, which is becoming increasingly important in the global economy. In addition, it is harder to design a conceptual simulation model than a manufacture supply chain does to the diversity of structures in the service sector. In this paper, we develop a service supply chain simulation models, based on object-oriented approach using UML analysis/design tools and ARENA® simulation language. We use a conceptual model of simulation because of its diversity and changeability. Objective-oriented modeling has demonstrated an effective simulation design and the practical process. The two main features of the proposed procedure are the definition of a systematic conceptual procedure to design service supply chain models and of a set of rules for the conceptual model translation in an ARENA® simulation language. The goal of this research is to improve the knowledge on service supply chain management with the developed model and support the simulation model development on service supply chain efficient.

Keywords: Service supply chain management, Objective-oriented, Simulation.

1 Introduction

The expansion of service industries in the world economy has already surpassed the manufacturing industry in developed country. However, the existing supply chain management literature has been focused on the manufacturing industry applicable and has limited application to service industries.

Korea's mobile manufacturing industry has had great success in the early 21st century. But manufacturing-centered mobile phone industry is difficult that can lead to continued success in the current corporate environment. Korea's cell phone company had struggled for the market to move to smartphones. Because, the previous market has be competed in performance of product but now mobile industry requires quality of products and services, innovation.

The study of supply chain service is made more difficult by the lack of a formal logical structure. Design a standardized model of service supply chain results from a

* Corresponding author.

J.-H. Kim et al. (Eds.): AsiaSim2011, PICT 4, pp. 268–280, 2012.

variety of service systems. Therefore, the first step of the study of supply chain service is to include a variety of service systems. According to previous studies of service were targeted as a single service system. However, the service industry also has a complex supply chain structure similar to manufacturing industries. Therefore, research must be worked in terms of the supply chain rather than only considering a single service system. Service supply chain and manufacturing supply chain are part of a very small share in many parts. In particular, the service characteristics of accounted for in Manufacturing supply chain is models that describe inventory / replenishment.

In this paper, we develop a service supply chain logical structure and simulation models, based on an object-oriented approach. Service supply chain takes advantage of programming an objective-oriented model to build up a conceptual model of simulation because of its diversity and changeability. Objective-oriented modeling has demonstrated an effective simulation design and the practical process. Especially, it is based on the UML analysis/design tools and on the ARENA® simulation language.

The main contributions of this paper exist in three areas: 1) Conceptual model of service supply chain defines for development of simulation, 2) Logical structure of service supply chain simulation model is based on object-oriented approach, 3) We propose to design a simulation model of service supply chain by using ARENA® simulation language.

This paper is organized as follows. The service supply chain deals with literature in section 2. In section 3, service supply chain explains conceptual model. In section 4, general model of the service supply chain explains procedure and implementation. In section 5, general model of the service supply chain is applied to the tourism sector, followed by the conclusions and insights in section 6.

2 Literature Review

2.1 Service Supply Chain

The characteristics of services: intangibility, perishability, simultaneity, heterogeneity and customer participation. These characteristics of service make it difficult to develop the general framework that is about the structure of service supply chain. However, there have been some attempts. Armistead and Clark (1993) have applied to the service supply chain through the concept of the value chain. Ellram et al. (2004) proposed service supply chain framework that is included the process of managing and coordinating in the service industry from the first supplier to the final customer. Zhang et al. (2009) developed service supply chain conceptual model based on business process. His conceptual model is composed service providers, service integrators and consumer's demand.

Another challenge to building a framework is that unlike manufacturing, service supply chain does not have a standardized Performance evaluation of rating scale. Ahn and Lee (2010) developed service supply chain's performance measure framework using Fuzzy-AHP.

Although earlier researches have made a contribution of concept of the service supply chain, such studies have failed to provide a formal model. For example, they cannot precisely define and the role of the elements in service supply chain.

2.2 Supply Chain Simulation

Supply chain Simulation is a dynamic model combining the characteristics of the quantitative, mathematical and computer model. Although simulation has mathematical characteristics, the mathematical model with closed form does not provide the solution. The main area of simulation studies are as follows:

Spreadsheet simulation is very easy to use and popular. Sounderpandian (1989) implemented MRP using Spreadsheet simulation. Towill (2003) studied the VMI (vendor managed inventory) model in supply chain using this methodology.

System dynamics (SD) was used to demonstrate the bullwhip effect in supply chain. Forrester (1961) classified supply chain that is constructed retailer, wholesaler, distributor and factory. This study analyzed the reactions in a link to each area of supply chain. Forrester (1961) explained through experimental variation of the sensitivity of customer demand. After which Lee et al. (1997) by the amplification of these fluctuations defined bullwhip effect. Higuchi and Troutt (2004) studied the Tamagotchi's supply chain by using system dynamics.

The most widely used simulation methodology in the supply chain is discrete event simulation. This method is most appropriate to express the dynamic supply chain system. Vollmann et al. (1997) was making researches MRP, ERP in the lower part of the supply chain system.

Object-oriented simulation has proven to be powerful techniques which are reusability and scalability. Rossetti and Chan (2003) studied the supply chain simulation framework by using object-oriented approach. This study established a manufacturing-oriented supply chain based on object-oriented conceptual model. Alfieri and Brandimarte (1997) modeled the supply chain logistics network through object-oriented techniques.

3 Object-Oriented Simulation Model for Service Supply Chain

3.1 Service Supply Chain Structure

The most important part of simulation design is from the physical environment to the process of converting of a conceptual model. Conceptual model design of the study is the first step of the simulation and it is the most influential in all phases.

This research proposes service supply chain structure. The structure is to include the wide range of services supply chain system. Service supply chain simulation consists of the four modules which are Customer module, Service Agent module, Service Operator module and Service Provider module.

Demand module represents the information of customer demand and customers receive the ordered services. Customers can choose the type of service, or be selected in consideration of their state and environmental. For example, customers can choose the tour package in tourism service but in case of hospital and product service center, they are being received in consideration of consumers' health status or condition of the product.

Service agent module checks service capacity according to the customer's service order and it distributes the optimal allocation of service activities.

Service operator module is responsible for the design of the service activities to respond to the needs of customers. For example, service operator role is to design a tour package in tourism service and assigns service resource to each service activity.

Service provider module supplies to provide services service activity with service resource. Each area of the object of the service supply chain can use to design simulation's conceptual model. Service supply chain simulation framework is shown in Figure 1.

3.2 Procedure for Developing Simulation Model

In this study, in order to develop the simulation model uses object-oriented methodology. Object-oriented methodology reduces the complexity of simulation modeling by scalability and reusability.

This study is based on the UML analysis/design tools and on the ARENA® simulation language. ARENA® simulation package has been made by the SIMAN language, it express feature of animation, input and output analysis in GUI(graphical user interface) environment. UML(unified modeling language) take the blueprint to develop simulation model in various types of software (Booch G. et al., 1999). UML diagram has Class diagrams, object diagrams, use case diagrams, sequence diagrams, etc. UML express each module of the object definitions, parameters, operations in object-oriented service supply chain simulation model.

The two main features of the proposed procedure are the definition of a systematic conceptual procedure to design service supply chain models and of a set of rules for the conceptual model translation in an ARENA® simulation language. The following is a simulation model development process.

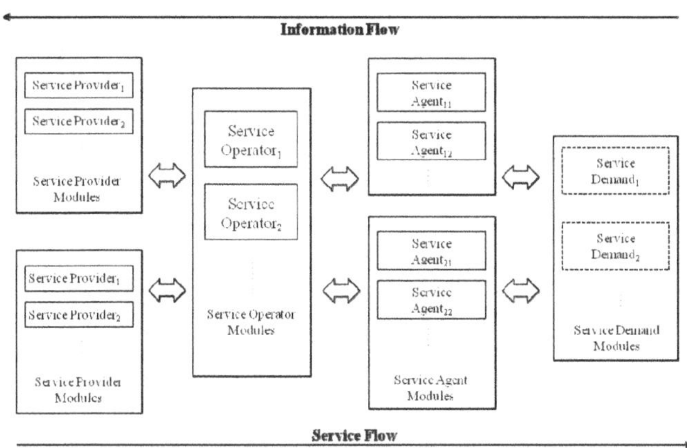

Fig. 1. Service supply chain simulation framework

The first step is the design of functional model. The purpose of this step is the analysis phase of service supply chain modules for UML conceptual modeling phase of the simulation. The design phase functional model represents UML use case diagram through the definition of a component. In other words, this phase analyzes role of demand for service, service agent, service operator and service provider, defines message exchange procedure between each module.

The second step is the design of dynamic model. The purpose of this step is the interaction of components for the messages and actions represents through the UML sequence diagram. Messages sent from one object to another object means to move the lifeline of one object to another object's lifeline in a sequence diagram.

The third step is the design of object model. In this step, the components inside the structure and the relationship through the UML class diagram collect the appropriate class group. The Class diagram is clearly the internal structure of the object that defines each object's name and attributes, relationships with other objects.

The last step is the implementation of the general model by UML and ARENA®. This step classifies to express UML classes in classes used in the simulation package ARENA®.

4 General Model for Service Supply Chain

The diversity and complexity of supply chain services makes it difficult to define a general model. However, the definition of a general model is necessary. General model of the service supply chain simulation gives an outline of the basic components and features.

4.1 Functional Model

In general, functional model shows the flow of information within the system. The Internal flow of information relates to resources, targets, and activity. The objective of this phase expresses abstraction of system components through the use case diagram. The use case diagram gathers system's functional requirements. The use case diagram is hierarchically organized so that system's functional requirements represent phase of the abstract level (upper level). This use case diagram provides the basis of design and development. As a result, this step is an important step that reflects the perspective of developers in the areas of applied simulation. Figure 2 shows a use case diagram of the service supply chain.

In Figure 2, Service Supply Chain System Actors are customer, Service agent, Service operator and Service provider, system use case is Service Demand, Service Type, Service Processes, Service Procedure, Service Activity, Service Resource and Resource. Of these, Service Procedure, Resource is expanded use case of Service Processes and Service Resource.

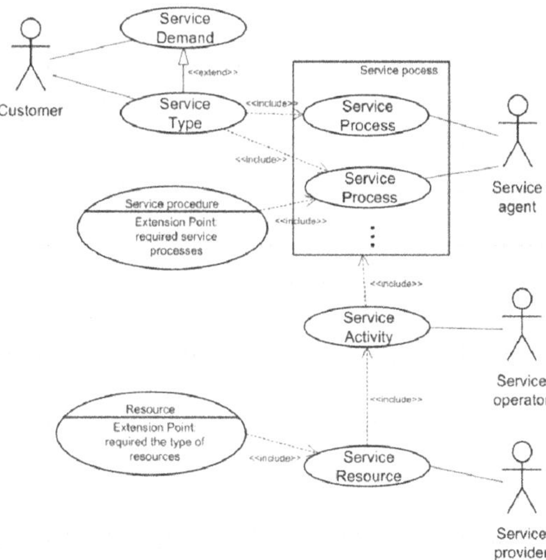

Fig. 2. Services of the supply chain use case diagram

4.2 Dynamic Model

The use case diagram explains service supply chain components and their relationships. The dynamic model can explain the components of each area in service supply through the sequence diagram (Roff, 2003).

The message is passed to an object from an actor and this diagram shows system control flow. The sequence diagram represents the interactions between instances of the class and it shows in detail the flow of processing. Thus, a sequential diagram represents all possible paths or single path. Figure 3 shows a sequential diagram of the service supply chain.

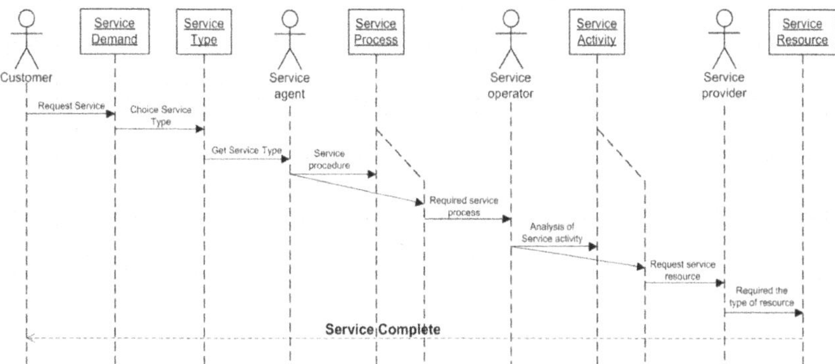

Fig. 3. Services of the supply chain sequential diagram

4.3 Object Model

In this step, Depending on the simulation software, developers should consider the type of class related to software. It is important that defines the relationship between classes corresponding to the software. Object model design is based on analysis of the following three categories:

The class identification: Class definition depends on the considerations of each developer. There is no standard method for finding all the classes in a system. In this study, the definition of the class proposes to find through the use case diagram.

The class relationship identification: Defining relationships between classes make easy model generalization that reduces complexity of the actual implementation of the simulation.

The class structure identification: Attribute and operation for each class define through functional model and dynamic model. Attribute is defined by use case diagram, operation is defined by sequence diagram.

Figure 4 shows four areas of supply chain services for the class diagram. Four areas of module classes represent the operations and attribute in service the supply chain.

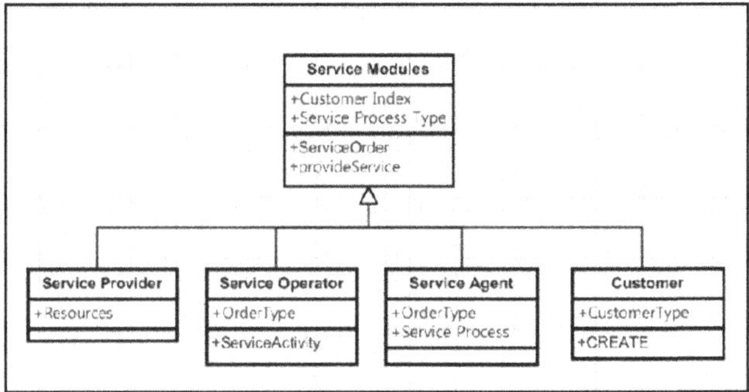

Fig. 4. Conceptual model of supply chain services

ARENA® consists of two structures, one is the EXPERIMENT(static structure), and another one is MODEL (dynamic structure) . ELEMENTS panel is a tool of the EXPERIMENT, BLOCK panel is the process of running the tool by taking the information of ELEMENTS.

In this study, the general model of the service supply chain uses the template of the ARENA. The object class of the static structure uses ELEMENTS panel, and dynamic structure uses BLOCK panel and another panel. ARENA® mapping table for converting shown in Table 1.

Table 1. ARENA mapping table

Template Panel	ARENA Class	Modules Class			
		Customer Demand	Service Agent	Service Operator	Service Provider
ELEMENTS Template Panel	Entities	•			
	Expressions	•			•
	Attributes	•	•		
	Resources				•
	Sequences		•		
	Stations		•		
BLOCK Template Panel	Create	•			
	Assign	•			
	Route			•	
	Dispose	•			
	Size				•
	Delay				•
	Release				•
Advanced Transfer Template Panel	Stations			•	

Based on this table, General model of service supply chain modules expresses class diagram in UML taking into ARENA® Templates. Simulation of the general model is shown in Figure 5. Its aim is to converts for ARENA simulation model.

In this UML class diagram, customer demand module uses Entities, Expressions, Attributes in ELEMENTS template, and Create, Assign and Dispose in BLOCK template. Entities generates an event in the simulation model. Demand information of the object is expressed in the Expressions module. Attributes module defines variable property of object.

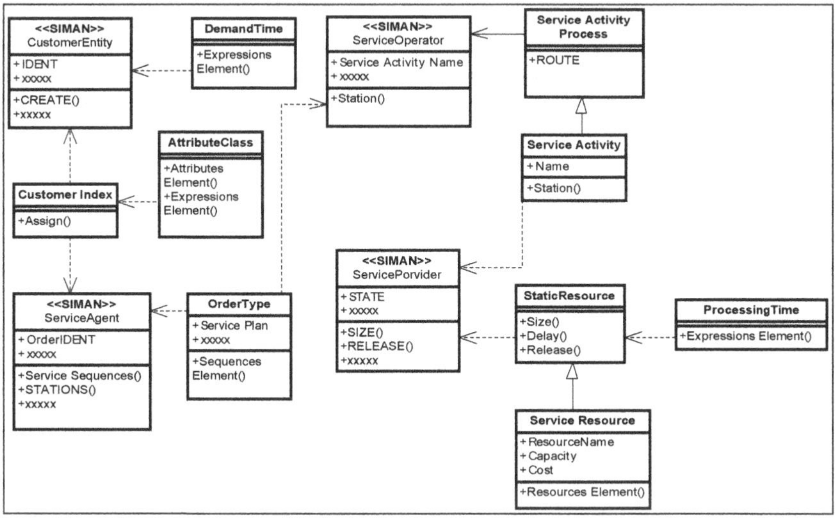

Fig. 5. General model of service supply chain simulation

Service agent module uses Attributes, Sequences and Stations in ELEMENTS template, and Stations and Sequences in BLOCK template. Service operator uses Expressions, Route, Station. Service provider uses Resources in ELEMENTS template, and Create, Size, Delay, Release in BLOCK template. Information exchange between classes is shown in Table 2.

Table 2. Information exchange class

	Modules Class		
	Customer Demand - Service Agent	Service Agent - Service Operator	Service Operator - Service Provider
Information Sharing Class	Customer Index	Order Type	Service Activity
Related ARENA Class	Assign Expressions Element	Sequences	Station Route

5 Applying to Tourism Services Supply Chain

Tourism service is one of service industry that includes hotel, passenger transport and sightseeing. Service supply consists of various service components in a value service network. Zhang Ying-zi [10] proposed that tourism serve supply chain is a bundle of tourism service product which is oriented to tourists and is based on the order of travel agents arrangement and uninterrupted time. However, most research on the service sector is focused on marketing. Sinclair and Stabler(1997) emphasized the importance of the tourism industry in terms of the supply chain. Page (2003) proposed that supply chain has to be the premise in the tourism industry.

5.1 The Structure of Tourism Services Supply Chain Simulation Model

Tourism service supply chain consists of customer, travel agent, tour operator, service provider on the basis of developed general model in this study.

5.2 Conceptual Model

In this study, the conceptual model of tourism services supply chain is based on a general model presented in section 4. Figure 6 is a conceptual model of tourism supply chain services describes how the demand module generates an event and gives unique characteristics to each customer. This model shows the flow where the Tour agent module confirms customer order and allocates is the travel package process. Tour operator module allocates the service activities to customer. Service provider module supplies service resources.

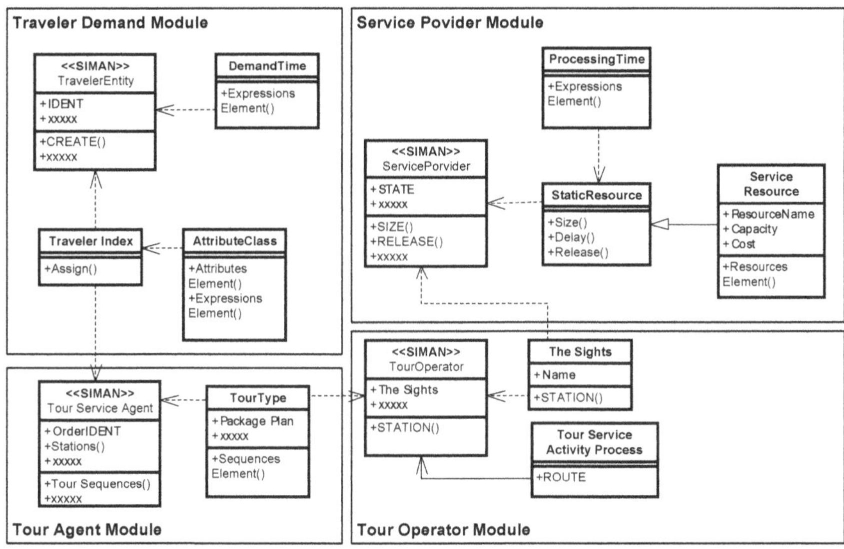

Fig. 6. Conceptual model of tourism services supply chain

5.3 Implementation Using ARENA®

Tourism services supply chain consists of six travel package for implement of ARENA®. The process for each travel package is shown in Table 3. Customer's arrival time interval considers expo (6.5) and expo (10). Warm-up period is 2000 hours (EXPO(6.5)), 1000 hours(EXPO(10)). Travel customers to choose the ratio is 13%(Package 1), 12%(Package 2), 10%(Package 3), 20%(Package 4), 23%(Package 5) and 22%(Package 6). Moving time of customer was not considered. Simulation length is 365 days, and repetition is five times. Resources and processing time for each service activities are shown in Table 4.

Table 3. Travel Packages process

Tour Type	Plan				
Package 1	Sight 1	Sight 2	Sight 3	Sight 4	End of Tour
Package 2	Sight 2	Sight 3	End of Tour		
Package 3	Sight 1	Sight 2	Sight 4	End of Tour	
Package 4	Sight 4	Sight 5	Sight 6	End of Tour	
Package 5	Sight 6	Sight 5	Sight 1	End of Tour	
Package 6	Sight 3	Sight 4	Sight 5	Sight 6	End of Tour

Table 4. Resources and processing time for each service activities

Type	Sight 1	Sight 2	Sight 3	Sight 4	Sight 5	Sight 6
Resource	Hotel 1 Transportation 1 Restaurant 1	Hotel 2 Transportation 2 Restaurant 2	Transportation 3 Restaurant 3 Hotel 3	Transportation 4 Excursion 1	Hotel 5 Transportation 5 Excursion 2 Restaurant 4	Hotel 4 Restaurant 3 Excursion 3 Restaurant 5
Processing Time	48	24	36	12	24	36

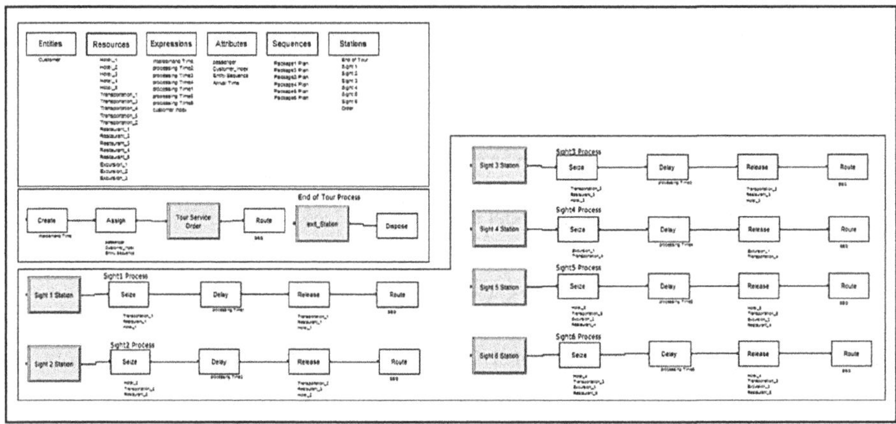

Fig. 7. Object-oriented simulation model of tourism service supply chain in ARENA

Figure 7 is tourism services supply chain simulation model by object-oriented in ARENA environment.

5.4 Experimental Results

Service supply chain simulation model compare results of expo(6.5) with results of expo(10) for model validation.

Customer demand's customer data, service agent's service process data, service operator's activity data, provider's resource utilization of the Factor analysis according to changes in demand factors do not affect each output data.

In addition, the validation of simulation model is that it only generates one of the objects and traces the flow of an object.

6 Conclusion

Until now, the service industry has been primarily studied in the fields of management such as marketing. Supply chain simulation research has focused on manufacturing supply chain. Recently, service industry is being important day by day, and service system is made up supply chain.

The purpose of this study develops service supply chain simulation model by object-oriented under the ambiguous conceptual model of service supply chain. This research progress defines the property and structure of service supply chain and then, simulation model implement in ARENA®.

The result of study, first, structure of service supply chain was defined for development of simulation, and the role of each module in service supply chain was proposed.

Second, this study propose general service supply chain model using objet-oriented programming for development of service supply chain simulation model. This general model can apply the future research for service industry. Finally, this study proposes to design simulation model based on object-oriented in ARENA®.

In future research, we will develop a hybrid model that uses mathematical optimization and simulation models are mixed.

Acknowledgment. This research was supported by Basic Science Research Program through the National Research Foundation of Korea(NRF) funded by the Ministry of Education, Science and Technology (2011-0002993).

References

1. Anglani, A., Grieco, A., Pacella, M., Tolio, T.: Object-Oriented Modeling and Simulation of Flexible Manufacturing System: a Rule-Based Procedure. Simulation Modeling Practice and Theory 10, 209–234 (2002)
2. Armistead, C., Clark, G.: Resource Activity Mapping: the Value Chain in Service Operations Strategy. The Service Industries Journal 13(4), 221–239 (1993)
3. Biswas, S., Narahari, Y.: Object Oriented Modeling and Decision Support for Supply Chains. European Journal of Operational Research 153, 704–726 (2004)
4. Booch, G., Rumbaugh, J., Jacobson, I.: The Unified Modelling Language User Guide. Addison-Wesley, MA (1999)
5. Disney, S.M., Towill, D.R.: MRP on Spreadsheets: A Do-It-Yourself Alternative for Small Firms. Production and Inventory Management Journal, Second Quarter, 6–11 (2003)
6. Ellram, L.M., Tate, W.L., Billington, C.: Understanding and Managing the Services Supply Chain. Journal of Supply Chain Management 40(4), 17–32 (2004)
7. Forrester, J.W.: Industrial Dynamics. MIT Press, MA (1961)
8. Fry, T.D., Cox, J.F.: Manufacturing Performance; Local versus Global Measures. Production and Inventory Management Journal 30(2), 52–56 (1989)
9. Lee, H.L., Padmanabhan, V., Whang, S.: The Bullwhip Effect in Supply Chains. Sloan Management Review 38, 93–102 (1997)
10. Liu, J., Wang, W., Chai, Y., Liu, Y., Easy, S.C.: A Supply Chain Simulation Tool. In: Proceedings of the 2004 Winter Simulation Conference, pp. 1373–1378 (2004)

11. Nasereddin, M., Mullens, M., Cope, D.: The Development of A Reusable Simulation Model for the Modular Housing Industry Using ProModel and Visual Basic. In: Proceedings of Industrial Engineering Research Conference (2002)
12. Page, S.J.: Tourism Management: Managing for Change. Butterworth-Heinemann, Oxford (2003)
13. Robinson, S.: Simulation: The Practice of Model Development and Use. John Wiley & Sons, England (2004)
14. Roff, J.: UML: A Beginner's Guide. McGraw-Hill, NY (2003)
15. Rossetti, M.D.: A Prototype Object-Oriented Supply Chain Simulation Framework. In: Proceedings of the 2003 Winter Simulation Conference, pp. 1612–1620 (2003)
16. Sinclair, M.T., Stabler, M.: The Economics of Tourism. Routledge, London (1997)
17. Smith, S.L.J.: The Tourism Product. Annals of Tourism Research 21(3), 582–595 (1994)
18. Sounderpandian, J.: MRP on Spreadsheets: A Do-It-Yourself Alternative For Small Firms. Production and Inventory Management Journal 30(2), 6–11 (1989)
19. Vieira, G.E.: Ideas for Modeling and Simulation of Supply Chains with Arena. In: Proceedings of the 2004 Winter Simulation Conference, pp. 1418–1427 (2004)
20. Vollmann, T.E., Berry, W.L., Whybark, D.C.: Manufacturing Planning and Control Systems. McGraw-Hill, NY (1997)
21. Zhang, R., Chen, R., Zhang, Y.: The Conceptual Model of the Service Supply Chain Research Based on Business Processes. Management and Service Science, 1–4 (2009)
22. Zhang, X., Song, H., Huang, G.Q.: Tourism Supply Chain Management: A New Research Agenda. Tourism Management 30(3), 345–358 (2009)

Power Saving in Wireless LANs by Adapting the Listen Interval

Kazushi Shinomiya, Ahmad Bazzi, and Yoshikuni Onozato

Graduate School of Engineering, Gunma University
1-5-1 Tenjin-cho, Kiryu, Gunma, 376-8515 Japan

Abstract. We are increasingly relying on wireless networks to access different Internet and Intranet services. With the current battery technologies, the manufacturers of portable wireless devices have to compromise between portability and battery life. The IEEE 802.11 standard includes a power saving mode (PSM), which is fixed and does not adapt to the varying traffic patterns. In this paper, we propose an enhancement to the standard PSM through the implementation of an adaptive listen interval that changes in accordance with traffic patterns. Our simulation tests show that the proposed technique provided better power saving compared to the standard PSM.

Keywords: Power saving, PSM, Listen Interval, 802.11.

1 Introduction

Recent years have seen a remarkable development of mobile systems, accompanied by an increasing reliance on mobile and wireless devices in everyday lives. However, the increase in the usage duration and rate also leads to an increased power consumption. Unfortunately this is posing a challenge with the current battery technologies.

In normal working conditions, the relatively high power consumption can lead to mere user inconvenience if the battery power is drained quickly for example. However, in an emergency situation, such as those caused by natural disasters, the battery power is the only way to stay connected to the network. Moreover, the power reserve in backup power supplies becomes the only chance to keep the network alive. In such cases, communications can help facilitate the work of rescue teams and might eventually save human lives. Hence, we can see the increasing importance of power saving in communication systems.

The objective of our study is to improve the power-saving in wireless communications by improving the Power Saving Mode (PSM) standardized in IEEE 802.11 [2]. PSM greatly reduces the power consumption of user terminals during communications; however, it still wastes power when no data frames are being sent to the user for a relatively long time. In this paper, we concentrate on the power-saving capability of PSM, and we introduce a technique for decreasing the power consumption beyond that of the current standard PSM.

J.-H. Kim et al. (Eds.): AsiaSim2011, PICT 4, pp. 281–290, 2012.

In this paper, we are concerned with the access point side. We tackle this power saving problem by implementing an adaptive Listen Interval (LI) and we show through various simulations that our approach is efficient and can better fit the wireless communication with the access point.

2 Previous Works

In the past years, there have been various works attempting to improve the power saving of 802.11 wireless networks. The published research shows diverse approaches to improve power saving in wireless networks by modifying the client software, while others concentrated on the access point.

We start by presenting some of the works concerned with reducing power consumption through modifying the client:

Krashinsky and Balakrishnan proposed Bounded-Slowdown (BSD) protocol [3]. BSD protocol aims to minimize energy consumption without letting the round trip time (RTT) exceed a certain configurable limit, eventually a compromise can be reached between power saving and speed. This protocol relies on past network activity in order to adapt the sleep duration, while ensuring that the RTT is not stretched by more than a chosen factor p. Consequently, the network interface will wake up less often to listen for the beacon during idle times. The authors provide different simulations when BSD is used in the case of HTTP requests issued to a web server.

Qiao and Shin proposed Smart Power Saving Mode (SPSM) [5], which aims to find the best action sequence to enter the wake and doze states in order to guarantee a limited delay with minimum energy consumption. An action sequence is what tells the network interface to stay awake or sleep after a request is sent in. In order to achieve this optimal action sequence, SPSM uses a two-step approach. First, it interprets the desired delay performance using a generic penalty function. Second, it feeds this penalty function to their proposed algorithm in order to produce the optimal action sequence. According to their simulations, SPSM provided and ensured the same responsiveness as BSD while achieving additional energy saving.

Nath et al. proposed a Dynamic Beacon Period (DBP) [4], where the access point uses different beacon periods for the different clients. The beacon period is calculated by the client based on the RTT of the client's current HTTP connection. Once chosen, it notifies the AP of its choice and the AP in turn uses this beacon period to communicate with the client. Hence this is different from the standard case where the same beacon period is used uniformly across all the clients.

3 Power Saving Mode

Generally speaking, in order to reduce power consumption in 802.11 wireless LANs, the users need to use the available power management functions standardized by IEEE 802.11. PSM and Active Mode (AM) are designed as the operational modes of these power management functions.

The states of the client in each mode are defined by:

1. The doze state with minimal power utilization
2. The awake state which can transmit and receive frames

The PSM function manages the power supply through these two states. On the other hand, during the AM, the client remains in the awake state, so it can transmit and receive packets any time. When the client switches the operational mode, the access point recognizes this switch through the management bit in the frame control field of the frame transmitted by the client.

3.1 Basic Function of PSM

When a client uses PSM, it switches to the doze state and only enters the awake state according to the LI signifying a period of receiving beacons. Because the client in the doze state cannot transmit and receive frames, the AP stores the incoming frames in its own buffer until the client enters the awake state again. When the AP starts storing frames directed to a certain client, it notifies that client with a TIM (Time Indication Map) beacon generally once in a 100 millisecond period. Then the AP transmits the frames to the client once the latter enters the awake state. The client remains in the awake state until the AP finishes transmitting all the stored frames. Once done, the client switches back to the doze state. This way the client that uses PSM can reduce power consumption by using the doze state most of the time except when transmitting or receiving frames.

3.2 Limitations of PSM

Although PSM improves the client's power saving, it suffers from the several drawbacks:

Fixed Listen Interval. Because the LI has a fixed length, there is a wasted awake time when there are no frames to be sent to the client for a relatively long time. Consequently, this negatively affects the efficiency of the PSM. Although a user can change the LI, this is a difficult task for the average user. For practical reasons, we can assume that the developer has previously set a relatively long value of LI. When LI is long, useless switching decreases when the frame arrival interval is large, and the client remains in the doze state for a longer period. Unfortunately, this approach would amplify the weakness presented in the next point.

Delays and TCP. Another limitation of PSM is that it might hinder the performance of the TCP protocol. Because the transmitted frame is stored in the AP, when switching to PSM, the ACK packet might be delayed and the TCP connection would suffer a time-out before the client can receive and process it. Eventually, this might lead to unnecessary retransmissions and the client would eventually consume additional electric power.

4 Adaptive Listen Interval PSM

As described in the previous section, PSM can improve the power consumption, but it has some limitations. Therefore we introduce the Adaptive LI PSM (ALIPSM), where the AP automatically and adaptively updates the client's LI using the frame arrival time. To solve the previous two problems, the AP in ALIPSM uses some additional functions as follows:

4.1 Adaptive Updating of the LI

In PSM, LI is constant; moreover, it is difficult to be modified by the average user. Consequently, there is very little that the user can do to avoid unnecessary mode switching in order to save battery power. Moreover, even if the user is familiar with PSM, it is clear that any user effort to modify LI is inefficient as the traffic pattern keeps changing. Therefore, we implement a new function where the AP monitors and estimates the traffic flow, updates the LI automatically and notifies the relative client. In ALIPSM, the traffic estimation is based on the *average frame arrival interval.* We must note that the time frame used to calculate the average arrival interval should not be too long otherwise the AP won't make a useful estimation.

If the AP uses long term average of the arrival interval for the AI, then it is clear that the estimation calculated by the AP will not work when the arrival interval is too high or too low. However, even if the estimation is good, there will always be some frames that arrive at an interval different than those originally estimated and if we were to rely only on them, then we will make unnecessary mode switching just like the case of PSM. To avoid this pitfall, the value of LI estimation should not be updated based on one frame arrival interval, but rather upon multiple frame arrival intervals. In this paper, to estimate the arrival interval in an efficient manner, we follow the following procedure: First, for the initial value of AI, we rely on a number of collected frames in PSM. Then we update this value based on the packets that arrive during each doze state before it switches to the awake state.

Finally, the AP often sends several frames while the client is in the awake mode. We label this number, the Collected Frames (CF). In order to adjust the LI, the CF must be suitable as well. Our approach can be described as follows, the CF decreases if the client receives the packets during its awake interval. Moreover, the CF increases if the client doesn't get its packets during that time. This approach is similar to congestion control.

If the traffic is slowly changing, using the average interval will work perfectly. But this technique has limitations when the arrival interval changes abruptly. In this case, the client would suffer useless switching at the next awake because the previous average differs greatly from the next average interval.

4.2 Automatic Reply

Another limitation of PSM is that it can negatively affect the function of the TCP protocol. Because the transmitted frame is stored in the AP when switching to PSM, the ACK packet might suffer time-out before the client gets the chance to receive and process it. Therefore in ALIPSM, we assume that the AP replies with a suitable ACK packet in order to avoid such useless retransmissions [1].

When the client uses ALIPSM, the client switches to the doze state, and the AP allocates a private buffer for the client as part of its own buffer. The AP measures the arrival time of the frame sent to each client and then, using these values, it calculates the frames' average arrival time for each of the clients. When the client enters the awake state, the AP starts transmitting the frames to the client along with the calculated Average Interval (AI), which specifies the next LI of the client. In other words, the LI of the client is replaced by the received AI. The value of the AI is calculated by multiplying the average arrival time by the CF — the CF is the size of the virtual buffer allocated for the client on the AP. And the CF must be smaller than the buffer of the client. Because, in ALIPSM, the AP sends the frames while the client is in awake state; moreover, it has to store the frames in the client buffer when the client is in doze state. If the size of the CF is bigger than the buffer for that client, buffer overflow would occur.

Consequently, ALIPSM can reduce power consumption more efficiently than PSM by using an adaptive LI and taking into consideration the particular wireless network communication in progress.

5 Simulation Model and Parameters

In this section, we describe our simulation model in order to compare the performance of ALIPSM with that of PSM. Note that we carry out the performance assessment by adjusting the LI so that the AP replies with an ACK packet in the PSM state as well. In this model, packet loss and data errors are not considered.

The sender who transmits the data that the client requests only has the simple ability to send packets to the client. Similarly, the client only has the simple ability to receive packets and to change LI depending on AI. Because the purpose of this simulation is to study the situation where the arrival interval of the packets is bigger than LI, the traffic between the sender and the AP is described only in terms of the arrival interval of the packets.

Because we assume that the AP in ALIPSM doesn't have information about arrival interval at first, the AP using ALIPSM follows PSM until it receives 5 packets and can calculate the average interval.

It is worth noting that the time is normalized to be measured in steps, where a step is the time required for a frame to reach the destination from the AP. The purpose of this approach is to avoid incorporating the device performance and environment, while concentrating on evaluating the performance.

The number of users is considered to be only one. Moreover, in this paper, the devices are considered to be fixed and immobile. The number of packets which the sender transmits to the client is 100. We assume that if the buffer size of the AP is large enough, we won't suffer any packet loss. Useless switching takes 1 step, LI in PSM is 10 steps. Power consumption during the doze state is assumed 1 mW/step and power consumption during the doze state is assumed to be 10 mW/step. CF is set to be 10 or 50. We describe the traffic by the median of arrival interval and the difference with median. It means that arrival intervals of the packets which the AP receives are ($median \pm difference$) steps. Median of the arrival interval is set as 10 to 100 steps and the difference with the median and the arrival interval is set to be less than 5 to 100 steps. Each simulation is repeated 10,000 times and the average result is used. Table 1 summarizes the simulation parameters used.

Table 1. Simulation parameters

Number of users	1
Number of simulations	10,000
Number of frames	100
Movement speed of the terminal	0 m/s
PSM's LI	10 steps
Power consumption during the doze state	1 mW/step
Power consumption during the awake state	10 mW/step
CF	10, 50
Median of the arrival interval	10 to 100 steps
Difference between median and arrival interval	up to 5 or up to 100 steps

5.1 Total Reception Steps

Total reception steps refer to the total number of steps which the client needs to receive all the packets which it requested. If this value is small, the client gets all the packets in short time. ALIPSM improves power efficiency by changing LI where it can be longer than of PSM; therefore, the client might need longer time to get all the packets compared to using PSM. Hence, we need to provide a comparison between ALIPSM and PSM regarding this factor.

5.2 Reception Power Consumption

Reception power consumption refers to the total power which the client consumes until it receives all the packets form the sender. If this value is small, the client achieves higher power efficiency. In this paper, power efficiency is the most important factor because the purpose of this study is to improve PSM's current power efficiency.

5.3 Packet Delay

Packet delay is the incurred delay for the AP to send a packet to the client and it is compared with the case of AM. If this value is small, the client gets the packet practically as soon as the AP gets it. The AP using ALIPSM or PSM stores some packet until the client switches to the awake state. It means that the client needs more intervals to get a packet than when using AM. Moreover, ALIPSM's LI is calculated longer, so it is assumed that using ALIPSM makes more delay than using PSM.

6 Simulation Results

In this section, we describe the numerical simulation that we implemented in C language in order to compare the performance of ALIPSM with that of PSM. We carry out the performance assessment by adjusting the LI to adapt to the traffic; moreover, the AP replies with an ACK packet in the PSM state as well. It is worth noting that the time is standardized to be measured in step, where a step is the time required for a frame to reach the destination from the AP. Note that the difference between the median and the arrival interval is less than 5 steps in figures 1–4 except for figure 3. In addition, ALIPSM'X' in the following figures means using ALIPSM with CF set to the value 'X'.

Figure 1 shows the number of the total reception steps until the client gets all its packets. We can see that the number of the total steps of PSM is only 9% less than that of ALIPSM. In other words, ALIPSM only needs 9% longer time. If the difference between the median and the arrival interval is small, then the difference between the real average interval and the average interval calculated by the AP should be small as well. Moreover, if the CF buffer is small, the number of the total number of steps of ALIPSM would be shorter than when using a large CF buffer because the value of the AI is directly proportional to the size of the CF buffer. So in this case, the client using ALIPSM can receive the last frame within a minimal time, while avoiding unnecessary power consuming cycles.

Figure 2 shows the total power consumption that the client needs to receive all the packets. We can see that ALIPSM is more efficient than PSM with respect to power consumption. Following our objective that we described in section 3, we think that reducing useless awake states should improve the overall efficiency. Moreover, useless switching is increased as the difference between PSM's LI and arrival interval increases. Hence, ALIPSM is more efficient as the difference between PSM's LI and arrival interval decreases.

Figure 3 shows the power consumption when the difference between the median and arrival interval changes. We can see that large CF buffer may lead to an unstable performance. If the AP uses a large CF buffer, the AP may calculate an AI that is much longer than that when using a small CF. Hence, it may take useless doze intervals to get the last small packet number. Consequently, using large CF may cause additional power consumption than when using a small CF.

Figure 4 shows the average delay time from the AP to the user terminal when buffering packets in comparison with continuous awake mode. As we can see in this figure, PSM keeps a steady delay; this is because the transmission interval in PSM only depends on LI. In other words, the AP receives packets during sleep mode, so the delay is always smaller than LI, where the average delay is LI/2. On the other hand, the delay in ALIPSM increases as the value of the packet arrival

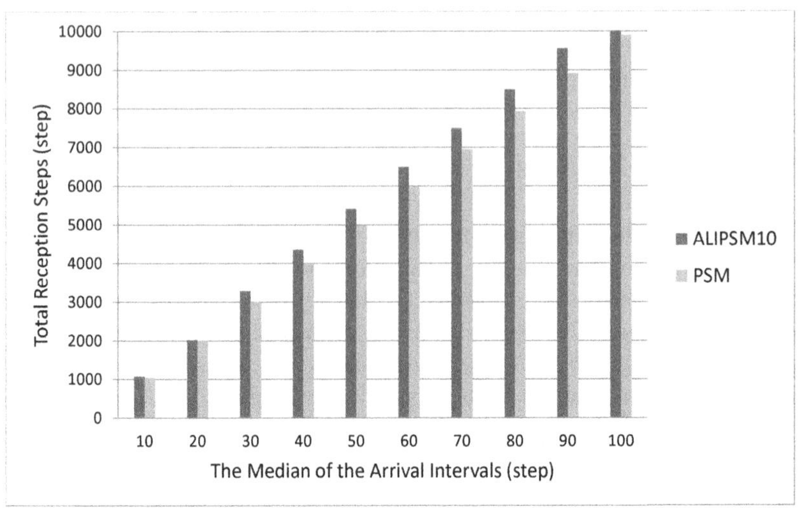

Fig. 1. Total reception steps versus the median of the arrival intervals when the difference between the median and the arrival interval is less than 5 steps

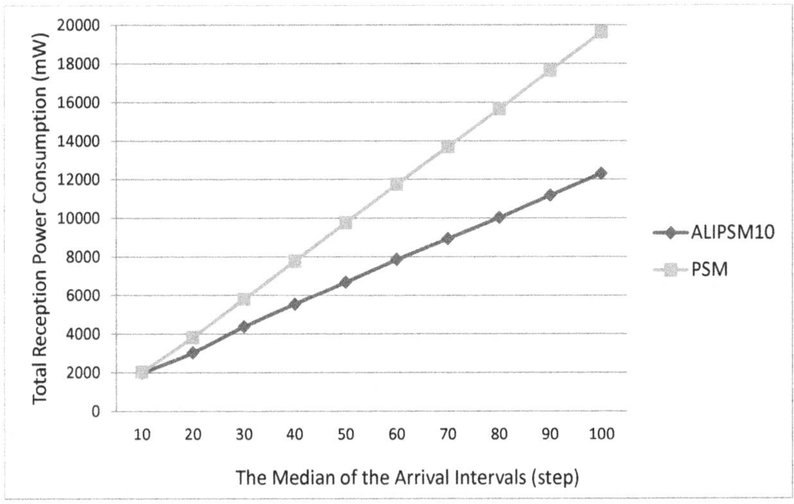

Fig. 2. Total reception power consumption versus the median of the arrival intervals when the difference between the median and the arrival interval is less than 5 steps

interval and CF increase. The transmission interval in ALIPSM is calculated by the arrival time and CF. Hence, the transmission interval in ALIPSM grows as the arrival interval and CF grow.

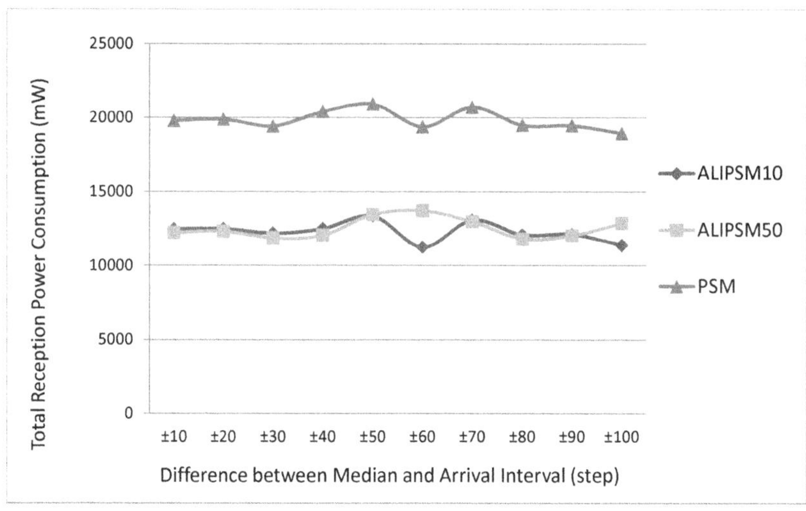

Fig. 3. Total reception power consumption versus deviation from the median

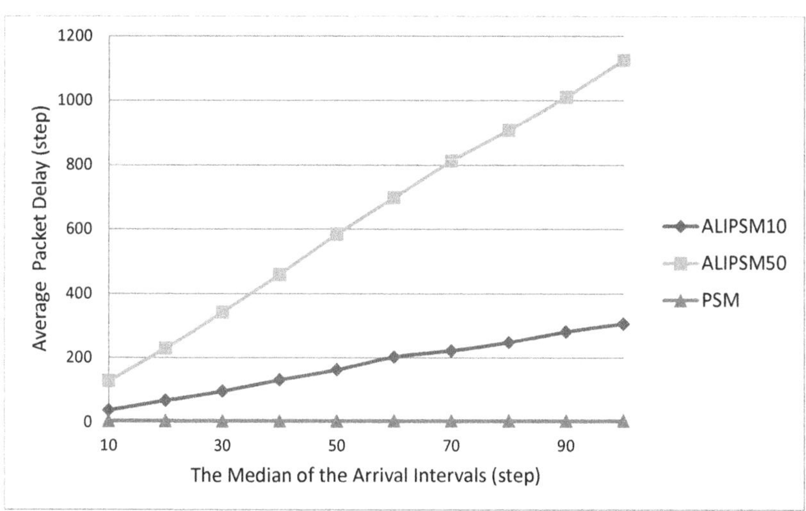

Fig. 4. Average packet delay versus the median of the arrival intervals when the difference between the median and the arrival interval is less than 5 steps

7 Conclusion

Based on our simulations, ALIPSM allows a more efficient power consumption compared to PSM. Moreover, the AP does not have to use large CF buffer for the clients to achieve this reduction in power consumption. On the other hand, ALIPSM leads to further delay in order to receive the packets, while still improving the power saving efficiency. The AP doesn't need to have a large buffer size when the traffic is steady. The decrease in power consumption leads to a longer up-time on backup power sources, and this can be of great value when facing certain disasters where reliable power sources become unavailable.

References

1. Hashimoto, Y., Imai, N., Morikawa, H., Aoyama, T.: Improving Throughput of IEEE 802.11 Power Saving Mode with TCP Proxy. IEICE Technical Report, IN2002-207, NS2002-234 (2003)
2. IEEE Computer Society: IEEE Standard for Information technology — Telecommunications and information exchange between systems — Local and metropolitan area networks — Specific requirements — Part 11: Wireless LAN Medium Access Control (MAC) and Physical Layer (PHY) Specificatio. In: LAN/MAN Standards Committee (ed.) IEEE 802.11-2007, chap. Part 11: W, pp. C1–1184. IEEE Computer Society (2007)
3. Krashinsky, R., Balakrishnan, H.: Minimizing energy for wireless web access with bounded slowdown. In: Proceedings of the 8th annual International Conference on Mobile Computing and Networking, pp. 119–130. ACM (2002)
4. Nath, S., Anderson, Z., Seshan, S.: Choosing beacon periods to improve response times for wireless HTTP clients. In: Proceedings of the Second International Workshop on Mobility Management & Wireless Access Protocols, pp. 43–50. ACM (2004)
5. Qiao, D.: Smart power-saving mode for IEEE 802.11 wireless LANs. In: 24th Annual Joint Conference of the IEEE Computer and Communications Societies, INFOCOM 2005, Proceedings IEEE, vol. 3, pp. 1573–1583 (2005)

Study on the Artificial Synthesis of Human Voice Using Radial Basis Function Networks

Yuuki Naniwa[1], Takaaki Kondo[1], Kyohei Kamiyama[1], and Hiroyuki Kamata[2]

[1] Graduate School of Science and Technology, Meiji University, Kawasaki, Japan
{ce01067,ce11039,kamiyama}@meiji.ac.jp
http://www.isc.meiji.ac.jp/~kamata/lab_kamata/
[2] School of Science and Technology, Meiji University, Kawasaki, Japan
kamata@isc.meiji.ac.jp

Abstract. In this study, we introduce the method of reconstructing more natural synthetic voice by using radial basis function network (RBF) that is one of neural network that is suitable for function approximation problems and following and synthesizing vocal fluctuations. In the synthetic simulation of RBF, we have set the Gaussian function based on parameters and tried to reconstruct the vocal fluctuations. With respect to parameter estimation, we have adopted to nonlinear least-squares method for making much account of the nonlinearity of human voice. When we have reproduced the synthesized speech, we have tried to reconstruct the nonlinear fluctuations of amplitude by adding normal random number. We have made a comparison the real voice and the synthetic voice obtained from simulation. As a consequence, we have found that it was possible to synthesize the vocal fluctuations for a short time.

Keywords: RBF network, K-means, Nonlinear least square method, Fluctuation of vocal cord.

1 Introduction

In recent year, there are many methods centered upon text-to-speech synthesis because of the performance of computers that handle large amount of speech database being improved. In the study of text-to-speech synthesis, especially in recent, hidden Markov models (HMM) has often been used in some way[1].These studies have human voice as a recording medium. So, when we set the rule of connecting each syllable, phoneme, waveform of 1 cycle and that of controlling prosodic information such as pitch and amplitude precisely, we able to reproduce portable synthetic voice and it is also commercialized.

However, in terms of humanity about generated voice, it is still not quite express. Therefore, the technique of speech production for the machine model has become a hot topic. Sawada has developed a robot that captured the appropriate speech techniques by associating not only leaning of the correspondence between acoustic

J.-H. Kim et al. (Eds.): AsiaSim2011, PICT 4, pp. 291–300, 2012.

features and vocal tract shape, but also leaning of the amount of motor control [2]. In regard to speech synthesis on software, Kanda has constructed and validated the vowel acquisition model by segmenting continuous acoustic signal and articulatory movement based on the vocal tract production model by Maeda and the neural model [3, 4]. The following thing has been mentioned in various papers, and we think that it is extremely meaningful in the field of speech recognition and speech synthesis that we make clear language acquisition board as a basic flow of the process of language acquisition from infant[5]. In addition, we believe it is necessary to try to synthesize self-organized speech like with vowel acquisition through voice imitative interaction in speech production.

In this study, we try to create the new artificial voice but in a short time, not voice recording medium by focusing on the human vocal cords. By using the radial basis function neural network (for pattern classification and function approximation problem), we aim to synthesize vocal fluctuations and to produce artificial sounds that are similar to the real voice by keeping track of real voice as much as possible.

Therefore, we will cope with the following three points.

(1) The reconstruction of the fundamental frequency with chaos in short time.

(2) The verification of the synthesized fluctuations improved by adding the vocal fluctuations to the smoothed data in the input.

(3) The reconstruction of the intensity of each period by Box-Muller-method.

In keeping with above points, we attempt to synthesize the nonlinear vocal fluctuations by using RBF network with the nonlinear approximation of parameters.

2 Fluctuation of Vocal Cords

At first sight, the human speech waveform seems to have a periodic structure. However, the shape of the waveform is changing for each cycle and the cycle width subtle changes accordingly[7]. The difference in width of the period is called the fluctuation of the vocal cords.

2.1 Sampling of the Vocal Fluctuations

With regard to the method of sampling vocal fluctuations, we determine the initial value $x_{max}(k)$(the number of sample $t_0 = 0$) arbitrarily from the maximum (minimum) value of voice waveform, and starting from there, number of sample until the next peak is 1 component of vocal fluctuations. In the case of k fluctuations, the following formula is given.

$$y(k) = x_{max}(k) \tag{1}$$

Visualizing t_k as the total number of samples, we seek in term of the width of each period $x_{max}(k)$.

$$x_{max}(k) = t_{k+1} - t_k \tag{2}$$

We collect the components of vocal fluctuations for the simulation by repeating this procedure. In addition, we set up a threshold to make the exact maximum (minimum) value and are to be used to normalize each of it.

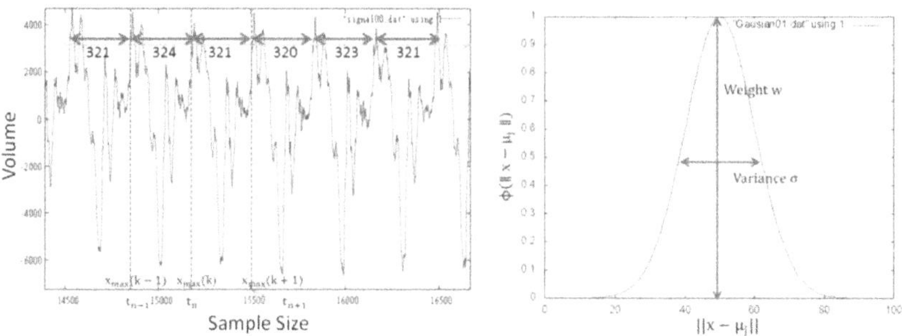

Fig. 1. The extraction method of vocal fluctuations and Gaussian function of RBF

3 RBF Network Including the Smoothing Components

RBF(Radial Basis Function Network) is a kind of deformed network across the three-layers, and it was considered as the way to complement desired arbitrary function by overlaying the localized basis functions. With respect to basis functions, there are several functions about Simlate-spline, bell-shaped, but we use Gaussian function (see Fig.1) in this study.

3.1 Structure

The formula that represents the structure of RBF network that obtains 1-dimensional output f(x) to K-dimensional inputs is as follows. In this study, we will try to validate the multi-dimensional output too [8, 13].

$$f(k) = \sum_{i=0}^{M} \omega_i \varphi_i (r) \tag{3}$$

Here, $\varphi_i(r)$ is the basis function, ω_i is the coupling coefficient of it, M is the amount of it. Gaussian function used as RBF $\varphi_i(r)$ is shown below.

$$\varphi_i(r) = \exp\left(-\frac{||x_k - \mu_i||^2}{2\sigma_i^2}\right) \tag{4}$$

Here, x_k is the input vector and φ_i is the variance of basis function.

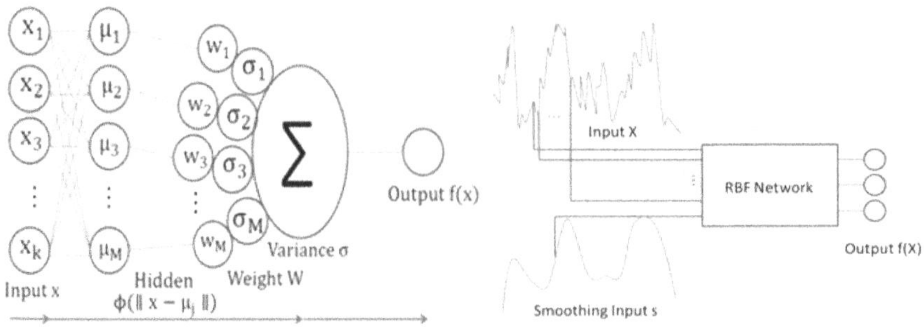

Fig. 2. Structure of RBF and Input of RBF

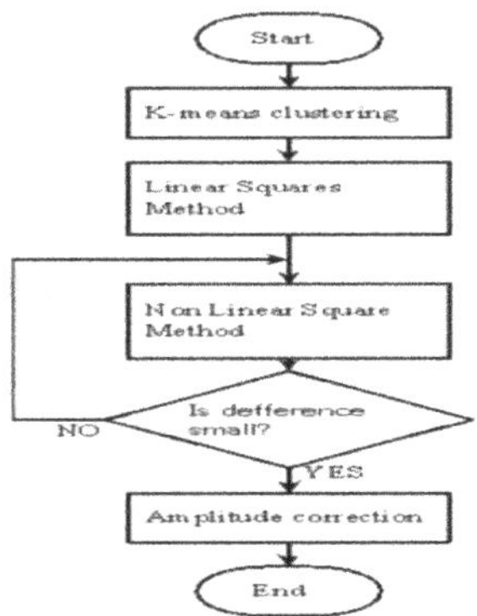

Fig. 3. Flowchart of RBF network

μ_i is the i th center vector and determines the number of it according to the number of RBF. The interior of $\|\cdots\|$ also represents the Euclidean norm.

Depending on how to give this vector μ_i and the number of RBF M, the accuracy of predictive value varies significantly. In this study, we have focused on the variance σ_i included in Gaussian function. In most of the studies using RBF, they always have treated the variance σ_i as a constant value, but we treat it as a variable and considered the changing value according to the center vector μ_i to enhance the reproducibility.

The weight w_i is obtained by using least squares from M-input data and Eq.(3) from the general formula of RBF. In addition, utilizing the obtained weight w_i, we seek the variance σ_i by using the Gauss-Newton method from Eq.(4).

We aim to reconstruct the predictive value closer to the actual data from two tasks of this. Fig. 2 shows the structure of RBF that obtains one to several outputs against K-inputs, and Fig. 3 summarizes the flow of RBF.

3.2 K-means Method

(1) We give appropriate initial values of M from the actual data of the vocal fluctuations.

(2) We calculate the distance between the center of the cluster and all the data, and classify the cluster most similar to the Euclidean distance.
(3) We seek the center from the formed cluster.
(4) We repeat the step of *(2), (3)* until there is no change in the center of the cluster.

For simplicity, we put the image of the algorithm of the K-means method by using the scatter plot on two-dimensional planes.

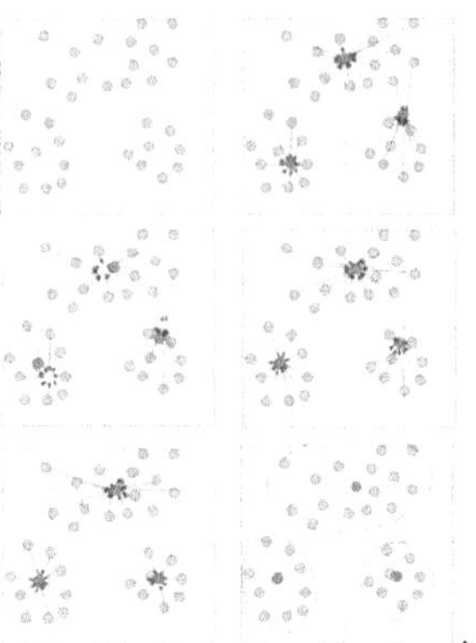

Fig. 4. K-means clustering

3.3 Nonlinear Least-Squares Method

In seeking the variance σ_i included in Gaussian function of the nonlinear equation $p(x) = \sum_{i=0}^{M} \omega_i \varphi_i(r)$, we use the Gauss-Newton method of being least squares approximation of Newton's method that we approximate the curve of function p(x) by means of the tangent by as a way to approximate more accurate results by little computation.

3.3.1 Gauss-Newton Method

We use Gauss-Newton method in M-dimensional input to approximate the variance σ_i included in RBF [12].

In determining the number of M-parameters, given the n data $(x_i, y_i)(i = 1, 2, \cdots, n)$, sum of squared errors of real data y_i and nonlinear equation $p(x_i, a_0, a_1, \cdots, a_M)$is shown as below.

$$f(a_0, a_1, \cdots, a_M) = \sum_{i=1}^{n} (y_i - p(x_i, a_0, a_1 \cdots, a_M))^2 \tag{5}$$

In this equation, we determine the value of the parameter. When we determine a_i parameter, the follow matrix is calculated by making use of differential and Taylor developing.

$$\begin{bmatrix} \sum_{i=0}^{n}\left(\frac{\partial p_i}{\partial a_0}\right)^2 & \sum_{i=0}^{n}\left(\frac{\partial p_i}{\partial a_0}\frac{\partial p_i}{\partial a_0}\right) & \cdots & \sum_{i=0}^{n}\left(\frac{\partial p_i}{\partial a_0}\frac{\partial p_i}{\partial a_M}\right) \\ \sum_{i=0}^{n}\left(\frac{\partial p_i}{\partial a_1}\frac{\partial p_i}{\partial a_0}\right) & \sum_{i=0}^{n}\left(\frac{\partial p_i}{\partial a_1}\right)^2 & \cdots & \sum_{i=0}^{n}\left(\frac{\partial p_i}{\partial a_1}\frac{\partial p_i}{\partial a_M}\right) \\ \vdots & \vdots & \ddots & \vdots \\ \sum_{i=0}^{n}\left(\frac{\partial p_i}{\partial a_M}\frac{\partial p_i}{\partial a_0}\right) & \sum_{i=0}^{n}\left(\frac{\partial p_i}{\partial a_M}\frac{\partial p_i}{\partial a_1}\right) & \cdots & \sum_{i=0}^{n}\left(\frac{\partial p_i}{\partial a_M}\right)^2 \end{bmatrix}$$

$$\begin{bmatrix} \delta a_1 \\ \delta a_2 \\ \vdots \\ \delta a_M \end{bmatrix} = \begin{bmatrix} -\sum_{i=0}^{n}\frac{\partial p_i}{\partial a_0}R_i \\ -\sum_{i=0}^{n}\frac{\partial p_i}{\partial a_1}R_i \\ \vdots \\ -\sum_{i=0}^{n}\frac{\partial p_i}{\partial a_0}R_i \end{bmatrix} \tag{6}$$

By solving the matrix equation above, it is necessary to calculate the following equation iteratively to determine the true value a_i obtained from the ters $\delta a_1, \delta a_2, \cdots, \delta a_1$.

$$a_i = a_i^{(0)} - \delta a_i \tag{7}$$

Here, $a_i^{(0)}$ is the initial value set arbitrarily. When the convergence condition is satisfied, we will terminate the iteration.

$$\sum_{i=0}^{n}(f(a_0, a_1, \cdots, a_M))^2 < \delta^2 \tag{8}$$

At this time, the obtained parameters are the true values.

4 Amplitude Correction of Synthesized Speech Signal

It is obvious that human voice contains a variety of nonlinear elements in addition to vocal fluctuation. Then, now we focus on this nonlinear change in amplitude, and try to vary amplitude for each cycle with random number. For random number generation, we collect the difference between the maximum and minimum of real voice signals (the width of amplitude), and we generate the random number conforming to Gaussian distribution by Box Muller method because distribution of the data relatively close to it [6]. Then, we use the random number as a multiplication of each period of real speech signal..

4.1 Box-Muller Method

Random numbers with normal distribution (mean μ, variance σ^2) η_1 and η_2 can be expressed as follow using independent uniform random number ξ_1, ξ_2 in [0, 1].

$$\eta_1 = \mu + (\sqrt{-2\log(\xi_1)}\cos 2\pi\xi_2)\sigma \tag{9}$$

$$\eta_2 = \mu + (\sqrt{-2\log(\xi_1)}\sin 2\pi\xi_2)\sigma \tag{10}$$

5 Simulation Conditions

The measurement condition of voice signal used in the simulation is shown in Table 1 bellow.

Table 1. Measurement conditions

The subject	Japanese adult man (22)
Measurement signal	A vowel /a/
Sampling frequency	44.1[kHz]
Bit rete	16[bits]
The number of samples (voice signal)	About 50000[points]
The number of the vocal fluctuations (period)	154[periods]

6 Synthesis Results

In Fig. 5, against the original waveform of vocal fluctuations of /a/, we show a comparison of the synthetic vocal fluctuations in the pattern of the proposal method that we have added the one to several smoothed data in the input. In the Gauss-Newton method used in the approximation of variances σ_i, it is considerably variable whether the initial values converge or not depending on the value. So we compared it to second decimal places strictly.

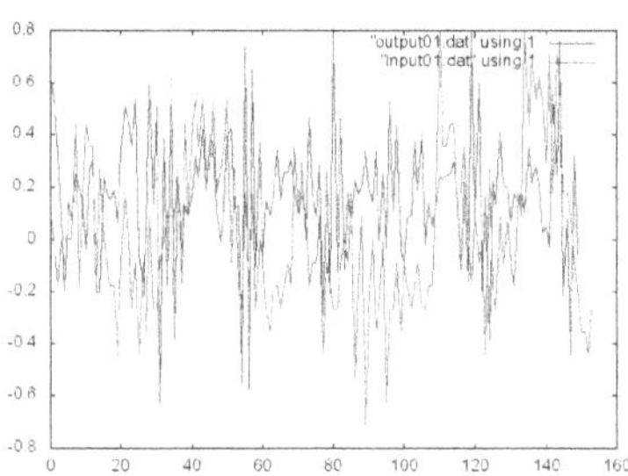

Fig. 5. Comparison of the real vocal fluctuations /a/(input00.dat: Real vocal fluctuations), (outpu01.dat: Synthesized vocal fluctuations with smoothed data)

In the conventional method, with respect to the reconstruction of high parts of fluctuations, there was something lacking fidelity [14]. However, the proposal method in Fig. 5 shows that it is relatively better. Also, we succeeded in shortening the error between real vocal fluctuations and synthesized vocal fluctuations than the conventional method. We have been able to get the best results since we adopted the nonlinear approximation. However, there was no significant improvement about the high frecuency components. To workaround this, we think that it is necessary to increase the degree of smoothed fluctuations in input or to apply LPF (low-pass filter) to the output.

After that, we compare between the real speech waveform and the synthesized speech waveform that we have faithfully reconstructed based on the synthesized vocal fluctuations in Fig. 7.

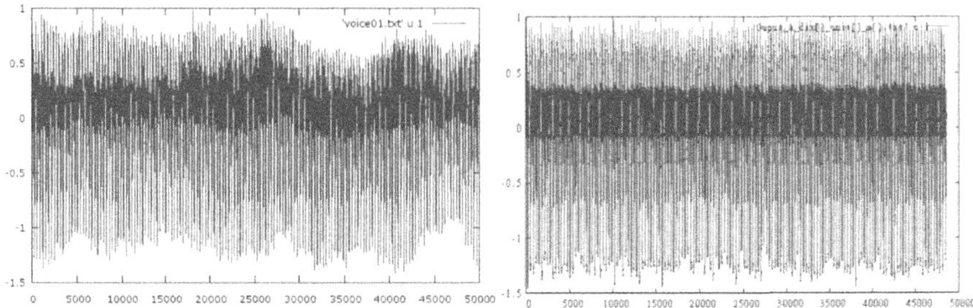

Fig. 6. Left: Real voice waveforms /a/ and Right: Synthetic voice waveforms /a/

About the wave of details, it is almost identical to the real speech waveform. We have also succeeded in improving the randomness in fluctuations of amplitude by expanding the variance of normal random variable. However, it hasn't become smooth like real speech waveform. So we have tried to reconstruct by adding the ratio of the maximum (minimum) value of the real speech signal and the synthesized speech signal for each period.

Next, we examined the behavior in frequency domain. Relating to the verification, it is considered to be 300Hz~3.4 kHz frequency of human voice, and we made FFT in the range of 0~4096Hz. In comparison with Fig. 7, the frequency spectrum of real speech signal, mainly in the range of 0~500 Hz, generally concentrate at a certain frequency. On the other hand, that of the synthesized speech waveform has some slight dispersion in Fig.7.

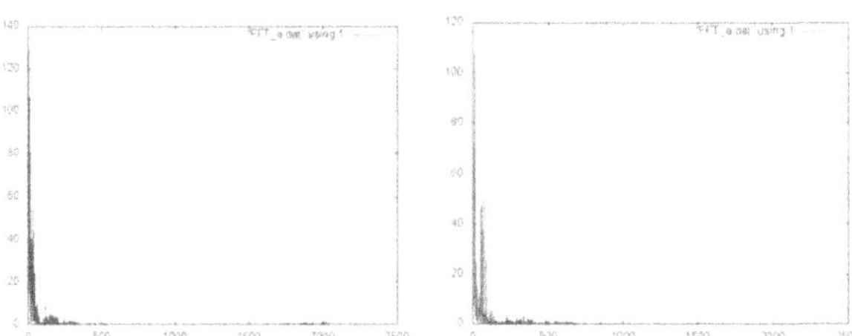

Fig. 7. Left: Frequency spectral density of real voice /a/ [2sec], Right: Frequency spectral density of Synthetic voice /a/ [2sec]

Then, in the field of time series analysis, human voice and the fluctuations of vocal cords have been known to be included in the chaotic[7,9,10]. Therefore, we have also decided to estimate the value of the maximum Lyapunov exponents. Regarding estimation, we have carried out the embedding dimension estimation by using the FNN (False Nearest Neighbor) method. This method enabled to reconstruct an attractor with objectivity and accuracy. And then, we have carried out the time-delay estimation based on the mutual information. This will seek Lyapunov exponents for a reconstructed attractor. As a result, in the largest Lyapunov exponents, there are one or more positive values in both that of the synthesized speech and of the synthesized vocal fluctuations. Therefore, it was found that they have chaos.

Table 2. Lyapunov spectrum of voice signal

	Vowel of /a/ (Embbeding dimention, Delaytime)	Lyapunov spectrum
Voice /a/	(3 , 5)	$\lambda_1 = -0.333104$ $\lambda_2 = -0.259747$ $\lambda_3 = 0.28725$
Synthetic Voice /a/	(3 , 4)	$\lambda_1 = -0.157536$ $\lambda_2 = -0.050309$ $\lambda_3 = 0.31957$
Synthetic Voice with smoothed data /a/	(5 , 5)	$\lambda_1 = -0.630296$ $\lambda_2 = -0.041686$ $\lambda_3 = 0.001254$ $\lambda_4 = 0.020488$ $\lambda_5 = 0.669256$

Table 3. Lyapunov spectrum of vocal fluctuations

	Vowel of /a/ (Embbeding dimention, Delaytime)	Lyapunov spectrum
Voice /a/	(2 , 2)	$\lambda_1 = -0.172612$ $\lambda_2 = 0.28751$
Synthetic Voice /a/	(2 , 2)	$\lambda_1 = -0.657639$ $\lambda_2 = 1.11770$
Synthetic Voice with smoothed data /a/	(2 , 5)	$\lambda_1 = -0.729565$ $\lambda_2 = -1.62867$

7 Conclusion

In this simulation, we have focused on vocal fluctuations of real speech waveform /a/ to reconstruct more natural synthetic voice. To reconstruct these nonlinear vocal fluctuations, by RBF neural network, and by adding the smoothed vocal fluctuations to input, we tried to reproduce more natural synthesized speech. In synthesis of vocal fluctuations, we enable to reconstruct the synthesized vocal fluctuations in fewer dimensions than previous, and the measurement error is improved. In addition, regarding synthesized voice, it close to human voice by complementing voice data according to the synthesized vocal fluctuations and reconstructing fluctuations of amplitude. So we have succeeded in achieving the desired value from sound materials for shorter intervals. However, adding smoothed data of vocal fluctuations to the input didn't seem to directly link to ease the high frequency components.

Finally, throughout, in the field of speech synthesis, we were able to get a limited success by the parameter estimation with nonlinear least square method and improvement of input. But in terms of human voice, we believe that more information is needed to reconstruct the synthesized speech that was aware of the individual components [11]. Therefore, in the future, we put the multidimensional map of the real speech that includes a lot of more information as the materials, and we hope to efforts to improve the reproducibility by fitting it into the nonlinear function.

References

1. Tokuda, K.: Fundamentals of Speech Synthesis Based on HMM. IEICE 100(392), SP2000-74, 43–50 (2000)
2. Kitani, M., Hara, T., Sawada, H.: Autonomous Voice Acquisition of a Talking Robot Based on Topological Structure Learning by Applying Dual-SOM. Transactions of the Japan Society of Mechanical Engineers Series C 77(775), 1062–1070 (2011)
3. Kanda, H., Ogata, T., Takahashi, T., Komatani, K., Okuno, H.: Simulation of Babbling and Vowel Acquisition based on Vocal Imitation Model using Recurrent Neural Network. In: IPS 2009, March 10, pp. 2-133–2-134 (2009)
4. Maeda, E., Arai, T., Saika, N.: Study of mechanical models of the human vocal tract having nasal cavity. IEICE 103(219), 1–5 (2003)
5. Minematsu, N., Nishimura, T., Sakuraba, K.: Consideration on infants' speech mimicking and their language acquisition based on the structural representation of speech
6. Golder, E.R., Settle, J.G.: The Box-Muller Method for Generating Pseudo-Random Normal Deviates. Journal of the Royal Statistical Society. Series C, Page 19 of 12–20 (1976)
7. Dai, S., Hirohku, T., Toyota, N.: The Lyapunov Spectrum and the chaotic property in speech sounds. IEICE Technical Report. Speech 99(576), 37–43 (2000)
8. Ogawa, S., Ikeguchi, T., Matozaki, T., Aihara, K.: Time Series Analysis using Radial Basis Function Networks. IEICE Technical Report. Neurocomputing 95(505), 29–36 (1996)
9. Suzuki, T., Nakagawa, M.: Fluctuashion of the vocal sound and its chaotic and fractal analyses. IEICE Technical Report. Nonliniea Problems 104(334) (2004)
10. Koga, H., Nakagawa, K.: Chaotic Properties in Vocal Sound and Synthesis Model. IEICE Technical Report, NLP99-120 (November 1990)
11. Wang, X., Niu, Y.: Adaptove synchronization of chaotic systems with nonlinearity inputs
12. Hartley, H.O.: The Modified Gauss-Newton Method for the Fitting of Non-Linear Regression Functions by Least Squares. American Statistical Association and American Society for Quality 3(2), 269–280 (1961)
13. Watanabe, T.: Consideration of Prediction Accuracy of Chaos Time Series Prediction by RBFN. The Research Reports of Oyama Technical College 39, 107–111 (2007)
14. Naniwa, Y., Kondo, T., Kamiyama, K., Kamata, H.: The exact reproduction in the voice signal of radial basis function network. IEICE Technical Report 110(387), 199–204 (2011)

A Business Entity Correlation Discovery Method between Cloud Services in Cloud Manufacturing System

Hua Guo, Lin Zhang*, Yilong Liu, Yongliang Luo, and Fei Tao

School of Automation Science and Electrical Engineering, Beihang University, Beijing 100191, P.R. China
zhanglin@buaa.edu.cn

Abstract. In order to overcome the bottlenecks of traditional network manufacturing, a new service-oriented manufacturing model, i.e., cloud manufacturing (CMfg), is proposed currently. Cloud service composition (CSC), which can realize the added value of cloud service (CS), is the core to implement CMfg. Considering that there always exist business entity correlations among CSs, so how to find the business entity correlation between two CSs and apply them to CSC is a key issue. This paper presents the formalized description for business entity correlation, and designs discovery algorithms to find business entity correlation between two CSs. A simulation experiment indicates the advantage of considering correlation discovery.

Keywords: Cloud manufacturing (CMfg), Cloud service composition, business entity correlation, Discovery algorithm.

1 Introduction

Although lots of research results has been obtained in theoretical researches of networked manufacturing, there are still no good solution for the lack of open architecture, common standards and specifications, without realizing intelligent embedded access of underlying physical manufacturing resources as well as service encapsulation of manufacturing resource and capability, etc., which result in the widen application and development of manufacturing modes hindered. Driven by the technologies of the cloud computing, Internet of things, CPS (Cyber-physical systems), high performance computer (HPC), virtualization, semantic web, etc., a new service-oriented networked manufacturing model-cloud manufacturing (CMfg) [1, 2] has been proposed and widely accepted both in industrial and academic communities. CMfg is a new manufacturing paradigm based on networks. It transforms manufacturing resources and manufacturing capability into manufacturing services, which can be managed and operated in an intelligent and unified way to enable the full sharing and circulating of manufacturing resources and manufacturing capability.

In CMfg, request of user can be divided into task request of single cloud service (CS) and multi-CSs. For task request of single CS, the best CS must be selected from lots of

* Corresponding author.

J.-H. Kim et al. (Eds.): AsiaSim2011, PICT 4, pp. 301–307, 2012.

candidate CSs to execute the task, i.e., the optimal-selection of CS. For task request of multi-CSs, the CS for each subtask will be composed to composite CSs according to a certain flow and logical relation to execute the task, i.e., cloud service composition (CSC) [3, 4, 5, 6]. When employ CSC to complete a task, a group of CSs with better quality of service (QoS) [7, 8] are needed to be selected from lots of CSs with the same function, i.e. the optimal-selection of CSC. CS and CSC can form manufacturing CS and capability of different granularity, which can provide the individuation CS for users.

However, the actual network CSs are not mutually independent, and there always exist correlations among them, where business entity correlation (BuC) [3] can influence QoS of CSC badly. For example, in the CSC flow of "online purchasing automobiles parts (OPAP)" [3], there are five elementary CSs constructed as CSC to complete the task request of ordering automobile parts. Here, if supplier select service (SSS) and delivery arrangement service (DAS) are arranged for two companies with a certain business entity correlation, these two companies are likely to offer favorable price for users, which will make the quality of CSC improved.

Although there exist business entity correlation between CSs, how to employ the discovery method to find out them, i.e. the business entity correlation discovery, such researches are still not too many. The existing works mainly focused on the mining framework and the technology details of discovery process [9, 10, 11]. Even if some works designed certain discovery algorithms, most of them focus on composable correlation [12, 13, 14, 15], and lack of the works for business entity correlation. Therefore, in order to accurately and comprehensively discovery the business entity correlation between CSs, this paper presents discovery algorithms to find business entity correlations between CSs base on the formalized description of business entity correlations.

This paper is organized as follows. Section 2 presents formalized description of business entity correlation. Section 3 designs the discovery algorithm of business entity correlation. Section 4 is the simulation experiment. Section 5 concludes the whole paper.

2 Formalized Description of Business Entity Correlation

Business entity correlation refers to the correlation that two or more providers (or enterprises) of elementary CSs in the business alliance maximize their own benefits through business cooperation or competition, so that influence QoS of CSC.

In order to realize business entity correlation discovery, business entity correlation should be described as a form of easy to find. Adjie-Winoto [16] proposed International Naming System (INS) that can be used to discovery resources and locate services, which mainly describes services through transforming description of services into service-tree with name-specifiers of attribute-value. But this kind of name-specifiers is only suitable for simple basic description of services, which cannot support QoS description for business entity correlation.

In order to realize the formalized description of business entity correlation, this paper extends the name-specifiers of attribute-value in INS to description specifications that can adapt to business entity correlation. The main description specifications based on CS-tree are as follows:

Rule 1: The description specifications include five kinds of nodes, i.e., basic property node, basic property-value node, condition property node, condition property-value node and QoS-value node.

Rule 2: For the basic characteristics of CSs, use ○ and ● to denote basic property and basic property-value of CS respectively. When there exist discovery conditions, use ✧ and ◆ denote condition property and condition property-value respectively. QoS value is denoted as■ .

Rule 3: Each property node corresponds to one or multiple property-value nodes. Each property-value nodes corresponds to one property node. Property nodes and property-value nodes are alternately in CS-tree.

Rule 4: If the relationship among one property node and multiple property-value nodes is relationship of "and" or "or", then it is denoted as "and-tree" or "or-tree" (see Fig.1).

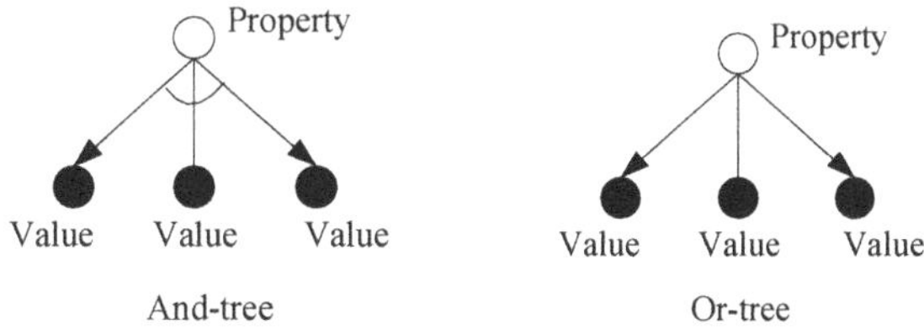

Fig. 1. And-Or-tree based on property and property-value

Rule 5: When multiple discovery conditions are needed to be satisfied, these discovery conditions will be placed on one branch until a leaf node.

3 The Discovery Algorithm of Business Entity Correlation

According to the description of CS based on CS-tree in section 2, the process of business entity correlation discovery is actually that a mapping process from the condition that a certain QoS value depends on to property and property-value of CSs.

The flow of discovery algorithm of business entity correlation is shown in Fig. 2, and the process can be described as follows.

Step 1. For any two CSs S_i and S_j, based on the description specification of CS-tree in section 2, the description information of S_i and S_j is transformed to the form of CS-tree.

Step 2. For a certain QoS property of S_i, if only default exists, then it is not needed to test whether S_i and S_j are business entity correlation. If there are other branches, then for each condition property node P_{ik} in these branches, the corresponding same property node P_{jl} will be searched in the CS-tree of S_j.

Step 3. In a certain branch, if there is condition property node of S_i that cannot find corresponding same property node in CS-tree of S_j, then this branch will be no longer considered. If all condition property nodes in this branch can find corresponding same property nodes, then testing whether the same property node P_{ik} and P_{jl} has the same property-value node d_{ik} and d_{jl}, until up to leaf node.

Step 4. If a certain couple of property-value nodes is not equal, then this branch is not considered. For those branches whose all condition property-value nodes are also equal, comparing their QoS values of leaf nodes in the branches and extracting the best QoS property-value as QoS value of S_i corresponding to this kind of QoS property.

Step 5. Employ the same method to search QoS value of S_j corresponding to this kind of QoS property.

Step 6. If anyone or both two of S_i and S_j can find QoS property-value except default, then they have business entity correlation.

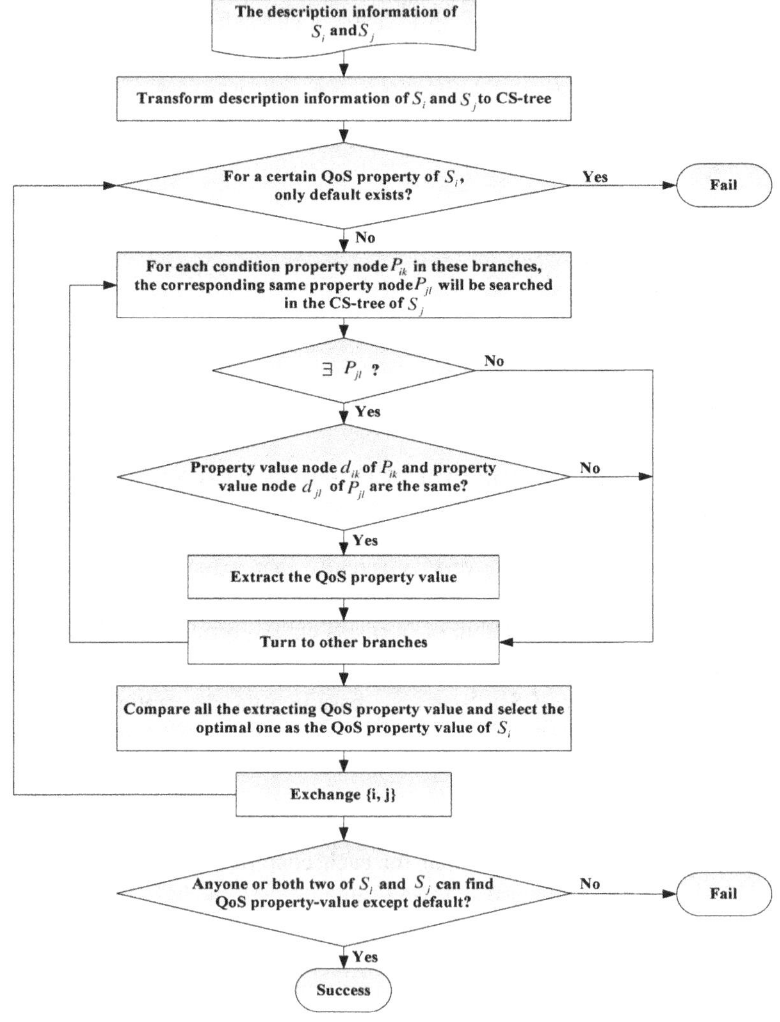

Fig. 2. The flow of discovery algorithm of business entity correlation

4 Simulation Experiment

In this section, it will be tested that employing discovery algorithm to find correlations between CSs is helpful to find more CSC plans with correlations.

It is assumed that there are five elementary CSs in the flow of CSC, and each elementary CS has n candidate CSs. In the flow of "OPAP", there will be n^5 kinds of CSC plans that can execute the task of purchasing automobiles parts. According to a certain ratio $r_{corelation}$, some CSC plans are selected as having correlations between CSs.

Let the ratio of CSC plans with correlations to all CSC be $r_{corelation} = 1/4$. Randomly select $r_{plans} = 1/3$ CSC plans from all CSC plans. Considering when n transforms from 3 to 30 according to a certain proportion, the proportion $p_{discovery}$ and $p_{no\text{-}discovery}$ of the found CSC plans with correlations to all CSC plans with correlations when employing discovery algorithm or no employing discovery algorithm will be separately calculated. The average of 50 times implementation is shown in Fig. 3.

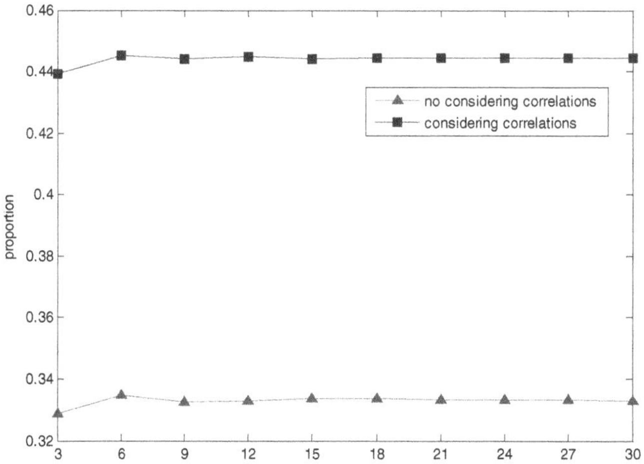

Fig. 3. Comparison of changing of $p_{discovery}$ and $p_{no\text{-}discovery}$ when n increases

The following conclusions can be seen from Fig. 3.

(1) With the number n of candidate CSs increases, the number of CSC plans also increases, while the proportions $p_{discovery}$ and $p_{no\text{-}discovery}$ of the found CSC plans with correlations to all CSC plans with correlations remain unchanged.

This is because when no employing discovery algorithm, since r_{plans} CSC plans are randomly selected, the theoretical proportion of the found CSC plans with correlations to all CSC plans with correlations is

$$p_{no\text{-}discovery} = \frac{r_{plans} \times n^5 \times r_{correlation}}{n^5 \times r_{correlation}} = r_{plans}$$

When no employing discovery algorithm, the theoretical proportion of the found CSC plans with correlations to all CSC plans with correlations is

$$p_{discovery} = \frac{r_{plans} \times n^5 \times r'_{correlation}}{n^5 \times r_{correlation}} = r_{plans} \frac{r'_{correlation}}{r_{correlation}}$$

where $r'_{correlation}$ is the actual proportion of the found CSC plans with correlations.

It has been set that $r_{corelation} = 1/4$ and $r_{plans} = 1/3$, so $p_{no\text{-}discovery} = 1/3$ will not change with n increasing. Seeing that the discovery ability of discovery algorithm is fixed, $r'_{correlation}$ is fixed, and then $p_{discovery} = 4r'_{correlation}/3$ is fixed too.

(2) No matter how n changes, comparing with the proportion of the found CSC plans with correlations to all CSC plans with correlations when no employing discovery algorithm, the proportion of that when employing discovery algorithm is larger. This is because when employing discovery algorithm, more CSC plans with correlations will be found out. At this time, $r'_{correlation} > r_{correlation}$, so $p_{discovery} > p_{no-discovery}$.

5 Conclusions

In this paper, the issue of business entity correlation discovery between two CSs in CMfg is considered. The formalized descriptions for business entity correlation were presented first, and then business entity correlation discovery algorithm was designed. The advantage of researching on business entity correlation discovery is demonstrated through simulation experiment. In the future, we will further develop the methods of how to apply correlation discovery to automatic construction and optimal-selection of CSC path.

Acknowledgment. This work is partly supported by the 863 program and the National Science Foundation in China (NO.61074144 and NO.51005012), the Doctoral Fund of Ministry of Education (NO.20101102110009), the Fundamental Research Funds for the Central Universities, and the Innovation Foundation of BUAA for PhD Graduates.

References

1. Li, B.H., Zhang, L., Wang, S.L., et al.: Cloud manufacturing: A service-oriented new networked manufacturing model. Computer Integrated Manufacturing Systems 16(1), 1–9 (2010)
2. Tao, F., Zhang, L., Venkatesh, V.C., Luo, Y.L., Cheng, Y.: Cloud manufacturing: a computing and service-oriented manufacturing model. Proc. ImechE Part B: J. Eng. Manufact. 225, doi:10.1177/0954405411405575

3. Guo, H., Tao, F., Zhang, L., Su, S.Y., Si, N.: Correlation-aware web services composition and QoS computation model in virtual enterprise. International Journal of Advanced Manufacturing Technology 51, 817–827 (2010)
4. Tao, F., Zhao, D.M., Hu, Y.F., Zhou, Z.D.: Correlation-aware resource CSC and optimal-selection in manufacturing grid. European Journal of operational Research (2009)
5. Tao, F., Zhao, D., Hu, Y.F., Zhou, Z.D.: Resource CSC and its optimal-selection based on particle swarm optimization in manufacturing grid system. IEEE Transactions on Industrial Informatics 4(4), 315–327 (2008)
6. Guo, H., Tao, F., Zhang, L., Laili, Y.J., Liu, D.K.: Research on measurement method of flexibility of Resource CSC in service–oriented manufacturing. International Journal of Computer Integrated Manufacturing (2011), doi: 10.1080/0951192X.2011.596572
7. Zeng, L.Z., Benatallah, B., Ngu, A.H.H., Dumas, M., Kalagnanam, J., Chang, H.: QoS-aware middleware for Web services composition. IEEE Transactions on Software Engineering 30(5), 311–327 (2004)
8. Tao, F., Hu, Y.F., Zhou, Z.D.: Study on Manufacturing Grid & Its Resource Service Optimal-Selection System. International Journal of Advanced Manufacturing Technology 37(9-10), 1022–1041 (2008)
9. Chen, S., Feng, Z., Wang, H.: Service relations and its application in services-oriented computing. Chinese Journal of Computers 33(11), 2068–2083 (2010)
10. Qianhui, L.A., Steven, M., Jenyao, C.: Service mining for web CSC, pp. 470–475 (2005)
11. George, Z., Athman, B.: Service mining on the web. IEEE Transactions on Services Computing 2(1), 65–78 (2009)
12. Li, B., Min, L.: A fuzzy-set based semantic similarity matching algorithm for web service. In: Proceeding of International Conference on Service Computing, pp. 529–532 (2008)
13. Zhou, B., Hui, S.C., Chang, K.: A formalized concept analysis approach for web usage mining
14. Peng, D., Huang, S., Wang, X., Zhou, A.: Management and retrieval of web services based on formalized concept analysis. In: Proceedings of the Fifth International Conference on Computer and Information Technology, CIT 2005 (2005)
15. Jiang, F., Fan, Y.: Web relationship mining based on extended concept lattice. Journal of Software 21(10), 2432–2444 (2010)
16. William, A.-W., Elliot, S., Hari, B., Jeremy, L.: The design and implementation of an intentional naming system. In: Proceedings of 17th ACM Symposium on Operating Systems Principles (SOSP 1999), Published as Operating Systems Review 34(5), 186–201 (1999)

Interactive ACO Algorithm toward Practical IEC Application Fields

Ryoji Tanabe, Tad Gonsalves, and Kiyoshi Itoh

Department of Information and Communication Sciences,
Faculty of Science and Technology, Sophia University,
7-1 Kioicho, Chiyoda-ku, Tokyo, 102-8554, Japan
{tanabe-r,t-gonsal,itohkiyo}@sophia.ac.jp

Abstract. Interactive Evolutionary Computational (IEC) algorithms have to be able to search robustly under the conditions of lower population number and fewer generations. This paper presents a novel IEC called interactive Ant System (iAS) which is interactively used by an user to determine the optimized combination of jacket, T-shirt, trousers and shoes. The iAS is the first of its kind in extending the well-known ACO to interactive optimization. Our experiments show that the iAS is capable of optimizing the user's requests in real time.

Keywords: Interactive Evolutionary Computation, Ant Colony Optimization, Combinatorial Optimization of Fashions.

1 Introduction

Interactive Evolutionary Computation (IEC) deals with the optimization of an objective function that is subjectively evaluated by a constant interaction with the user [1]. IEC is computationally a difficult process as it involves the users' subjective judgment in the process of optimization. Unlike the conventional EC optimization procedures in which the objective function is evaluated by the system using a pre-established fitness function, in interactive optimization the evaluation of the fitness function is left to the subjective judgment of the user.

The critical problem in IEC application areas is the load on the user because the iterative interaction between the human user and the system to evaluate the solutions produces stress and fatigue to the human user. Because of this, the IEC algorithms are required to have faster convergence than the general EC algorithms. Additionally, in IEC the number of population for search is lower than the general EC algorithms due to the constraint of the interface such as the display size. Therefore, the IEC algorithms have to be able to search robustly under the conditions of lower number of population and fewer generations for search. Furthermore, to reduce the user's stress which appears as the user evaluates the solutions, the subjective evaluation is generally the rough discrete evaluations such as 2 and 5 evaluation levels in IEC. This causes the evaluation of solutions to include noise and makes it more difficult for the IEC algorithms to proceed with the search.

J.-H. Kim et al. (Eds.): AsiaSim2011, PICT 4, pp. 308–315, 2012.

Traditionally, interactive Genetic Algorithm (iGA), interactive Genetic Programming (iGP) and so on are widely used in IEC [1]. Recently, interactive Particle Swarm Optimization (iPSO) [2] and interactive Differential Evolution (iDE) [3] are proposed and show a successful result.

Ant Colony Optimization (ACO) borrows its analogy from the stigmergy mechanism of a colony of ants searching for food [4]. Starting with random paths, the colony manages to determine the shortest path from the colony to the food source by means of a pheromone-based communication. The ACO algorithm is used in several combinatorial optimization problems like the Traveling Salesman Problem (TSP) [5], Quadratic Assignment Problem (QAP) [6], Flow-shop Scheduling Problem (FSP) [7] and so on. The ACO algorithms can be applied to a wide variety of problems same as the other EC algorithms. However, to the best of our knowledge, interactive ACO algorithm has not yet been developed. Although there are studies related to the use of ACO algorithm in music composition [8], in painting art works [9], to the best of our knowledge, there is no interactive ACO algorithm to date.

In this paper, we introduce an interactive Ant System (iAS) which is based on the ACO algorithm. Section 2 briefly explains the Ant System (AS) that is the basic ACO algorithm. Section 3 describes the iAS proposed method with an appropriate test problem. Section 4 shows experimental results of the iAS. Section 5 concludes this paper.

2 Ant Colony Optimization

In this section, we explain the AS by applying it to TSP which is one of the typical combinatorial optimization problems. Given a set of cities and the edge weight between them, TSP is the problem of finding the shortest tour visiting the cities exactly once.

In the beginning of the AS algorithm, the pheromone trails on all the edges is initialized. In the solution construction procedure, each agent is given a start city with random or uniform selection and constructs a closed tour that satisfies the constraints of TSP. The probability that agent k ($k = 1, \ldots, m$) in city i selects city j to move to the next city is given by the following formula:

$$p^k_{ij} = \frac{[\tau_{ij}]^\alpha [\eta_{ij}]^\beta}{\sum_{l \in N^k} [\tau_{il}]^\alpha [\eta_{il}]^\beta} \tag{1}$$

where τ_{ij} is the pheromone between city i and j, η_{ij} is the heuristic information that is the inverse of the distance between cities i and j, N^k is the set of cities that agent k has not visited, α and β control the relative importance of the pheromone versus the heuristic. This selection rule is the roulette wheel selection based on the pheromone density τ_{ij} and the heuristic information η_{ij}. This process is repeated until agent k completes the closed tour.

After all the agents have constructed their tours, these are evaluated and the pheromone trails are updated as follows:

$$\tau_{ij} = \rho\tau_{ij} + \sum_{k=1}^{m} \Delta\tau^k_{ij} \tag{2}$$

$$\Delta\tau^k_{ij} = \begin{cases} Q/L^k & [if\,(i,j)\in T^k] \\ 0 & [otherwise] \end{cases} \tag{3}$$

where ρ is a pheromone evaporation rate, $\Delta\tau^k_{ij}$ is the incremental amount of pheromone deposited by the agent k, Q is a constant value, T^k is a set of edges included in the tour of agent k, and L^k is the tour length of agent k. The pheromone evaporation allows the algorithm to decrease the pheromone density on undesirable tour previously constructed. The amount of pheromone trail on the edges depends on the tour length of the agents. They increase per each tour length. So, more pheromone is deposited on the edges belonging to the shorter tour.

3 Interactive Ant System

3.1 The Interactive Problem of Our Study

We constructed the interactive fashion combinatorial problem to which the iAS is applied. Fig. 1 shows the system's interface and the user evaluating the outfits generated by the iAS. The system shows several outfits, each being a combination of four items: jacket, T-shirt, trousers and shoes. The problem is typically a combinatorial optimization problem as it involves dealing with a large number of dress items stored in the apparel database. The system follows a recommend-evaluate cycle in aiding the user in determining the combination of the four clothing items that matches the user's preferences.

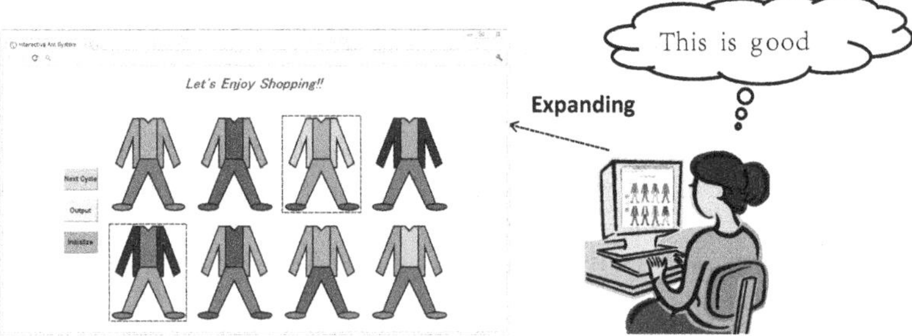

Fig. 1. The system's interface and the user who evaluates the outfits generated by the iAS. The user selects a minimum of one item of his/ her liking and requests the iAS to generate a new set of clothing by combining the selected items.

Considering the "user interaction load" which emerges when the user evaluates the solutions displayed by the system, we have designed an interface allowing the user to exercise his/her option only by selecting the item of his/her liking from a set of items displayed by the system. The user selects a minimum of one item and requests the system to produce a new set of clothing by combining the selected items. The user continues to select one or more dress items from those displayed by the system, till he/she is satisfied with the apparel displayed by the system.

3.2 Interactive Ant System Algorithm

The algorithm of the iAS, on the whole, is not very different from the AS algorithm previously explained in Section 2. But the evaluation of the solutions is replaced by subjective human evaluation.

Initially, the pheromone density on all items is set the same value. Then each ant constructs a solution, which is a combination of the four dress items: jacket, T-shirt, trousers and shoes. The probability of the agent k allotting the jth item to the ith clothing type is given by:

$$p_{ij}^{k} = \frac{\tau_{ij}}{\sum_{l=1}^{n} \tau_{il}} \tag{4}$$

where, τ_{ij} is the pheromone density of the jth item being allotted to the ith dress category. The formula (4) is based on the formula (1). Because there is no available heuristic information η_{ij} in our problem, we did not use it.

In general, to improve the performance of the EC algorithm, we need to maintain a balance between intensification and diversification. But there is the user's "interaction load" problem in IEC application, because the user becomes tired due to evaluating the solutions during a number of iterations. Therefore, the IEC algorithms are required to have a faster convergence than EC algorithms to terminate the interaction between the user and the system as fast as possible. With this in mind, we have adopted the elite strategy in preserving the agents selected by the user in the next generation for faster convergence. So, the number of the newer outfits generated by agents using formula (4) is the result of subtracting the total number of agents from the number of agents selected by the user in the previous cycle. In addition, to maintain the minimal diversification of the outfits displayed by the system, we use the random selection [10] with a fixed probability.

After the evaluation step, the pheromone density on items is updated based on the user's evaluation. In general, the definition of the pheromone depends on the problem which the ACO algorithm is trying to solve. Inappropriate definition of the pheromone could lead to poor solutions. In TSP the definition of the pheromone τ_{ij} is generally the desirability of selecting city j from city i in which the agent stays [5]. In the case of QAP, that is generally the desirability of assigning facility j to location i [6]. In this iAS study, pheromone density is determined by the degree to which the user desires item j to be chosen in cloth category i. Since the problem in our study is essentially different from the typical combinatorial problems such as TSP and QAP, we cannot use the pheromone matrix as the data structure to store the pheromone density. So, we store the pheromone density of each item in each array.

The updating of the pheromone density is done using the following formulae. When items preferred by the user are included in the selected clothing, the pheromone intensity increases.

$$\tau_{ij} = \varrho\tau_{ij} + \sum_{k=1}^{m} \Delta\tau_{ij}^{k} \tag{5}$$

$$\Delta\tau^{k}_{ij} = \begin{cases} 1 & [if\,(i,j) \in user\ selected] \\ 0 & [otherwise] \end{cases} \tag{6}$$

4 Experimental Results

We have performed several experiments to test the accurate working of the iAS. The number of items in each clothing category is as follows - jackets: 35, T-shirts: 40, trousers: 40 and shoes: 30. Each item in a given clothing category is different from the others in the category in its color. The items cover a broad range of colors typically available in an apparel store. The iAS experiment parameters are as follows: number of agents = 8, ρ = 0.4, random selection probability = 0.05. These control parameter values are obtained by performing several pilot interactive optimization experiments.

A sequence of pictures from Fig. 2 to Fig. 6 show the transitional interfaces by the interaction of the user and the system until the 5 cycle. The items in dotted lines are those selected by the user to obtain a new combination of clothing. In Fig. 2, the first interface, all outfits are randomly generated because initially the pheromone density on all items of each dress category is exactly equal.

In Fig. 4 we can see that the number of items that include the outfits selected by the user in Fig. 3, such as the pink trousers and the khaki T-shirt, is increased. This tendency is easy to notice in Fig. 5. The cause of this is that the pheromone density on a particular item that the user desires to be chosen in a given category tends to be augmented. On the other hand, the pheromone density on the item including outfits that do not match the user's preferences is decreased. This makes the probability of selecting the items resembling the user's taste increase. In Fig. 6 of the 5 cycle interface, we can see all the outfits have a unity because the search of agents has converged to some extent. In spite of this situation, random selection makes the outfits displayed by the iAS include the new item by ignoring the pheromone density.

Through the experimental results, the iAS can generate the outfits that match the user's preferences in spite of the characteristic nature of IEC such as human evaluation, rough level evaluation, the low number of agents and a few cycle for search. We can verify the iAS can display outfits that match the user's preferences with a little interaction between the user and the system. This means that iAS is a promising IEC algorithm.

Fig. 2. The interface of 1 cycle

Fig. 3. The interface of 2 cycle

Fig. 4. The interface of 3 cycle

Fig. 5. The interface of 4 cycle

Fig. 6. The interface of 5 cycle

5 Conclusion

In this paper, we have proposed an interactive Ant System (iAS) using the well-known ACO algorithm. To the best of our knowledge, iAS is the first of its kind in the area of Interactive Evolutionary Computation. To illustrate the working of the iAS, we have introduced a clothing combinatorial problem. Initially the system displays a set of clothing images. Each clothing image consists of a T-shirt, jacket, trousers and a pair of shoes. The user selects one or more images of his/her liking and requests the iAS to display a new set of clothing images combining the selected items. In the next run, the system displays items resembling the user's taste. The interaction optimization cycle continues till the user feels satisfied with the color-clothing combination suggested by the iAS. The practical example verifies that the iAS is successful in producing items that match the user's preferences in real-time.

If using pictures of real fashion items, the problem we used in this paper can be made available on an online shopping site as a real-world application. Furthermore, the iAS has the potential of applying to a broad utility. So, if the problem we are trying to solve has a combinatorial nature such as the decision on a filtering sequence [11] and the montage systems [12], then iAS can be applied. As further research, we intent to compare the performance of iAS with other interactive optimization algorithms like iGA on this problem with actual human user or a simulated pseudo-IEC user.

References

1. Takagi, H.: Interactive Evolutionary Computation: Fusion of the Capabilities of EC Optimization and Human Evaluation. Proceedings of the IEEE 89(9), 1275–1296 (2001)
2. Nakano, Y., Takagi, H.: Influence of Quantization Noise in Fitness on the Performance of Interactive PSO. In: Proceedings of the IEEE Congress on Evolutionary Computation (CEC 2009), pp. 2416–2422 (2009)
3. Takagi, H., Pallez, D.: Paired Comparison-based Interactive Differential Evolution. In: Proceedings of the first World Congress on Nature and Biologically Inspired Computing (NaBIC 2009), pp. 475–480 (2009)
4. Dorigo, M., Stützle, T.: Ant Colony Optimization. MIT press (2004)
5. Dorigo, M., Maniezzo, V., Colorni, A.: The Ant System: Optimization by a colony of cooperating agents. IEEE Transaction on Systems, Man and Cybernetics-Part B 26(1), 29–41 (1996)
6. Gambardella, L.M., Taillard, É.D., Dorigo, M.: Ant Colonies for the quadratic assignment problem. Journal of the Operational Research Society 50, 167–176 (1999)
7. Stützle, T.: An Ant Approach to the Flow Shop Problem. In: Proceedings of the 6th European Congress on Intelligent Techniques and Soft Computing (EUFIT 1998), vol. 3, pp. 1560–1564 (1998)
8. Guéret, C., Monmarché, N., Slimane, M.: Ants Can Play Music. In: Dorigo, M., Birattari, M., Blum, C., Gambardella, L.M., Mondada, F., Stützle, T. (eds.) ANTS 2004. LNCS, vol. 3172, pp. 310–317. Springer, Heidelberg (2004)

9. Aupetit, S., Bordeau, V., Monmarché, N., Slimane, M., Venturini, G.: Interactive Evolution of Ant Paintings. In: Proceedings of the IEEE Congress on Evolutionary Computation (CEC 2003), vol. 2, pp. 1376–1383 (2003)
10. Nakamichi, Y., Arita, T.: Diversity control in ant colony optimization. Artificial Life and Robotics 7(4), 198–204 (2004)
11. Mutoh, T., Komagata, N., Ueda, K.: An experimental study for automatically generating image filter sequence by using simulated breeding. In: Workshop on Interactive Evolutionary Computation (Fukuoka, Japan), pp. 7–12 (1998) (in Japanese)
12. Caldwell, C., Johnston, V.S.: Tracking a Criminal Suspect Through "Face-Space" with a Genetic Algorithm. In: Proceedings of the 4th International Conference on Genetic Algorithm (ICGA 1991), pp. 416–421 (1991)

Developing a Practical Machine Scheduler for Worker-Involved Systems

Jinwoo Seo, Sumin Han, and Jinwoo Park

Dept. of Industrial Eng. Automation & Systems Research Institute, Coll. of Eng., Seoul National University, Daehak-dong, Gwanak-gu, Seoul, Korea
{jinoo02,hans8501,autofact}@snu.ac.kr
http://mailab.snu.ac.kr

Abstract. There have been numerous researches in machine scheduling, but rarely applied to fields of industry, because previous studies tend to focus on schedule generation phase, neglecting input and execution phases before and after it. A worker plays important rolls during those phases so worker's abilities and needs should be investigated and reflected to make scheduler practical; actually used as well as resulting good performance. This study aims at developing practical machine scheduler especially for worker-involved machine cell. In this study worker's rolls and characteristics are analyzed in scheduling aspect, and requirements on scheduler are derived. After that how to reflect those requirements are proposed and actual machine scheduler has developed upon it.

1 Introduction

A machine scheduler determines machine level resource allocation and is the lowest level scheduler in terms of closest to actual execution. It may schedule either a single or multiple machines and may exist embedded in a machine itself or apart from the machine as an external controller. To create a schedule, a production plan and master data are needed from the outside and system statuses and conditions from the inside as input. Machine schedule, created by machine scheduler, requires high level integrity according to which production equipments are directly controlled.

Today's manufacturing system usually consists of workers and machines. Though automation technology has advanced, most manufacturing systems still need workers in whole or some parts of their operations or tasks, especially in many types and small volume production industry. Therefore, machine scheduler should consider workers as well as machines.

Schedulers and workers are different in level of knowledge and in preference. A scheduler is well in procedural knowledge whereas a worker is good at synthetic knowledge. A scheduler seeks apparent measures such as utilization, makespan, and due date fulfillment, while a worker additionally demands schedules which are safe, comfortable and easy to execute.

Most previous schedulers have neglected these differences and demanded workers to just follow given schedule. However, the worker suffers from physical and psychological problems.

J.-H. Kim et al. (Eds.): AsiaSim2011, PICT 4, pp. 316–325, 2012.

On the other hand, a worker often breaks the schedule both intentionally and unintentionally. This is not only because the schedule is confusing and burdensome but also because the worker judges that the schedule is not fit to current situation. In the latter case the worker makes real-time decision and releases an action based on his/her knowledge regardless of existing schedule. Besides, the worker often makes unintentional error such as wrong input. All these leads the schedule meaningless.

This conflict between worker and scheduler usually ends up with disuse of scheduler. Considering the time and cost for scheduler development, it is very undesirable. Note that schedule-less system is also likely to be unpredictable and uncontrollable. Problem in lower level system usually propagates to higher level system.

To avoid such waste and problem, the scheduler should cooperate with the worker; i.e. the scheduler should consider worker's ability and need and should request what is acceptable to the worker. In this study, requirements for scheduling worker-involved system are derived and how to implement those requirements is investigated. Based on solutions derived, a machine scheduler is developed and incorporated into a real system. A test and evaluation of the implemented system are also included.

Contents of the remaining sections are as follows. In the next section the target system is introduced and results of previous research and limitation are discussed. After that worker's rolls and requirements in scheduling aspects are analyzed. Next, actual implementation of scheduler is explained. Finally, the evaluation and future direction of study are discussed.

2 Target System: Reconfigurable Manufacturing Cell

Reconfigurable manufacturing cell (RMC), as representative worker involved production system, is selected as a target system. RMC is a production system which can change rapidly its hardware and software configurations in order to catch up with changes in production requirements, i.e. product mix and volume. RMC is considered the next generation of production system following FMS or FMC providing more flexibility to shop floor. A conventional RMC consists of four elements: Through loading/unloading(L/U) stations, parts are inserted into and removed from the system. In machines(CNCs and auxiliary machines) operations such as machining and washing are performed. Buffer or storage lack plays as an intermediate storage preserving parts before and after processing. Finally material handling system mediates among those elements delivering part into/from them.

Part loading and unloading is performed by worker outside. Automation may be possible, but it rarely happens because of unfixed and complex routing between production systems. To load/unload part in L/U station, it takes about tens of minutes.

In scheduling aspect machine schedule is affected by loading/unloading schedule, both machines and L/U stations should be scheduled as resources. Therefore RMC can be viewed as a 2-staged flexible flow shop(FFS) with each stage consisting of non-identical processors [1].

Besides the characteristics of conventional FFS, RMC has an additional constraint about fixture and pallet and such that every part should be mounted on a fixtured pallet to enter and to be processed in RMC. Each part type may need a different type of fixture and The number of pallets fixtured with a certain type of fixture is limited. This constraint is different from that of [2][3] where pallets are not distinguished each other, while here pallets are divided by types according to fixture types installed on them. This is rather similar with that of [4].

As for objectives primary objective is minimizing completion time of whole system. As secondary objective minimizing of L/U stations' completion time is considered excluding machine completion times, for L/U stations' loading processes are performed by worker and worker efficiency and working hour have more importance than those of machines.

Our goal in this research is developing a practical machine scheduler for RMC. RMC has both characteristics of worker-involved system and schedule-required system. So it is expected that the result of scheduler development of RMC may be generalized to any worker-involved production system which necessitates scheduling.

3 Related Work

RMC is closely related to flexible manufacturing system (FMS) or Cell (FMC). These has been numerous research about FMS/C. Luh et al.(1990) used lagrangian relaxation to decompose the original problem and proposed heuristics to solve each problem [5]. Gultekin et al.(2007) suggested a cyclic schedule considering material handling time for FMC which produces identical parts and is operated by a robot [6].

There also exist research about scheduling flexible (often hybrid) flowshop. Sriskandarajah and Sethi(1989) proposed a heuristic for k-staged flexible flowshop with purpose of minimizing makespan and analyzed the worst and average case performance[7]. Lee and Vairaktarakis(1994) presented an $2 - 1/\max\{m_1, m_2\}$ approximation algorithm for a 2-staged flexible flow shop minimizing makespan [8], where m_i is the number of machines for stage i.

There exist relatively few research considering pallet-constrained shop. Sethi et al.(1999) proposed an algorithm for determining the optimal number of pallets in 2-staged flow shop [2]. Mashaei et. al.(2009) also developed an pallet number optimization algorithm for cyclic manufacturing plant, and verified the result using Colored Petri Net [3].

Previous research has focused on the schedule generation phase regardless of input and actual phase. This causes rare use of scheduling results though their performance itself is quite good. Meanwhile Lin and Lee (1997) discussed to use Colored Timed Petri-nets to resolve discordance problem between scheduling and control phases[9].

4 Roles of Worker in Scheduling

In this section the roles and needs of worker in the aspect of scheduling are investigated. Worker plays various roles such as an user of scheduler, scheduling resource, and an executor and supervisor of schedule.

4.1 Worker as an User of Scheduler

Here an user means one who prepares for scheduling. To create schedule, information should be prepared. Even though ERP, MES, and other techniques of information system such as sensor network are advanced, still many systems depend on human worker's manual input. It tends to be evident as in low (control) level systems.

Required information is like followings: At first production plan, i.e. what and how much to produce is needed. It may or may not include due dates. As second, specifications of those to be produced, often called master data, are necessary. They consist of routings i.e. orders and appropriate resources of operations, and additional information such as processing times and NC program numbers. Finally resource information such as the number of resources along type and availability(resource schedule) and status for each resource.

Worker may input all of this information or part of it. It should be noted that worker is often not skilled in this task, rather worker usually has expertise in handling machine and workpiece. Worker may feel confused or exhausted if the task is complex and burdensome. Thus input process and interface should be designed to minimize confusion and discomfort by making contents simplified and by using terminology friendly to worker.

In another hand, possibility should be considered that worker may make error. It is difficult to completely sure there is no error, though process and interface are well designed. Types of error and impact on system should be investigated. If input has some errors, then resulting schedule may not be consistent with actual system status and this may lead to fatal consequence such as collision. Therefore every time a schedule is executed, consistency must be checked. In our system this is realized by another module beside scheduler which will be explained in the later section.

Another problem with input error is performance degradation of system; the execution may be far away(usually to bad side) from what the scheduler expected. To make schedule 'honest', error should be fixed. In this study an input error detection and automatic and manual interpolation scheme is designed.

4.2 Worker as a Scheduling Resource

After information is prepared, the scheduler generates a schedule. The scheduler focuses on a good schedule with respect to objectives. But the resulting schedule may not be satisfactory to worker. This is because the schedule is hard for worker to follow or even it may be hard to understand. In those cases the worker may not follow the given schedule whether it is consciously or unconsciously. This may cause a dangerous situation as well as performance degradation. So it's better to create a schedule considering the worker's need and ability. This means additional constraints are imposed to scheduling problem and achievable performance may decrease. But an acceptable schedule is better than a misread or discarded schedule. It should be investigated that the trade-off between satisfying worker's need and seeking scheduling performance. What the worker needs should be converted to constraints to make it incorporated in the scheduler. The scheduler should select the essentials among all works' explicit and implicit needs.

In cases of this research, the worker feel confused when a certain pallet is assigned for more than two types of parts, though the pallet can support all of them. So scheduler has considered setting a pallet dedicated to a single part type. This, however, may result in considerable performance degrading when the number of pallets is not sufficient with respect to its production volume. And in contrast some pallets may remain idle for a long time after finishing processing all parts. So finally it is decided that each pallet is dedicated to a single part, until production of that part is finished. After that, the pallet is reassigned to another type of part. This is considered still acceptable to worker.

4.3 Worker as an Executor and Supervisor of Schedule

Created schedule, even though input was correct and worker's needs are carefully considered and reflected in scheduling, may not be executed as scheduled. This is because variances and disturbances. For various reason worker may not follow schedule. Scheduler may not know what worker knows. System should be designed to process unexpected actions and can verify reason of it and have information synchronized among worker and control system.

Worker may stop resource from usage and renew resource so that it can be used again, though scheduler cannot understand reason, these conditions should be reflected in schedule. Worker may bread schedule for urgent order comes or existing order is modified or cancelled. Scheduler should support this kind of exceptions and should receive external information from worker. With updated information and condition scheduler regenerate schedule and with that schedule scheduler may expect worker to follow it as long as additional exception happens. By doing so schedule may be executed as close to real as possible.

Information from worker even help scheduler response disturbance. When machine breakdown happens, and reschedule, how much the breakdown will last is a very important information, however it is usually hard to estimate it by machine itself so usually a unit time is assumed previously. Seo et al(2010) reported this may result in bad schedule and verified difference between two cases when MTTR information is given and not [10]. Workers judgment can be a good estimation for these periods. At least it is better than no information or simple assumption. So scheduler should support functions of receiving information from worker when disturbances happens.

5 Implementation

Developed machine scheduler is operated with three phases. First in the input phase worker inputs a production order and master data into system and the scheduler receives those information together with system status information from the dispatcher. Then the dispatcher requests the scheduler to schedule and the scheduling phase starts. The created schedule then delivered to the dispatcher and the dispatcher interprets, verifies and gives instructions to worker and machines in the execution phase. Any inconsistency is detected between the current schedule and system information over a certain limit, rescheduling starts.

In each phase, requirements derived in previous section are reflected. From the following sections detailed information and explanation are presented.

5.1 Overall Structure

Physical system can be operated when combined with control system. Overall structure including both physical and control parts of the system is depicted in Fig. 1. Scheduler determines what and when to process for each resource. It does not schedule the material handling system(MHS), since too much variation it has and so it is almost impossible to control MHS with schedule. MHS is directed to serve the schedule for the other resources(L/U stations and machines) and the dispatcher decides the movements of MHS based on the schedule.

Let's go further about the function of the dispatcher. Besides scheduling, information gathering and consistency assuring are required to operate the system in a stabilized manner. Information used in scheduler may not reflect the current status of system by some reasons. In this case, the schedule may demand resources to perform wrong actions such as mounting a new part on a machine currently in processing another part. This may cause collision and leave irrecoverable damage into the system.

Therefore Some should check validity of given schedule and coordinate it if necessary not to cause any error during execution. This is what the dispatcher does. It requests the scheduler to create a new schedule providing current status information of physical system, interprets and converts the schedule received for execution, then checks whether it is consistent with current status reflecting possible changes in the system, and finally send signals to equipments via PLC to make them perform corresponding actions. If the schedule has problems, it suspends to execute it and behaves separately with the schedule in accordance with pre-determined logic. The dispatcher usually request scheduler periodically, i.e. rescheduling so that the resulting schedule may be applicable for execution with synchronized information.

Based on the structure described above, the roles of worker in the system can be defined. The first role was indicated at the middle of Fig. 1 where worker input production information into scheduler, using that information the scheduler can create schedule. As the second role, worker performs tasks in L/U station i.e. loading/unloading workpiece to/from the system. These worker-performing tasks requires considerable time and unless they are completed, production cannot proceed. As the third, worker administers the system during execution and release responsive actions when necessary. This role appears implicitly in the structure. Worker can determine whether the system submits to given schedule or just follows worker's empirical rules or customs, or takes the compromise. Worker also detects problems during production, may set certain resources or part not to be used or processed until the problem is solved.

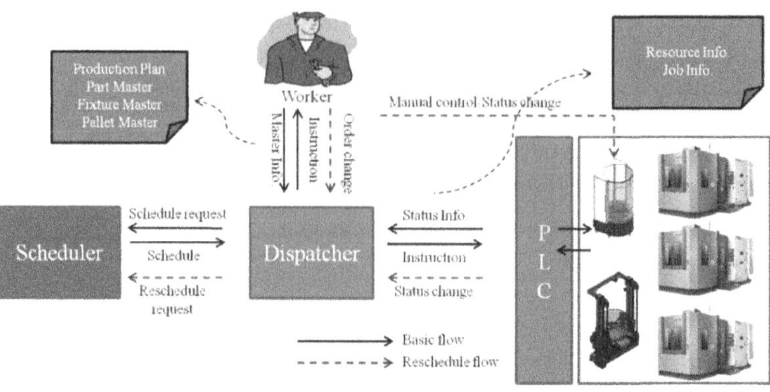

Fig. 1. Overall structure of machine scheduler

Based on these roles defined, issues in developing practical machine scheduler are discussed in the next section.

5.2 Input Phase

Requested input from worker is as follows (key fields are underlined):

<Production plan>

Part Name	*Volume*	*Due Date*

Production plan includes names, amount and due date of parts to produce. Part name is often called part type.

<Part information>

Part Name	*Proc. No.*	*Fixture Type*	*Proc. Time*	*Program No.*	*L/U stations Available*	*Machines Available*	*Auxiliaries Available*

To produce a part, a workpiece should go through several processes including loading/unloading, machining and washing (in auxiliaries). For further processing fixture change may be required thus the unfinished workpiece may be taken away from the pallet currently equipped with, and mounted on a proper pallet with required fixture. Program number is not directly related to scheduling but is referred when processing starts. Processing of each part-process is often restricted to certain resources only, with reason such as difference in resource capability, This is included in resource availability parts.

<Fixture information>

Fixture Type	*Part Name*	*Proc. No.*

Each type of fixtures can support several parts and operations. Fixture-parts-operations combinations may often be many even for a single fixture (as all five constitute key fields), only combinations worker has inputted are considered.

<Pallet information>

Pallet No.	Fixture Type	Part Name

The number of pallets are limited and each fixture can support with limited parts and processes, what fixture each pallet is equipped with may have much impact on scheduling performance. However pallet fixturing i.e. installing a fixture on a target pallet is done by worker and it takes considerable time to remove and reinstall a fixture. So if the scheduler directs it, it may be very burdensome and even unnecessary. So it is concluded to commit fixturing to worker and to receive fixturing result from worker.

Though a fixture type can support various part types, worker may need a pallet dedicated to a single part type to avoid confusion. In this case worker may set part name of each pallet. But this is not desirable because this pallet allocation to part also affects scheduling performance. So unless there is special reason, this field is left blank as default and filled during scheduling phase.

<Resource maintenance schedule>

Resource No.	Start Time	End Time

Maintenance schedule may include workers break, meal time and so on. During these period, corresponding resource cannot be allocated.

5.3 Scheduling Phase

How are processes followed and what techniques does each process use and why.

In this phase, a schedule is generated based on system information given from worker and resources.

Scheduling start from allocating each of unassigned pallet to a part called pallet-to-part allocation. In the developed scheduler following rule is applied:

$$N_i = \frac{O_{if} \times V_i \times P_f}{\sum_j O_{jf} \times V_j} \tag{1}$$

Where N_i is the number of pallet assigned to part type i, O_{if} is the number of processes(operations) of part type i which required fixture type f and P_f is the number of total pallets fixtured with fixture type f.

After pallet-to-part allocation, conventional scheduling i.e. resource allocation to jobs can be done. Because scheduling RMC is an NP-hard problem. Scheduling is done using dispatching rules and for further improvement is employed multi-pass scheduling approach, i.e. applying several rules and create a schedule for each rules compare those schedule and determine the best one as a final schedule. Scheduling result, information to be delivered to dispatcher for execution is as follows:

<Job Lists>

Start Time	Pallet No.	Fixture No.	Resource No.	Part Name	Proc. No.	Proc. Time

A schedule consists of job lists. Each job list represent one resource allocation.

Here fixture number seems redundant, since it is already recorded in pallet information, but this is not true because pallet information is not fixed i.e. a pallet may be fixtured with different fixture afterward. So fixture information in each point of scheduling execution should be maintained.

5.4 Execution Phase

Fig. 2 indicates a display of a generated schedule delivered to the dispatcher. When any resources becomes available, the dispatcher searches in the schedule table a nearest job list from current point of time and check it is executable with regard to current system status. If so and it is L/U operation, the dispatcher informs worker that there is a job to perform and worker can select to follow it (called auto-call mode) and if it is machining or auxiliary operation, the dispatcher at once release instruction to corresponding resource through PLC.

5.5 Rescheduling

When instructions delivers to worker, worker may reject to follow suggested operation. In that case worker can decide job for him/her to perform and request system to follow his action i.e. bringing a workpiece to L/U station or just bring load workpiece worker has mounted on the L/U station.

Sometimes worker may want to change resource status as locking certain resources or allowing some resources(e.g. pallet) only be used for certain purpose(e.g. machining). Worker may also receive instruction from outside the system such as additional urgent order or order cancellation/modification. In another case delay or disturbance may happen in the system itself.

Whatever case are among cases mentioned, it results in discordance of information between scheduler and system. The dispatcher checks it periodically and when the degree of discordance exceeds a certain limit, the dispatcher request rescheduling, then three phases have been discussed are repeated again. During rescheduling information is synchronized and system consistency is re-guaranteed.

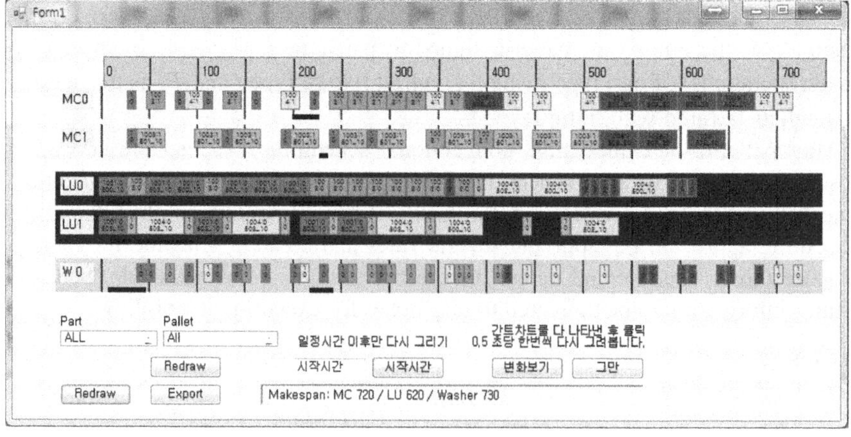

Fig. 2. Display of the scheduling result

6 Conclusion and Future Work

In this study, workers' needs and corresponding requirements in scheduling aspects are derived and a machine scheduler is developed based on those requirements. When a scheduler understands and cooperates with the worker, the scheduler may realize its potential to the maximum extent.

The developed scheduler still has limitation. Performance degradation from imposing additional constraints to make the scheduler practical should be investigated. And beyond just obeying the workers' needs and sacrificing performance, a way how to get aid from workers to improve the schedule is now in consideration.

Acknowledgements. This research is supported by the project, 'Development of integrated operation technology for autonomous reconfigurable production system' funded by the MKE (Ministry of Knowledge Economy), Republic of Korea. We would like to acknowledge the support by the ASRI (Automation and Systems Research Institute) in Seoul National University in the form of resources and administrative support.

References

1. Park, J., Seo, J., Park, J.: RMC Scheduling Considering Setup and Pallet Constraints. In: Proc. of 2010 KIIE/KORMS Joint Conf. (2010)
2. Sethi, S., Srishkandarajah, C., Velde, S., Wang, M., Hoogeveen, J.: Minimizing Makespan in a Pallet-Constraint Flowshop. Journal of Scheduling 2, 115–133 (1999)
3. Mashaei, M., Lennartson, B., Sannehed, F., Abbestam, G.: Optimal Number of Pallets for Reconfigurable Cyclic Manufacturing Plants. In: Proc. of 2009 Emerging Technologies & Factory Automation Conf., pp. 906–913 (2009)
4. Solot, P.: A Heuristic Method to Determine the Number of Pallets in a Flexible Manufacturing System with Several Pallet Types. The International Journal of Flexible Manufacturing Systems 2, 191–216 (1990)
5. Luh, P., Hoitomt, D., Max, E., Pattipati, K.: Schedule Generation and Reconfiguration for Parallel Machines. IEEE Transactions on Robotics and Automation 6, 687–696 (1990)
6. Gultekin, H., Akturk, M., Karasan, O.: Scheduling in a Three-machine Robotic Flexible Manufacturing Cell. Computers and Operations Research 34, 2463–2477 (2007)
7. Sriskandarajah, C., Sethi, S.: Scheduling Algorithms for Flexible Flowshops: Worst and Average Case Performance. European Journal of Operational Research 43, 143–160 (1989)
8. Lee, C., Vairaktarakis, G.: Minimizing Makespan in Hybrid Flowshops. Operations Research Letters 16, 149–158 (1994)
9. Lin, J., Lee, C.: A Petri net-based Integrated Control and Scheduling Scheme for Flexible Manufacturing Cells. Computer Integrated Manufacturing Systems 10, 143–160 (1997)
10. Seo, J., Park, J.: Applicability of Variability-Forecasting Technology in Scheduling. In: Korean Society for Precision Engineering 2010 Spring Conf., pp. 107–108 (2010)

Battlefield Data Quantization Method for War-Game Optimal Path Generation in Distributed Simulation

Hyunchang Oh and Jong Sik Lee

School of Information Engineering
Inha University #253, YongHyun-Dong, Nam-Ku
Incheon 402-751, Republic of Korea
windbust@hotmail.com, jslee@inha.ac.kr

Abstract. Modern War became to obtain information about a battlefield in real-time according to increase of intelligence and this information is used in creating the optimal path of units under operation. When battlefield data becomes diversified and detailed, reliability of the optimal path increases. But this situation also increases a possibility to cause trouble while transmitting these expanded data to each node and the time which is consumed to process battlefield information becomes the cause of declining performance. This study had applied quantization with an objective to increase performance as decreasing battlefield data and to maintain reliability. The study confirmed that only about half messages of DCPM [6] were created through experiments and proved that performance of the system was improved. The study also showed reliability of the created routes was not so different from routes, which DCPM is generated.

Keywords: Quantization, Distributed simulation, War-game, Optimal path.

1 Introduction

In War-game simulation [1], existing methods of path generation [2] were creating optimal path with using single processor in fixed battlefield data. However, in actual war, battlefield situations change moment by moment and numerous objects move simultaneously. If multiple objects request path simultaneously, operation with single processor is bound to face a delay of path generation. And if a unit movement is delayed with this problem, units are likely to experience loss of fighting power due to failure to acquire advantageous terrain and evade risk. A solution to cope with such a problem is creating paths with using multiple clients through distributed processing in War-game simulation [3] [4].

However, distributed processing requires that clients, which create optimal paths, should remain synchronized with the same battlefield data. Nodes, which play the role of a client, generally remain separated geographically and communicate through network. Because situations of battlefield are changed in real-time, the battlefield data is required to be transmitted to all the clients on the network in short period. But the more the battlefield enlarges, the more the amount of data increases in geographical progression. Not only receiving battlefield data can take more time than path generation time, but also data cannot be transmitted smoothly depending on network condition.

J.-H. Kim et al. (Eds.): AsiaSim2011, PICT 4, pp. 326–335, 2012.

In order to shorten synchronizing time of battlefield data, it is necessary to reduce the amount of data, which is transmitted to client, or adjust frequency of transmission efficiently. In existing researches, all battlefield data was transmitted periodically to renew clients' battlefield information. In this paper, the amount of data, which is transmitted to clients, was reduced through applying of the quantizing technique [5]. The technique divided the battlefield data into several segments. The clients received only battlefield data, which change more than threshold that was setup. Efficiency of quantizing battlefield data was evaluated through experiments. Time efficiency with the reduced battlefield data was measured. Difference between a path, which was generated with a quantization applied system, and a path, which was generated with DCPM [6], were also analyzed to evaluate usefulness of the quantization technique.

2 Related Works

2.1 Optimal Path Generation System for Moving Objects in War-Game

There are many path finding algorithms because of long period of research. Each algorithm has different characteristics. However, there are special characteristics in movement of military strategy. In military strategy, an algorithm requires considerations on concealment and covering different from general movement and locations of enemy forces and risk elements should be considered. To meet such special requirements, in this study, optimal path was created in War-game simulation using Genetic algorithm [2]. The battlefield data is composed of various information like terrain, altitude of the entire battlefield, locations of enemy force and abnormal symptom etc. The more the types of data increase, the more reliable the optimal path, which is generated, becomes. Because Genetic algorithm generates optimal paths through procedures of selection, crossover, evolution and mutation, this algorithm can make suitable optimal paths in War-game simulation.

2.2 Distributed Processing System Based on Client-powered Model (DCPM)

In battlefield, due to the fact that numerous objects move simultaneously, requests of optimal path generation also should be served simultaneously. In single processor

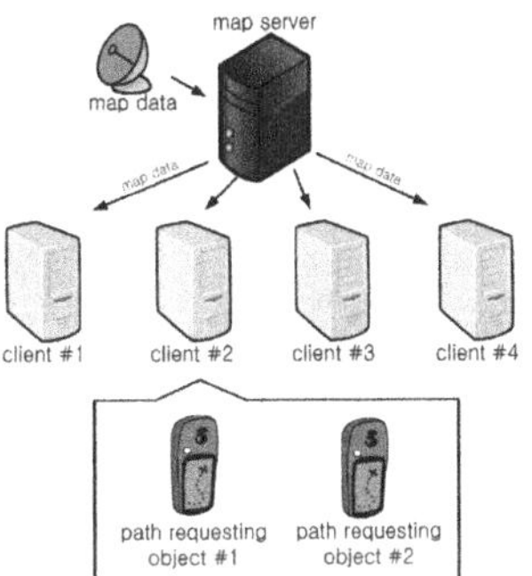

Fig. 1. A structure of DCPM

environment and central processing method, optimal paths are generated with one processor or node. In this situation, if optimal path generations are increasingly requested, satisfactory performance can hardly be expected. A solution to such this problem is to simulate on distributed processing environment [7].

Fig. 1 shows a structure of optimal path generation system using distributed processing based client-powered model (DCPM) [6]. In this structure, path requests of moving object are processed separately with multiple clients, so the same number of requests may be processed in shorter time than central processing method. However, in this structure, the battlefield data, which is obtained through various equipments, should be transmitted to all clients for being synchronized. And battlefield information should be transmitted again immediately when the battlefield condition is changed in order to effectively deal with actual battlefield situations, so extra overhead occurs. A purpose of distributed simulation [7] is to improve system performance. Therefore, the overhead, time delay, should be solved.

3 Quantizing Battlefield Data to Improve System Performance

Quantization technique divides battlefield data into several segments. Then, this technique reduces size of message, which is transmitted to clients, with excluding changes of the battlefield data that is lower than a threshold.

3.1 Quantizing Battlefield Data

As shown in Fig. 2, operation areas are divided into grid as much as the required level of precision. Each data, which is applicable to each area, is input. The battlefield data is made with this process. 'D' means anticipated degree of risk with enemy force, 'h' means height and 'α' means weighted value of moving speed with area type, slope or obstacle. All clients receive the same battlefield data from a map server and generate optimal paths based on such information. Battlefield situations are changed in real-time. If the latest data show difference between the current data, the map server makes recent battlefield data as a message and transmits to the clients.

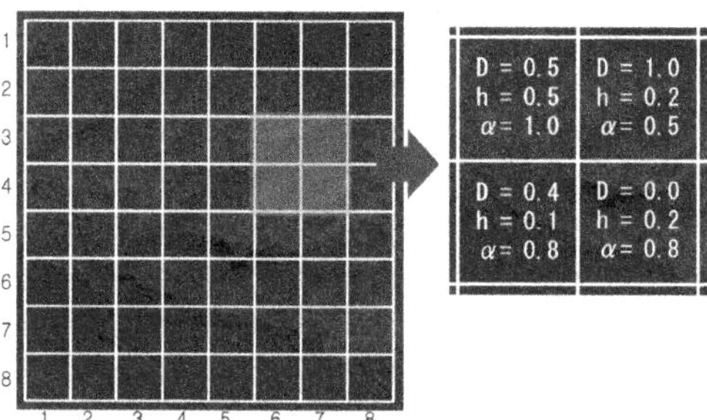

Fig. 2. Input battlefield data, which is used for path generation

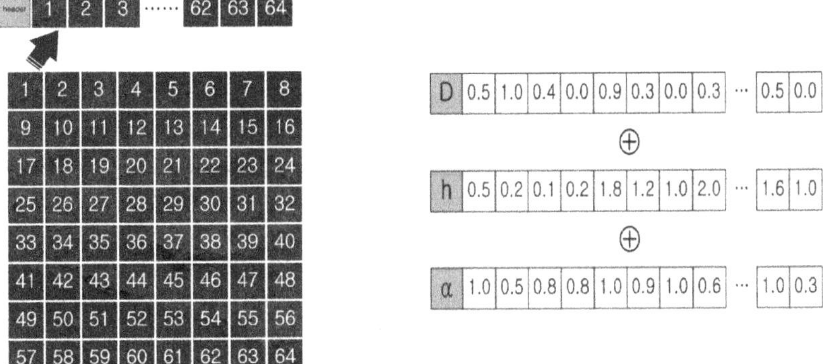

(a) Converting battlefield data to message (b) Message structure

Fig. 3. Message of DCPM

Fig. 3 (a) shows the method that DCPM converts battlefield data to a message. All the data of the battlefield is composed of one long message. The clients do not need to consider for where coordinate the data located, because the system is designed to apply the message sequentially.

A structure of the message is shown in Fig. 3 (b). The message is classified into three selections of data elements 'D', 'h' and 'α', which are used for generating optimal paths, and transmitted as one long character string. The bigger size of the battlefield and the more types of data increase significantly the battlefield data. This type of the message structure causes overhead when the clients receive the message and process to generate optimal paths. In addition, even a slight change of the battlefield causes transmission of the entire battlefield data to all the clients. The massive size message has a high error probability during the transmission.

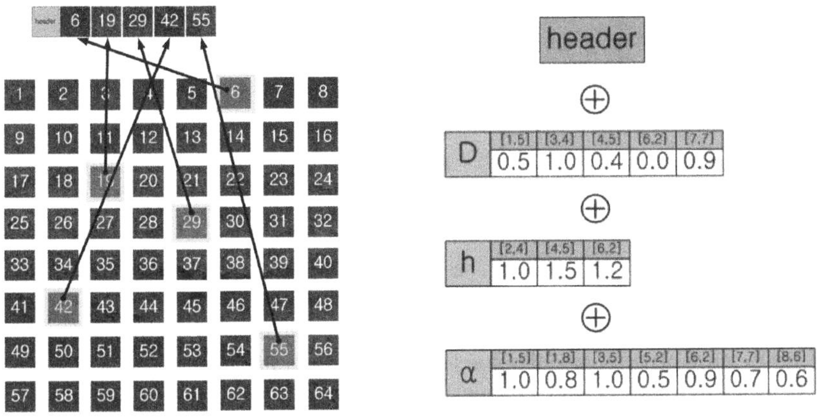

(a) Converting battlefield data to message (b) Message structure

Fig. 4. Message of quantization applied system

Fig. 4 (a) shows the method that a quantization applied system converts battlefield data to a message. In this process, only changed parts of the battlefield information are composed of the massage. Therefore, the message should have additional information about coordinates, where the data is located.

A structure of the message, which is applied quantization technique, is shown in Fig. 4 (b). The length of the message is not uniform because when some of the elements (D, h, α) change values, these elements are only included in the message. Therefore, there are three types of headers in this message structure. A message header should have additional information on the length of the message itself, and three element headers also need to inform each element boundary. These two types of headers that inform of data boundary are negligible when seen among the entire length of the message and are the fixed length. On the other hand, coordinate headers, which is right in front of each element and have information on coordinates of each element, increases in proportion to the changed battlefield data. So, the coordinate headers are the greatest overhead with quantization technique.

3.2 Reduction of Battlefield Data with a Threshold

Quantization technique [5] reduces the size of data, which is transmitted to clients. When a map server composes a message from battlefield data, it selects some data, which may affect to change an optimal path. Among the battlefield information, altitude and shape of terrain change barely if special situations do not happen. Usually, the battlefield data changes very slowly. And, most of small changes are not able to modify an optimal path. A threshold is needed to filter out these negligible changes. A change of the data, which is below the threshold, is not included in the message to be transmitted to the clients. The change of the battlefield data is ignored.

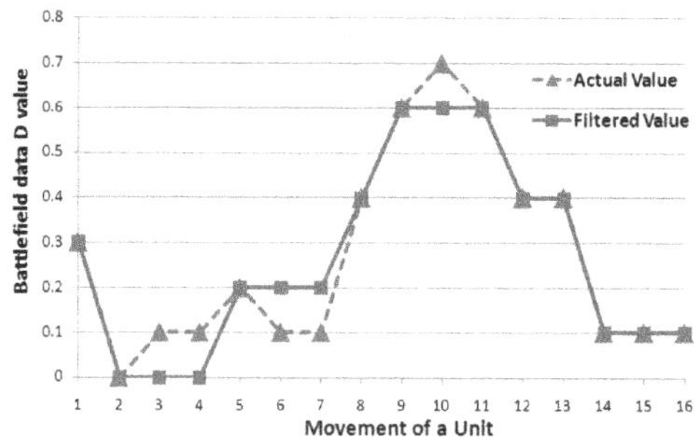

Fig. 5. Difference between actual and filtered value

Fig. 5 shows the graph of difference between an actual and a filtered 'D' value in a specific location. The threshold value is 0.2. The horizontal axis of the graph indicates a movement of units, which request optimal paths. The vertical axis of the graph indicates 'D' value of battlefield data at that time. 'D' changes periodically, and clients receive the changed value after the units move one segment (1Km) along the path. Minute changes are not transmitted with the threshold 0.2. So, the battlefield information, which the clients maintain, does not match information of the map server.

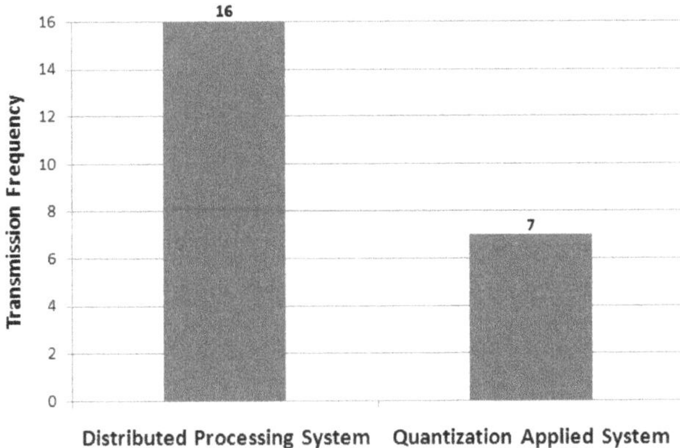

Fig. 6. Transmission frequency of battlefield data

Fig. 6 graph shows that how frequent 'D' value of Fig. 5 is included in the message to be sent to the clients. The graph has each result of two different systems, DCPM and the quantization applied system. Across entire experiments, the units move 16 segments, and 'D' changes 11 times. In the map server of DCPM, which is not applied any technique, 'D' is included in the regardless of whether the value changes or not. Therefore, 'D' is transmitted 16 times as same as the number of the unit movements. On the other hand, in the map server of the system, which is applied quantization technique, 'D' is only included in the message if the value exceeds the threshold. Therefore, 'D' is transmitted just 7 times while 16 segments of the unit movements. 'D' changes 11 times actually, but the graph shows the quantization applied system ignores 4 times of changes. The battlefield data, which changes larger than the threshold, is transmitted selectively if quantization technique is applied to optimal path generation systems. So, the size of a message may be reduced significantly.

4 Simulation Design and Results

4.1 Experiment Environment

A configuration of an experiment is shown in Fig. 7. Fig. 7 (a) is the distributed processing path generation system based on Client-powered model [6]. Simulation environment is organized with total five PCs. Four PCs act clients, and one PC acts a map server. The clients process twelve optimal path requests simultaneously. Once the clients create optimal paths, units, which send path requests, move 1Km along the optimal paths. After that, the units repeat sending requests of optimal paths to a destination again. The map server transmits battlefield data, which changes periodically, after the clients receive requests. When the clients receive a message, which has the battlefield information, they generate optimal paths immediately. The map server modifies the battlefield data at certain probability after transmitting the message. And again, the map server makes a new message and prepares to transmit it. In the experiment, the battlefield data elements 'D' and 'α' are only used to generate optimal paths because height 'h' barely changes.

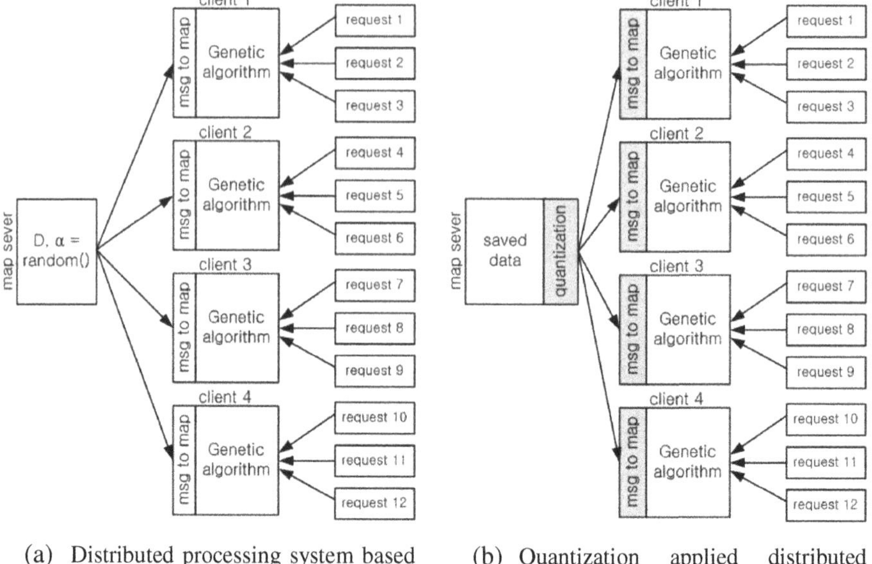

(a) Distributed processing system based on Client-powered model [6]

(b) Quantization applied distributed processing system

Fig. 7. Experimental configuration

Fig. 7 (b) is the new system; the quantization applied distributed processing path generation system. This system has mostly the same basic configuration of Fig. 7 (a). Changes of battlefield data should be the same as changes of the battlefield data in Fig. 7 (a). So, a map server uses saved data, which was used in Fig. 7 (a) instead of modifying the battlefield data in the experiment process. One of difference with Fig. 7 (a) system is a quantization module. This module is added to the map server. This module selects the battlefield data, which a change happens. Then, the selected data is filtered with a threshold. After these processes, the module composes a message as shown in Fig. 4 (b) from remaining data. According to the change of the message structure, the clients are also modified to be able to accept the new type message. These two systems, Fig. 7 (a) and Fig 7 (b), differ only the message composing module of the map server and the receiving message part of the clients. The rests of all functions are completely the same. Both the systems are modeled in DEVS methodology [8].

Three are measured to compare system performance through the experiments Fig. 7 (a), (b). First of all, the amount of messages, which the clients of each system receive at each period, is compared. Then, processing time from the moment, when the clients receive a message, until generation of an optimal path is completed is measured. At the last, difference between paths from Fig. 7 (a) DCPM, and paths, which are generated with reduced battlefield date in Fig. 7 (b) the quantization applied system, are compared.

4.2 Experiment Results

In this chapter, results of the experiments show how better the quantization applied system is. Fig. 8 shows the graph of the total amount of the messages that the map server transmitted to the clients. The total amount of the message is the cumulative amount while the units move to the destination. The horizontal axis of the graph indicates the movement of the units. And, the vertical axis of the graph indicates the total amount of the messages transmitted to the clients. The clients receive the message from the map server immediately after the units move 1Km along optimal paths, which is generated in the each system. The graph shows the amount of the message transmitted to the clients while the units move 15 times. Because the clients of DCPM receive the accurately same size of a message each time, the graph shows constantly rising shape. However, the clients of the quantization applied system receive the inconstant size of a message each time since the quantization module reduces messages. As a result, the quantization applied system produces approximately messages of 50% compared to DCPM.

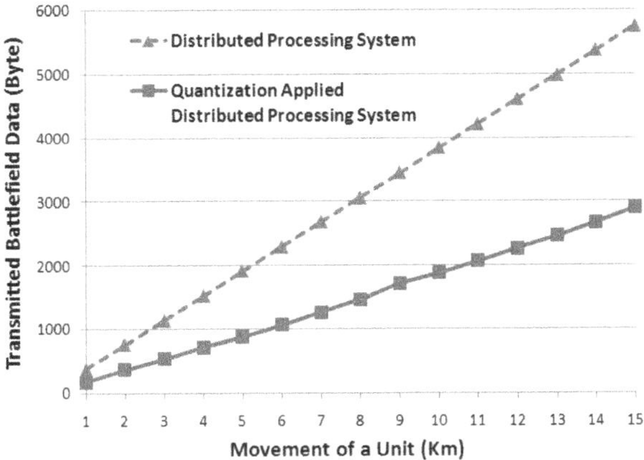

Fig. 8. The total amount of transmitted battlefield data

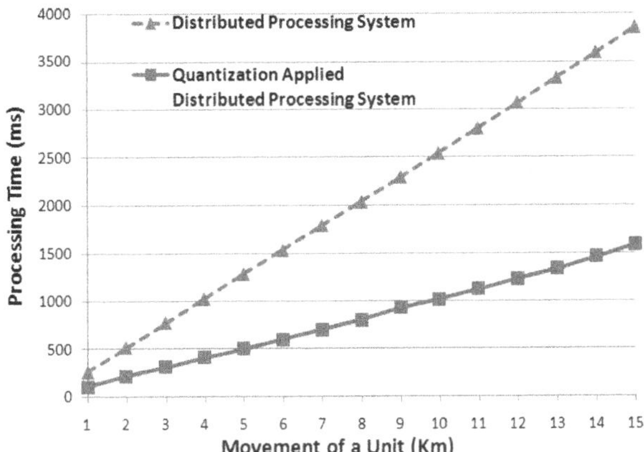

Fig. 9. Path generation time

Fig. 9 shows the graph of processing time from the moment, when the clients receive a message, until generation of an optimal path is completed. The processing time is measured in millisecond and the cumulative time. The horizontal axis of the graph indicates the movement of the units. And, the vertical axis of the graph indicates the cumulative processing time of the path generation in the clients while the units move 15 times. Difference of the processing time between both the systems occurs while the clients convert the message to the battlefield data, which is used to generate optimal paths. Difference of the processing time is not related to the path generation time. Because the clients of DCPM always receive the same size of a message each time, the processing time also shows within 16ms (250ms~266ms) difference without significant changes at every path generation. On the other hand, the clients of the quantization applied system receive the inconstant size of a message each time. Because of the various size of messages, the processing time shows relatively large difference of 50ms (82ms~132ms). But, the processing time is reduced about 60% compared to DCPM due to reduced messages.

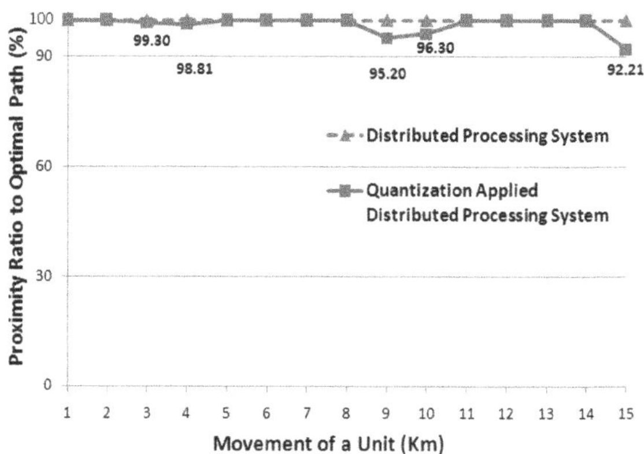

Fig. 10. Proximity ratio to optimal path

Travel time using an actual optimal path: T

Travel time using a path generated with quantization applied system: t

$$\text{Proximity ratio to an actual optimal path (\%)} = \frac{T - |t - T|}{T} \times 100 \qquad (1)$$

If the map server filters out the battlefield data with the threshold when composing a message, slight difference occurs between actual battlefield information and information, which the clients has. Fig. 10 shows difference between paths from DCPM, and paths, which are generated with the reduced messages in the quantization applied system. The horizontal axis of the graph indicates the movement of the units. And, the vertical axis of the graph indicates the proximity ratio of the paths generated in each system to an actual optimal path. The proximity ratio can be obtained as in equation (1). This ratio becomes 100% if a generated path takes the same time as the actual optimal path does. Because the map server of DCPM composes a message without any filtering, the system always generates the same paths as actual optimal paths. The graph shows that the paths, which are generated in the quantization applied system, are very close to the actual optimal paths even though the total amount of messages decreases to 50% level. During the units move 15 times throughout the experiments, there is average error of 1.21%. However, at 10 times, which is equivalent to 2/3 of 15 entire path generations, generated paths are exactly the same as the actual optimal paths. And, the worst path that is generated at 15th movement, also shows only 7.79% difference with the actual optimal path. So, paths, which are generated with the quantization applied system, are reliable.

5 Conclusion

In this paper, battlefield data quantization method was proposed to improve the performance of optimal path generation system of War-game moving object on distributed simulation environment. According to the experiment results of chapter 4, the amount of messages was reduced to approximately 50% level compared to DCPM if changed battlefield information beyond the threshold was only transmitted with quantization method. The reduced messages also decreased a probability of network transmission error and transmission time. In addition, the reduced messages shortened the message processing time of the clients, so system performance was improved about 60%. During 15 movements of the units occurred throughout the experiments, few errors happened. The results showed low average error rate, 1.21%. And 2/3 of the generated paths were as the same as the optimal paths.

Acknowledgments. This work was partially supported by Defense Acquisition Program Administration and Agency for Defense Development under the contract. (UD110006MD).

References

1. Vlahos, M.: The Naval War College and the American Mission, pp. 8–14. Naval War College Press (1981)
2. Lee, S.Y., Jang, S.H., Lee, J.S.: Modeling and Simulation of Optimal Path Considering Battlefield-situation in the War-game Simulation. Journal of the Korea Society for Simulation 19(3), 27–35 (2010)
3. Fullford, D.: Distributed interactive simulation: it's past, present, and future. In: Proceedings of the Winter Simulation Conference, pp. 179–185 (1996)
4. Weatherly, R.M., Wilson, A.L., Griffin, S.P.: Alsp - Theory, Experience, and Future Directions. In: Proceedings of the Winter Simulation Conference, pp. 1068–1072 (1993)
5. Zeigler, B.P., Cho, H.J., Kim, J.G., Sarjoughian, H.S., Lee, J.S.: Quantization-based Filtering in Distributed Discrete Event Simulation. J. Parallel Distrib. Comput. 62, 1629–1647 (2002)
6. Oh, H., Lee, J.S.: Optimal Path Generation System Using Distributed Simulation for War-game Moving Objects. In: The Korea Society for Simulation, Spring Conference, pp. 52–57 (2011)
7. Fujimoto, R.M.: Parallel and distributed simulation. In: Proceedings of the 1999 Winter Simulation Conference, pp. 122–131 (1999)
8. Zeigler, B.P., Praehofer, H., Kim, T.G.: Theory of Modeling and Simulation: Integrating Discrete Event and Continuous Complex Dynamic Systems. Academic Press (2000)

A Methodology for Effectiveness Analysis of Future Weapon System Using a PLAF Based Simulation System*

Seung Keun Yoo[1], Jae Shin Lee[2], and Doo-Kwon Baik[1,**]

[1] Dept. of Computer and Radio Communications Engineering, Korea University, 1, 5-ga, Anam-dong, Seongbuk-gu, Seoul, Republic of Korea
{iori27,baikdk}@korea.ac.kr

[2] Center for Military Planning, Korea Institute for Defense Analyses, San 5-7, Cheongryangri-dong, Dongdaemoon-gu, Seoul, Republic of Korea
jslee86@kida.re.kr

Abstract. Newly emerging, latest aspects of war have made a lot of nations to accelerate in developing future weapon systems. However, there are only a handful of methods to analyze and estimate their effectiveness. Conventional simulation systems are too ineffective and limited in testing the unique nature of future weapon systems. Instead, a more flexible PLAF (Product Line Architecture Framework) based simulation is required to take into account new characteristics of these weapon systems. We therefore propose a new methodology for effectiveness analysis of future weapon system using the newest simulation system, OneSAF, and seek reliability of this method by analyzing the results for our test simulation.

1 Introduction

The latest wars in Iraq and the Middle East shows us the new characteristics which differ to that of past conventional wars. Wars have transformed from linear wars where battles are fought between large scale troops to non-linear wars where guerrillas fight a team of special force or where few types of new weapon systems battle against numerous conventional weapons.[1] This shift in war trend has emphasized the importance of acquiring new weapon systems across many nations, especially in developed nations such as the U.S. and European countries. They have invested significant amounts of their national defense budget in developing future weapon systems. In comparison to the fast developing rate of new weapon systems, however, the means of measuring and analyzing the effectiveness of these newly developed weapon systems are quite insufficient.

Many M&S systems performed well in reflecting the characteristics of conventional weapon systems in conventional war environment until now, but they have shown great limitations in testing new traits of future weapon systems which incorporates

---

* This research was supported by Second Brain Korea 21 Project.

** Corresponding author.

J.-H. Kim et al. (Eds.): AsiaSim2011, PICT 4, pp. 336–344, 2012.

concepts such as stealth, unmanned systems, and multi-purposed functions. Therefore it is crucial to develop new M&S concepts and methodologies that take these new features into account for effectiveness analysis of future weapon systems.[2]

This paper proposes a new methodology to analyze the effectiveness of future weapon systems using a flexible and PLAF based simulation in order to simulate the unique characteristics of future weapon systems. It also tests the new methodology using a simulation system, compares the test result with the methodology for the effectiveness analysis using an existing simulation system, to demonstrate the usefulness of the new method. For our experiment, we used OneSAF, a PLAF based simulation system created by the U.S. Army, and OICW (Objective Individual Combat Weapon) as the choice of future weapon system to be tested for our proposed methodology.

This paper consists of five parts. In Chapter 1, we briefly explain the background and purpose of this study and the proposed methodology briefly, and move on to Chapter 2 to describe the characteristics and strength of a PLAF based simulation system. In Chapter 3, we propose a new methodology to analyze the effectiveness of future weapon systems using this simulation system, and test this methodology, show the test results, and compare it with the results from an existing methodology in Chapter 4. Chapter 5 summarizes the whole contents of this paper and emphasizes the outcome and meaning of our proposed methodology.

2 PLAF Based Simulation Systems

2.1 The Limitations of Existing Simulation Systems

Future weapon systems have differentiated itself from conventional weapon systems been developed and defined by sets of concepts such stealth, unmanned system, multi and complex function. The objective of these new weapon systems is to minimize the damage of personnel and equipment on our side while effectively damaging the enemy. New weapon systems have special functions and attributes that the conventional weapon systems don't have.

Traditional simulation systems models characteristics of weapon systems according to the type of military units such as infantry, artillery, armor, and engineer and implements conventional functions of warfare such as maneuver, detection, direct fire/area fire, damage assessment, combat service support etc. However, there are great limitations to model future weapon systems with these conventional features alone. In order to analyze future weapon systems through simulations, existing simulation systems must be tweaked to incorporate their functions.[3]

2.2 The Characteristics of a PLAF Based Simulation System

PLAF means the architecture framework based on the Product Line concept. A simulation system with it has a flexible, scalable, evolutional structure to overcome the limitations of the existing simulation systems. Unlike previous systems where a single simulation program was built by bringing together multiple warfare functions, PLAF based systems are divided into functional blocks. A specific warfare environment could be created by selecting certain blocks. It can compose some function components already implemented in case of simulation of conventional weapon systems and implement the new component which fits to a new weapon system in case of simulation of future weapon systems. In fact, a PLAF based simulation system is so expandable, that it can simulate many heterogeneous future weapon systems together if they can be modeled and implemented. It makes the capability of the simulation system to be improved evolutionally.[4]

2.3 OneSAF Simulation System

OneSAF is a PLAF based, entity level, and next generation simulation system. It is developed to provide the integrated simulation service in various M&S domains such as ACR (Advanced Concepts and Requirements), RDA (Researches, Developments and Acquisitions) and TEMO (Trainings, Exercises, and Military Operations). Strictly speaking, it is not a single simulation model, but an integrated simulation system which includes all infrastructures for simulation. Its features differ from existing defense simulation systems as follows.[5]

CGF(Computer Generated Forces) Support. Just like the term indicates, forces are automatically generated and operated by the computer instead of users or training participants. OneSAF can support CGF in analysis and training simulation.

PLAF based Simulation. OneSAF doesn't have a single configuration as a single executable program but is able to construct various configurations with a number of components integrated systemically according to the user's objective.

Multiple Fidelity Support. OneSAF supports 3 levels of fidelity (Low, Medium, and High) depending on the degree of precision of simulated entity.

Urban Area Operation Support. OneSAF can thoroughly depict urban terrain through its UHRB (Ultra High Resolution Building) function which in return makes it possible to simulate urban area military operations such as counter-terrorism operations.

2.4 The PLAF Based Feature of OneSAF

OneSAF is composed of many components with PLAF to support the concept of component based system shown as Fig.1.[6]

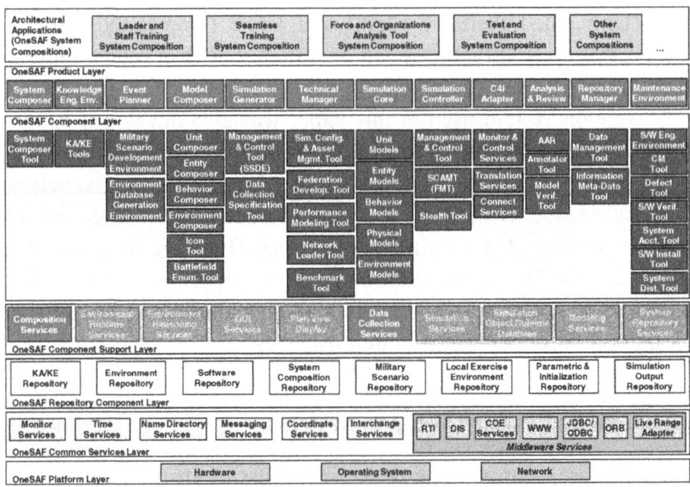

Fig. 1. PLAF in OneSAF

The various function components of OneSAF are structured as hierarchical function blocks. Users of OneSAF can produce desired functions by putting together these blocks of components. OneSAF can simulate a number of new future weapon systems because the component of each weapon system is in the same layer and users can get them easily.

3 The Methodology for Effectiveness Analysis of Future Weapon Systems

In this paper, we propose a new methodology for effectiveness analysis of future weapon systems by using PLAF based simulation system implemented in OneSAF. The method is divided into 6 steps, 'Characteristics Analysis' - 'Modeling' - 'Model Implementation' - 'Composition' - 'Simulation' - 'Result Analysis'. This chapter gives a detailed account of the contents from 'Characteristics Analysis' to 'Composition.'-More details of 'Simulations' and 'Result Analysis' are given on the following chapter.

3.1 The Characteristics Analysis of Future Weapon Systems

Future weapon systems are in pursuit of the most effective means of striking methods by endlessly combining the newest weapons and IT technologies such as wireless communication technologies at their fullest. For instance, the key role of the next generation JSF (Joint Strike Fighter), equipped with stealth function, is to precisely bomb key enemy bases and facility without being detected. Unmanned robot systems such as UAS (Unmanned Aerial System) and UGV (Unmanned Ground Vehicle) accomplish their missions without human operators which in return minimize loss of our forces.[7]

The OICW weapon system used in this paper was also developed to take out enemy effectively. OICW, with the help of advanced fire control system, can kill concealed enemies behind walls. Shooters can choose their firing options between normal bullets and grenade launcher depending on the characteristic of enemy hideout (type, number of enemy, degree of concealment, position). Therefore, in order to analyze the effectiveness of an OICW, the simulation must incorporate functions that consider dual caliber shots (between normal bullets and grenade launcher) and the ability to switch shooting options depending on circumstances.

3.2 The Modeling of Future Weapon Systems

The modeling of future weapon systems is the process of abstracting the key functions of future weapon systems and expressing them as mathematical expressions or logical expressions that could be used in simulations. The model for future weapon system exists in a single component in a PLAF based simulation system. This paper generates a single weapon system model for OICW in this process. Table 1 shows the modeling process for 3 key functions of OICW; fire control system, rifle, and grenade launcher

Table 1. The modeling process for OICW

Target Function	Modeling	Note
Fire Control System	Laser aim, Thermal detection, Automatic distance measure	Improves accuracy rate
Rifle	Shot by bullets, Point-to-Point fire damage assessment	To strike individual targets
Grenade Launcher	Shot of rounds, Area fire damage assessment	To strike group targets

3.3 The Implementation of Future Weapon Systems

The implementation of future weapon system is the process of converting the result of modeling process into the model for simulation through implementation. In this paper, we chose to use OneSAF as our simulation system. Therefore, the model was implemented as a component of OneSAF. Table 2 shows the results of implementation for OICW modeling into OneSAF components.

'Newly developed' in the note column indicates that the component was newly developed while 'Reuse' means that the component already exists in OneSAF. In the case of OICW, new functions such as the fire control system and the part related to grenade launcher were newly developed, whereas the existing components in OneSAF were reused for rifle function.

Table 2. The implementation process for OICW model

Modeling	Component (name) implemented	Note
Laser aim	Laser sight	Newly developed
Thermal detection	Thermal detector	Reused
Automatic distance measure	Device for automatic distance measure	Newly developed
Shots by bullets	Rifle	Reused
Point-to-Point fire damage assessment	Model for Point-to-Point fire damage assessment	Reused
Shot of rounds	Grenade launcher	Newly developed
Area fire damage assessment	Model for Area fire damage assessment	Newly developed

3.4 The Composition of Future Weapon Systems

The composition of future weapon systems is the process of integrating lower-level components into higher-level models just so it could be used in simulations. Table 3 shows the assembly process of a single OICW model used in OneSAF from its individual components.

Table 3. The composition process for OICW

Individual component	Higher-level component	Target component
Laser sight Thermal detector Device for automatic distance measure	Model for fire control system	Model for OICW
Rifle Model for Point-to-Point fire damage assessment	Model for rifle	
Grenade launcher Model for Area fire damage assessment	Model for grenade launcher	

4 Simulation and Test Results

'Simulation' and 'Result Analysis' are the last two processes in analyzing the effectiveness of future weapon systems. This chapter explains the two remaining processes, provide~~s~~ the ~~real~~ results for this test, and compares the results to the results from existing methodology using a conventional simulation system.

4.1 Simulation

In this section, we showed the process of simulating the composite components of future weapon systems using a PLAF based simulation system. OneSAF and OICW components were used to carry out our simulation. Table 4 shows the test scenarios for this simulation.

Table 4. Test Scenarios

Scenario	Blue force	Red force	Terrain	Operations
SCN#1	OICW (2), Individual rifle (7), Machine gun (1)	Individual rifle (10), Machine gun (1)	Open	Blue force as defense, Red force as offense
SCN #2	Same as SCN#1	Same as SCN#1	Open	Blue force as offense, Red force as defense
SCN #3	Same as SCN#1	Same as SCN#1	Hill	Blue force as defense, Red force as offense
SCN #4	Same as SCN#1	Same as SCN#1	Hill	Blue force as offense, Red force as defense
SCN #5	Same as SCN#1	Same as SCN#1	Mountain	Blue force as defense, Red force as offense
SCN #6	Same as SCN#1	Same as SCN#1	Mountain	Blue force as offense, Red force as defense

4.2 Test Results and Implications

Fig.2 shows the result of our test simulation (marked blue) in terms of Weapon System Contribution Ratio (WCR). WCR is the measure to be computed by taking the number of casualties incurred by the weapon system and dividing it by the number of total enemy casualties. We also provide the simulation results using Janus (marked red) model, a methodology developed by the U.S. Army to analyze effectiveness of weapon systems.

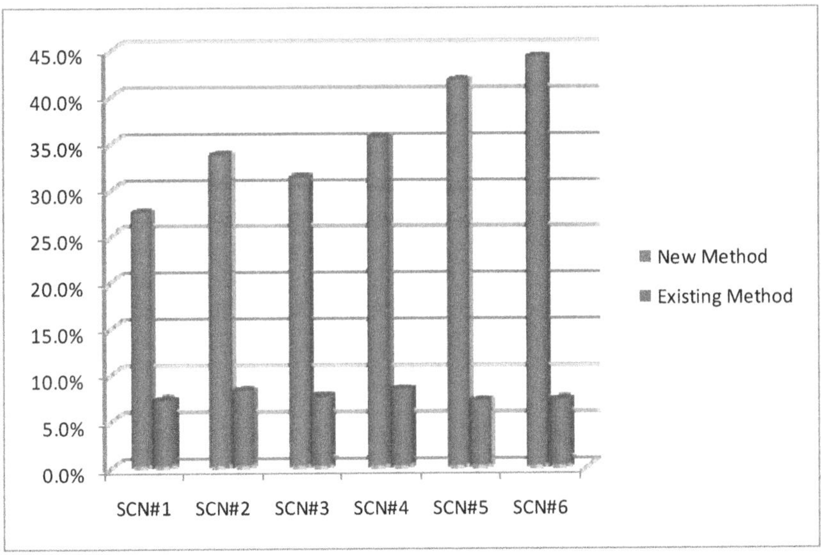

Fig. 2. Comparison of results between new and existing methodology for OICW

The results of simulation using Janus model is rather constant, regardless of the scenarios. After all, this is not so surprising because grenade launchers attached to OICW are treated as normal rifles in Janus. Our proposed methodology, on the other hand, takes into account the effectiveness of OICW when attacking groups of enemies. This is well reflected in our results by the varying WCR ratio in each scenario. The existing methodology using a conventional simulation system does not simulate the effectiveness of future weapon systems realistically. The result in this test simulation shows that the new methodology using a new PLAF based simulation system is more reliable than previous methods.

5 Conclusion

This paper proposes a new methodology to systemically and realistically analyze the effectiveness of future weapon systems which is being recognized to be one of the key capacities for ~~the~~ force improvements. While technologies for developing weapon systems have been rapidly improving with many nations actively investing billions of dollars in them, the improvement of M&S, which is the means to analyze and estimate the effectiveness of weapon systems, has been lagging in comparison. This gap between the two technologies has become so wide that existing M&S systems now are not apt for analyzing new future weapon systems. Accordingly, the effects of newly developed weapon systems are being evaluated and interpreted by qualitative aspects rather than quantitative aspects. We believe recently developed PLAF based simulation systems such as OneSAF and newly developed methods such as the one we have proposed in this paper, with further improvements, could not only overcome the problems of existing methodologies but also effectively analyze feasibilities of developments of weapon systems realistically at early stages. These quantitative data could also be helpful to the Ministry of Defense, who is constantly prioritizing on optimal investments under great budgetary pressure.

However, the methodology proposed in this paper requires robust infrastructure for detailed models before additionally adding models for a new future weapon system. Unless the infrastructure fails to incorporate enough conventional weapons systems and war environments, the number of new functions that could be composed would be quite limited. Therefore, more efforts must be made in consistently adding new function components and precise models for each newly components must also be developed in accordance. Further studies on modeling C4I and interoperability is also required to more realistically portray future network centric warfare environments in simulations.

References

1. Hammes, T.: The Sling and The Stone – On War in the 21st Century. Zenith Press, Minneapolis (2004)
2. Moon, H.K., Yoo, S.K.: Application of Next Generation Defense Simulation Model for Analysis of Future Combat System. In: The Korea Society for Simulation Spring Conference, KSS Suwon (2008)
3. Yoo, S.K., Moon, H.K., Lee, M.H.: OneSAF Localization (I). KIDA Press, KIDA Seoul (2010)
4. Wittman, R., Harrison, C.: OneSAF: A Product Line Approach to Simulation Development. In: OneSAF User Conference, PEO STRI Orlando (2001)
5. Surdu, J.: OneSAF Program Overview. In: OneSAF User Conference, PEO STRI Orland (2004)
6. Wittman, R.: The OneSAF Product Line Architecture – An Overview of the Products Process. In: OneSAF User Conference, PEO STRI Orlando (2002)
7. Son, T.J., Noh, H.: Network Centric Warfare. KIDA Press, KIDA Seoul (2009)

Simulation-Based SAM (Surface-to-Air Missile) Analysis in OpenSIM (Open Simulation Engine for Interoperable Models)

Kangsun Lee[1], Joonho Park[1], Chanjong Park[1],
Seogbong Kim[2], and Hyun Shik Oh[2]

[1] Department of Computer Engineering, MyongJii University,
38-2 San, Namdong, YongIn, Kyunggi-Do, South Korea
{ksl,allin917,pcj0824}@mju.ac.kr
[2] Agency for Defense Development
{sbkim,ohs}@add.re.kr

Abstract. Assessing MOE (Measure of Effectiveness) of a weapon system is important as we invest significant efforts to develop the weapon system. OpenSIM (Open Simulation environment for Interoperable Models) has been developed to help users perform complex tasks for simulation-based weapons effectiveness analysis. In this paper, we present tools and services of OpenSIM and illustrate how we can utilize the OpenSIM for weapons effectiveness analysis with an example of SAM (Surface-to-Air Missile) in AAW (Anti-Air Warfare).

1 Introduction

As modern weapon systems need significant efforts to develop, there is an increasing need to predict the effectiveness of the weapon systems before the actual development. Conventional approaches, such as mathematical and analytic solutions, become insufficient for this purpose, as the weapon systems are operated in non-deterministic network-centric warfare [1]. SBA (Simulation-Based Acquisition) aims to help decision makers by providing measurements of performance and effectiveness for weapon systems before their actual development and deployment. [2]

There are some challenging issues for the simulation-based weapons effectiveness analysis. Modern weapons themselves are complex in their dynamics, and therefore, we need complex modeling tasks to represent the dynamics. Also, we need to represent nature and operational tactics as well to realistically simulate environmental effects to the weapon systems. Moreover, we have to deal with legacy simulators, since weapons sometimes operate with legacy live, virtual and constructive simulators. All these complex tasks are hard to be done by non-M&S (Modeling and Simulation) experts. An integrated simulation environment can help users to partly automating complex M&S tasks for weapons effectiveness analysis.

In this paper, we present a SAM (Surface-to-Air Missile) simulator in AAW (Anti-Air Warfare). The SAM simulator has been constructed by our simulation

J.-H. Kim et al. (Eds.): AsiaSim2011, PICT 4, pp. 345–351, 2012.

environment, OpenSIM (Open Simulation engine for Interoperable Models). [3] The SAM Simulator is composed of four components – Aircraft, AA (Anti-Air) Radar, Launcher and Missile. *Aircraft* approaches to the target. *AA Radar* detects the aircraft if it's within the detection zone. *Launcher* receives the aircraft location from AA Radar and fires a missile accordingly. *Missile* starts inertial navigation and then homing guidance to strike the target. *Aircraft* may begin avoidance maneuver upon facing the incoming missile. These four components have been developed with the help of services and tools in OpenSIM. We measure a set of hit/miss rate of a SAM to find out effective ways of employing SAMs in AAW.

This paper is organized as follows. Section 2 introduces OpenSIM with its tools and services. Section 3 presents our SAM (Surface-to-Air Missile) simulator and illustrates how OpenSIM can help users to construct a simulator for weapon systems. Section 4 concludes this paper with summaries and future works.

2 OpenSIM

OpenSIM (Open Simulation engine for Interoperable Models) is an integrated simulation environment for weapons effectiveness analysis. OpenSIM provides a suite of tools and services for developing, executing and analyzing simulations of weapon systems, as shown in Figure 1 [3].

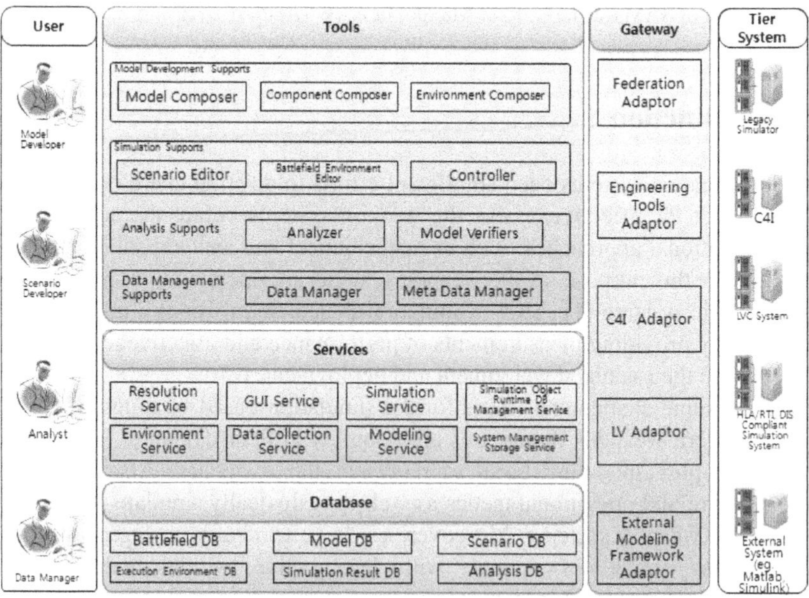

Fig. 1. Overview of OpenSIM

2.1 OpenSIM Services

OpenSIM provides a set of services that users can use to develop models and conduct simulations.

- **Modeling Services:** The modeling framework of OpenSIM has been devised based on DEVS (Discrete Event System Specifications) formalism. Details on DEVS can be found in Reference [4]. OpenSIM provides a set of services to define weapon components and their coupling information.
- **Simulation Services:** OpenSIM provides various simulation services including scheduling services, high-speed communication services, resolution services, journaling services, and message logging services. Scheduling services of OpenSIM manage time to support hybrid simulation. Internal high-speed communication services run models in either single- or multi-threaded modes to speed up simulation. Resolution services allow analysts to dynamically simulate models with different levels of abstraction. Journaling Services are used to sample the values of selected attributes during a simulation and save the sampled values to a specified file format in OpenSIM. Message Logging services write messages logged during a simulation to the corresponding output medium.
- **Linkage Services:** OpenSIM provides a set of services to link HLA federations, legacy LVC simulators, commercial engineering tools and web applications. Gateway services connect external HLA federations. A SOAP [5] interface enables various web applications to communicate with the OpenSIM models. OpenSIM includes LVC gateway services to facilitate integration with LVC standards, such as HLA, DIS (Distributed Interactive Simulation), and TENA (Test and Training Enabling Architecture) [6,7]. OpenSIM also provides a linkage to commercial engineering tools, such as. MATLAB/SIMULINK. Reference [9] provides an efficient mechanism to handle timing and data discrepancies between MATLAB/SIMULINK models and DEVS-compliant models. OpenSIM uses the mechanism proposed in Reference [9] and provides a set of services to realize the mechanism.

2.2 OpenSIM Tools

OpenSIM provides a set of tools to help modelers and analysts define models, simulation scenarios, and analyze simulations, as shown in Figure 2.

Modelers can construct weapon models, natural environment models and operational environment models with blocks and arcs on easy-to-use GUIs. Inside each block, modelers may insert detailed behaviors with mathematical equations or APIs provided in OpenSIM services. All these information are sent to the *Translator* of OpenSIM, where automatic translation into C++ codes takes place. Models can be further compiled either into DLL components or into C++ executables. All model components are stored in our internal reuse library. Modelers can facilitate the development process by searching reusable components in the reuse library, and putting the reusable pieces together according to simulation objectives.

Analysts can define/modify a set of simulation scenarios. OpenSIM also provides a simulation controller so that analysts can start, stop, pause and resume simulations. A set of analysis tools are additionally provided to help users view simulation results with reports and graphs.

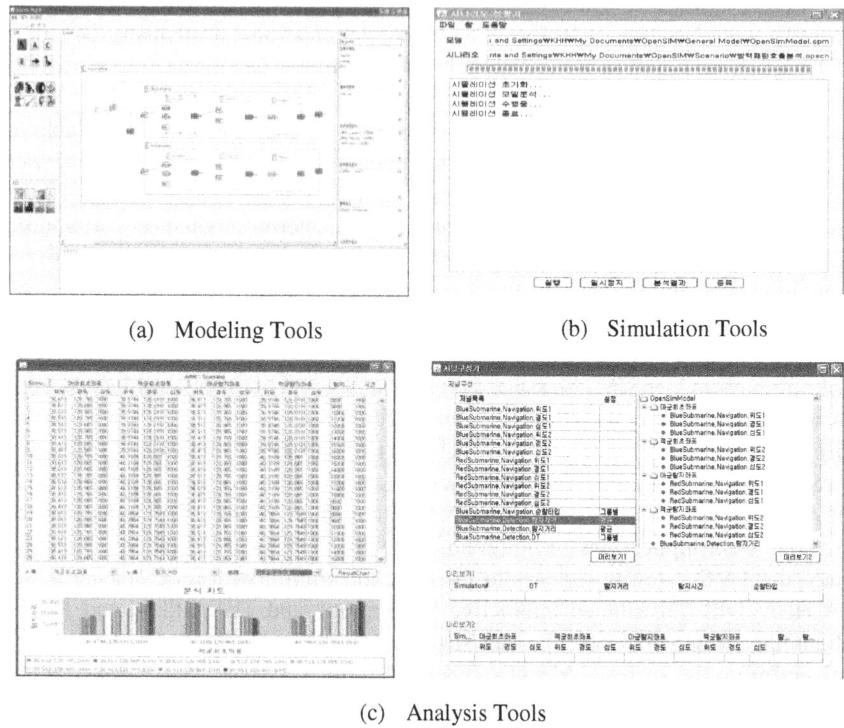

(a) Modeling Tools (b) Simulation Tools

(c) Analysis Tools

Fig. 2. OpenSIM Tools

3 SAM (Surface-to-Air Missile) Simulator

In order to analyze effectiveness of SAM, we need to consider not only its dynamics but also its natural environment (i.e. Atmosphere) and engagement environment (i.e. Anti-Air Warfare) as well. As shown in Figure 3, our SAM Simulator is composed of Aircraft, AA (Anti-Air) Radar, Launcher and Missile to address these considerations.

Aircraft approaches to the mission area. It employs a set of radars that can detect approaching missiles. Upon detecting the missiles, it starts horizontal balance turn based on the azimuth value, φ, obtained from the following equation.

$$\dot{\varphi} = -\frac{a_n}{V}$$

, where V is velocity, and a_n is acceleration for the horizontal turn. [11]

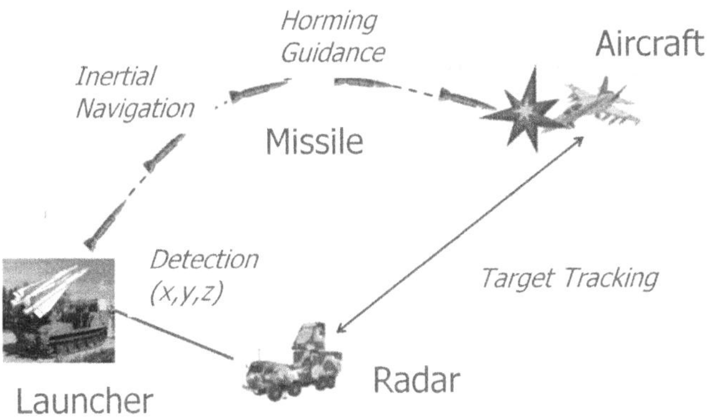

Fig. 3. SAM Simulator – Components [10]

AA Radar detects the aircraft if it's within the detection zone. *Launcher* receives the aircraft location from *AA Radar* and determines the initial inertial guidance position, (Xc, Yc, Zc) with the following equation.

$$X_c = \frac{X_{DR} + (V_{DR} * t_A)\sin \varphi_{DR}}{2}$$

$$Y_c = \frac{Y_{DR} + (V_{DR} * t_A)\cos \varphi_{DR}}{2}$$

$$Z = Z_{DR} + 100$$

, where (X_{DR}, Y_{DR}, Z_{DR}) is the aircraft location that AA Radar detects, V_{DR} and $\psi_{DR}\psi$ is the velocity and azimuth of the aircraft, respectively.

Missile flies to (Xc,Yc,Zc) with inertial navigation and then starts homing guidance to strike the target. [12]

Aircraft, AA Radar, and *Launcher* have been constructed by our research team. With the help of modeling services and tools in OpenSIM, we could successfully construct DEVS-compliant models for the three DLL components. Meanwhile, *Missile* has been developed by a separate research team from ADD (Agency for Defense Development) in South Korea and imported into OpenSIM as an external DLL component. Figure 4 shows the top- level coupled DEVS model for the SAM simulator. By using simulation services and linkage services in OpenSIM, these three OpenSIM components can communicate with external *Missile* component during the simulations.

Figure 5 shows a trajectory of *Missile* and *Aircraft*. The thick line represents the trajectory of Aircraft, while the thin line represents the one of *Missile*. As shown in Figure 5, Missile starts homing guidance right after it reaches the inertial guidance point to shot down the Aircraft. Upon seeing the missile, aircraft starts the horizontal turn to avoid the missile. Hit/Miss rates can be recorded as we change initial position and velocity of the aircraft. With total of 1,000 simulation trials, the SAM simulator exhibits 68.5% hit rate. These simulation results can help decision makers to figure out effective ways to operate SAMs in AAW.

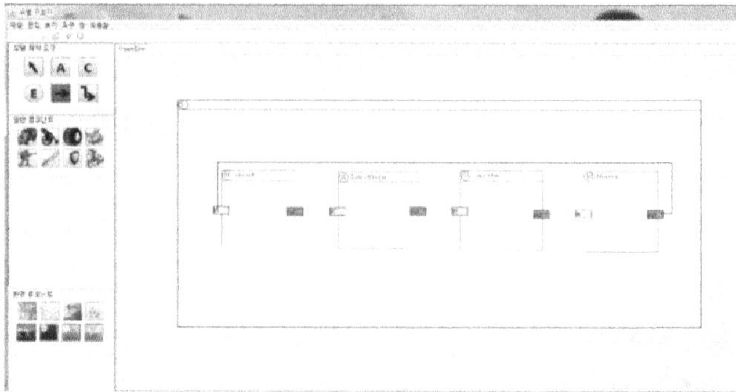

Fig. 4. Construction of SAM on OpenSIM

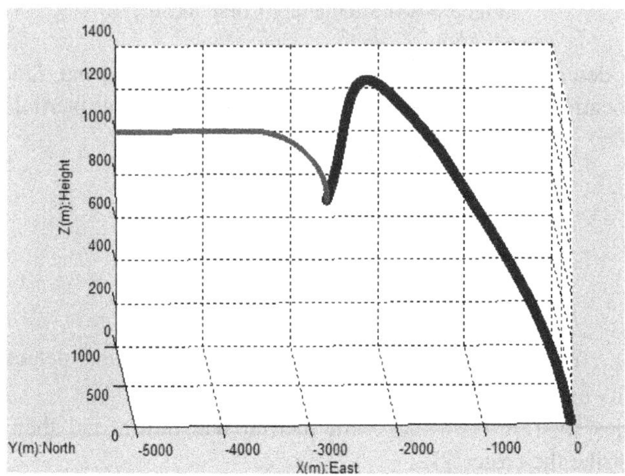

Fig. 5. A Sample Trajectory: Aircraft (Thin Line) and Missile (Thick Line)

4 Conclusion

In this paper, we have presented a simulation environment, OpenSIM. OpenSIM provides useful tools and services to aid simulation-based weapon's effectiveness analysis. SAM simulator has been constructed with four DLL components. Three DLL components – Aircraft, AA Radar and Launcher – have been constructed by using OpenSIM services. An external Missile DLL component has been separately developed and imported into OpenSIM. We have shown that the four DLL components could talk to each other seamlessly throughout the simulations with the help of the linkage services in OpenSIM.

We would like to add more components to SAM simulator for enhancing reality - Atmosphere component will soon be available in the SAM simulator in order to simulate atmospheric effects on the maneuver in Aircraft and Missile. Also, various types of aircraft models will be developed, so that we can analyze SAM's effectiveness with various avoidance tactics of the aircrafts. We also would like to continuously improve OpenSIM by addressing lessons learned from SAM simulator.

Acknowledgement. This work was supported by Defense Acquisition Program Administration and Agency for Defense Development under the contract UD080042AD, Republic of Korea.

References

1. Department of Defense Report to Congress, Network Centric Warfare (July 27, 2001), `http://www.dod.mil/nii/NCW/`
2. Sage, A.P., Olson, S.R.: Whither Simulation Based Acquisition?: Part II: Modeling and Simulation in Systems Engineering. Simulation, 283–285 (May 2001)
3. OpenSIM (Open Simulation Engine for Interoperable Models) for Weapons Effectiveness Analysis. In: Proceedings of the, International Conference on Modeling, Simulation & Visualization Methods, pp. 116–120 (2011)
4. Zeigler, B.P., Kim, T.G., Praehofer, H.: Theory of Modeling and Simulation, 2nd edn. Academic Press, New York (2000)
5. Möller, B., Dahlin, C.: A first look at the HLA Evolved Web Service API. In: Proceedings of Euro Simulation Interoperability Workshop (2006)
6. IEEE Std 1516-2000, IEEE standard for modeling and simulation (M&S) high level architecture (HLA) – federate interface specification (2000)
7. TENA – Test and Training Enabling Architecture, `https://www.tena-sda.org/display/intro/Home`
8. MATLAB® and SIMULINK®, `http://www.mathworks.com/products/`
9. Seo, K.-M., Song, H.S., Kwon, S.J., Kim, T.G.: Measurement of Effectiveness for an Anti-torpedo Combat System Using a Discrete Event Systems Specification-based Underwater Warfare Simulator. Journal of Defense Modeling and Simulation (2011)
10. Oh, H.S., Kim, D.H., Jo, S.J.: A Method To Evaluate Capabilities of Simulation Environments by Appling Engineering Models To Engagement Case. In: Proceedings of KIMST, pp. 617–620 (2011)
11. Roskam, J., et al.: Airplane Aerodynamics and performance, Univ of Kansas (1980)
12. Fleeman, E.L.: Tactical Missile Design, 2nd edn. AIAA (2006)
13. Siouris, G.M.: Missile Guidance and Control Systems. Springer (2004)

Evaluating the Effectiveness of Shoot-Look-Shoot Tactics Using Discrete Event Modeling and Simulation

Kyung-Min Seo, Changbeom Choi, and Tag Gon Kim

School of EECS KAIST
373-1 Kusong-dong, Yusong-gu, Deajeon, Korea 305-701
{kmseo,cbchoi}@smslab.kaist.ac.kr,
tkim@ee.kaist.ac.kr

Abstract. Modeling and Simulation (M&S) methods of analysis play an increasingly significant role in planning military operations and determining the optimal use of available defense forces to ensure that national security is maintained. M&S engineering enables the procurement of military equipment to become more efficient. This paper presents a case study that applies DEVS-based M&S technology to develop a simulation of shoot-look-shoot tactics. The developed simulation allows us to conduct a statistical evaluation of shoot-look-shoot tactics overall, and to provide an assessment of tactics development. In order to analyze the system effectiveness, we performed extensive combat experiments by varying certain parameters, including tactics and weapon performance. The experimental results show how these factors influence the effectiveness of the proposed system.

1 Introduction

Modeling and Simulation (M&S) methods of analysis play an increasingly significant role in planning military operations and determining the optimal use of available defense forces to ensure that national security is maintained [1]. These methods constitute a rational approach to exploring the different alternatives for force deployments, and to determining the best strategies and tactics in actual combat engagement. For the sake of enhancing the development of weapon systems, Modeling and Simulation (M&S) engineering has been widely employed for requirement analysis, development, test and evaluation, and weapon system training. M&S Engineering is a particularly appropriate approach for weapon systems in vehicles used for underwater warfare, such as torpedoes, decoys, or submarines, because the factors related to those vehicles have become more complicated and diverse. Applying M&S technique, simulation-based acquisition and development facilitates good decision-making about equipment procurements and tactics development.

This paper presents a case study that applies DEVS-based M&S technology to develop a simulation of shoot-look-shoot tactics. The purpose of this paper is to introduce and develop simulation models for simplified battlefield situations. The developed simulation allows us to conduct a statistical evaluation of these shoot-look-shoot tactics overall, and also to provide an assessment of tactics development. In

J.-H. Kim et al. (Eds.): AsiaSim2011, PICT 4, pp. 352–360, 2012.

order to analyze the system's effectiveness, we performed extensive combat experiments by varying certain parameters, such as tactics and weapon performance. The experimental results show how the factors influence the effectiveness of the proposed system.

This paper is organized as follows. Section 2 presents the simulation model and DEVS formalism. Section 3 explains the design of the overall model, while section 4 illustrates the effectiveness analysis using the experimental results from the shoot-look-shoot model. Finally, section 5 offers a conclusion.

2 Related Work

First, we introduce simulation modeling for underwater warfare combat. Second, we briefly introduce the DEVS formalism that we apply to modeling the underwater warfare system in this paper.

2.1 Simulation Modeling

Modeling is a way of solving problems that are encountered in the real world. It is applied whenever prototyping or experimenting with a real system is prohibitively expensive, or indeed impossible. Modeling allows us to optimize systems prior to their implementation. Modeling includes the process of mapping the problem from the real world to its model, model analysis and optimization, and mapping the solution back to the real-world system. We can distinguish between analytical and simulation models [2]. In an analytical model, the result functionally depends on the input (a number of parameters); it is possible to implement such a model in a spreadsheet. However, analytical solutions do not always exist, or may be very hard to find. In those instances, simulation modeling may be applied. A simulation model may be considered as a set of rules (e.g. flowcharts, state machines, cellular automata) that define how the system being modeled will change in the future, given its present state. Simulation is the process of model 'execution' that takes the model through state changes (either discrete or continuous) over time. In general, for complex problems where the time dynamics are important, simulation modeling is the superior option. In this paper, we employed simulation modeling rather than analytical modeling.

2.2 DEVS Formalism

The DEVS formalism is a general formalism for discrete event system modeling based on set theory, and it is one of the M&S theories that are applied to various military simulations [3]. The DEVS formalism supports the specification of discrete event models in a hierarchical and modular manner. The DEVS formalism exhibits the concepts of system theory and modeling, and with this formalism, the user can model the target system by decomposing a large system into its smaller components and identify the coupling scheme among them. There are two kinds of models in the formalism: the atomic model and the coupled model.

The atomic model is a specification of basic model behavior as a timed state transition. Formally, an atomic model can be defined by 7-tuples as follows:

$$M = < X, Y, S, \delta_{ext}, \delta_{int}, \lambda, ta >,$$

Where

X: a set of input;

Y: a set of output events;

S: a set of sequential states;

δ_{ext}: $Q \times X \rightarrow S$, an external transition function, where $Q = \{(s,e) | s \in S, 0 \leq e \leq ta(s)\}$ is the total state set of M;

δ_{int} : $S \rightarrow S$, an internal transition function;

$\lambda : S \rightarrow Y$, an output function;

$ta : S \rightarrow Real$, time advance function.

A coupled model is a specification of a hierarchical model structure. It enables the method of assembling atomic and/or coupled models to build the hierarchy of a complex system. Formally, a coupled model is defined as follows:

$$DN = < X, Y, M, EIC, EOC, IC>,$$

Where

X: a set of input;

Y: a set of output events;

M: a set of all component models;

$EIC \subseteq DN.X \times \cup M.X$: external input coupling;

$EOC \subseteq \cup M.Y \times DN.Y$: external output coupling;

$IC \subseteq \cup M.Y \times \cup M.X$: internal coupling;

SELECT: $2^M - \phi \rightarrow M$: tie-breaking selector.

3 Model Design

The exiting model is designed based on the DEVS formalism. In Section 3, we describe in simple terms the shoot-look-shoot tactics and the overall model design.

3.1 Shoot Look Shoot Tactics

The shoot-look-shoot assignment tactics for the attacker are based on the assumption that the command and control for the battle management has the ability to assess the consequences of each shot, i.e., whether the target was missed or destroyed, and to assign successive shots only to the surviving targets. This means that the kill assessment must be perfect [4]. A typical engagement for shoot-look-shoot tactics is illustrated in Figure 1. In spite of the precision of modern weaponry systems, shots fired at targets do occasionally miss or cause only partial damage. Consequently, a sequence of more than one shot may be directed at a particular target to increase the probability of a kill [5].

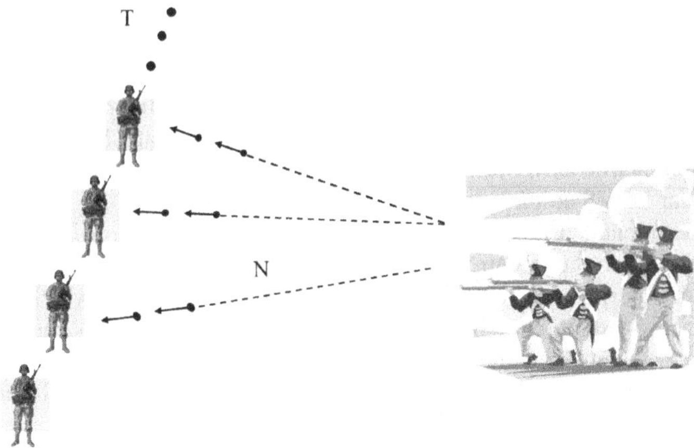

Fig. 1. Attacker's shoot-look-shoot assignments

3.2 Model Design

A model of a typical engagement for shoot-look-shoot tactics is illustrated in Figure 2. The entire simulation model consists of a system model and an experimental frame. The overall model architecture that has been developed is described in Figure 2, and consists of two main models: a simulation model for the user's system, and an experimental frame model for analysis of the user's system. The simulation model consists of blue force and red force models: a hit, a damage and a killManage model.

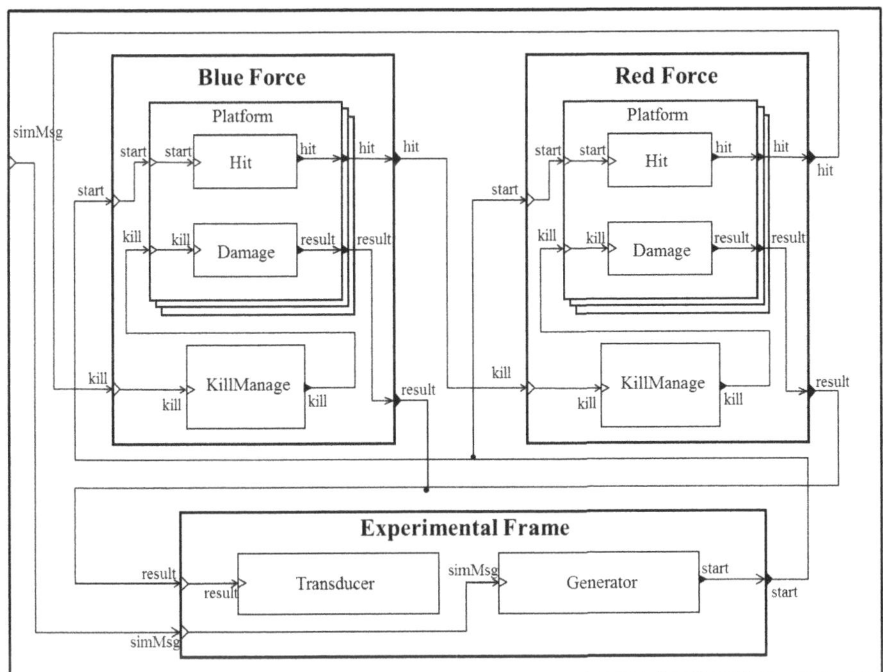

Fig. 2. Overall model of shoot-look-shoot tactics

The DEVS-coupled model describes the event message exchange relations among the DEVS atomic models. The following specification represents the DEVS-coupled model of the shoot-look-shoot model, and Figure 2 presents a diagram of the model.

Figure 3 and Figure 4 show the DEVS atomic model diagram for the maneuver model. In Figure **3**, the colored circle represents the initial state. Since the set-theoretic specification can be easily turned into the diagram, we will use this diagram to show the particular DEVS atomic model for this investigation.

An experimental frame typically has three components: a generator, which generates inputs to the system; an acceptor, which monitors an experiment to see that the desired experimental conditions are met; and a transducer, which observes and analyzes the system outputs. In practice, many experimental frames can be formulated for the same system. This means that we might have different objectives in modeling the same system. In this paper, we design a generator and transducer model, as illustrated in Figure 4.

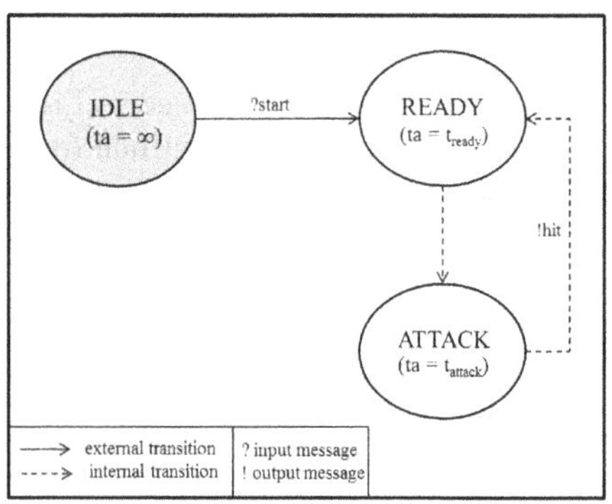

(a) Hit model

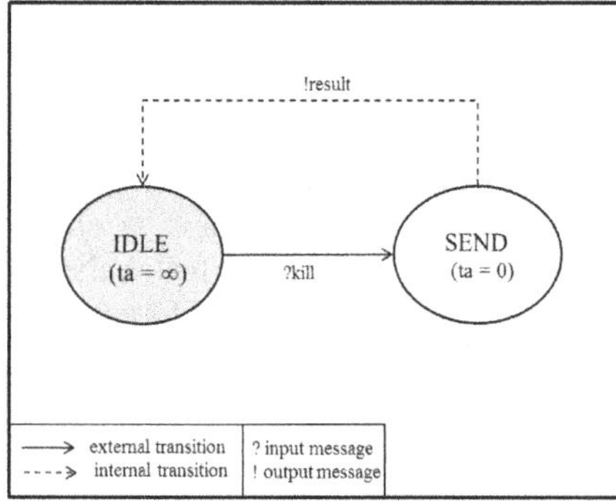

(b) Damage Model

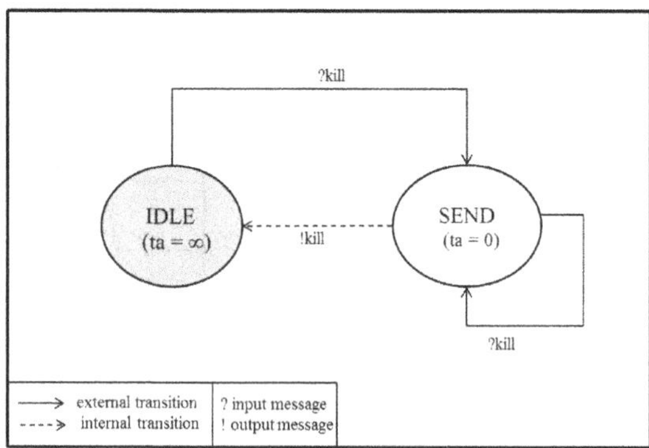

(c) HitManage Model

Fig. 3. Atomic model of system model (blue force)

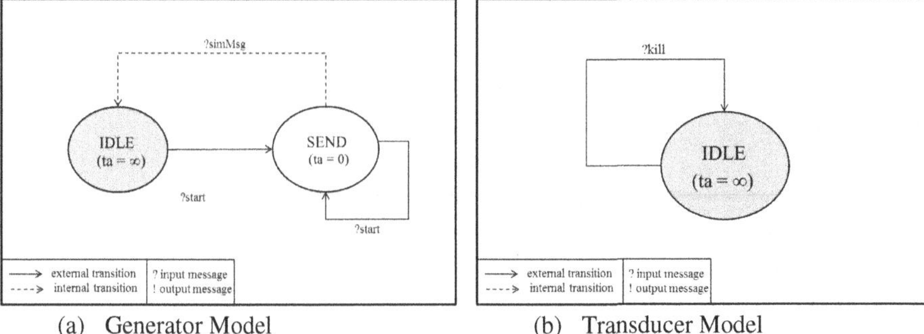

(a) Generator Model (b) Transducer Model

Fig. 4. Atomic model of experimental frame

A hit model assumes the role of attacking the enemy. The hit model transfers a 'hit' message to the hitManage model of the opponent's side. The hitManage model receives the message and decides whether its own model is killed or not. If its own model is killed, the hitManage model sends a 'kill' message to the damage model. The damage model receives the message and sends the result to the experimental frame.

4 Case Study – Shoot Look Shoot Tactics

All DEVS models and experimental frame models are implemented with the DEVSim++ library, which is a DEVS execution environment that uses C++ by SMSLab in KAIST [6]. This section illustrates the developed simulator, which is a shoot-look-shoot tactics simulator for effectiveness analysis. Analyzing the survivability of the blue force across diverse parameter values and tactics, we show how the simulator enables us to gather insights into equipment procurements and tactics development.

4.1 Model Simulation of Shoot-Look-Shoot Tactics

Now that we have completed the model implementation using C++, we will design the overall architecture of the shoot-look-shoot simulation.

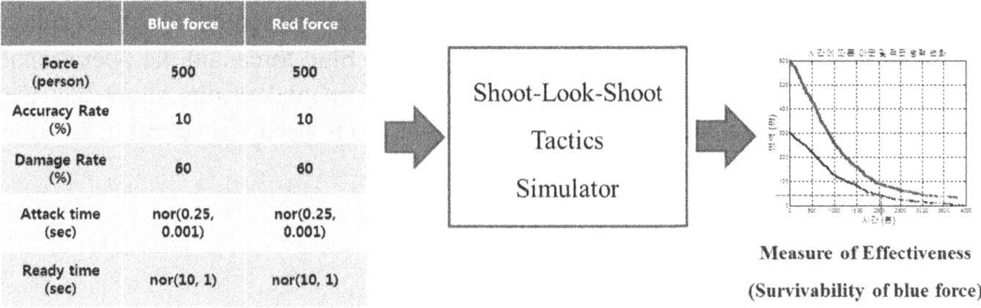

	Blue force	Red force
Force (person)	500	500
Accuracy Rate (%)	10	10
Damage Rate (%)	60	60
Attack time (sec)	nor(0.25, 0.001)	nor(0.25, 0.001)
Ready time (sec)	nor(10, 1)	nor(10, 1)

Fig. 5. High-level view of shoot-look-shoot system

Figure 5 presents the high-level view of the discrete event simulation for underwater warfare. The underwater warfare simulation is not a live simulation, but a constructive simulation. Generally, a constructive simulation has various simulation parameters.

Table 1. Scenario input parameters

Parameter name	Default value	Implications
Troops	500	Blue and Red force
Accuracy Rate	10%	The probability of accuracy of weapon system
Damage Rate	60%	The probability of damage of weapon system
Attack Time	0.25 sec	The time for attacking
Ready Time	10 sec	Reloading time

We identify these parameters as the scenario parameters and enumerate the input parameters in Table 1. As our goal is to evaluate the simulation's MOEs, such as the survivability of the blue force and the operational success rate for various situations, we will vary the input parameters in a later case study. In Table 1, "Troops" refers to the military strength of both armies, and "Accuracy Rate" of the weapon system refers to the percentage of target hits. "Attack Time" refers to the period in which the weapon flies toward the target, and finally, "Ready Time" refers to the time for reloading and aiming.

4.2 Experimental Results

We performed four experiments to analyze the MOE using this extended simulator. In each experiment, we varied the value of one input parameter, such as Troops, Accuracy Rate, Damage Rate, or Ready Time. We did not experiment by changing the Attack Time. The main purpose of these experiments was to evaluate the simulation's MOEs, such as the survivability of the blue force and the operational success rate in various situations.

As depicted in Table 2, we performed four experiments to analyzing the MOE using this extended simulator. The main purpose of these experiments was to evaluate the simulation's MOEs, such as the survivability of the blue force and the operational success rate in various situations. Basically, the experimental results show that the military strength of both sides and the gap between both strengths decrease as time goes on.

Table 2. Scenario parameters for experiments

Parameter name	Blue Force	Red Force
Troops (persons)	700	500
Accuracy Rate (%)	10	10
Damage Rate (%)	60	60
Attack Time (sec)	nor(0.25, 0.001)	nor(0.25, 0.001)
Ready Time (sec)	nor(10,1)	nor(10,1)

(a) Experiment 1

Parameter name	Blue Force	Red Force
Troops (persons)	500	500
Accuracy Rate (%)	20	10
Damage Rate (%)	60	60
Attack Time (sec)	nor(0.25, 0.001)	nor(0.25, 0.001)
Ready Time (sec)	nor(10,1)	nor(10,1)

(b) Experiment 2

Parameter name	Blue Force	Red Force
Troops (persons)	500	500
Accuracy Rate (%)	10	10
Damage Rate (%)	80	60
Attack Time (sec)	nor(0.25, 0.001)	nor(0.25, 0.001)
Ready Time (sec)	nor(10,1)	nor(10,1)

(c) Experiment 3

Parameter name	Blue Force	Red Force
Troops (persons)	500	500
Accuracy Rate (%)	10	10
Damage Rate (%)	60	60
Attack Time (sec)	nor(0.25, 0.001)	nor(0.25, 0.001)
Ready Time (sec)	nor(8,1)	nor(10,1)

(d) Experiment 4

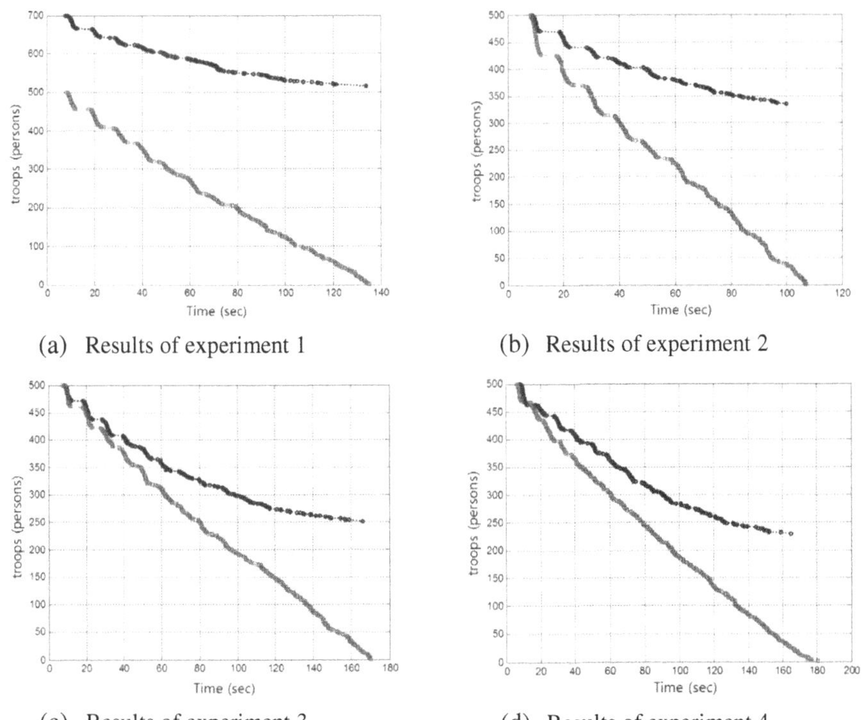

(a) Results of experiment 1

(b) Results of experiment 2

(c) Results of experiment 3

(d) Results of experiment 4

Fig. 6. Experimental Results

The results of experiment 1 show, basically, that having more troops is superior to having fewer troops. Experiments 2 and 3 show the troops that survive according to the performance of the weapon system, such as the accuracy rate and damage rate. As shown in (b) and (c) of Figure 6, when the weapon system performance of the blue side is superior to the opponent's force, more troops survive than in the opponent force as the battle progressed. Finally, experiment 4 shows that well-trained troops are superior to troops that have inadequate training, because the reloading time is short when troops are well trained.

5 Conclusion

This paper presents a case study that applies DEVS-based M&S technology to the development of a simulation of shoot-look-shoot tactics. The developed simulation allows us to perform a statistical evaluation of the overall shoot-look-shoot tactics, and also to perform an assessment of tactics development. Using the proposed simulation, we can determine how various factors, such as military strength, performance of weapons and physical training, influence effectiveness of the system. Experimental results support assessment of system effectiveness. It requires much effort in modeling in detail and developing various input parameters, yet that is out goal for future works.

Acknowledgements. This research is supported by Ministry of Culture, Sports and Tourism(MCST) and Korea Creative Content Agency(KOCCA) in the Culture Technology(CT) Research & Development Program 2009.

References

1. Seo, K.-M., Song, H.S., Kwon, S.J., Kim, T.G.: Measurement of Effectiveness for an Anti-torpedo Combat System Using a Discrete Event Systems Specification-based Underwater Warfare Simulator. The Journal of Defense Modeling and Simulation: Applications, Methodology, Technology (2011)
2. Borshchev, A., Filippov, A.: Form System Dynamics and Discrete Event to Practical Agent Based Modeling: Reasons, Techniques, Tools. In: Proceedings of The 22nd International Conference of the System Dynamics Society, England (July 2004)
3. Zeigler, B.P., Praehofer, H., Kim, T.G.: Theory of Modeling and Simulation, 2nd edn. Academic Press, London (2000)
4. Glazebrook, K., Washburn, A.: Shoot-Look-Shoot: A Review and Extension. Operations Research 52(3), 454–463 (2004)
5. Aviv, Y., Kress, M.: Evaluating the effectiveness of shoot-look-shoot tactics in the presence of incomplete damage information. Military Operation Research 3(1), 79–89 (1997)
6. Kim, T.G., Park, S.B.: The DEVS formalism: Hierarchical Modular systemsSpecification in C++. In: Proceedings European Simulation Multi-conference, York, United Kingdom, pp. 152–156 (1992)

Battle Space Model Based on Lattice Gas Automata for Underwater Warfare Simulation

Sol Ha[1], Namkug Ku[1], Kyu-Yeul Lee[2], and Young-In Nah[3]

[1] Department of Naval Architecture & Ocean Engineering,
Seoul National University, Seoul, Korea
hasol81@snu.ac.kr
[2] Department of Naval Architecture & Ocean Engineering and Research
Institute of Marine Systems Engineering, Seoul National University, Seoul, Korea
[3] Agency for Defense Development, Jinhae, Korea

Abstract. To simulate complex undersea engagement, many platforms, such as submarines and battleships, participate in underwater warfare simulation. To perform an underwater simulation with reasonable communication among the platforms and environmental factors, a middleware that can treat communication and environmental factors is needed. This paper presents the battle space model, which is capable of propagating various types of emissions from platforms in underwater warfare simulation, predicting interesting encounters between pairs of platforms, and managing environmental information. The battle space model has four components: the logger, spatial encounter predictor (SEP), propagator, and geographic information system (GIS) models. The logger model stores brief data on all the platforms in the simulation, and the GIS model stores and updates environmental factors such as temperature and current speed. The SEP model infers an encounter among the platforms in the simulation, and progresses the simulation to the time when this encounter will happen. The propagator model receives various emissions from platforms and propagates these to other "within-range" platforms by considering the propagation losses and delays. The battle space model is based on the discrete event system specification (DEVS) and the discrete time system specification (DTSS) formalisms. Especially, the propagator and GIS models are based on lattice gas automata for considering an underwater acoustic field and environmental space. To verify the battle space model, simple underwater warfare between a battleship and a submarine was simulated. The simulation results with the model were the same as the simulation results without the model.

1 Introduction

The sea and undersea environments in an engagement among platforms such as submarines and battleships have several components. A battlefield has some media (e.g., air and water) that influence platforms that participate in underwater warfare. For example, to identify the location of enemy submarines, sound signal parameters that pass through the seawater are determined. The sound signal intensity is reduced

J.-H. Kim et al. (Eds.): AsiaSim2011, PICT 4, pp. 361–376, 2012.

by the characteristics of seawater or is disturbed by the other platforms in the battlefield. Thus, the environment influences the interaction among the platforms. To perform underwater warfare simulation, all the platforms in the simulation should exchange information on a position and attitude, among others. To simulate underwater warfare, information on the environment must be stored, and the values of the factors that affect an acoustic signal must be calculated. This paper presents a battle space model that can manage information on the environment and platforms. Using such model, some advantages can be predicted, as follows: (1) information on the battlefield can be integrated and managed; (2) information on platforms, such as their positions and attitudes, can be managed efficiently; (3) platforms are independent of the environment and can be composed simply; and (4) the battle space model can predict interesting spatial relationships among platforms.

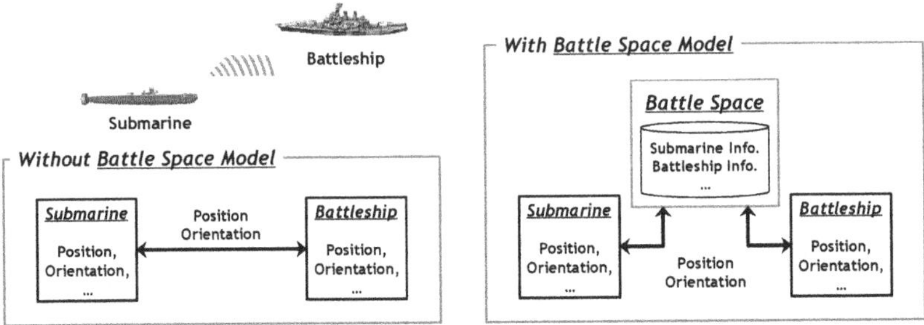

Fig. 1. Efficiency of the battle space model with two platforms

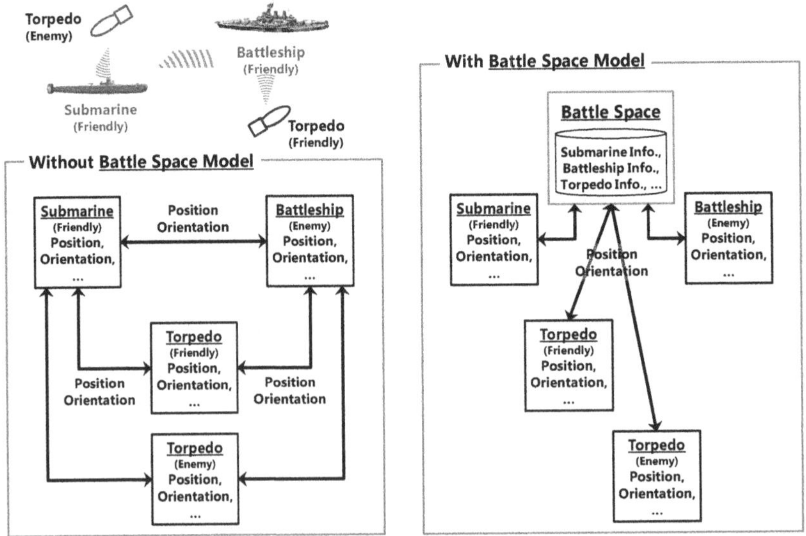

Fig. 2. Efficiency of the battle space model with four platforms

The rest of this paper is organized as follows. Section 2 presents the related works in the areas of the battle space model, lattice gas automata (LGA), and discrete event system specification (DEVS). Section 3 presents the concept of the battle space model. Section 4 shows the modeling of the battle space model based on LGA. In section 5, underwater warfare simulation is explained to verify the battle space model. Finally, section 7 presents some conclusions and future directions.

2 Related Works

2.1 Battle Space Model

Zeigler and Sarjoughian suggested the space model shown in Fig. 3 (Zeigler et al., 1999; Sarjoughian et al., 2000). The space model consists of the propagator, logger, spatial encounter predictor, and geographic information system models. Each component has a unique function, as follows:

- The propagator model calculates the propagation loss, which is a function of both the distance travelled and the nature of the environment.
- The logger model logs the platform's major information such as trajectory parameters, .
- The spatial encounter predictor (SEP) model predicts the interesting spatial relationships between the platforms, and furnishes such information to the platforms.
- The geographic information system (GIS) model contains the environmental information on the battlefield.

DEVS is a mathematical formalism for expressing discrete event models that support the discrete event approximation of continuous systems, and has an object-oriented substrate that supports model implementation and repository reuse (Zeigler, 2000). Based on the DEVS concept, the space model is defined as a coupled model. A coupled model in DEVS is a model that encapsulates other models.

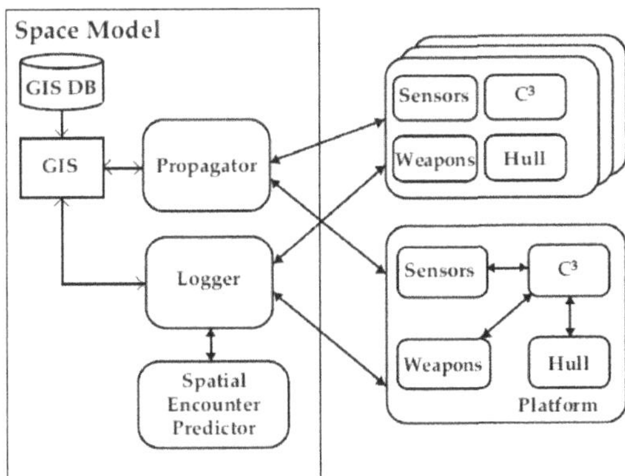

Fig. 3. Configuration of the space model (Zeigler et al., 1999; Sarjoughian et al., 2000)

By adding Selector and Environment, Chi and Jung suggest the modified space model shown in Fig. 4. Selector is a middleware that manages the data between the space model and a platform. Environment manages time-varying information, and GIS manages time-invariant information. All the components of the modified space model are defined based on DEVS.

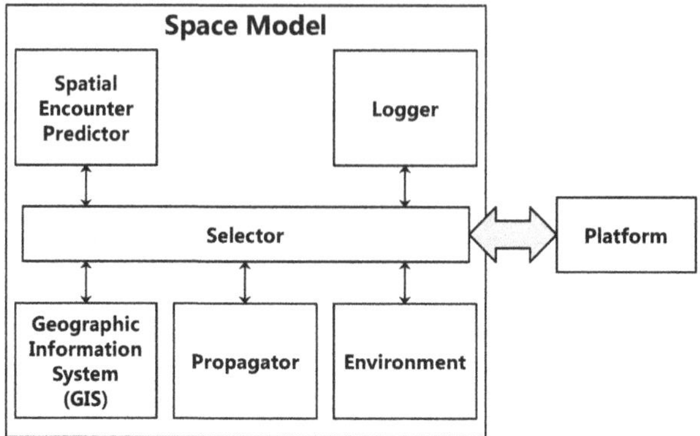

Fig. 4. Configuration of the modified space model (Chi et al., 2009; Jung et al., 2009)

2.2 Lattice Gas Automata and the Lattice Boltzmann Method

The purpose of LGA is to simulate the behavior and interaction of many single particles in a gas as simply as possible. LGAs can be seen as very simple molecular-dynamics methods. Macroscopic quantities such as particle density and velocity can be recovered from this microscopic scope, making it possible to study the macroscopic behavior of a fluid in different geometries with this model.

Gas is modeled as a multitude of hard spheres moving along a regular grid, with a discrete set of possible velocities $\vec{c}_i$ for each particle. Collision between particles is handled by a set of elastic collision rules that must conserve the system's quantities of mass m and momentum $\vec{p}$.

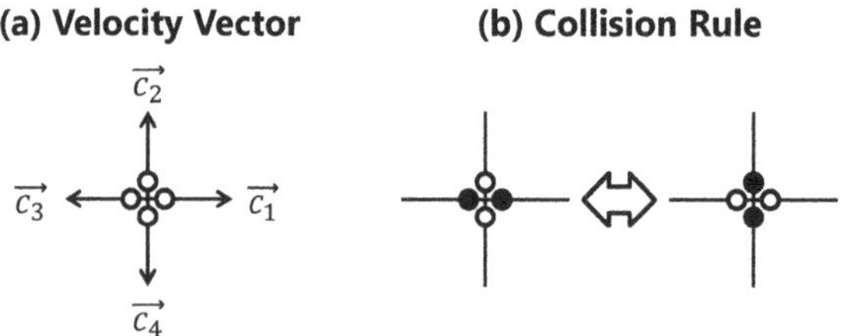

Fig. 5. (a) Velocity vectors of the HPP model. (b) Head-on collision rule for the HPP model.

The HPP model was proposed by Hardy, Pomeau, and De Pazzis (hence, the name of the model) as early as 1973 (Hardy et al., 1973). In the HPP model, the grid is two-dimensional and square, so that each node in the grid has four neighbors. The particles can have four possible velocities, $\vec{c_1} = (1,0)$, $\vec{c_2} = (0,1)$, $\vec{c_3} = (-1,0)$, and $\vec{c_4} = (0,-1)$, as shown in Fig. 5(a). For each time step, each particle is moved forward one step in the direction of its velocity. When two or more particles meet in the same node after a time step, a collision occurs. To conserve mass and momentum, the number of particles and the total velocity of all the particles in the node must be the same before and after the collision. When two particles collide head on, they are thrown out at right angles to their original velocities, as shown in Fig. 5(b). This conserves momentum, as the sum of the velocities of the two particles is zero in both configurations.

Possibly the greatest weakness of the HPP model is that it cannot achieve rotational invariance, which means that its behavior becomes anisotropic. As a result of this, HPP systems fail to behave in accordance with the Navier-Stokes equations. This weakness alone is crippling to the point that the HPP model is not useful for fluid simulations. The HPP model was abandoned for this reason in the late 1980s for the FHP model, which manifests an isotropic Navier-Stokes behavior by changing the shape of the lattice.

The FHP model was suggested by Frisch, Hasslacher, and Pomeau (hence, the name of the model) in 1986 (Frisch et al., 1986). It was shown that by moving from a square lattice to a hexagonal lattice, the incompressible Navier-Stokes equations could be recovered due to the extra rotational invariance afforded by the hexagonal lattice.

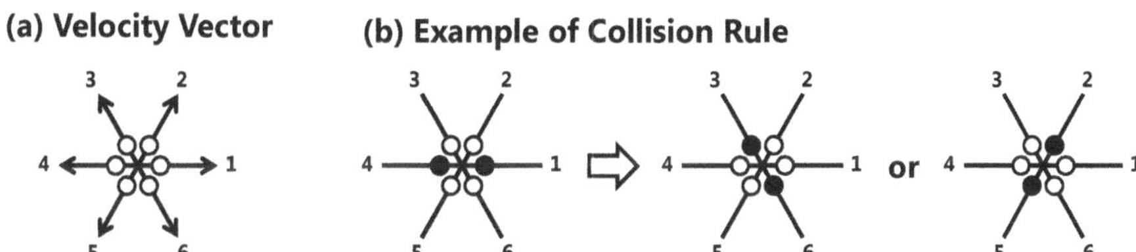

Fig. 6. (a) Velocity vectors of the FHP model. (b) Example of the head-on collision rule for the FHP model.

Lattice Boltzmann simulations were developed from the lattice gas approach as an attempt to overcome the noise problems. They can be regarded either as a "mean-field" lattice gas or as a slightly unusual finite-difference solver for the Navier–Stokes equations. Like the lattice gas, the lattice Boltzmann approach can be used to simulate streaming without having to calculate the forces produced by the wave. Once a sound wave is generated in the model, streaming should follow automatically. Haydock (2001) and Viggen (2009) applied the Lattice Boltzmann method to the acoustic fields.

2.3 Discrete Event System Specification Formalism

Praehofer and Zeigler proposed a modeling and simulation method that can handle simulation models of a discrete event and a discrete time (Praehofer, 1992; Zeigler et al., 2000). This method, called *discrete event system specifications (DEVS)/discrete time system specifications (DTSS) formalism,* is widely used as a standard for modeling and simulation. Using this model, a simulation engine changes the state of a model according to the events, and calculates the state variables of the model at each unit time when the model is in a specific state.

DEVS formalism consists of two kinds of models: the atomic and coupled models. The atomic model is the basic model and has specifications for the model's dynamics. The coupled model, on the other hand, provides the method of assembly for several atomic and/or coupled models to build a complex system hierarchy.

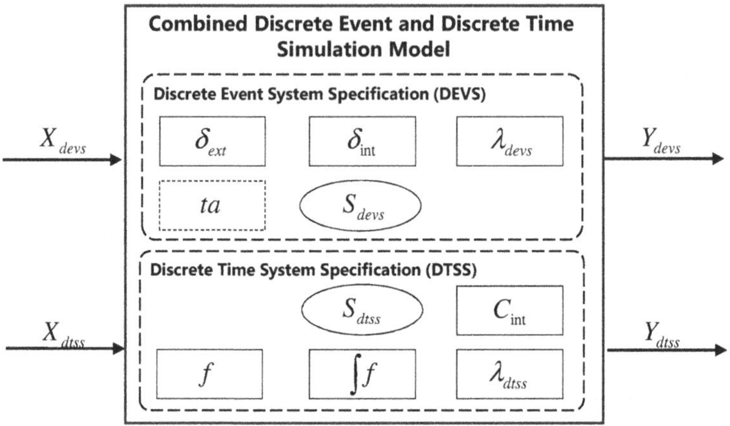

Fig. 7. Configuration of the combined DEVS/DTSS atomic model (Bang, 2006)

Bang and Cha developed a simulation framework based on the combined DEVS/DTSS concepts (Bang, 2006; Cha et al., 2009; Ha et al., 2009; Cha et al., 2010). To evaluate the efficiency and applicability of the simulation framework, they applied it to the block erection process in shipbuilding and underwater warfare simulation.

3 Battle Space Model Concept

In this paper, the battle space model is suggested. The proposed battle space model used the configuration of the space model suggested by Zeigler and Sarjoughian. All the components of the battle space model are defined based on the combined DEVS/DTSS formalism. To accurately model the characteristics of an undersea environment, the propagator and GIS are represented as LGA, and each node in the LGA is also defined based on the combined DEVS/DTSS formalism.

3.1 Configuration of the Battle Space Model

The platforms in Fig. 5 represent the systems that are to be evaluated. Platforms have a location in the world and can move their location. They have a spatial extent and can collide with other platforms. Platforms may actively or passively sense the world around them and are potentially detectable because they generate various kinds of emissions. They can launch subplatforms such as torpedoes. Multiple platforms can be instantiated in a given simulation. The platform model is a coupled model that contains subsystem models such as the command & control and the motion models.

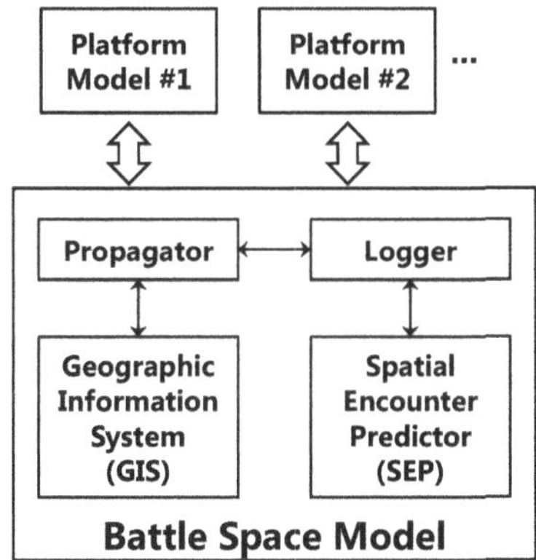

Fig. 8. Configuration of the battle space model

The battle space model helps simulate underwater warfare. First, it infers when interesting encounters will occur between pairs of moving platforms, and provides encounter information to the platforms. Second, it calculates the propagation losses and delays of various emissions from the platforms and propagates those to other "within-range" platforms. These functions are localized in the components of the battle space model, as illustrated in Fig. 6: the logger, SEP, GIS, and propagator models. All the models are defined based on the combined DEVS/DTSS concepts.

1) Logger model
The logger model manages the main information on platforms, such as their positions, attitudes, and detection radii. The platform is responsible for updating the logger model each time the platform's trajectory parameters change. The information is saved to the database. This information is also used to replay the simulation at the end. The logger model can provide the information to the propagator and SEP models.

2) SEP model
The battle space model predicts interesting spatial-relationship events and informs the various platforms so they will not unnecessarily attempt to interact. To make these predictions, the "trajectory" of a platform must be known by the battle space model. The SEP model currently knows the following kinds of trajectories. The SEP model can produce a number of spatial encounter predictions with what it knows about any two platform trajectories. These predictions currently include the time at which they will be at their closest point of approach (potentially a collision). The battle space model also predicts at what point in time, if at all, a platform will "run aground." Platforms subscribe to encounters of interest and are informed of these at the appropriate time.

When platforms effectively exploit this information, significant simulation efficiencies are achieved as a consequence of jumping over dead time intervals (when no platforms are interacting) and by limiting which platforms are time-stepping to those that are potentially interacting.

3) GIS model
The GIS model manages information on the environment, such as terrain and environmental information. The GIS database provides the terrain and environmental information to the GIS model. The GIS model provides the information to the propagator model and platforms.

4) Propagator model
The propagator model of the battle space model propagates emissions with their appropriate propagation losses, delays, and associated background noises. Platforms are limited to two forms of interaction: physical transfers (e.g., supplies or fuel) and emissions. Emissions cover pressure waves from explosions, acoustic noise (self-generated and reflected), and electromagnetic emissions (self-generated and reflected). The model allows emissions to carry semantic content as appropriate, as, for example, in radio communications.

Propagation loss (i.e., how much of the source signal strength remains when the signal arrives at the destination's location) is a function of both the distance traveled and the nature of the environment. Both factors (distance and environment) similarly influence the propagation delay (i.e., how long it takes for the signal to be transmitted from the source to the destination). Finally, noise (in which the signal is embedded) is a function not only of the environment but also of the other platforms in the region. These delays (especially in the acoustic realm) and losses can be significant.

4 Modeling of the Battle Space Model Based on Lattice Gas Automata

The proposed battle space model was developed based on the combined DEVS/DTSS concepts. The DEVS model in the battle space model receives data from platforms and sends encounter information, whereas the DTSS model updates environmental information and various emissions from platforms. The battle space model provides the user with simple platform model architecture and a simple connection among the platforms.

4.1 Overall Model Configuration

The overall configuration of the developed battle space model is shown in Fig 9. All the components of the battle space model are based on the combined DEVS/DTSS formalism. To express an undersea environment as a 3D space, the propagator and GIS models are represented as LGAs, and each node in LGA is defined as an atomic model. Each platform has only two connections with the battle space model: its own information and emissions.

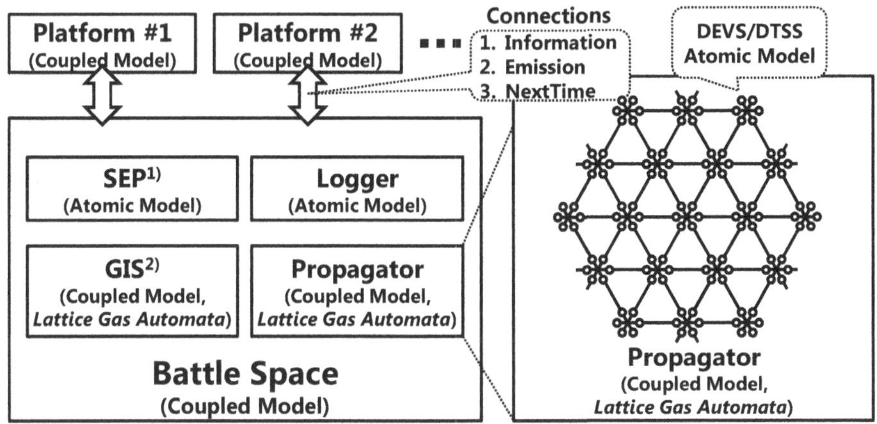

Fig. 9. Overall configuration of the battle space model

4.2 Propagator Model

4.2.1 Lattice Gas Automata for Acoustic Fields

The lattice gas automaton is based on a regular grid. Each node in the grid has several different variables associated with it $[f_i(\vec{x},t)]$. These variables represent the density of particles traveling in direction $\vec{c_i}$ at the node at $\vec{x}$ and time t. In the most commonly used two-dimensional lattice, there are six different velocity vectors $\vec{c_i}$, as shown in Fig. 6(a).

The macroscopic quantities of particle density and velocity at the node can be recovered from $f_i(\vec{x},t)$ by

$$\rho(\vec{x},t) = \sum f_i(\vec{x},t) \text{ and}$$

$$\rho(\vec{x},t)u(\vec{x},t) = \sum \vec{c_i} f_i(\vec{x},t),$$

where ρ is the particle density and $\vec{u}$ is the flow velocity. In the fluid nodes, the particles' collisions and propagations are handled according to

$$f_i(\vec{x}+\vec{c_i},t+1) = \left(1-\frac{1}{\tau}\right) f_i(\vec{x},t) + \frac{1}{\tau} f_i^{(0)}(\vec{x},t),$$

where τ is a time relaxation constant and $f_i^{(0)}$ is the equilibrium distribution for the particles. τ is related to the kinematic shear viscosity through

$$\nu = c_s^2\left(\tau-\frac{1}{2}\right),$$

where c_s is the speed of sound in the fluid, which is $1/\sqrt{3}$ in lattice units for the FHP lattice. Each node's equilibrium distribution of particles, $f_i^{(0)}$, is constructed from the macroscopic quantities of density and velocity at the node, and is given by

$$f_i^{(0)} = \rho t_i [1 + \frac{\vec{u}\cdot\vec{c_i}}{c_s^2} + \frac{(\vec{u}\cdot\vec{c_i})^2}{2c_s^4} - \frac{\vec{u}^2}{2c_s^2},$$

where t_i is a set of weights, one for each velocity vector.

4.2.2 Acoustic Point Source

Every platform has its own self-generated noise. This noise is propagated to other platforms with their appropriate propagation losses, delays, and associated background noises. A platform's self-noise was considered an acoustic point source. Viggen (2009) modeled an acoustic point source that sends out a sinusoidal signal. The simplest way to implement this is to lock a node's density to a sine function around the equilibrium density

$$\rho(\vec{x},t) = \rho_0 + \rho_s \sin\left(\frac{2\pi}{T}t\right),$$

where ρ_s is the point source amplitude and T is the period of the oscillation in lattice units. As wave propagation is a linear phenomenon and the Navier-Stokes equation can be assumed to be linear only for small disturbances, it is important that $\rho_s \ll \rho_0$ so that nonlinear wave effects can be avoided. Macroscopic velocity $\vec{u}$ at the point source is locked to a constant source velocity, typically $\vec{u} = 0$.

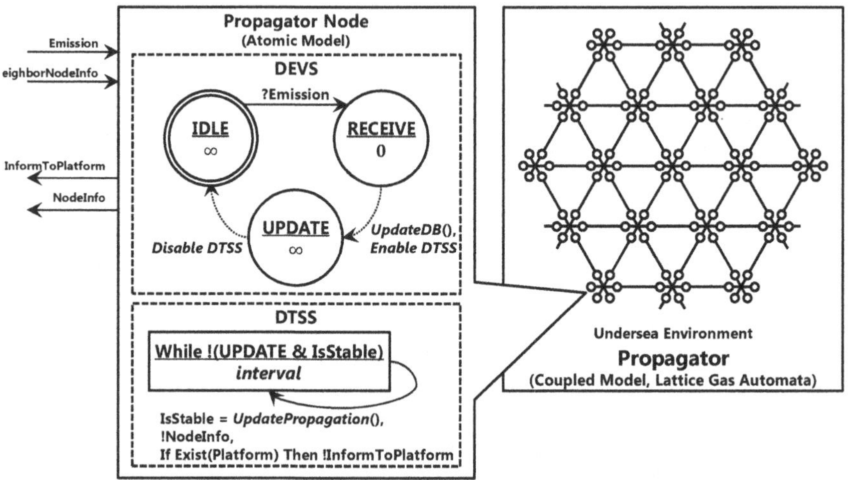

Fig. 10. Atomic-model diagram of each node in the propagator model

4.2.3 Modeling of the Propagator Model Based on the DEVS/DTSS Formalism

Fig. 10 shows the atomic-model diagram of each node in the propagator model. Each node updates its information through its rules for propagating and colliding at each unit time. Each node exchanges its own data with those of its six neighbor nodes.

4.3 Logger Model

The logger model stores each platform's information in the simulation. The major roles of the logger model are as follows:

- All platforms should send their information each time the trajectory changes.
- When the logger model receives information from the platforms, it should update the database immediately.
- If the database has some data at the same time, the data should be overwritten.

Fig. 11 shows the atomic-model diagram of the logger model. Here, the double circle means the initial state. If there is an input such as "Info" and "Emission," the logger model's state will change to the "UPDATE" state. Then the logger model will update the database with these input data and will return to the "IDLE" state.

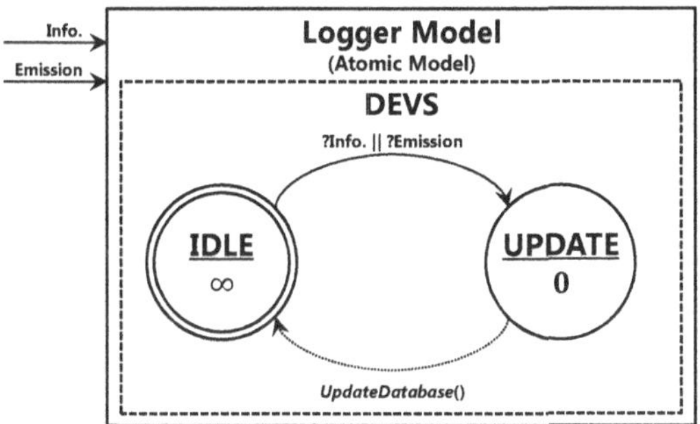

Fig. 11. Atomic-model diagram of the logger model

4.4 Spatial Encounter Predictor Model

The SEP model predicts interesting encounters between pairs of platforms. Fig. 12 shows the atomic-model diagram of the SEP model. The initial state of the SEP model is the "PREDICT" state. If there is no encounter among platforms in a simulation, the SEP model predicts the next encounter and progresses the time until the encounter occurs.

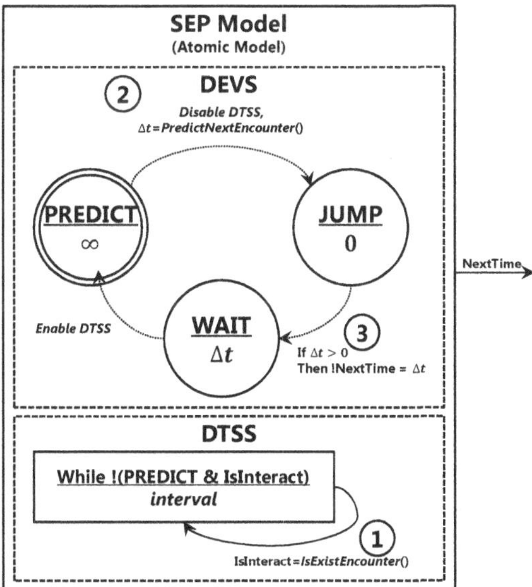

Fig. 12. Atomic-model diagram of the SEP model

4.5 Geographic Information System Model

Fig. 13 shows the atomic-model diagram of each node in the GIS model. Each node of LGA in the GIS model is defined as an atomic model. Each node updates its environmental factors (e.g., temperature and current speed) at each unit time. Each node exchanges its own data with those of its six neighbor nodes.

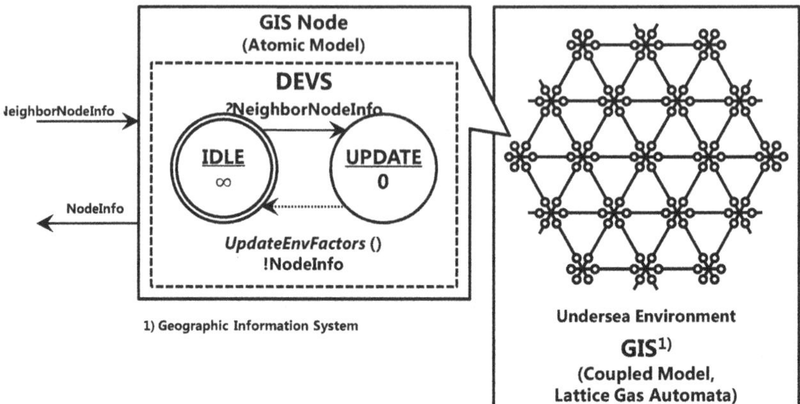

Fig. 13. Atomic-model diagram of the GIS model

5 Underwater Warfare Simulation Using the Battle Space Model

This section illustrates the underwater warfare simulation that was developed in this study to test the battle space model. Based on the combined DEVS/DTSS concepts, Ha (Ha et al., 2010) simulated underwater warfare between a battleship and a submarine. This simulation was reconfigured by adding the battle space model. The simulation results were compared with those of the previous work.

5.1 Scenario of Underwater Warfare Simulation

Mjelde studied the simulation of a torpedo engagement against a battleship (Mjelde, 1977). He concentrated on the relationship between a torpedo and a battleship. In the study, a battleship was selected as a target. The velocity of this battleship was 18-20 knots when cruising. When the submarine launches a torpedo, it conducts target motion analysis (TMA) using its sensor, and it keeps a certain distance from the target. Using equation (1), the firing angle (launching direction) can be calculated based on the target speed, direction, and torpedo speed. Fig. 14 shows the definition of *firing angle.*

$$\Psi_{DA} = \sin^{-1}\left(\frac{V_{TE}\sin\phi_{ASP}}{V_M}\right) \tag{1}$$

The launched torpedo steps into the search mode. Then it starts maneuvering with a set search pattern. The search pattern is usually snake search, a method that enlarges the search area via repetition.

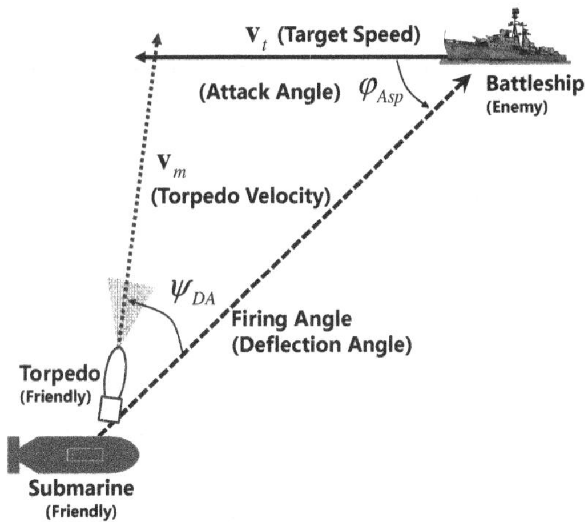

Fig. 14. Firing angle of the torpedo

In the search mode, if the torpedo succeeds in detecting a target, it steps into the attack mode and starts homing guidance. The attack mode is continued until the torpedo hits a target or loses its target. If the torpedo cannot find the detected target again, it regards its target as lost. In the attack mode, as the torpedo approaches a target, it steps into the final attack mode. In this mode, the speed and the course steering speed increase so that the torpedo can keep tracking the target and can hit it effectively. If the torpedo loses its target, it conducts a search with a re-attack search pattern, using circle search. If the torpedo loses a target in the attack mode, it goes back to snake search.

To simplify the problem, some assumptions were made in the simulation of this scenario. Such assumptions are as follows:

- Only surface targets are considered.
- The battleship remains on a steady course after firing a torpedo. As such, there is no anti-torpedo system in the battleship, and the battleship does not evade the torpedo.
- The target data estimated via TMA was used to solve the deflection angle problem.

5.2 Configuration of Models in Underwater Warfare Simulation

Fig. 9 shows the structure of the battle space model, which aims to simulate underwater warfare between a battleship and a submarine. The battleship, torpedo, and submarine models in Fig. are defined as the combined DEVS/DTSS models. To send a "firing" command to the torpedo, the battleship's position, orientation, and detection range are directly connected to the torpedo model. All the platforms are connected to the battle space model. Each platform updates the logger model anytime when its position, orientation, and detection range change. The platforms also exchange emissions with other platforms using the propagator model.

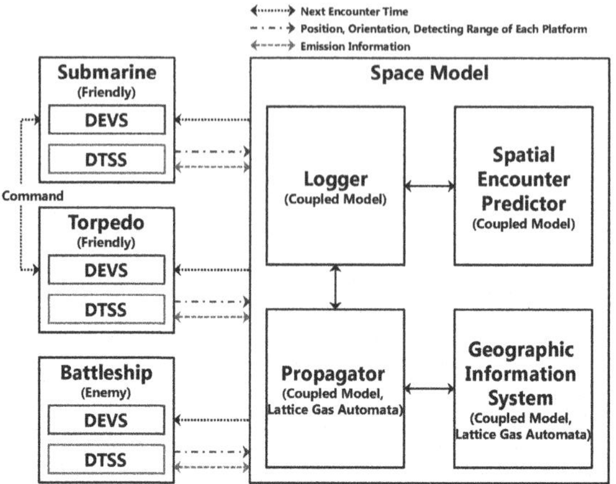

Fig. 15. Coupling of models in underwater warfare simulation

5.3 Simulation Results

Initial values were set for each platform model in Table 1, and the simulation was run. When the simulation ran, the logger model stored each platform's information each time, and the result was confirmed by drawing graphs.

Table 1. Initial values of models of underwater warfare simulation

Name	Value
Torpedo speed	40 knots
Search method	Snake search
Detection range	1,500 m
Ping repetition interval	2 sec
Beam width (transmission)	30 degrees
Beam width (reception)	40 degrees
Battleship speed	15 knots
Battleship dimensions	**80 × 7 × 7** m (**L × B × D**)
Firing distance	3,000 m
Turn rate	6 degrees/sec
Sweep angle	30 degrees
Simulation time step	0.05 sec

According to the assumptions in this simulation, the battleship has no anti-torpedo system and runs straight ahead at a given velocity. Fig. 10 shows that the battleship moved in a straight line from the starting location.

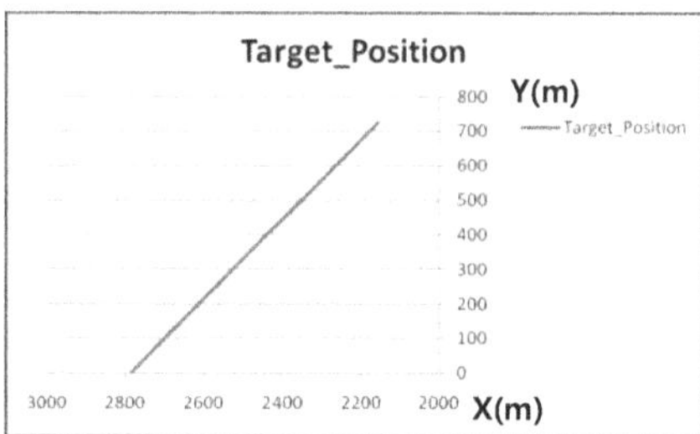

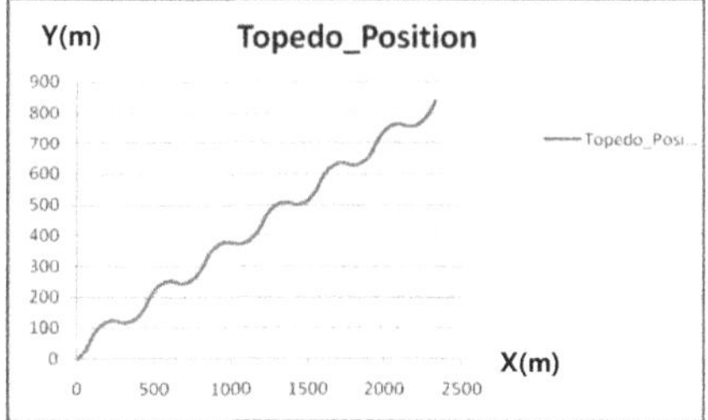

Fig. 16. Simulation results: left - battleship movement; right – torpedo movement

After the start of the simulation, the torpedo started maneuvering with a snake search pattern. This search pattern is a method that enlarges the search area by repeatedly searching at the right side and then the left side. Fig. 16 shows that the torpedo moved like a snake according to the sweep angle and the turn rate.

6 Conclusion

In this paper, the battle space model is proposed for managing information on the environment and platforms. The battle space model has two important functions. First, it can infer when encounters will occur among platforms, and can furnish such information to the platforms. Second, it can calculate the propagation losses and delays of various emissions from the platforms and can propagate them to other platforms within a range. The battle space model was designed using the combined discrete event/time system specifications formalism. To verify the battle space model, an underwater warfare simulation was conFigd with it. Using this simulation, the function of the battle space model was confirmed.

In this research, external environmental factors such as wind speed and ocean current were not considered. Therefore, the results of this study are not applicable to the correct handling of exact active or passive sensors. The latter requires much effort in modeling and developing multiple platforms and sensors as well as environmental factors. Nonetheless, that is the authors' goal in future studies.

Acknowledgement. This work was partially supported by SM-11 ("A Study on the Network-based Architecture of a Virtual System for the Simulation of Underwater Vehicles") of the Underwater Vehicle Research Center (Republic of Korea), "Measuring the Effectiveness of the Next-Generation Torpedo Using the Modeling & Simulation Framework" funded by the Agency for Defense Development (Republic of Korea), the Industrial Strategic Technology Development Program (10035331, Simulation-based Manufacturing Technology for Ships and Offshore Plants) funded by the Ministry of Knowledge Economy (MKE, Republic of Korea), the Brain Korea 21 program of the Marine Technology Education and Research Center of Seoul National University, the Research Institute of Marine Systems Engineering of Seoul National University, and the Engineering Research Institute of Seoul National University.

References

1. Bang, K.W.: Combined Discrete Event and Discrete Time Simulation Framework for Shipbuilding Process Planning. Master Thesis, Seoul National University, Seoul, Korea (2006)
2. Cha, J.H., Lee, K.Y., Ham, S.H., Roh, M.I., Park, K.P., Suh, H.W.: Discrete Event/Discrete Time Simulation of Block Erection by a Floating Crane Based on Multibody System Dynamics. In: 19th International Offshore and Polar Engineering Conference (ISOPE), vol. 3, pp. 678–685. Osaka International Convention Center, Osaka (2009)
3. Cha, J.H., Roh, M.I., Lee, K.Y.: Combined discrete event and discrete time simulation framework for the improvement of shipbuilding process planning. Journal of the Korean Society for Simulation 17(4), 71–80 (2008)
4. Chi, S.D., You, Y.J., Jung, C.H., Lee, J.S., Kim, J.I.: FAMES - fully agent-based modeling & emergent simulation. In: Proceedings of the 2009 Spring Simulation Multiconference, pp. 1–8 (2009)
5. Frisch, U., Hasslacher, B., Pomeau, Y.: Lattice-Gas Automata for the Navier-Stokes Equation. Physical Review Letters 56(14), 1505–1508 (1986)
6. Ha, S., Cha, J.H., Lee, K.Y.: Analysis of Detecting Effectiveness of a Homing Torpedo using Combined Discrete Event & Discrete Time Simulation Model Architecture. Journal of the Korean Society for Simulation 19(2), 17–28 (2010)
7. Hardy, J., Pomeau, J., de Pazzis, O.: Time Evolution of a Two-Dimensional Classical Lattice System. Physical Review Letters 31(5), 276–279 (1973)
8. Hardy, J., Pomeau, J., de Pazzis, O.: Time evolution of a two-dimensional model system. I. Invariant states and time correlation functions. Journal of Mathematical Physics 14(12), 1746–1759 (1973)
9. Haydock, D., Yeomans, J.M.: Lattice Boltzmann simulations of acoustic streaming. Journal of Physics A: Mathematical and General 34(25), 5201 (2001)
10. Jung, C.H., You, Y.J., Ryu, H.E., Lee, J.S., Kim, J.I., Chi, S.D.: Multi-Platform Warship M&S System Using the Hierarchical Multi-Agent System. Journal of the Korean Society for Simulation 18(4), 117–125
11. Mjelde, A.: A Homing Torpedo. The Effect of the Tactical Situation and the Torpedo Parameters on the Torpedo Effectiveness, Naval Postgraduate School Monterey Calif (1977)
12. Praehofer, H.: System theoretic foundations for combined discrete-continuous system simulation. VWGO (1992)
13. Viggen, E.M.: The Lattice Boltzmann Method with Applications in Acoustics. Master Thesis. Department of Physics - NTNU Norway (2009)
14. Zeigler, B.P., Hall, S.B., Sarjoughian, H.S.: Exploiting HLA and DEVS to promote interoperability and reuse in lockheed's corporate environmentp. Simulation 73(5), 288 (1999)
15. Zeigler, B.P., Praehofer, H., Kim, T.G.: Theory of modeling and simulation. Academic Press, New York (2000)

Stereoscopic 3D Model Viewer with Improved Depth Perception for Battlefield Visualization

Jiyoung Park[2], Hye-Jin Park[1], and Myoung-Hee Kim[1,2,*]

[1] Dept. of Computer Science & Engineering,
Ewha Womans University,
11-1 Daehyundong, Seodaemungu, Seoul, Korea
{lemie,clrara}@ewhain.net

[2] Center for Computer Graphics and Virtual Reality,
Ewha Womans University
11-1 Daehyundong, Seodaemungu, Seoul, Korea
mhkim@ewha.ac.kr

Abstract. We developed a stereoscopic 3D model viewer and constructed a prototype visualization system to provide an overview of the battlefield as well as the detailed view of a user-selected entity. For the overview of the battlefield, a battlefield visualization software, the SIMDIS, is used. The user is allowed to select an entity on the overview screen; the detailed view of this entity is displayed in the 3D model viewer. In order to provide the user appropriate depth perception under the condition of user movement, the distance of the screen from the user, which is one of the stereo parameters used in the 3D model viewer, is required to be set correctly. In order to update this parameter while the user is moving, we employed a depth camera and applied a real-time head tracking method to the acquired depth image frames. In the implemented system, we used a 2D display and a 3D TV to show the overview of the battlefield by the SIMDIS and the stereoscopic 3D model viewer, respectively.

1 Introduction

Advances in digital modeling and simulation have fostered the development of complicated structures with minimal design time and implementation cost [1]. In military applications, the use of constructive battlefield simulation software packages for effective strategic decision-making has increased as opposed to the use of physically simulated situations, which are dangerous and expensive. In particular, the emphasis on battlefield visualization and virtual-reality techniques is constantly gaining attention because these techniques offer the following advantages. They are capable of developing a realistic 3D model of the battlespace, and they impart a feeling of immersion to help commanders monitor the battlespace better evaluate the simulated operation in a more intuitive way.

[*] Corresponding author.

J.-H. Kim et al. (Eds.): AsiaSim2011, PICT 4, pp. 377–384, 2012.

In this work, we developed a virtual-reality-based battlefield visualization system by using a combination of two different visualization tools for obtaining different perspectives and different levels of detail. We used a battlefield visualization software, SIMDIS [2], for providing an overview of the battlefield, and we developed a stereoscopic 3D model viewer for a detailed model of the entity selected by the user in the SIMDIS view. This work contributes to the development of a separate view that shows the detailed entity model. This detailed model cannot usually be used in the battlefield overview for achieving a better performance. Moreover, our stereoscopic 3D model viewer improves depth perception by setting appropriate stereo parameters in real time. These parameters are set by employing user head tracking using a depth camera.

The aim of the proposed method is to enhance a common battlefield visualization tool by integrating it with a separate visualization tool for obtaining a detailed view. Therefore, the user can analyze and monitor the battlefield effectively by intuitively switching between the two views, each having a different level of detail. In a simulation scenario, this functionality is very useful for the user in monitoring the damage status and its detailed location.

The rest of the paper is organized as follows. Section 2 offers a brief review of a few examples of commercial constructive simulation software, with focus on the visualization of the battlefield and the entity model (e.g., battleship, submarine, torpedo, etc.). Section 3 presents an overview of the proposed system. Section 4 presents the four main components of our system and assignment of appropriate values to the stereo parameters. Section 5 discusses the implementation of the system and the obtained results. Finally, in Section 6, we conclude the paper by presenting the potential benefits and pitfalls of the proposed system and by outlining the imminent areas of research. We also provide a tentative direction for future work.

2 Related Work

Several battlefield simulation systems and solutions are currently available, including commercial off-the-shelf software packages such as VR-Forces (MAK) [3] and STAGE (Presagis) [4], and their capabilities are improving at a phenomenal rate. In particular, the advent of novel computer graphics and display technology has encouraged the development of a variety of 3D battlefield visualization tools. In general, battlefield simulation products are available in the form of software packages composed of several software modules, which can be substituted with other tools or sources. As an example for illustrating visualization, the 2D visualization result is provided to the user along with a military symbology; however, the user is allowed to run the simulation scenario and send the scenario data to a 3D visualization tool, which results in better situational awareness.

Apart from large-scale software packages, 2D/3D visualization-specific tools are gaining prevalence. These tools support various types of external simulation data sources, and they provide large-scale and detailed virtual scenes in real time by using a high-performance rendering engine. The Virtual Reality Scene Generator (VRSG) [5] from MetaVR, TView [6] from Tiltan, and the Dragon [7] and the SIMDIS from the Naval Research Laboratory are typical examples of these visualization tools. Several of these tools support stereoscopic 3D visualization, so the user can connect a 3D display to the tools, and the user can monitor the battlespace with improved depth perception by wearing special glasses. These glasses are not required for autostereoscopic displays. However, all the above-mentioned visualization tools have a limitation in that they do not show detailed information views of simulation entities in a simulation scenario.

In the present battlefield visualization, because the overview of the battlespace takes priority over the detailed view of each entity, the entity information is usually presented as simple text or as a simple 2D image sequence. However, the modeling of entities and weapon systems is becoming more complex, and several stand-alone simulation models are increasingly migrating to large-scale simulation techniques. Therefore, the detailed 3D visualization of a geometric model and the status of a certain entity in a battlefield situation are constantly gaining the attention of researchers. In the case of damaged status and ship sinking situations, the advent of engineering-level simulation techniques has yielded realistic visualization results at high frame rates [8, 9]. In addition, several efforts need to be directed toward the suggestion of visualization frameworks for the abovementioned simulation techniques.

Depth perception is a very important factor governing the effectiveness and usability of a virtual environment, and several real-time eye or head tracking techniques [10, 11] have been presented to calculate stereo parameters and generate appropriate stereoscopic images.

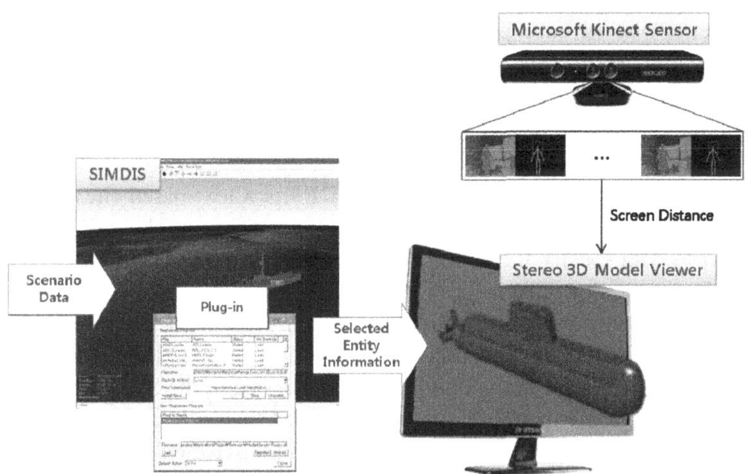

Fig. 1. System overview

3 System Overview

Our system overview is shown in Figure 1. The system is composed of the SIMDIS, the stereoscopic 3D model viewer, a plug-in to connect the viewer to the SIMDIS, and a head tracking software module of the Microsoft Kinect sensor [12].

The operation of the system starts with loading the SIMDIS with scenario data that are to be visualized. The user can play and stop the scenario by using the SIMDIS GUI, and the user is allowed to click on an entity for getting a more detailed view. Then, the underlying plug-in captures the click event and generates a message, which is sent to the stereoscopic 3D model viewer. The viewer displays the selected entity model in a stereoscopic mode. We have two 3D visualization models which differ in terms of the level of detail shown in the visualization. We use a simplified model for the SIMDIS, and we use a model having more details such as the inner structure of the entity for the model viewer. To reproduce the sense of movement parallax for a single viewer, we employ head tracking. In the current prototype, we use only the distance from the screen and extend it to 6DOF pose in the future work. The head tracking module obtains the depth image frames from the Kinect sensor and computes the user's head position in each frame in real time. Then, we calculate the distance from the display to the head position and send it to the viewer via a network connection. The viewer receives the screen distance value and uses it to update the corresponding stereo parameter.

4 Autostereoscopic 3D Visualization of Target Entity

In this section, we provide a detailed explanation of the four main components of our visualization system. First, we need to prepare the input scenario data and the 3D models to be visualized by the SIMDIS. Second, we need to implement the plug-in, which records the user input and connects the SIMDIS to the model viewer. Third, we need to set up the stereoscopic model viewer to visualize the selected entity model; finally, we implement the head tracking module with a depth camera.

4.1 Scenario and Model Preparation

The scenario data can be loaded to the SIMDIS as a text-based file, or they can be uploaded from an external data source connected via a network or through a plug-in. The scenario data specify the scenario entities and their time, space, and position, as well as the orientation information. All the 3D models of the scenario entities are required to exist in a predefined directory so that these models can be loaded to the SIMDIS. Generally, the models used in the SIMDIS are simplified for improved performance, and the file format is only applicable for use in the SIMDIS. Therefore, we generate a separate 3D model that is more detailed than that used in the SIMDIS, so the 3D model enables the user to navigate through the selected entity.

4.2 Plug-In Implementation

Once the user clicks on an entity in the SIMDIS view, the plug-in obtains the user ID from the entity under the cursor, and retrieves its model name. Then, the plug-in sends a message to our stereoscopic 3D model viewer, which includes the model name. The plug-in is implemented by using the SIMDIS Plug-in API.

Fig. 2. Screen image of the stereoscopic 3D model viewer (anaglyph stereo mode)

4.3 Stereoscopic Model Viewer

The main functionality of our stereoscopic model viewer is to show the detailed 3D model of the selected entity in the SIMDIS view. Figure 2 shows a screen image of the model viewer. The user can rotate the view clockwise or counterclockwise and perform zoom-in and zoom-out operations on the view with a mouse and keyboard. The model viewer is implemented using the OpenSceneGraph (OSG) toolkit [13] on which the SIMDIS is based. Because the OSG toolkit supports various formats for stereoscopic 3D outputs, it is possible to run our developed viewer on common passive and active stereoscopic 3D displays, as well as on autostereoscopic displays. We used an autostereoscopic 3D display, Pavonine Dimen G170A, which supports four different stereoscopic image formats—side-by-side, top-bottom, interlaced (field sequential), and frame sequential. We used the frame sequential mode and quad-buffered stereo mode for the display and viewer, respectively.

A stereoscopic image is generated by capturing two images from two different perspectives, one from the left eye and the other from the right. The images are captured from the view that would be seen by a person located inside the virtual scene. In order to obtain appropriate depth perception, we write the stereo code and assign appropriate parameter values to separate left and right eye views. Both the frustum and the model view are shifted to account for the separate eye views. Extreme parameter values that are either too large or too small would give most people a headache.

To achieve the right amount of adjustment, the OSG requires the initialization of the parameters—the user eye separation, the distance from the eyes to the screen, and the height of the screen [13]. The OSG uses default values of 0.05 m, 0.5 m, and 0.26 m for the user eye separation, the distance from the eyes to the screen, and the height of the screen, respectively. These values are assumed to be reasonable defaults for most workstation configurations. In general, we measure these values and set them up via the environmental variables, and then, the views obtained ideally provide improved depth perception. However, the current used display of 47 inches is very large so that it happens often that the user moves further than the default value. Therefore, it is not appropriate to leave the parameter default with no regard to user movements.

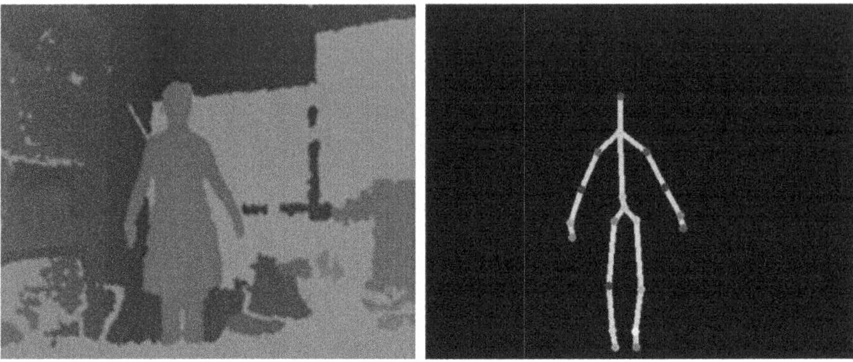

Fig. 3. Example of a depth image (left) from the Kinect sensor and its skeletal tracking result (right)

4.4 Updating of Stereo Parameters Using a Depth Camera

In order to obtain the best stereo effects for a moving user, we track the user's head position and update stereo parameters such as the screen distance in real time. We use the Microsoft Kinect sensor and its software development kit (SDK) for head tracking. The Kinect sensor has a built-in depth camera that acquires depth data; these data indicate the distance of the object from the Kinect sensor. The image frames of the depth map are 640 × 480, 320 × 240, or 80 × 60 pixels in size. Each pixel represents the Cartesian distance (in millimeters) from the camera plane to the nearest object at that particular Cartesian coordinate (x, y). The Kinect SDK enables the acquisition of depth data and skeletal tracking of the images of one or two persons, who are within the Kinect sensor's field of view. The Kinect SDK provides information about the location of up to two players standing in front of the Kinect sensor, with detailed position and orientation information. An example of a depth image and the result of the skeletal tracking of the image are shown in Figure 3. The data are provided to the application code as a set of points that compose a skeleton. Among the other skeleton joint positions, we only use the head position because we need to calculate the distance from the display to the user's head.

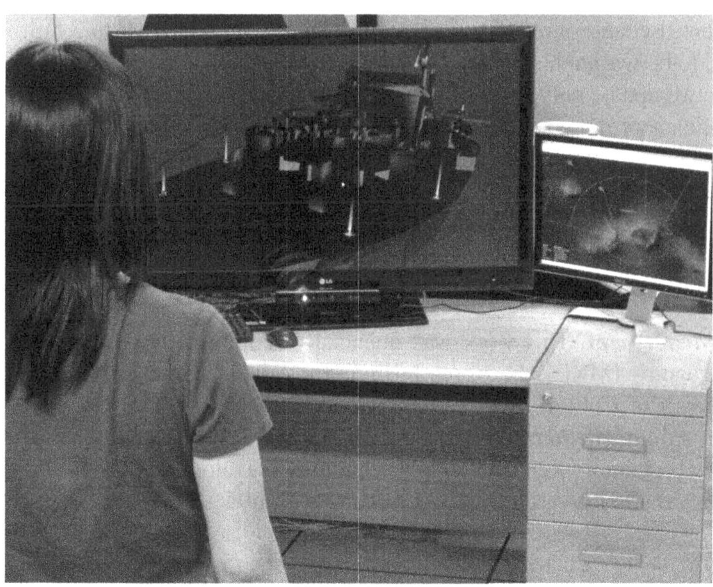

Fig. 4. Stereoscopic 3D model viewer in use. SIMDIS provides an overview of the battlefield. The 3D model viewer shows the detailed view of an entity in a separate display.

5 Result

Our visualization system is shown in Figure 4. We used two different displays: a 2D monitor for the SIMDIS view and a 3D TV for the stereoscopic 3D model viewer. The user can see the stereoscopic images by wearing shutter glasses. The SIMDIS and the model viewer run on a standard workstation with Intel dual CPUs running at 3.2 GHz, and having a RAM of 4 GB. The workstation runs on the Windows XP Professional operating system and is equipped with an NVIDIA graphics board. Because the Kinect sensor and the SDK require a development environment based on Windows 7, we used a separate PC for connecting to the Kinect sensor and for running the skeleton tracking software. The resolution of the acquired depth image was 320 × 240 pixels. A user can interact with the SIMDIS and the model viewer by using a mouse and keyboard. The skeletal tracking module sends the head position at around 30 frames per second, and the entire system operates in real time.

6 Conclusion

We developed a stereoscopic 3D model viewer and constructed a prototype to simultaneously visualize the overview of the battlefield and the detailed view of a certain entity selected by the user. A depth camera and a real-time head tracking module were employed to update the screen distance, which is one of the stereo parameters used in the model viewer, to provide appropriate depth perception under the condition of user movement. Because the depth range of the Kinect sensor is 1.2 to 3.5 m, the user is required to be within this range.

Since we update only the distance from the display, the user movements in the other directions are not considered for providing movement parallax. For making future improvements, we intend to calculate 6DOF pose of the user and apply it to updating views. Currently, the developed viewer can visualize a single static 3D model. However, to visualize dynamic models such as those showing the detailed damage status in the physical surface of the entity, we need to implement additional functionalities such as the time synchronization of the model with the SIMDIS.

Acknowledgement. This work was financially supported by the Ministry of Knowledge Economy (MKE), the Korea Institute for Advancement of Technology (KIAT), the Ministry of Culture, Sports and Tourism (MCST), and the Defense Acquisition Program Administration (DAPA) through the Human Resource Training Project for Strategic Technology, the Strategic Technology Development Program, and the Underwater Vehicle Research Center (UVRC) Program.

References

1. Anderson, P., Kenny, T., Ibrahim, S.: The Role of Emerging Visualization Technologies in Delivering Competitive Market Advantage. In: Proceedings of the 2nd International Conference on Total Vehicle Technology (2002)
2. The SIMDIS, https://simdis.nrl.navy.mil
3. MAK VR-Forces, http://www.mak.com
4. Presagis STAGE, http://www.presagis.com
5. MetaVR Inc., http://www.metavr.com
6. Tiltan Systems Engineering Ltd., http://www.tiltan-se.co.il
7. Durbin, J., Swan, J. E., Colbert II, B., Crowe, J., King, R., King, T., Scannell, C., Wartell, Z., Welsh, T.: Battlefield Visualization on the Responsive Workbench. In: Proceedings of the Conference on Visualization 1998, pp. 463–466 (1998)
8. Rosenblum, L., Durbin, J., Obeysekare, U., Sibert, L., Tate, D., Templeman, J., Agrawal, J., Fasulo, D., Meyer, T., Newton, G., Shalev, A., King, T.: Shipboard VR: From Damage Control to Design. IEEE Comput. Graph. Appl. 16(6), 10–13 (1996)
9. Varela, J.M., Soares, C.G.: A Virtual Environment for Decision Support in Ship Damage Control. IEEE Comput. Graph. Appl. 27(4), 58–69 (2007)
10. Su, C.-H., Chen, Y.-S., Hung, Y.-P., Chen, C.-S., Chen, J.-H.: A Real-time Robust Eye Tracking System for Autostereoscopic Displays Using Stereo Cameras. In: Proceedings of ICRA 2003, pp. 1677–1681 (2003)
11. Andiel, M., Hentschke, S., Elle, T., Fuchs, E.: Eye-tracking for Autostereoscopic Displays Using Web Cams. In: Proceedings of SPIE, vol. 4660, pp. 200–206 (2002)
12. Microsoft Kinect, http://www.xbox.com/en-US/kinect
13. OpenSceneGraph, http://www.openscenegraph.org

Effectiveness Analysis of Anti-torpedo Warfare Simulation for Evaluating Mix Strategies of Decoys and Jammers

Se Jung Kwon, Kyung-Min Seo, Byeong-soo Kim, and Tag Gon Kim

Department of Electrical Engineering, KAIST
291 Daehak-ro, Yusong-gu, Deajeon, Korea 305-701
{sjkwon,kmseo,bskim}@smslab.kaist.ac.kr,
tkim@ee.kaist.ac.kr

Abstract. Modeling and Simulation (M&S) engineering has been widely used for design and evaluation of underwater warfare systems. M&S engineering enables more efficient procurement of military equipment by analyzing these systems with simulations. This paper extends previous work, which used an anti-torpedo simulator with only decoys. To facilitate more diverse simulations of anti-torpedo combat, we design and add a simple jammer model, because the jammer is one of the most effective counter-measures against torpedo attacks. Utilizing this proposed model, we collect experimental data about the survivability of surface-ships that use anti-torpedo strategies and analyze the proper parameter values of jammers that satisfy the required Measure of Effectiveness (MOE). The experimental results show both the required performance of jammers and the efficiency of jammers with decoys. The results can be utilized to support the decision-making process for future equipment procurement.

1 Introduction

For the sake of developing weapon systems effectively, Modeling and Simulation (M&S) engineering has been widely employed for requirement analysis, development, test and evaluation, and weapon system training. M&S Engineering is an especially suitable approach for the weapon systems of vehicles designed for underwater warfare, such as a torpedo, a decoy or a submarine, because the factors of those vehicles have become more complicated and diversified. Applying M&S technique, Simulation-Based Acquisition (SBA) and development facilitate decisions about equipment procurements and tactics developments.

In past years, there have been some efforts to develop the underwater warfare simulation for academic purposes as well as commercial tools. For example some simulation tools, such as ODIN[1] and ORBIS[2], were developed for commercial uses, and some papers about simulation tools were published for academic purposes [3][4].

To contribute to these diverse efforts, this paper uses an anti-torpedo warfare simulator for developing the weapon systems of a surface ship [5]. This simulator is designed for an effectiveness analysis of weapon systems and uses Discrete EVent

J.-H. Kim et al. (Eds.): AsiaSim2011, PICT 4, pp. 385–393, 2012.

Specification (DEVS) formalism. By virtue of DEVS, the simulator can be extended with less cost, and the models of this simulator are reusable for various models.

In addition to the merits of DEVS formalism, it applies the collaborative modeling methodology [6]. This methodology decomposes the model into sub-models from a top-down perspective and partitions the sub-models into the two modeling levels of the object: discrete event-model (DEM) and detailed object model (OM). In the simulator, each platform is described as abstract DEM and has detailed OMs, which are algorithms for decision-making or calculation that involve domain-specific knowledge. Therefore, two different experts can develop their own models collaboratively.

This paper features extended content for evaluating mix strategies of decoys and jammers, whereas the existing simulator dealt with tactics of only decoys for counter-measures against a torpedo. The present study concentrates on jammers, as they are one of the most effective anti-torpedo counter-measures. Jammers generate air bubbles or noise to prevent the sonar of torpedoes from detecting our forces. While the jammer confuses the sonar of the opponents, the surface-ship can evade the opponents. In this paper, we introduce our design of a simple jammer model and revision of the simulator for mix strategies. We analyze the required parameter values of our jammer two-dimensionally for the sake of supporting the decision-making process for future equipment procurement.

This paper is organized as follows. Section 2 presents several of the related works and DEVS formalism. Section 3 explains the overall design and its additional components, and section 4 illustrates the effectiveness analysis with the experimental results from the anti-torpedo simulator with a jammer. Finally, section 5 concludes this paper.

2 Related Work

We firstly introduce surveyed papers about simulation tools of underwater warfare combat. Secondly, we briefly introduce the aspect of DEVS formalism that we apply to our modeling of the underwater warfare system in this paper.

2.1 Previous Research

In recent years, some efforts have been made to develop M&S techniques for underwater warfare. Among these efforts are two papers that deal with evaluating mix strategies of anti-torpedo combat including jammers.

One of them is by Cho et al. [3]. They developed all simulation models for an anti-torpedo simulation of submarines that uses DEVS formalism. By executing the simulation models in DEVSim++ [7], they examined maximum speed, acceleration and countermeasure systems capabilities. However, this research presents a disadvantage in that it is difficult to apply to other tests of candidate tactics of underwater platforms due to the low reusability of their study's models. This research is the most similar to the present paper with the difference that Cho et al. focused on submarine evasion, not surface-ship evasion.

Another study similar to our own is Liang and Wang's study [4]. This study included candidate tactics of underwater platforms. Unlike our study, Liang and Wang did not base their study on mathematical formalism. Consequently, complicated model development or functional extension of existing models was restricted in their study, and the cost of these works was expensive.

2.2 DEVS Formalism

The DEVS formalism, which is a set of theoretic formalisms, specifies a discrete event system in a hierarchical and modular form [8]. The DEVS formalism provides the framework for modeling, and its framework has several modeling advantages, such as completeness, verifiability, extensibility, reusability and maintainability. In addition to these advantages, the formalism is known to be compatible with the object-oriented view. The modeler can specify a system easily by decomposing a large system into smaller component models.

The DEVS formalism consists of two types of models, coupled models and atomic models. The coupled model provides the method of assembly of several atomic and/or coupled models, and the atomic model describes the basic behavior of models.

3 Model Design

The existing model is designed based on DEVS formalism. This paper adds a simple jammer model to the existing overall model. In this section, we provide an overview of model design and describe the model of the jammer as well as revisions of other models.

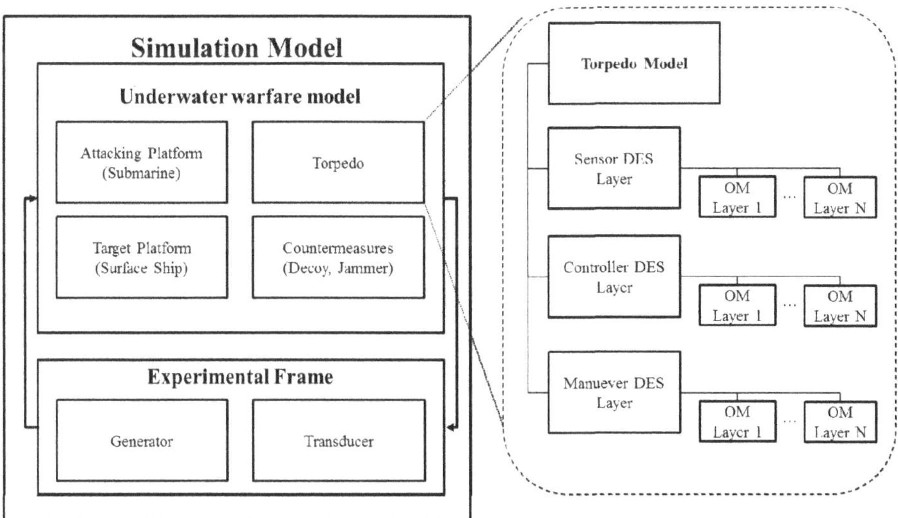

Fig. 1. Overall model design and an example model (a torpedo model) [5]

3.1 Overall Model Design

The existing model design is depicted in Figure 1. The whole simulation model consists of an underwater warfare model and an experimental frame; the underwater warfare model simulates the combat systems and the experimental frame is a controller of the underwater warfare model for analysis.

Each underwater warfare model consists of 3 DES layers: Sensor, Controller and Maneuver. Each layer has object models for detailed behaviors. As we mentioned, this simulation model offers a high degree of flexibility and reusability on account of these model structures.

3.2 Design of Added Component

A jammer is the underwater device that makes underwater noise or air bubbles for masking the sound of underwater vehicles during the performance of evasion maneuvers [8][9]. We design this jammer as a simple DEVS model to offer a new countermeasure component and to extend the objective strategies of the previous model. The behavior of our jammer model is simple. After the jammer is fired and is settled on the proper position, it sends jamming signals to a sensor model of the torpedo. As a result, the sensor of the torpedo cannot detect the target while its sensor is being attacked by the jamming signal.

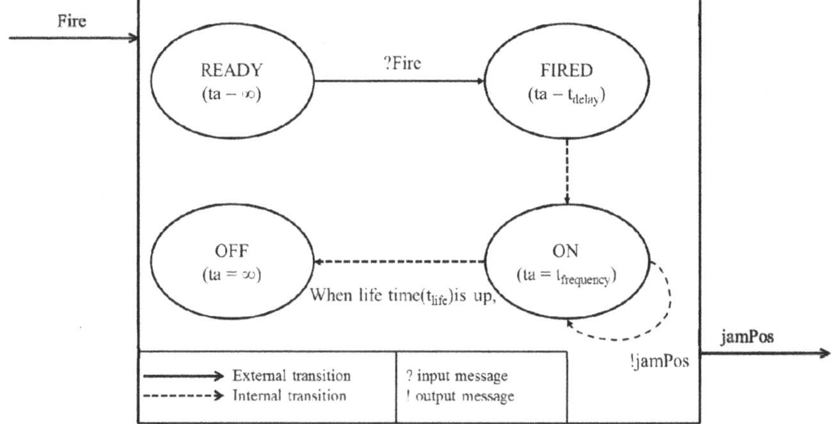

Fig. 2. Design of the jammer model (t_{delay} is the time of the firing delay. $t_{frequency}$ is the interval time required to send output messages. t_{life} is the total lifetime of the jammer).

Figure 2 demonstrates that the new jammer model has a simple design. In its initial state, 'READY', the controller of the surface ship orders 'Fire'. Then, the jammer is launched to a proper position on the surface of the water during t_{delay}. The next state after 'FIRED' is the 'ON' state. During the 'ON' state, the jammer transmits the jamming signals to the 'jamPos' port periodically. When the jammer's lifetime is up, the jammer seizes up.

It is the torpedo model's role to decide whether the jamming signal is effective or not. After receiving messages regarding the positions of the jammer, the torpedo's sensor makes a decision by using some parameter values. If the signals turn out to be effective signals, the torpedo's sensor ignores the renewal of position message that comes from the other underwater warfare model.

There are two object models in the jammer model. One model organizes the decision process regarding the effectiveness of the jamming signals. Effectiveness is determined only according to the distance between the jammer and the torpedo with no consideration of the particular frequency band involved. The other model shows how to launch the jammer in a proper position, and the position of the jammer is simply fixed to a short distance in front of the surface-ship.

The processes in this paper are simple but they are performed in a detailed object model, not in the DEVS model. When a more complex jammer is needed, these functions of the jammer are readily changeable.

4 Case Study - Anti-torpedo Warfare Simulator

All DES layers and experimental frame models are implemented with the DEVSim++ library, which is a DEVS execution environment that uses C++ by SMSLab in KAIST [7]. The rest of the simulation models, which are the object models representing detailed functions, are implanted in C++ language. The shared library technique ensures the effectiveness of these functions.

This section illustrates the developed simulator, which is an extended anti-torpedo simulator for effectiveness analysis that has a new jammer model. By analyzing the survivability of the surface-ship on diverse parameter values and tactics, we show how we gather insights for equipment procurements and tactics development by the simulator.

4.1 Torpedo Engagement

The attacking platform is a submarine and the target platform is a surface ship. The latter uses jammers and decoys for counter-measures. The brief scenario is illustrated in Figure 3. The surface-ship launches counter-measures (decoys and jammers) according to stored strategies against the torpedo's possible paths.

The objective of the experiment is to measure the survivability of the surface-ship as our new jammer model's MOE (Measure of Effectiveness). By changing the parameter of the jammer repeatedly, we analyze the survivability of the ship to support the decision making process for future equipment procurements.

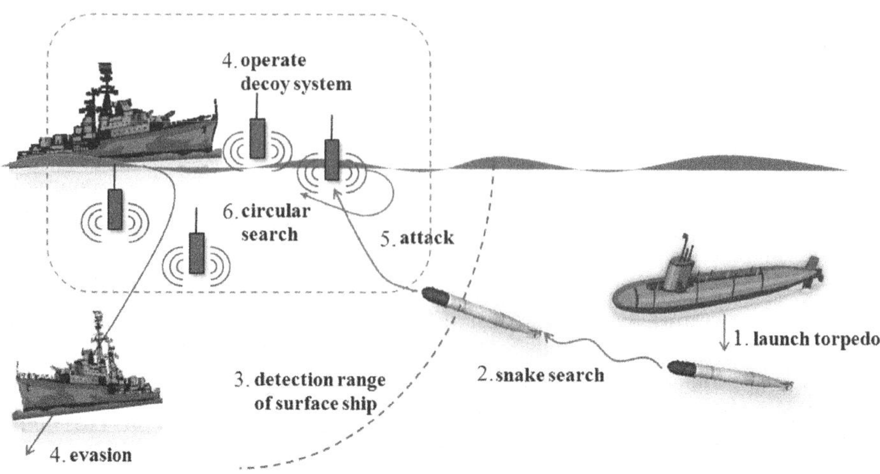

Fig. 3. Brief scenario of anti-torpedo warfare

Table 1 lists the major parameters for the jammer model in this paper. Parameters of other simulation models are listed in our previous paper [5].

Table 1. Major parameters for the jammer model

Parameter name	Default value	Implications
Lifetime	250s	Time period from launch to expiration
Jammed range	2500m	The range of the jamming effect
Reliability	90%	The probability of working normally

4.2 Experimental Results

We performed two experiments to analyze MOE using this extended simulator. The main purpose of these experiments is to discover the parameter pairs that satisfy required MOE. At first, we compare the survivability of the extended simulator to experimental results of our previous work. Choosing one example strategy from this result, we executed the simulator repeatedly by changing the parameters of the jammer model.

Table 2. Four patterns of the decoy operating system from the previous work [5]

Pattern#	Implications
1	4 static decoys (2 decoys at the front of surface ship and the others at the rear)
2	4 mobile decoys (2 decoys at the front of surface ship and the others at the rear)
3	2 static decoys at the front of surface-ship and 2 mobile decoys at rear
4	2 mobile decoys at the front of surface-ship and 2 static decoys at rear

The first experiment tests four patterns of counter-measures by using decoys. Table 2 shows detailed information of the patterns. We add a jammer to these four patterns and conduct 100 simulations of each mix strategy.

Table 3 shows the experimental results about the first experiment, comparing survivability of extended simulator to previous work.

Table 3. Results from the first experiment

Pattern #	1	2	3	4
Only decoys	7	32	46	3
Decoys with jammer A Lifetime = 200s, Jamming Range = 1500m	10	61	70	59
Decoys with jammer B Lifetime = 300s, Jamming Range = 2500m	91	78	81	78

The first row shows the results of the strategies that have only decoys. The rest of the table shows the results of the strategies with an added jammer. A comparison of the first row with the others shows that the survivability of the surface-ship is better with new counter-measure, the jammer. The increases in the survivability of the first pattern and the fourth pattern are the most dramatic. As a result, we can conclude that a jammer is a good option for anti-torpedo counter-measures in patterns 1 and 4.

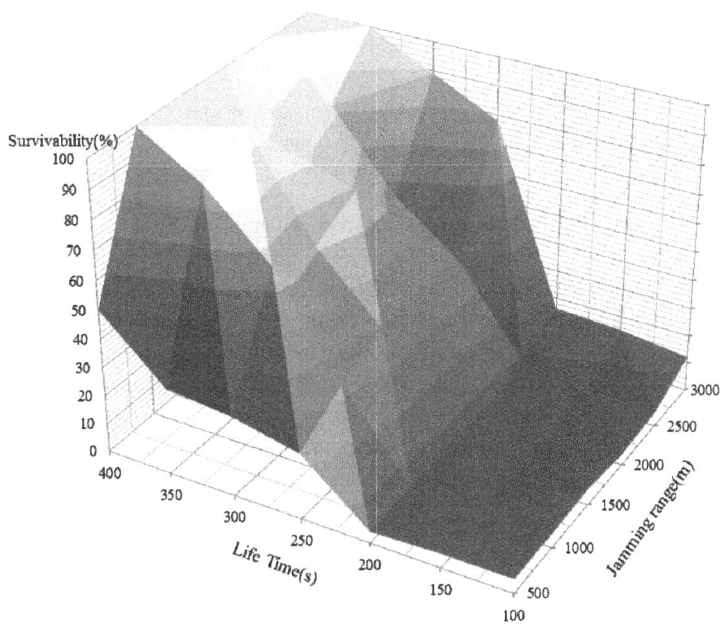

Fig. 4. Experimental result of survivability of the first pattern with a jammer

From the above result, we choose one example strategy to collect data for procurements. We chose pattern 1, because the survivability of pattern 1 increased considerably whenever pattern 1 was given an additional jammer. Figure 4 is a 3-dimensional graph that represents the survivability of pattern 1 with a jammer according to lifetime and jamming range.

This graph shows that survivability becomes better as the values of the parameters increase. Since one purpose of this experiment is to determine the parameters that satisfy the required MOE, we redraw this graph 2-dimensionally from the top of this 3d graph, and we set up 80% as the required MOE.

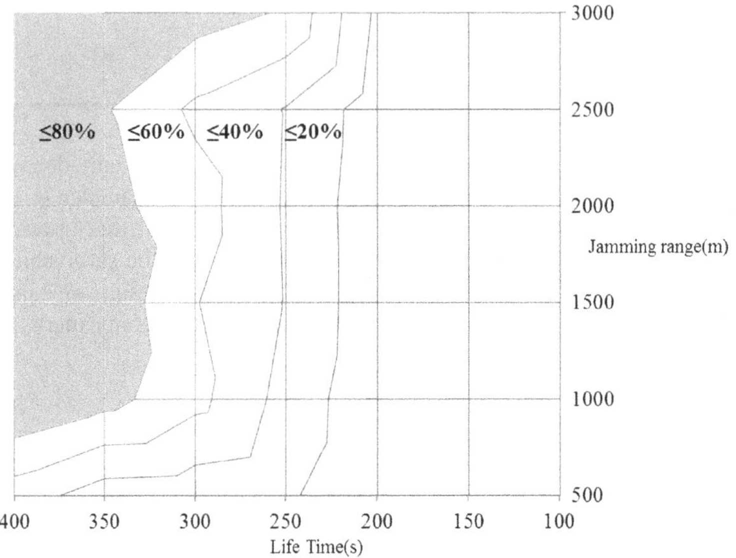

Fig. 5. Two-dimensional redrawn graph of Fig. 4

The gray territory of Figure 5 satisfies the required MOE. Thus, we can gather proper pairs of parameters that satisfy the required MOE: (300, 3000), (350, 1500 ~ 3000), (400, 1000 ~ 4000). In addition, we can obtain one more insight: when the lifetime of a jammer is over 350, the strategy of the counter-measure is effective regardless of the jamming range.

If experiments for other patterns or new patterns with different values are conducted, we could gain the proper parameter values from the above analysis.

5 Conclusion

This paper proposes an anti-torpedo simulator with a simple jammer model to analyze the MOE of more diverse counter-measures. We have successfully gathered knowledge supporting the procurement process of jammers and can continue to do so through analyzing survivability with this simulator.

Although an evident weakness of this simulator is that our jammer model is extremely simple and not based on real specifications, military experts can nevertheless apply this simulator and analysis methodology to real jammer models, because our jammer model offers a detailed and changeable object model. Future work should consider the addition of more complex jammer models.

Acknowledgement. This research is supported by Ministry of Culture, Sports and Tourism(MCST) and Korea Creative Content Agency(KOCCA) in the Culture Technology(CT) Research & Development Program 2009.

References

1. Robinson, T.: ODIN – an underwater warfare simulation environment. In: Proceedings of the 2001 Winter Simulation Conference, pp. 672–679 (2001)
2. Higgins, T.M., Turriff, A.E., Patrone, D.M.: Simulation-based undersea warfare assessment. Johns Hopkins APL Technical Digest 23(4), 396–402 (2002)
3. Cho, D.-Y., Son, M.-J., Lee, K.-Y.: Analysis of a submarine's evasive capability against an antisubmarine warfare torpedo using DEVS modeling and simulation. In: Proceedings of the Spring Simulation Multi-conference, pp. 307–315 (2007)
4. Liang, K.-H., Wang, K.-M.: Using simulation and evolutionary algorithms to evaluate the design of mix strategies of decoy and jammers in anti-torpedo tactics. In: Proceedings of the 2006 Winter Simulation Conference, pp. 1299–1306 (2006)
5. Seo, K.-M., Song, H.S., Kwon, S.J., Kim, T.G.: Measurement of effectiveness for an anti-torpedo combat system using a discrete event systems specification-based underwater warfare simulator. The Journal of Defense Modeling and Simulation: Applications, Methodology, Technology (2011)
6. Sung, C.H., Kim, T.G.: Collaborative modeling process for the development of domain-specific discrete event simulation systems. IEEE Transactions of System, Man, And Cybernetics Part C: Applications & Reviews (under submission)
7. Kim, T.G., Park, S.B.: The DEVS formalism: hierarchical modular systems specification in C++. In: Proceedings from the European Simulation Multi-conference, York, United Kingdom, pp. 152–156 (1992)
8. Zeigler, B.P., Praehofer, H., Kim, T.G.: Theory of modeling and simulation, 2nd edn. Academic Press, London (2000)
9. Hill, F.K., et al.: Underwater accoustal jamming apparatus, United States Patent, No. 3799094 (1974)
10. Watts, A.J.: Jane's underwater warfare system, 11th edn. Jane's Information Group Limited, UK (1999-2000)

The Analysis Method of Command and Control Measures of Effectiveness (C2MOE)

SeungHun Cha[1], DongJun Lee[1], JaeHo Na[1], ByungIn Cho[2], JongHyeon Jin[2], DongKuk Ryu[2], and HyungGyoon Kim[2]

[1] Constructive Model Team, M&S Research & Development Center
702 Sampyeong-dong, Bundang-gu,
Seongnam-city, Gyeonggi-do, South Korea 463-400
{seunghun.cha,dongjun.lee,jaeho.na}@lignex1.com
[2] Agency for Defense Development, Songpa P.O. Box 132,
Seoul-city, 138-600 South Korea

Abstract. Presently as development of information technology, combat systems are transformed from traditional PCW(Platform Centric Warfare) to NCW(Network Centric Warfare). In this regard several analysis framework and methods of Command and Control Measures Of Effectiveness (C2MOE) have been discussed and advanced. The analytical and conceptual frameworks are presented but how to analyze and be applied to MOE are not present clearly. The objective of this paper is to show new analysis framework of C2MOE, experiment procedure to apply items of MOE to simulator, how to decide items of MOE. An analysis method and a analysis framework presented in this paper are used and applied to C4ISR or other systems also.

1 Introduction

Earlier many researcher [1,2,3,4,5,6,7,8,9,10] have attempted to resolve and define various issues for Measures Of Effectiveness of C2 system. These issues are becoming important as systems are complex and networked together so called Network Centric Warfare Environment.

The combat system has been changed from individual combat system based PCW(Platform Centric Warfare) to complex combat systems based NCW(Network Centric Warfare) concept that has the advantages of information superiority, improved accuracy of decision making, Conducting speed increment by sharing information, knowledge and precision strike.

C2 MOE analysis for PCW is to evaluate the physical characteristics of the single platform and to summate the evaluation results of each platform. But it is difficult that the method of C2 MOE analysis for PCW is applied to networked system, that is, NCW because C2 in complex systems closely is related to and reflected by communication, sensors, and shooters. Thus the needs of new evaluation methods in terms of qualitative and quantitative measures of C2 in complex systems have emerged.

J.-H. Kim et al. (Eds.): AsiaSim2011, PICT 4, pp. 394–402, 2012.

Generally C2 MOE is to analyze the results obtained by shooters, that is, direct effectiveness of, such as enemy damage probability or hit probability is extracted from shooting results in which system delay time is reflected. [9]

In other words, destruction effectiveness is classified as direct effectiveness by PGM in C4I, ISR and PGM such as complex system or system of systems, system delay time, target analysis, decision of shooting specification, target allocation and information set obtained by detecting asset are classified as indirect effectiveness.

Ultimately, though C2MOE is represented by direct effectiveness like engagement success rate, hit probability, enemy damage probability, they are reflected by the items of indirect effectiveness of C2 such as information fusion processing time, decision times, information fusion accuracy etc.

Presently NCW based NCO-CF that is the conceptual framework for the analysis of C2MOE was published in America. NCO-CF emphasizes the conceptual aspect of C2 system and since NCO-CF is used to static evaluation method that means the extraction of quantitative value through stakeholders' interview, it is inadequate to apply NCO-CF to the analysis of C2MOE. In this paper, we don't discuss NCO-CF[13].

This paper presented new method and framework of C2MOE analysis to evaluate a synergistic effect of combat power due to C2 system suitable for NCW environment.

At the chapter 2, we introduced general explanation, various frameworks, the analysis methods of C2MOE, and understanding of what is MOE. At the chapter 3, experiment procedure for the analysis of C2MOE was presented in M&S (Modeling & Simulation) aspect, new type of MOE framework, method and the items of C2MOE. Finally, we suggested the conclusion of C2MOE and future works.

2 What Is MOE and the Objective of MOE Analysis?

There are numerous types of definition that have been used to explain MOE and C2 MOE in the past. First, before entering the analysis of C2 MOE, we present the definition of command and control Measures Of Effectiveness(C2MOE), C2 structure, and C2 system.,

Various researchers have represented the definition of MOE, but here we suggest a few ones among many definitions of MOE.

2.1 The Definition of MOE

First, Goode and Machol [11] provide the definition of MOE to the 1964 report by the Weapon System Industry Advisory Committee (WSEIAC). The definition is simple and straightforward [2].

"The measure of effectiveness is the criterion by which solutions will be judged-proposed solutions, solutions under test, or solutions in being".

Sproles (2002)' refocused the discussion of effectiveness back to the more general question of "Does this meet my need?" and hence defined Measures Of Effectiveness (MOE) as

"*standards against which capability of a solution to meet the needs of a problem may be judged. The standards are specific properties that any potential solution must exhibit to some extent. MOEs are independent of any solution and do not specify performance or criteria*".

Dockery's (1986) MOE definition to compare systems and to produce a rank ordering of effectiveness is following that

"*A measure of Effectiveness is any mutually agreeable parameter of the problem which induces a rank ordering on the perceived set of goals*".

Smith and Clark's (2004) MOE definition that is not to derive absolute measures as they do not support the making of comparisons between disparate systems whose measures may be based on totally different characteristics and produce values with different ranges and scales is following that

"*A measure of the ability of a system to meet its specified needs (or requirements) from a particular viewpoint. This measure may be quantitative or qualitative and it allows comparable systems to be ranked. These effectiveness measures are defined in the problem-space. Implicit in the meeting of problem requirements is that threshold values must be exceeded*".

The definition of some components of C2 by Malerud[12] is following that

C2 Concept: A set of characteristics of a C2 system describing how it reaches its objective.
C2 Structure: An assembly of personnel, organization, procedures, equipment and facilities arranged to meet a given objective, and within fixed economical limits.
C2 System: An assembly of personnel, organization, procedures, equipment and facilities organized to accomplish C2 related functions. A C2 system comprised three main components: C2 tasks, C2 functions and a C2 structure.

Here we introduce two analysis framework of C2MOE; one is VFT (Value Focused Thinking) with MUAT (Multi attribute Utility Theory) and BN (Bayesian Network)[1]. MAUT is used to formulate and evaluate utility function and is unsuitable for measuring effectiveness in complex domain, like NCW. VFT is used to determine the fundamental objective and value hierarchy used to derive the utility function. VFT provides a well grounded, consistent mathematical framework to analyze and model the problem domain to produce a value hierarchy and a means-end network. BN allows the rigorous propagation of effectiveness values between causally connected nodes and support the use of both qualitative and quantitative effectiveness measures.

The characteristics of a simple thought experiment [1] are to need to capture the problem domain requirements and map the solution back to this problem domain. This mapping will not always be direct and is constrained on the validity.

The properties of C2MOE using Newton's Second Law [10] is as follows. Combat power as the key factors that affect battle in terms of military is regarded as power, command and control times is regarded as time. Thus the values of combat power and command and control times are applied to Newton's Second Law, and we can obtain the results of C2MOE.

2.2 The Objective of MOE Analysis

MOE (Measures Of Effectiveness) in accordance with the objective can be divided into four cases such as strategic, combat/mission, engagement, system objectives. MOE analysis of Theatre level is applied to war game level, MOE analysis of combat/mission level is obtained by the degree of accomplishment of troop's mission. The one of engagement level is applied to engagement of N to N, N to 1, and 1 to 1. The one of engineering level is to predict the key parameters that affect the performance of system.

And in a different meaning, MOE is used for at least two purposes – to analyze systems in existence and to predict the effectiveness of future systems. Existing systems allow for direct measurement. For example, we extract optimal system requirements as varying key factors and performance of the system elements. Whilst future systems are measured against models developed with input from subject matter experts. The comparison of between Legacy system and new system, the proposal of alternative systems to replace the legacy system is equivalent.

3 The New Analysis Method of C2MOE

The analysis framework and method of C2MOE mentioned in the previous chapter can be used when the prerequisites meet, simply consider and provide the processing and quantifying for measuring effectiveness. The main difference between legacy analysis methods and new method in this paper is to reflect the characteristics of C2system that suits the purpose of analysis, not to reflect the overall characteristics of the real C2 system. This paper provided full steps and framework for evaluating the analysis method of C2MOE.

3.1 The Experimental Procedures for Analyzing the C2MOE

The whole process for the analysis of C2MOE is divided into four steps as follows: Experiment plan, preparation, performance, and analysis of results.

At the Experimental procedure for the analysis of MOE, we first construct the experimental design in the stage of experimental plan. The experimental design determines the analysis method and items for the analysis of C2MOE/MOP by using the analysis framework. We determined models to be simulated, the range of MOE analysis, the level of simulator to be developed based on established operational concept through the analysis framework. Thus experimental design is the most important stage during the experimental processing. If the experimental design is invalid, the results of simulation must be unuseful.

At the stage of experimental preparation, a simulator is developed and items of MOE/MOP, the analysis method of C2MOE are established. We perform the simulation for producing the MOE analysis data by using the developed simulator. Also extracted items of C2MOE on stage of the experimental preparation are reflected to simulation.

At the final stage, simulated data is collected and analyzed. Analyzed results are compared with the requirements of C2 system and C2MOE. If the analyzed results don't satisfy the purpose of experiment, what the reasons (simulation scenario, MOE items, the level of models so on) is must be investigated and extracted. After applying the reasons to the experimental design, simulation start, and simulated date are analyzed. These procedures will be repeated until the simulated results satisfy the purpose of experiment.

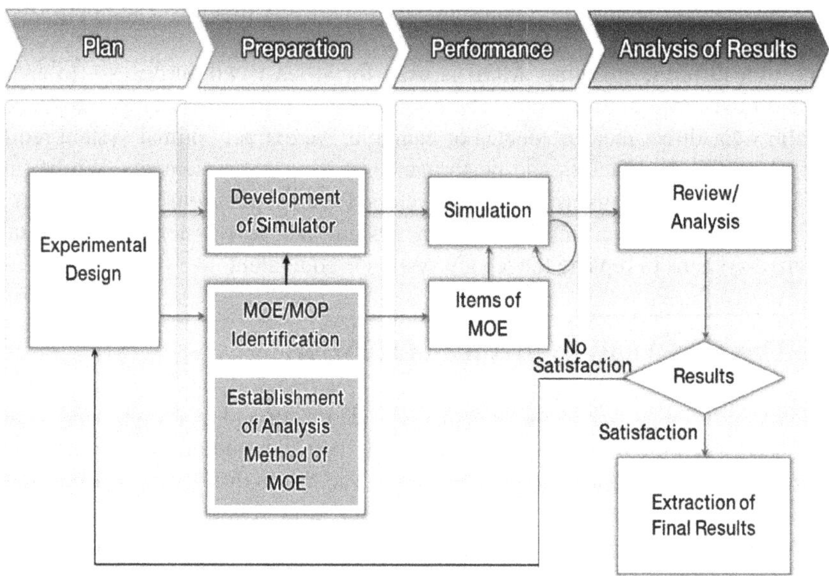

Fig. 1. The experimental Process presents how to analyze and evaluate C2MOE and what are the needs of C2MOE

3.2 The Analysis Framework of C2 MOE

We exhibited the analysis framework of C2MOE in the previous section. First, general analysis framework was described and we suggested the new framework of C2MOE.

The general analysis framework of C2MOE based on U.S. Field Manual 5.0 METT-TC considerations is following that

There are the objective and method of MOE analysis, the capabilities and performance for object of effectiveness analysis, Target and Threat, the environment (Terrain & weather, civil considerations) which surrounds these factors. Here Time elements represent simulation time when M&S(Modeling and Simulation) is used to the method of MOE analysis. According to the objective of MOE analysis, simulation scenario is constructed in time sequence.

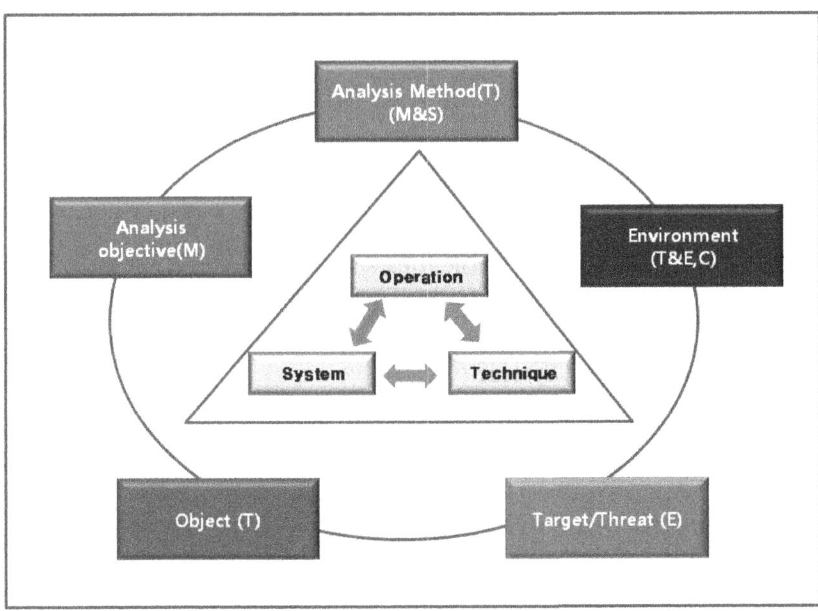

Fig. 2. The general analysis framework of C2MOE based on U.S. Field Manual 5.0 METT-TC considerations

The analysis framework of C2MOE in this paper established operational concept of system from the materials such as output of MND-AF, ROC, RFP, OCD, SSS etc on the stage of operational concept study. Operational scenario based on the operational concept was created. We verified that completed scenario is suitable for the purpose of analysis, that is, whether operational concept is faithfully reflected, or whether the results of analysis can be derived.

After verification of operational scenario, tactical required models and technical required model are determined. They are part of models which make up operational scenario. Tactical required models represent models reflected doctrine, tactics of foe and friend, technical required models indicate physical models, environment model.

Input/output parameters were determined through the analysis of C2 characteristics and operational scenario. If operational concept, operational scenarios, constructive models and characteristics, and input/output parameters are established, according to the analysis procedure, that is, it is a part of the experimental procedure simulation was performed and the results of analysis were extracted from the simulation data. These processes are repeatedly performed until the desired result is obtained.

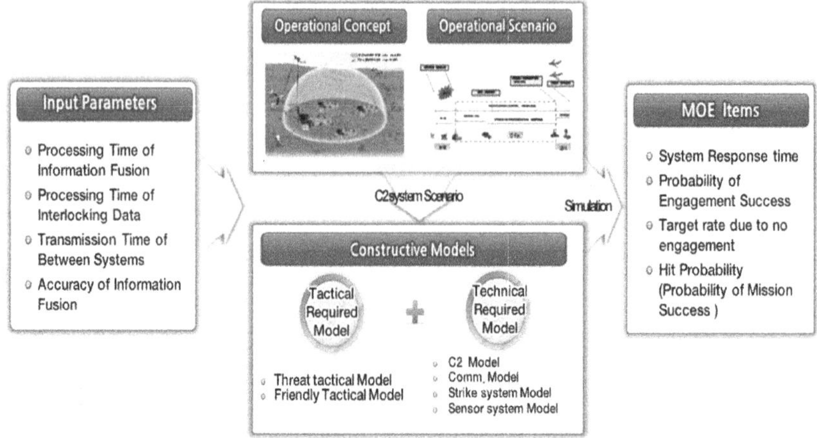

Fig. 3. The analysis framework of C2MOE extracts operational concept, operational scenario, constructive models, input/output parameters, and simulation procedures

3.3 The Structure of C2 System

To extract the items of MOE analysis as mentioned earlier, we must comprehend the constitution of command and control system.

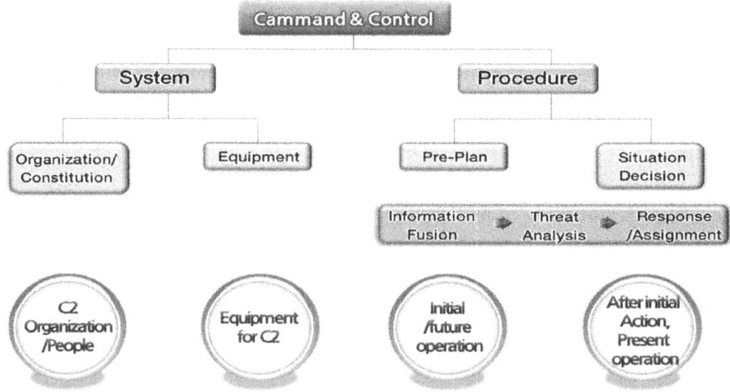

Fig. 4. The structure of command and control represents the constitution of organization and people, equipment, C2 procedure, and roles of each configuration

The command and control consists of system and procedure. System is composed of equipment of C2, organization and structure that is made up formation and the number of people. Procedure consists of planning ahead (pre-plan) like the initial or the future military action and after initial action, present operation that is made up a situation decision. Also procedure roles of performs information fusion, threat analysis, and response and assignment of command to sensor & strike system.

The following explanation of C2MOE will help in understanding what affects C2MOE. The analysis of C2MOE is to evaluate indirect effectiveness of system and procedure that compose command and control system.

C2MOE can be measured based on time, accuracy, and combined time and accuracy. Time based C2MOE presents Planning the work, alternative, recognition of event, system response time, time for achieving the goal state, time rate of staying on target, timeliness of response.

Accuracy based C2MOE is classified as accuracy and precision of mission achievement, the probability of error occurrence, the quality of decision making.

Combined time and accuracy based C2MOE is related to the time of error occurrence, recovery time when error is occurred.

4 Conclusion and Future Work

We introduced legacy analysis methods of C2MOE that enumerate simply items of MOE analysis and determine the measure by the quantitative formulation. Thus these methods didn't affect the practical analysis and method of C2MOE. But we suggested how to determine operational concept, the creation of operational scenario, the extraction of MOE items, Model construction for simulation, and the level and range of simulator development in this paper. This new method must help the analysis of C2MOE and is applied to various systems.

To measure effectiveness of command and Control system and to analyze how C2 affects the war are the difficult trial. Especially it is too difficult to measure effectiveness of C2 system that includes human element like commander and his staff. Thus in the future agent based C2 model will be developed. Also the complex model of Human and Machine will present meaningful indicators for the study of C2.

References

1. Smith, N., Clark, T.: A Framework to Model and Measures System Effectiveness. In: 11th ICCRTS, paper 54 (2006)
2. Smith, N., Clark, T.: An Exploration of C2 Effectiveness – A Holistic Approach. In: 9th ICCRTS 2004 Command and Control Research and Technology Symposium (2004)
3. Sproles, N.: Formulating Measures of Effectiveness. System Engineeirng 5(2) (2002)
4. Sproles, N.: Establishing Measures of Effectiveness for Command and Control. A System Engineering Perspective, DSTO report DSTO.GD-0278
5. Bornman Jr., L.G.: Command and Control Measures of Effectiveness Handbook. Technical Document TRAC-TD-0393 (1993)
6. Stone III, G.F., Jones I, O.J.: Applying Command and Control Measures of Effectiveness tot the Eagle Combat Simulation Model. A technical report of the operations research center U.S Military Academy (February 1992)
7. Green, J.M., Johnson, B.W.: Toward a Theory of Measures of Effectiveness. In: CCRTS (2002)
8. Bayne, J.S., Paul, R.: Performance Measurement in C2 Systems. In: 8th International Command & Control Research & Technology Symposium (June 2003)

9. Kim, C., Kim, I.: A Study of Influencing Factor Upon Using C4I System: The Perspective of Mediating Variables in a Structured Model. Asia Pacific Journal of Information Systems 19(2) (June 2009)
10. Lee, Y., Jung, W., Kim, Y., Lee, J.: A Proposal of New MOE to Acess the Combat Power Synergistic Effect of Warfare Information System. IE Interfaces 22(3), 205–213 (2009)
11. Goode, H.H., Machol, R.E.: System Engineering. In: An Introduction to the Design of Large scale Systems New York, McGraw-Hill, New York (1957)
12. Malerud, S., Feet, E.H., Enemo, G.: Assessing the Effectiveness of Maritime C2 System. In: 1998 CCRTS Processings, June 29-July1. Naval Postgraduate School, Monterey (1998)
13. Garstka, J.: Network Centric Operations Conceptual Framework Version 1.0, Office of Force Transformation (November 2003)

Data Management and Time Synchronization in PlugSim: A DEVS-Based Framework for Interoperation of Simulations

Jang Won Bae[1], Kyung-Min Seo[2], and Tag Gon Kim[2]

[1] Department of Industrial and Systems Engineering
KAIST, Daejeon, Republic of Korea
jwbae@smslab.kaist.ac.kr
[2] Department of Electrical Engineering
KAIST, Daejeon, Republic of Korea
kmseo@smslab.kaist.ac.kr, tkim@ee.kaist.ac.kr

Abstract. As modern systems are increasing in complexity, modeling and simulating them are also becoming difficult. In particular, modeling and simulation (M&S) with one modeling method reveals the limitations of many simulation purposes. In order to tackle this limitation, the interoperation concept has been developed. For the interoperation of simulations, data management and time synchronization are indispensable. In this paper, we explain how the data management and time synchronization are performed in PlugSim environment. PlugSim is the DEVS-based framework for interoperation of simulations and its theoretical basis depends on the DEVS-BUS. PlugSim helps developers create an environment of interoperation of simulations. For better understanding, we provide an example and show the result of simulation in the case study.

1 Introduction

As modern systems are increasing in complex, modeling and simulating them are also becoming difficult. The systems are classified into two groups according to complexity. The first is the view of the number of sub-models in the system; the second thing is the view of various sub-systems in the system [1]. In the first case, modeling and simulating the system is not a difficult problem. However, in the second case, this is hard to tackle, because many systems (e.g. continuous system (CS), discrete event system (DES)) have their own characteristics.

For this reason, researchers have developed new methods for presenting the systems in the second case, such as multi-formalism and interoperation of simulations [2] [3]. In the second case, Sub-systems should be modeled using the most appropriate formalism and tool. In the multi-formalism approach, however, a single formalism is identified into which each of the sub-systems may be symbolically transformed [1]. Interoperation of simulations can be used with the most appropriate formalism and tool for the sub-systems.

J.-H. Kim et al. (Eds.): AsiaSim2011, PICT 4, pp. 403–412, 2012.

There has been a great deal of researches regarding the representation of complex systems. For example, High Level Architecture (HLA) is a standard in IEEE which describes a specification for interoperation between heterogeneous simulators. Run-Time Infrastructure (RTI) is software that is the result of implementing HLA [7] [8] [9]. HLA/RTI supports interoperability and reusability of simulators. However, interoperation of simulations is inconvenient using HLA/RTI. In order to use HLA/RTI, users should understand the APIs of HLA/RTI and create interfaces that can handle HLA/RTI services. Another example is the Virtual Laboratory Environment (VLE) [4]. VLE is the framework for the integration of heterogeneous formalisms using Discrete Event System Specification (DEVS) formalism and associated extensions adapting a multi-formalism method.

PlugSim is the DEVS-based framework for interoperation of simulations [6]. It supports the interoperation of various simulators. Therefore, the models of various sub-systems are modeled and implemented by appropriate methods in PlugSim environment. The theoretical background of PlugSim is on DEVS-BUS [5], the interface that can communicate over the RTI. However, PlugSim is a complete environment for interoperation of simulations, not using RTI.

In this paper, we introduce the internal structure and algorithms of PlugSim in detail using an example. Data management and time synchronization in PlugSim are our main consideration. Then, we refer to the advantage of using PlugSim environment with the case study.

The rest of this paper is organized as following: In section 2, we briefly introduce the DEVS as our background. Section 3 shares concepts and components in PlugSim and explains how data and time are managed in PlugSim using torpedo engagement as an example. Section 4 shows our case study in PlugSim environment. We then conclude this paper in Section 5.

2 Background – DEVS Formalism

PlugSim is based on the DEVS formalism. This section briefly explains the DEVS formalism. The DEVS formalism specifies discrete event models in a hierarchical and modular form. There are two kinds of models in the formalism: atomic model M and coupled model DN, as follows [10] :

$M = < X, Y, S, \delta_{ext}, \delta_{int}, \lambda, ta >$,
Where
X : a set of input;
Y : a set of output events;
S : a set of sequential states;
$\delta_{ext} : Q \times X \rightarrow S$, an external transition function,
where $Q = \{(s,e) | s \in S, 0 \le e \le ta(s)\}$ is the total state set of M;
$\delta_{int} : S \rightarrow S$, an internal transition function;
$\lambda : S \rightarrow Y$, an output function;
$ta : S \rightarrow Real$, time advance function.

$DN = < X, Y, M, EIC, EOC, IC >$,
Where
X : a set of input events;
Y : a set of output events;
M : a set of all component models;
$EIC \subseteq DN.X \times \cup M.X$: external input coupling;
$EOC \subseteq \cup M.Y \times DN.Y$: external output coupling;
$IC \subseteq \cup M.Y \times \cup M.X$: internal coupling;
SELECT: $2^M - \phi \rightarrow M$: tie-breaking selector.

3 PlugSim: DEVS-Based Framework for Interoperation Simulation

In this paper, we focused on the data management and time synchronization performed in PlugSim. Before we handle these subjects, we briefly present the concept and components of PlugSim.

3.1 Concept and Component of PlugSim

Figure 1 shows the conceptual design of PlugSim. PlugSim provides data management and time synchronization services, which are the most important services in the interoperation simulation to users. Using these services and Interoperation Object Data (IOM), which is created with user-defined data information, users can join their models to the interoperation of simulations. Moreover, users can change their models during the simulation through PlugSim environment.

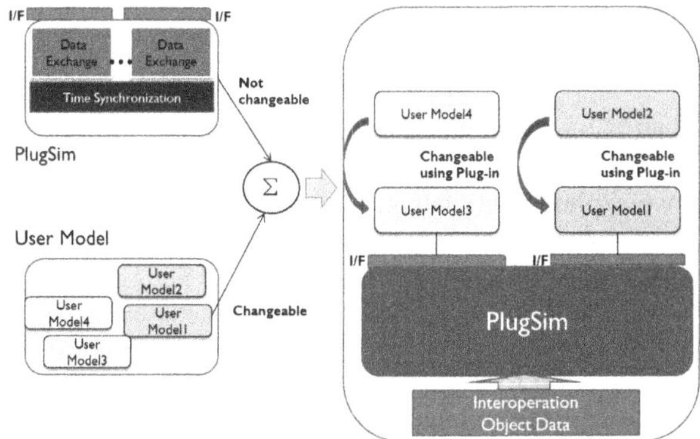

Fig. 1. Conceptual Design of PlugSim

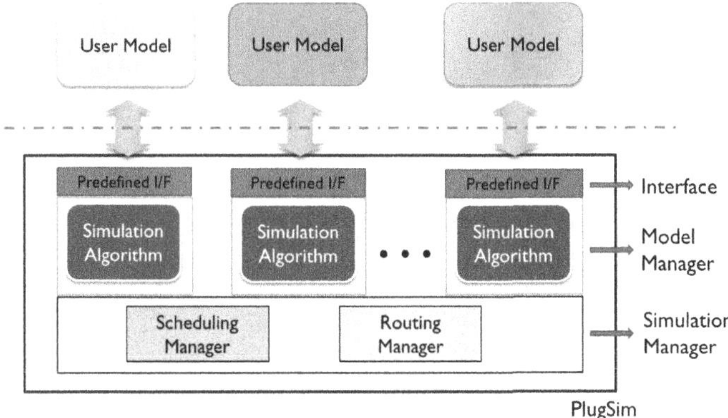

Fig. 2. Components in the PlugSim

In order to provide this environment, PlugSim consists of three main components: simulation manager, model manager and interface. Figure 2 shows the components of PlugSim. Roles of each component are as follows:

- Simulation Manager: The simulation manger consists of the scheduling manager and routing manager. The simulation manager synchronizes overall simulation time and gathers the information of data exchange then makes the IOM.
- Model Manager: The model manager has a discrete event based simulation algorithm. Using this algorithm, one model manager simulates a user model. It handles time and data information of the user model.
- Interface: The Interface allows the model manager to communicate with a user model. Through the interface, model manager can simulate a user model and obtain time information for the user model. According to the type of user model (i.e., DEVS model or non-DEVS model), the kind of interface employed can be different.

A detailed description of PlugSim, how to make the user model and other information can be found in [6].

3.2 Data Management and Time Synchronization in PlugSim

In this section, we explain data management and time synchronization of in PlugSim in detail. To aid in understanding, we use a specific example. In the following subsection, we briefly introduce the example.

3.2.1 Example: Torpedo Engagement [11]

We use the concept of torpedo engagement as our example. In this torpedo engagement, the engagement scenario and specification of model are defined.

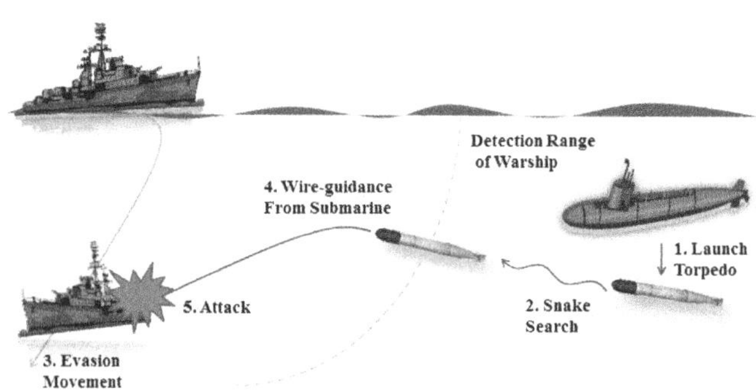

Fig. 3. Scenario of torpedo engagement

Figure 3 depicts a brief scenario of torpedo engagement. We explain this scenario in detail as followings.

1. A submarine launches a torpedo toward the warship
2. The torpedo searches for the position of the warship using its searching algorithm
3. When the warship detects the torpedo, the warship performs an evasion movement according to the direction of the torpedo
4. The submarines control the torpedo toward the warship using wire-guidance
5. As the control from the submarine, the torpedo goes to the warship and attacks it.

In the torpedo engagement simulation, we used only two models, a warship model and a torpedo model. The submarine model in the figure 3 is abstracted, because it does not affect the result of the simulation. The warship model and the torpedo model are separated by three parts: maneuver, sensor, control models. The maneuver model is based on a continuous system, while the others are based on a discrete event system, according to the characteristics of each model.

3.2.2 Time Synchronization

Time synchronization is an essential part of the simulation, particularly in the interoperation simulation, in which the various types of models can be joined. In order to synchronize simulation models by the simulation time, PlugSim should get the next simulation time of the models.

Before the simulation has begun, all the joined models notify their next simulation time to their model manager. Then, each model manager forwards the next simulation time to the simulation manager. The simulation manager decides the model, which has the minimum next simulation time, to simulate until its next simulation time. After model simulation concludes, the model notifies updated next simulation time to its model manager again. With iterating these steps, PlugSim can maintain the time synchronization of the interoperation simulation.

The time synchronization method, using in PlugSim, is based on the coordinator algorithm of DEVS [10]. In the coordinator algorithm, the coordinator obtains the subordinate simulators' next event times. The coordinator defines the minimum next event time among all next events times. Then, the coordinator sends a message to the simulator to simulate the atomic model.

We give an example of time synchronization in PlugSim using torpedo engagement. The warship and torpedo models are joined in the interoperation simulation using PlugSim. The warship and Torpedo models consist of three sub-models. However, according to the coordinator algorithm, it is not considerable in the time synchronization. Before beginning the simulation, two models send next simulation times to their model managers. When the simulation begins, the model managers serve the next simulation time to the simulation manager. The simulation manager defines the minimum next simulation time, then, lets the model manager simulate its model (for example, the warship model). The warship model will be simulated until the next simulation time. At the next simulation time, the warship model can define the next simulation time by its model specification. The model manager can obtain the updated next simulation time and forward it to the simulation again. Figure 4 shows this example as a sequence diagram.

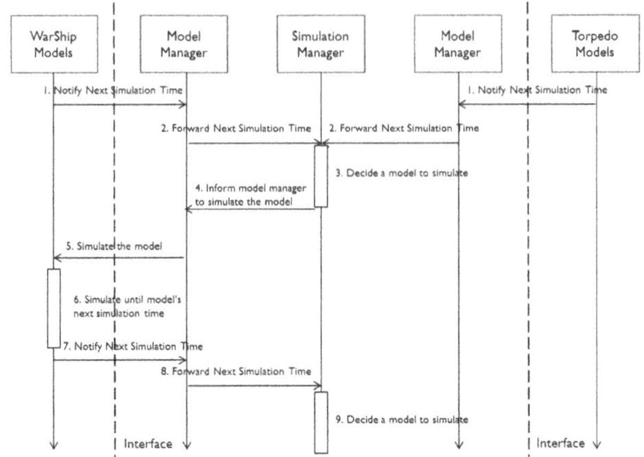

Fig. 4. Sequence Diagram of Time Synchronization in the example

The Discrete Event model (like the DEVS model) should retain its next event time. Therefore, the model manager can receive the next simulation time of its model. However, in a non-DEVS model case, for example continuous system model, there is no specification about the next schedule simulation time.

In the form of continuous system model, it is impossible to obtain the next simulation time of the model. There needs to be an additional method to the model. In [12], the interface which converts continuous time information to discrete event time information is well described. Continuous models, which are joined in PlugSim, can use that interface.

3.2.3 Data Management

In order to exchange data between user models, it is essential to define the transferred data first. We define this transferred data form as an Interoperation Object Model (IOM). Using user-defined IOM, the simulation manager makes the RoutingInfo which is used as a reference table in exchanging data. When the data is in transit from or to model, the model manager can transfer the data to/from the target model using the RoutingInfo. In the following subsection, we explain data management in PlugSim in detail.

*3.2.3.1 Interoperation Object Model (IOM).*The IOM describes all the transferred data among user models. The IOM consists of local data set (LDS) of user models. LDS is the set of data that represents input/output in a model. In order to join to PlugSim, users should define the LDS of the joining model. However, the LDS format differs by the kind of user model (i.e. DEVS or non-DEVS model).

In the DEVS model case, the LDS format consists of the number of publish/subscribe (output/input) data and the name of the publish/subscribe data. However, in the non-DEVS model case, the LDS format is more complicated than the DEVS model case. Packet ID, field name and type of packet are additionally needed. This is because non-DEVS models communicate with PlugSim by packet. Figure 5 shows an example of LDS in the torpedo engagement. The torpedo control model is based on a discrete event system and torpedo maneuver mode is based on a continuous system.

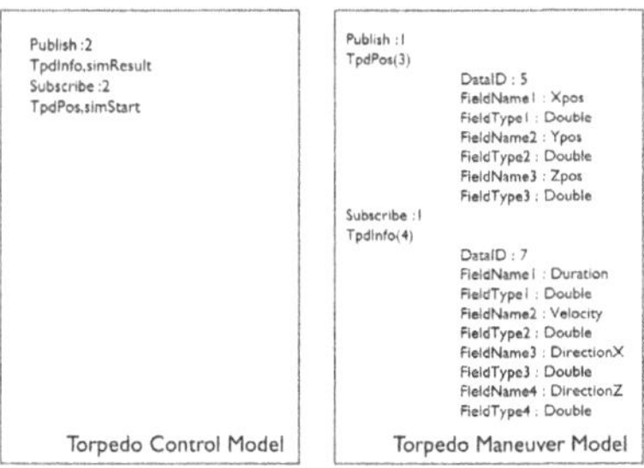

Fig. 5. Examples of LDS (L: CS model, R: DES model)

3.2.3.2 RoutingInfo. RoutingInfo is the table which represents from which the data is generated and to which it is forwarded. Using the user-defined LDS, the simulation manager defines the RoutingInfo table. For example, the RoutingInfo table indicates that 'TpdPos' data comes from the Torpedo Maneuver model and goes to the Torpedo Control model (in Fig. 6). When the RoutingInfo has been defined, the simulation manager sends the RoutingInfo to each model manager.

3.2.3.3 Data Exchange. Data exchange is performed among the model managers. We explained the detailed step of exchange data as follows.

When the data has been generated in the user model, it is forwarded to its model manager. The model manager finds target model referring the name of the output data to the RoutingInfo. The model manager forwards the output data to the model manager of the target model. Then, the model manager forwards the data to the target model. When the target model receives the data, it handles the data according to its specification. After the end of data handling, the target model updates its next simulation time and forwards it to the simulation manager through its model manager. Moreover, the model manager of the target model informs the model manager of the source model of response of the data. When the source model is informed of response of the data through its model manager, the source model updates its next simulation time and forwards it to the simulation manager. Figure 6 describes the process of data exchange using torpedo model and warship model in torpedo engagement example.

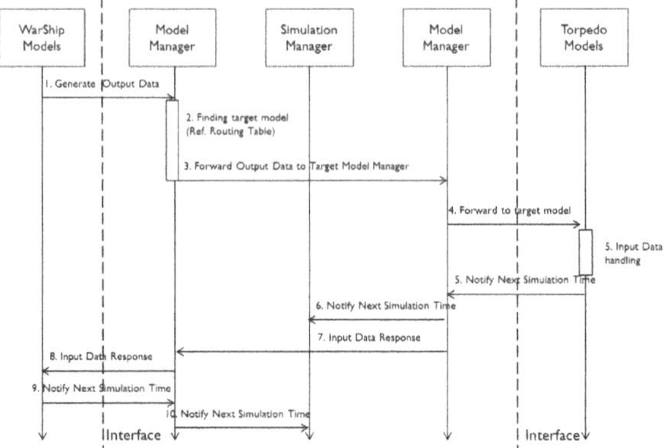

Fig. 6. Sequence diagram of data exchange

4 Case Study: Torpedo Engagement

Figure 7 shows the experimental design of the case study. The torpedo engagement simulation consists of the warship and the torpedo models. Each model is separated into three parts: sensor model, control model and maneuver model. These sub-models are developed as DEVS models (sensor, control model) and MATLAB model (maneuver model). For easy management of the simulation, we added Experimental Frame (EF) parts. The scenario Editor and statistical estimator are components of the EF. With separating EF and simulator, we can easily manage and obtain statistical data [10]. In particular, we use SIMDIS as a display tool for the simulation results [13]. Moreover, with SIMDIS, we can figure out the simulation status during the simulation at once. The IOM, which is used in the interoperation simulation, is represented in the Fig. 8. Scenario of this case study is same as that found in 3.2.1.

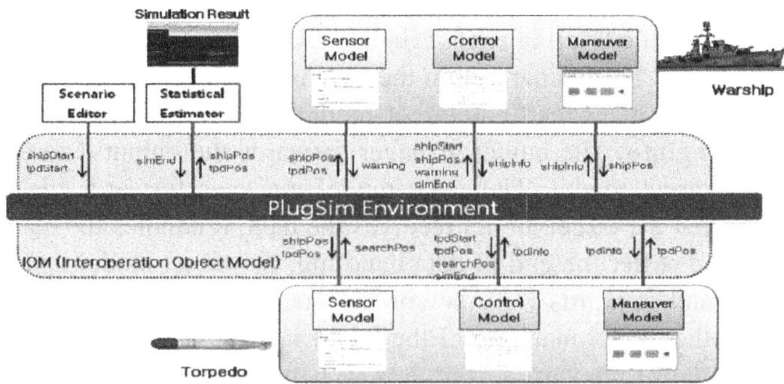

Fig. 7. Experimental design of the case study: torpedo engagement

With this case study, we have attempted to demonstrate two things: first, the interoperation simulation, which consists of the different kinds of models, can be conducted in PlugSim environment; second, changing simulation models during the simulation is effective in some simulations. For the first aspect, we designed the case study with various kinds of models (the DEVS model and the MATLAB model).For the second, we designed the scenario of case study to emphasize the changing models during the simulation. In the wire-guidance torpedo, maneuver algorithm of the torpedo is changed by the submarine. In PlugSim Environment, maneuver model of the torpedo model can be changed during the simulation. Therefore, using PlugSim, It is very similar to the wire-guidance torpedo.

Figure 8 depicts changing the maneuver model of the torpedo model during the simulation and the simulation result of that changing on the SIMDIS. We checked that the movement of the torpedo changed by SIMDIS, when the maneuver model of the torpedo model changed.

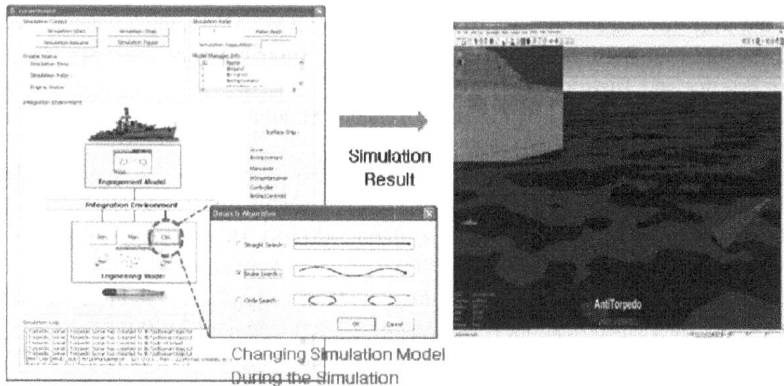

Fig. 8. Changing the maneuver model of torpedo during the simulation

5 Conclusion

In this paper, we introduce PlugSim environment. PlugSim supports interoperation simulation, which consists of various kinds of models (both DEVS and Non-DEVS model). Moreover, in PlugSim environment, simulation models can be changed during the simulation. In particular, we represent the data management and time synchronization in PlugSim, which are essential services in the interoperation simulation. We explain these services using torpedo engagement as an example. In the case study, we design the experiment using PlugSim and show that PlugSim environment can support the interoperation simulation and change the simulation models during the simulation. Changing simulation models during the simulation is particularly useful in some simulations, such as the development of tactics of the wire-guidance torpedo.

In the further work, we plan to develop PlugSim as a hybrid environment. For the hybrid environment, the time synchronization algorithm and interfaces need to be modified. Moreover, we plan to use Inter-Process Communication (IPC) as the communication method for efficient matter.

Acknowledgement. This research is supported by Ministry of Culture, Sports and Tourism (MCST) and Korea Creative Content Agency (KOCCA) in the Culture Technology (CT) Research & Development Program 2009.

References

1. Vangheluwe, H.L.M.: DEVS as a common denominator for multi-formalism hybrid system modeling. In: Proceedings of the 2000 IEEE International Symposium on Computer-Aided Control System Design, Anchorage, Alaska, USA, September 25-27 (2000)
2. de Lara, J., Vangheluwe, H.: *AToM*^{3}: A Tool for Multi-formalism and Meta-modelling. In: Kutsche, R.-D., Weber, H. (eds.) FASE 2002. LNCS, vol. 2306, pp. 174–188. Springer, Heidelberg (2002)
3. Dahmann, J.S., Morse, K.L.: High Level Architecture for Simulation: An Update. In: Proceedings of 2nd International Workshop on Distributed Interactive Simulation and Real-Time Applications, Montreal, Que., Canada, July 19-20, pp. 32–40 (1998)
4. Quesnel, G., Duboz, R., Ramat, E.: The Virtual Laboratory Environment – An Operational Framework for multimodeling, simulation and analysis of complex dynamic systems. International Journal of the Federation of European Simulation Societies: Simulation Modeling Practice and Theory 17, 641–653 (2009)
5. Kim, Y.J., Kim, J.H., Kim, T.G.: Heterogeneous Simulation Framework Using DEVS Bus. Simulation 79(1), 3–18 (2003)
6. Bae, J.W., Kim, T.G.: DEVS Based Plug-in Framework. In: Spring Simulation Multiconference 2010, Orlando, FL, USA, pp. 147–153 (April 2010)
7. IEEE standard for modeling and simulation (M&S) High level architecture (HLA) - framework and rules, IEEE Std 1516-2000, pp. i–22 (September 2000)
8. IEEE Standard for Modeling and Simulation (M&S) High Level Architecture (HLA) - Federate Interface Specification, IEEE Std 1516.1-2000, pp. i–467 (2001)
9. IEEE standard for modeling and simulation (M&S) High level architecture (HLA)-object model template (OMT) specification, IEEE Std 1516.2-2000, pp. i–130 (2001)
10. Zeigler, B.P., Praehofer, H., Kim, T.G.: Theory of Modeling and Simulation. Academic Press (2001)
11. Seo, K.-M., Song, H.S., Kwon, S.J., Kim, T.G.: Measurement of Effectiveness for an Anti-torpedo Combat System Using a Discrete Event Systems Specification-based Underwater Warfare Simulator. The Journal of Defense Modeling and Simulation: Applications, Methodology, Technology 8(3), 157–171 (2011)
12. Sung, C.H., Hong, J.H., Kim, T.G.: Interoperation of DEVS Models and Differential Equation Models using HLA/RTI: Hybrid Simulation of Engineering and Engagement Level Models. In: 2009 Spring Simulation MultiConf., San Diego, CA, USA (March 2009)
13. U.S. Naval Research Laboratory. SIMDIS user's manual (2006)

Generic Simulation Models to Evaluate Integrated Simulation Environment

Hyun Shik Oh and Do Hyung Kim

Yuseong P.O. Box 35, Agency for Defense Development,
Joint Modeling and Simulation Directorate, Daejeon, Korea
ohs@add.re.kr
http://add.re.kr

Abstract. This paper presents a method to evaluate ISE, Integrated Simulation Environment, on which engineering models can be interoperate with each other. A set of generic simulation models is suggested to evaluate ISE. Some models of ground-to-air engagement are defined and developed as follows: aircraft, detect radar, launcher, and surface-to-air missile (SAM). Each model represents to be an engineering level model of the specific weapon system, which are used to analyze the MOP (measure of performance) of each system during the system development phase. Their functional properties, such as program structure and input/output for interoperation, are similar to those of real system's simulation. Using these models, three types of engagement simulation can be tried: collecting based on source codes, linking them on a HLA/RTI federation as federates, and applying to ISE. Comparing the performances of these three simulation types, we will evaluate ISE in qualitative and quantitative manner. And, hopefully these models can be a reference for evaluating the next ISEs. This paper contains the beginning of this whole study. The detailed descriptions of each model and the results of the first one, source code collection, are presented.

1 Introduction

According to the dramatic improvement of computing capability these days, Modeling & Simulation takes an important role of Defense R&D area.[1] Defense M&S can be classified in several levels by resolution and complexity. There are theater, mission, engagement, and engineering levels. Each level has models which have different resolution and fidelity. In Korea, Theater and mission level M&S's have been used for war games and battle experiment by JSC, Joint Staff of Commanders, and the forces of Korea. Engineering and engagement level M&S's have been used by ADD, Agency for Defense Development, and defense industries for the weapon systems R&D. However, in the advanced countries, there were many efforts to perform engagement, mission level M&S using engineering models directly.[2] With this idea some academic studies are taken place nowadays in Korea.[3] And, a key technology development of ISE, Integrated Simulation Environment, is being conducted also in

J.-H. Kim et al. (Eds.): AsiaSim2011, PICT 4, pp. 413–424, 2012.

ADD. The name of this ISE is AddSIM, Adaptive distributed parallel Simulation environment for Interoperable and reusable models. The details of AddSIM were presented in the 2011 Conference of KIMST.[4] In the project, it is needed to evaluate the capability of AddSIM. To satisfy the needs, an evaluation method is invented and presented in this paper.

At first, a ground-to-air engagement situation is imagined. These are identified the participants as follows: aircraft, detect radar, launcher, and surface-to-air missile (SAM). The four simulation models are developed. Each model pretends to be an engineering level simulation of a specific weapon system, which is used to analyze the MOP, Measure-Of-Performance, of each system during the system development phase. It has the similar program structure and interface for inter-operation to those of the real system's simulation. They can be handled as legacy models.

To simulate the engagement situation, it can be used some commercial software such as EADSIM[5], TESS[6], etc. which are specialized for engagement level simulation. These are good tools for the purpose but, if we already have engineering level models fully proven during the development phase, it is another problem that making a new model which can run on EADSIM or TESS with same performance of the old one.

For that reason, the studies for engagement simulations with engineering level models are needed. So, three candidates were considered. The first candidate is collecting the source code of them. The second one is constructing an HLA/RTI federation with them as federates. And the last one is applying them to a ISE such as AddSIM.

To demonstrate the performance of AddSIM, the candidates for the engagement situation with the four engineering level models are made and compared with the others quantitatively. Furthermore, the models will do a role of a test bed for the follow-on ISEs.

This paper describes the forepart of our study as follows: establishing the evaluation concept, developing the four models and, making reference result with the collective model based on source code.

2 Engagement Scenario

A ground-to-air engagement situation is supposed as an engagement scenario. The concept of the scenario is shown on Fig 1. When an aircraft is coming into the engagement area, a detect radar detects it and sends the target information to a launcher. Receiving the information, the launcher calculates the launch direction and the mid-course way-point of inertial navigation guidance, injects the way-point to a missile, and gives launch command to the missile. After taking off, the missile flies to the way-point with inertial navigation system. Reaching the way point, the missile starts homing guidance with its own seeker. If the aircraft detect the missile coming with its own radar, it does an evasive flight within its maximum maneuverability.

To simulate this situation, four models, such as aircraft, detect radar, launcher, and missile, are needed.

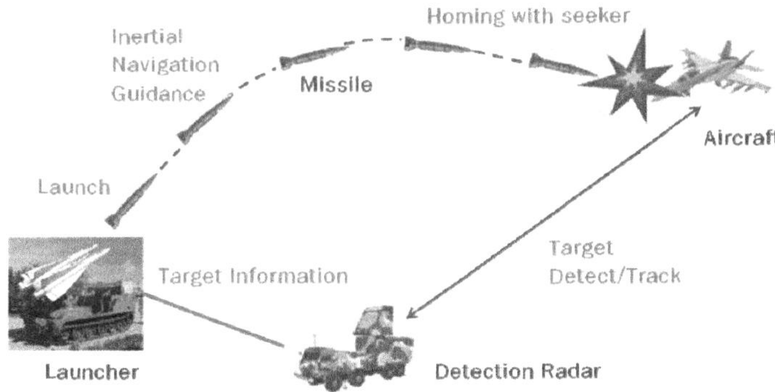

Fig. 1. Concept of Engagement Scenario

3 Engineering Models

3.1 Aircraft Model

The behavior of an aircraft is modeled in the aircraft model based on the scenario. The aircraft should fly into the engagement zone, detect the incoming missile, and, do evasive maneuver. The aircraft model, named PAircraft, is developed according to the flowchart in Fig 2. In this flowchart, there is a missile model which generate incoming missile trajectory to check the evasive function of the aircraft. The red outlined blocks are the models of aircraft functions.

Assuming that the aircraft can fly on the same horizontal plane and the same height, the dynamics of aircraft can be described as equation (1). [7]

$$\dot{\psi} = -\frac{a_n}{V} \tag{1}$$

Where, ψ means azimuth angle of aircraft

a_n means normal acceleration of aircraft

V means velocity of aircraft

The aircraft trajectory can be calculated from integrating the equation. To integrate it numerically, Runge-Kutta 4th Order, RK4, is used. [8]

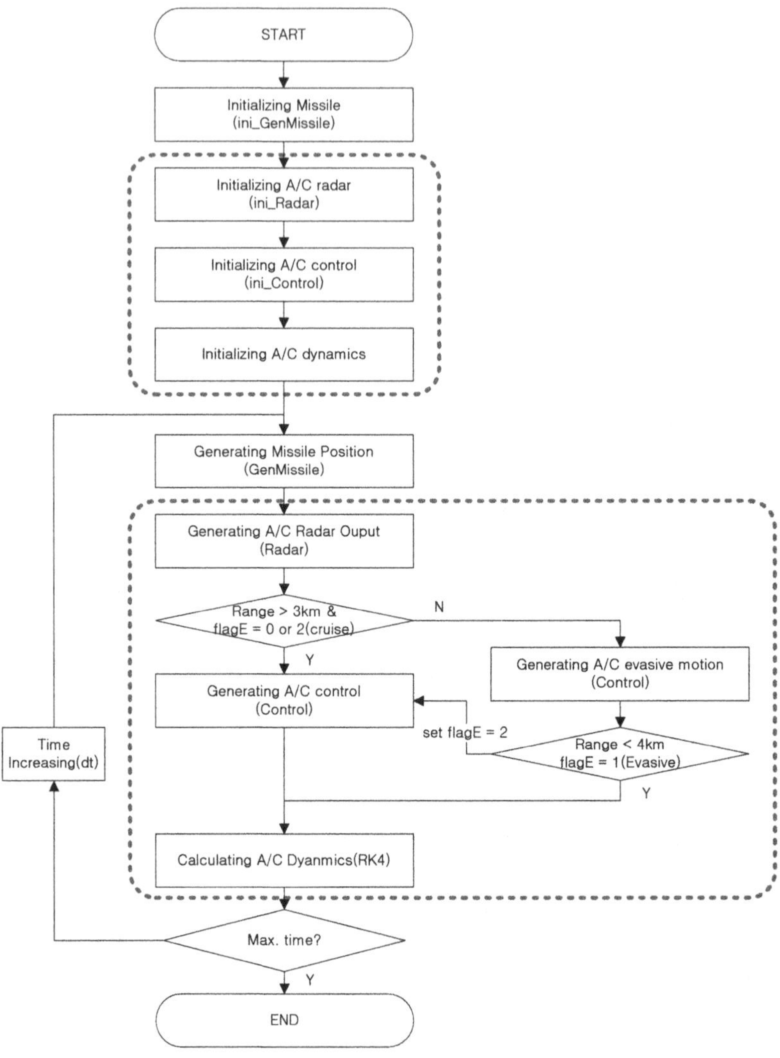

Fig. 2. Flowchart of PAircraft

An example of PAircraft simulation result is shown in Fig 3. The initial position of missile is on 0 km east, 0 km north, and 0 km height and that of aircraft is on -5 km east, 0 km north, and 1 km height. The aircraft cruises in 200m/s speed with 90 degree azimuth, toward westward direction. The aircraft starts the evasive motion when it notices that a missile comes up within 3 km. After that, when the aircraft goes away from the missile more than 4 km, it returns to cruise flight mode.

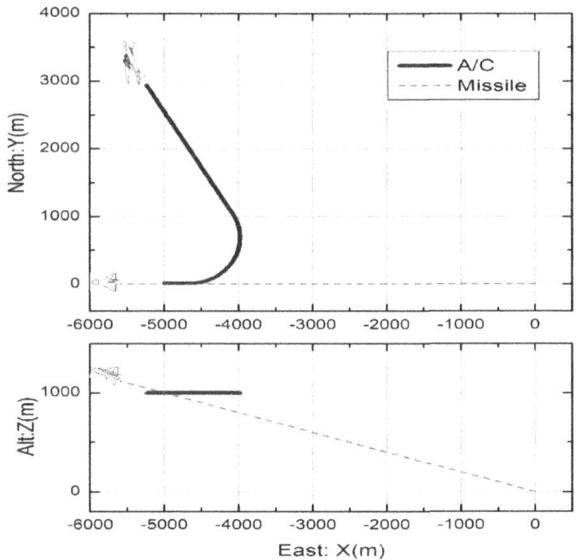

Fig. 3. An example of PAircraft simulation

3.2 Detect Radar Model

The detect radar model, named PDetectRadar is designed to monitor the position of aircraft and generate target information when the aircraft comes up in the detection range of it. Its flowchart is shown in Fig. 4.

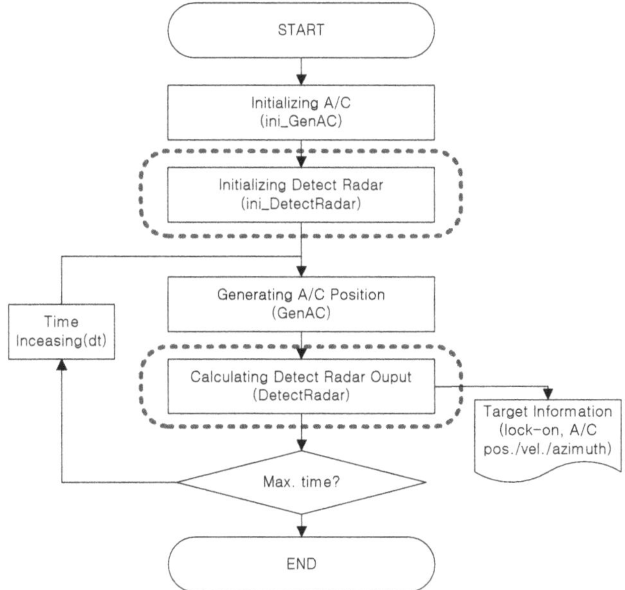

Fig. 4. Flowchart of PDetectRadar

For example, the detection range is set as 4.5km. The result of simulation is plotted in Fig 5. The lock-on flag changed from 0(search mode) to 1(lock-on mode) when the range between aircraft and detect radar is less than 4.5 km.

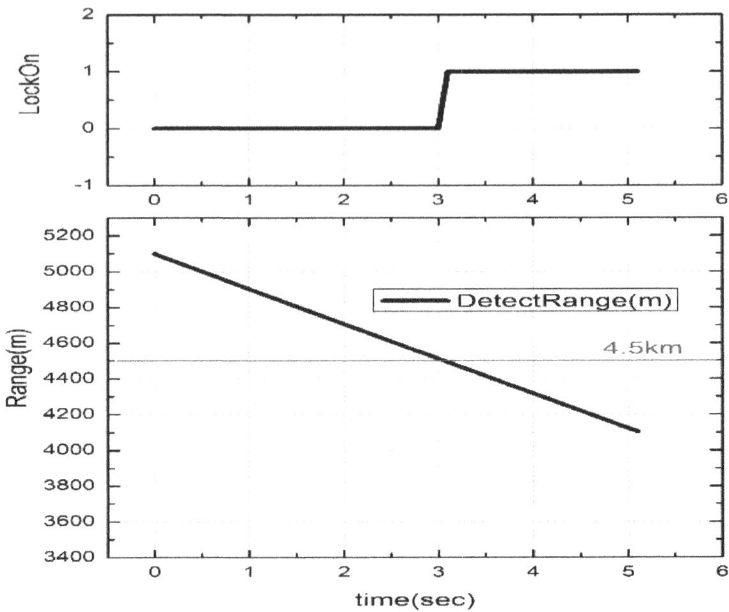

Fig. 5. An example of PDetectRadar simulation

3.3 Launcher Model

The launcher model is designed to represent the behavior of missile launcher which receives the target information from detect radar, calculate the way-point, set the missile to the launch azimuth, inject the way-point to the missile, and give the launch command to the missile.

To calculate the way-point, the engagement position should be predicted with the aircraft information and missile's performance. The horizontal range (R_l) can be determined with equation (2).

$$R_l = \sqrt{(X_{DR} - X_l)^2 + (Y_{DR} - Y_l)^2} \quad (2)$$

Where, DR means the target information detected by Detect Radar

l means launcher

X means East position

Y means North position

Assuming the mean speed of missile ($\hat{V}_M$), the time of collision (t_A) can be estimated with equation (3) and the way-point position (X_C, Y_C, Z_C) with equation (4).

$$t_A = \frac{R_l}{V_{DR} + \hat{V}_M} \tag{3}$$

$$\begin{aligned} X_C &= \frac{X_{DR} + V_{DR} t_A \sin \psi_{DR}}{2} \\ Y_C &= \frac{Y_{DR} + V_{DR} t_A \cos \psi_{DR}}{2} \\ Z_C &= Z_{DR} + 100 \end{aligned} \tag{4}$$

The flowchart of the launcher model, named PLauncher, is shown in Fig 6.

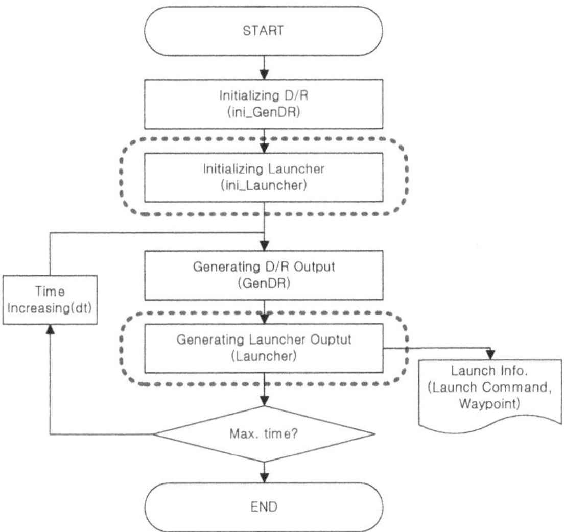

Fig. 6. Flowchart of PLauncher

3.4 Missile Model

The missile model represents the behavior of the missile system like receiving the way-point information, launch attitude, and launch command from launcher. After take-off, the missile flies to the way-point with inertial navigation. Passing the way-point, the missile changes the guidance mode to homing target with seeker information. To define the missile model, atmosphere, aerodynamics, gravity, mass, propulsion and dynamics models are needed. The expressions of these models are borrowed from ref. [9] and [10].

Also the embedded equipment models such as navigation, seeker, guidance, and autopilot are included the missile model. The structure of the missile model, named PMissile, is shown in Fig 7.

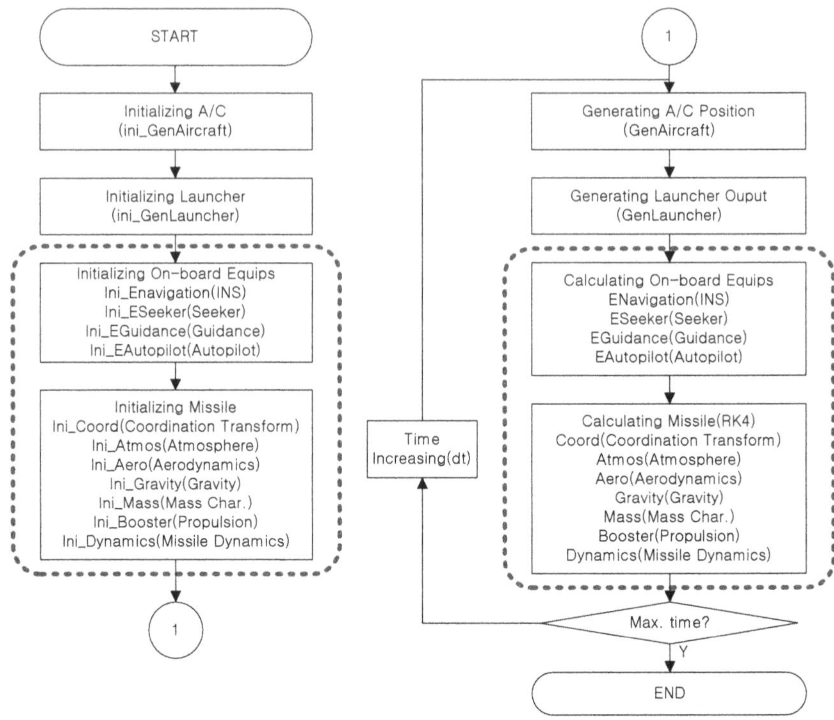

Fig. 7. Flowchart of PMissile

Fig 8 shows the sample result of PMissile simulation. A trajectory of aircraft which flies straight from the initial position (4.5km west, 1km high) to east is generated to check the missile's capability. When the missile intercepted its target, the final miss-distance, the closest distance between aircraft and missile, was 2.7 m.

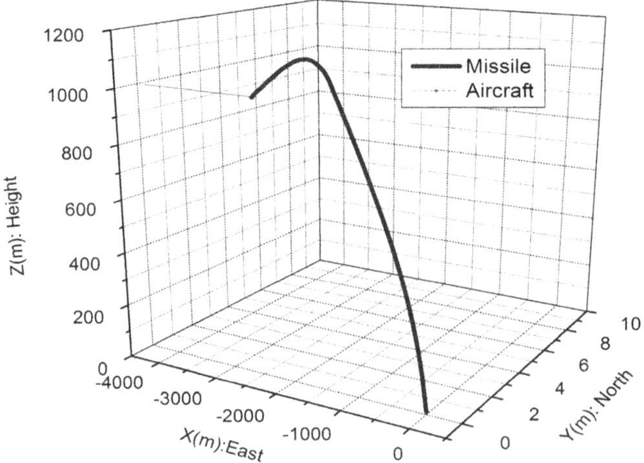

Fig. 8. An example of PMissile simulation

4 Source Code Integration Results

PMissile model runs based on the interception scenario. But it is easy case for missile because the aircraft just flies straight. That is not a fare game. So the participation of best aircraft model is needed. It can be agreed that the engineering model is the best model because of its high fidelity and high resolution. So, how the PAircraft model can participate in the engagement simulation? As the mention in the introduction section, there can be 3-types of simulation to make the ready-made engineering models inter-operated in same simulation. These are collecting the source codes, constructing an HLA/RTI federation and applying them to an ISE such as AddSIM.

The construction and sample result of the first one, source code integration, are presented in this section.

At first, the engineering model parts, which are marked with red dash lines in Fig 2, 4, 6 and 7 are extracted and collected in one source code, named CollectAll, as the following flowchart in Fig 9.

For the integration, some codes should be changed not to conflict with others. The changes are marked with red characters in Fig 9. Furthermore, the additional initializing code for missile is added because the missile should not be calculated during the aircraft model running.

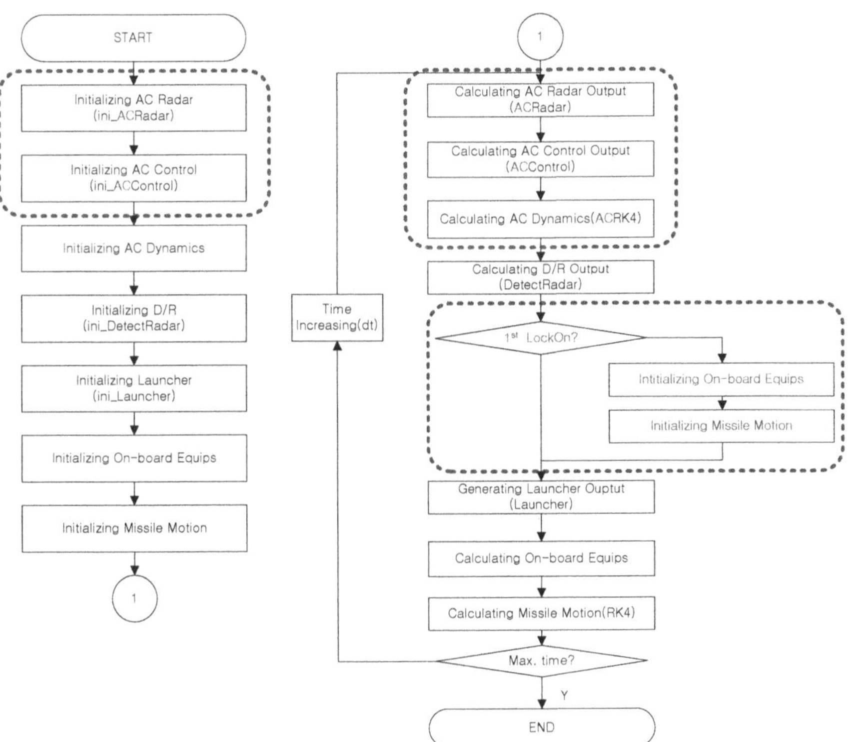

Fig. 9. Flowchart of CollectAll

Using the CollectAll, the engagement results of the scenario in Fig 10 are displayed with SIMDIS[11], a GOTS tool for visualizing data, in Fig 11.

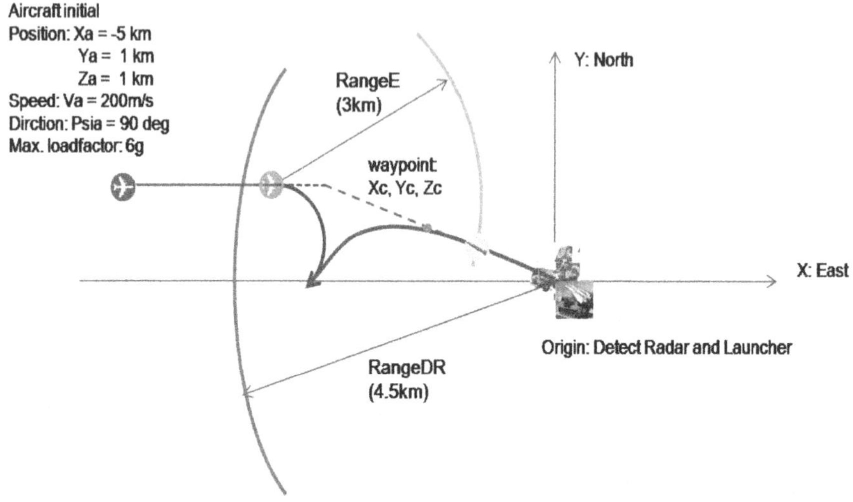

Fig. 10. A sample scenario of single simulation

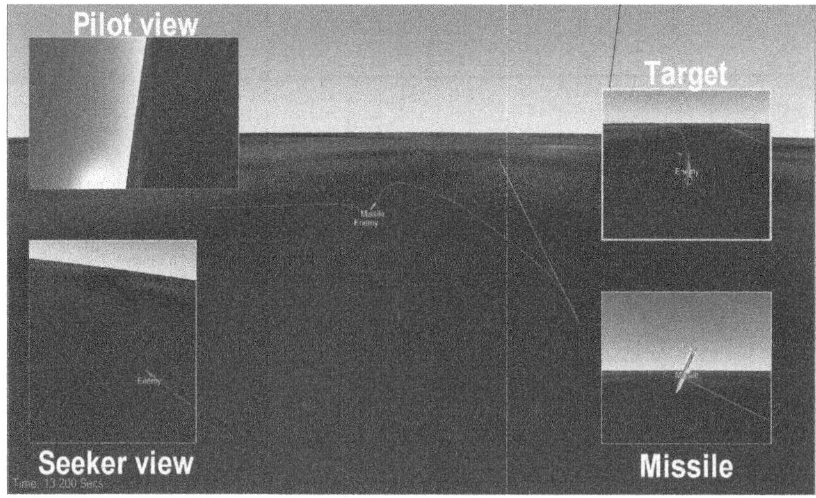

Fig. 11. A sample result displayed with SIMDIS

After that, a batch simulation with the scenarios in Fig 12 can be done.

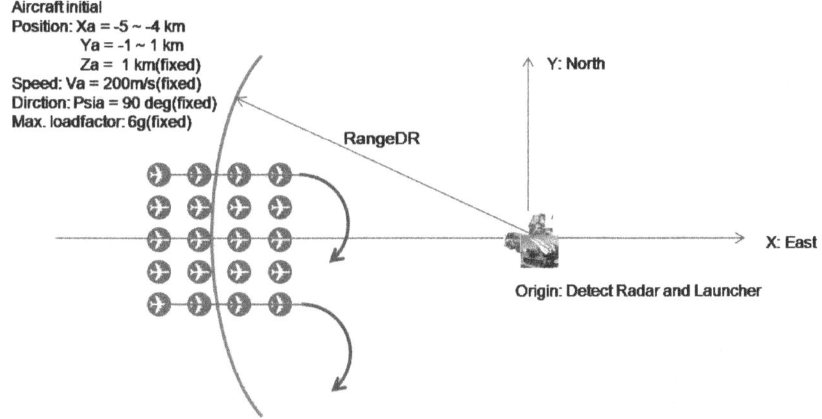

Fig. 12. Sample scenarios of batch simulation

Fig 13 shows the results of the final miss-distance of each simulation with batch simulation.

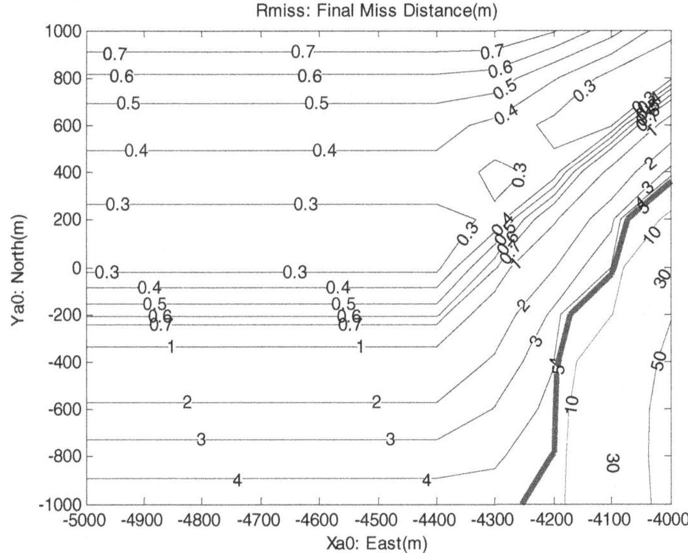

Fig. 13. Final miss-distance contour chart of the batch simulation sample

Assuming that the effective range of the missile's warhead is 5 m, the aircraft which starts with initial position on the left side of red line, the final miss-distance is 5m, is intercepted by the missile and that on the right side is not.

5 Conclusion

This paper describes the developing a method to evaluate capabilities of ISE. To achieve this goal, the generic models are suggested and the four engineering models such as PAircraft, PDetectRadar, PLauncher, and PMissile, which can be applied to engagement simulation, are developed.

The candidate of source code base collection is built. The result of it is presented. It will be handled as a reference data for comparing with that of other candidates of HLA/RTI federation and AddSIM. With comparing them in several ways, the capabilities of AddSIM can be evaluated.

And, hopefully, these four models can be the reference models of evaluating next ISEs.

References

1. Systems Acquisition Manager's Guide for the use of Models and Simulations. In: DSMC (1994)
2. McQuay, B.: The History and Significance of JMASS. Wright-Patterson AFB, OH (2000)
3. Lee, K.S., et al.: Defense M&S Technology Research Center, Simulation Engine Lab. The 1st phase end report, ADD (2010)
4. Lee, S.J., et al.: A Study on Construction of Component Based Distributed Simulation Environment (Korean language). In: Proceedings of the 2011 KIMST Conference (2011)
5. Bourassa, N.R.: Modeling and Simulation of Fleet Air Defense Systems Using EADSIM, p. 128. Naval Postgraduate School, Master Monterey (1993)
6. User's Guide: Tactical Engagement Simulation System RF Version 4.3 Master Interface, Tactical Technologies Inc. (2010)
7. Roskam, J., et al.: Airplane Aerodynamics and performance, Univ of Kansas (1980)
8. Numerical Recipes in C : The Art of Scientific Computing, 2nd ed. W. H. Press, Cambridge University Press (1992)
9. Fleeman, E.L. : Tactical Missile Design, 2nd edn. AIAA (2006)
10. Siouris, G.M.: Missile Guidance and Control Systems. Springer (2004)
11. SIMDIS Manual, U.S. Naval Research Lab (2010)

A Study on the Requirements for Designing Agent-Based Computer Generated Force

Yong-Jun You[1], Jang-Se Lee[2],
Sung-Do Chi[1], and Jae-Ick Kim[3]

[1] Department of Computer Engineering, Korea Aerospace University,
200-1, Hwajeon-dong, Deogyang-gu, Goyang-si, Gyeonggi-do, Korea
{ilog21c,sdchi}@kau.ac.kr
[2] Division of Information Technology, Korea Maritime University,
Dongsam 2-dong, Yeongdo-gu, Busan, Korea
jslee@hhu.ac.kr
[3] Agency for Defense Development, Jinhae, Korea
jaeick@add.re.kr

Abstract. CGF (Computer Generated Forces) is known as a representation of constructive forces in a simulation environment which attempts to model human behavior. Currently, CGF deals with some extent on lower levels of command as well as weapon system, however, their behavior is stereotyped so as to support only simple level of decision-making. In this paper, we propose the level of agent-based CGF by integrating various viewpoints of requirements for supporting the systematic design.

1 Introduction

At the end of the Cold War, M&S can be used as a tool to support the development of new concepts and systems for the future. M&S also help to better train and use existing forces and equipment and to improve operations in a new environment. Emerging technologies will have a great impact on the implementation and on the military use of such simulation systems in the future. CGF (Computer Generated Forces) as representations of forces in simulations which attempts to model human behavior play a main part in this development[1].

CGFs are still used to some extent on lower levels of command and on weapon system level. However their behavior is stereotyped and only very simple levels of decision-making is addressed[1].

The definition of CGF is mainly based on the definition given in the US DoD Modeling and Simulation (M&S) Master Plan[2]. This definition read as:

"A generic term used to refer to computer representations of entities in simulations which attempts to model human behavior sufficiently so that the forces will take some actions automatically (without requiring man-in-the-loop interaction)".

J.-H. Kim et al. (Eds.): AsiaSim2011, PICT 4, pp. 437–443, 2012.

CGF could be divided into two classes; SAF (Semi-Automated Force) and FAF (Fully-Automated Force) [3]. In this paper, however, CGF is divided into 4 levels; LAF (Low-Automated Force), SAF, HAF (Highly-Automated Force), FAF followed by their own autonomy[4]. The LAF, SAF, HAF and FAF are capable of remote-control, hazard avoidance, cooperation and fully autonomy, respectively.

This paper proposes the requirements of designing each CGF level. To do this, the various viewpoints for modeling an agent are introduced.

This paper is organized as follows. First it summarizes a few agent modeling viewpoints. This is followed by the requirements for modeling CGFs, and conclusion.

2 Agent Modeling Viewpoints

2.1 Autonomy

Agents are autonomous entities. The level of autonomy was defined as shown in Table 1 in U.S. military for Future Combat System[4]. According to this definition, there are 10 levels; from level 1- remote control to level 10 - fully autonomy including cooperative, collaborative operation. The SAFs of OneSAF, currently representative CGF system, have the 3~5 level autonomy; hazard avoidance, etc.

Table 1. Level of Autonomy

Level of Autonomy	Description
1	Remote control
2	Remote Control with vehicle state knowledge
3	External preplanned mission
4	Knowledge of local and planned path environment
5	Hazard avoidance or negotiation
6	Object detection, recognition Avoidance or negotiation
7	Fusion of local sensors and data
8	Cooperative operations
9	Collaborative operations
10	Full autonomy

2.2 Capability

Capability is the ability to make changes in given environment. As shown in Table 2, there are a few capabilities of agents[5]. A single agent can react to environment, adapt to its changes, plan for changing the environment. Multiple agents can cooperate for making bigger changes; they are too much for a single agent.

Table 2. Capabilities of Agent

Capabilities	Description
Reactivity	The ability to respond immediately to environmental changes
Adaptability	The ability to update own knowledge through feedback on the results of acts
Deliberability	The ability to plan for achieving a given goal
Cooperation	The ability to achieve joint objectives
Sociability	The ability to request and provide assistance
Autonomy	The ability to decide and act by itself

2.3 Architecture

The kinds of multi-agent architecture are layerd[6], blackboard-based[7], hierarchical [8] architectures, etc. as you can see Table 3. While the layered architecture is proper to represent the strict vertical relationship, the blackboard-based architecture is appropriate to express the randomly organization consisted of experts. The hierarchical architectures cannot be represent only the vertical relationship, but also the horizontal relationship flexibly.

Table 3. The Kinds of Multi-agent Architectures

Architectures	Description
Layerd	It is proper to represent the strict vertical relationship.
Blackboard-based	It is proper to express the randomly organization consisted of experts.
Hierarchical	It is proper to represent both the vertical relationship and the horizontal relationship flexibly.

2.4 AI Algorithm

A variety of AI algorithms have been used in multiple agents to perform its role[9]. The algorithms and its utilization are shown in Table 4. The proper AI algorithm has to be chosen by the level of modeling CGF.

Table 4. AI Algorithms

AI Algorithm	Description
ANN	Finding patterns in data
Fuzzy	Data Abstraction
A*	Pathfinding and Graph Traversal
Expert System	Solving complex problems
Learning	Local Evolution
GA	Global Evolution

2.5 Function

An agent has several functions in order to act by itself[5]. Table 5 shows the several functions of an agent. Because the functions of an agent can be implemented by AI algorithms, they have to be decided by the level of modeling CGF, too.

Table 5. Several Functions of an Agent

Function	Description
Perception	Awareness of the environment
Control	Response to the environment
Planning	Process of achieving the goal
Diagnosis	Discovery of what is wrong
Cooperation/Coordinator/Negotiation	Process of achieving the group goal
Task-assignment	Dividing up into subtasks

2.6 Specification

Intelligent agent has knowledge of real object. Between the knowledge and the object should have a relationship with a homomorphism. In order to represent homomorphic knowledge, system specification level is necessary as like Table 6[10,11]. Level 0, 1, 2, observation, I/O (Input/Output) relation (Script-based), I/O function (State Machine-based) are influenced by I/O values and initial state not time effect. To model higher autonomous agent, level 3, 4 has to be considered, because they can be applied time effect (internal state transition followed by time; DEVS-based and AEEE-based [12]).

Table 6. System Specification Level

Specification Level	Description
Level 0: I/O observation	Script-based
Level 1: I/O relation	Script-based
Level 2: I/O function	State Machine-based
Level 3: I/O system	DEVS-based
Level 4: Structured system	AEEE-based (Hierarchical DEVS-based)

2.7 Data Type

To cope with complex objectives, an agent-based system requires integrating of symbolic and numeric data. Number crunching ability is typically a requirement of traditional models but symbol manipulation capability is essential to agent-based simulation characterized by intelligent agents. The data type used in intelligent agents is shown in Table 7.

Table 7. Data Types

Data Type	Description
Numeric	data measured or identified on a numerical scale
Symbolic	data expressed by means of symbols
Abstracted Symbolic	extracted data from symbolic data

3 Integrated Viewpoint: Level of Agent-Based CGF

Agent-based CGF with the higher autonomy is essential in order to describe a modern complex engagement followed by development of advanced weapon systems. It is necessary to develop intelligent agent technology systematically. To do this, this paper proposes the requirements for designing the level of agent-based CGF integrating viewpoints introduced in chapter 2 as shown in Table 8.

Table 8. The Requirements for Designing the Level of Agent-based CGF

CGF Levels	Autonomy	Capa-bility	Archit-ecture	AI Algorithm	Function	Speci-fication	Data Type
LAF	Remote Control	Reactivity	Layered	ANN	Perception /Control	Script-based	Numeric
SAF	Preplanned Mission/Hazard Avoidance	Adaptability/ Deliberability	Blackboard	Fuzzy/A*	Planning /Diagnosis	SM-based	Symbolic
HAF	Cooperative Operations	Cooperation /Sociability	Hierarchical	Expert System/ Learning/GA	Cooperation/ Coordinator/ Negotiation	DEVS-based /AEEE-based	Abstracted Symbolic
FAF	Full Autonomy	Autonomy	Hierarchical	Expert System/ Learning/GA	Task-Assignment	DEVS-based /AEEE-based	Abstracted Symbolic

Figure 1 shows the requirements for designing of agent-based CGF using Table 8. Currently, a representative CGF system, OneSAF has SAFs. If someone wants to design the higher CGF level than SAF, the following requirements are needed.

- In order to describe the cooperative, collaborative operation without human intervention in a complex engagement, CGF with autonomous level 8 degree, sociability, and task-assignment function is required.
- In defense M&S, in order to communicate between vertical and horizontal members organically, the hierarchical architecture is required.
- In order to expect the emergent phenomena followed by assessing and evolving its strategies, evolutionary AI algorithm (ex. Genetic Algorithm) is required.
- In order to show such a high autonomy, the knowledge of CGF is homomorphic, system specification level 3, 4 degree, DEVS-based and AEEE-based, is required.

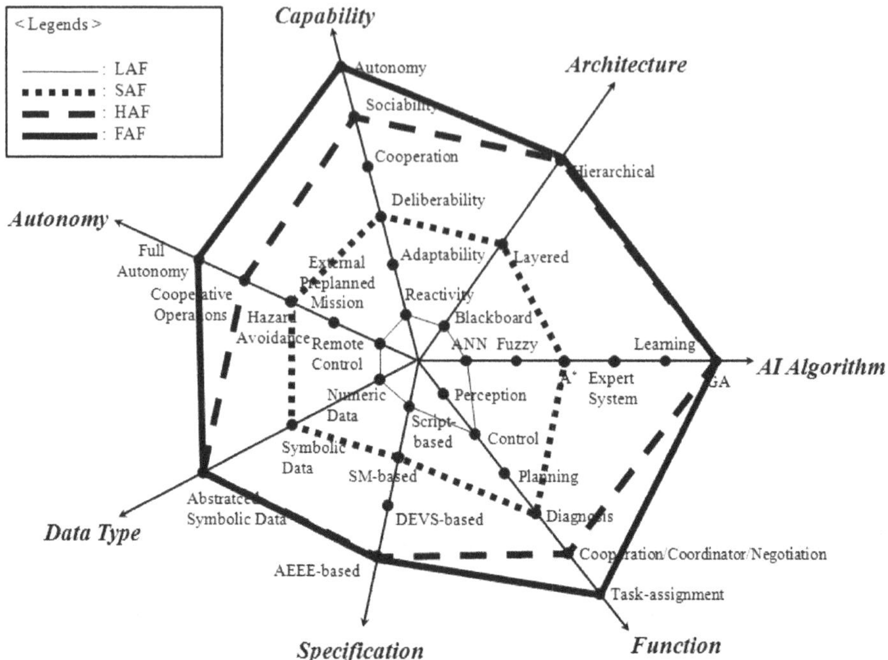

Fig. 1. The Requirements for Designing Agent-based CGF

4 Conclusion

In this paper, we analyzed the requirements for designing agent-based CGF to represent the complex battlefield environment. In order to improve the usability and affordability of defense simulation systems, one of the central goals of future M&S training should be to reduce the footprint of manned response cells by creating FAF that can replace many of the tasks currently done by human operators. Through more systematic design of agent-based CGF followed by the requirements of this paper, more various and sophisticated analysis is expected to available in defense M&S.

Acknowledgement. This research was supported by DAPA (Defense Acquisition Program Administration) and ADD (Agency for Defense Development) in Korea (UD100004KD).

References

1. Dompke, U.: Computer Generated Forces – Background, Definition and Basic Technologies. RTO SAS Lecture Series on Simulation of and for Military Decision Making, Rome, Italy, October 15-16; Stockholm, Sweden, October 18-19; Virginia, USA, October 23-25 (2001)
2. DoD 5000.59-M (January 1998)

3. Furness, Z., Tyler, J.G.: Fully Automated Simulation Forces(FAFs): A Grand Challenge for Military Training. MITRE Technical Papers (2001)
4. LTC Warren O'Donell, USA, Office of the Assistant Secretary of the Navy (Acquisition, Logistics, and Technology), Future Combat Systems Review, presentation to the committee (April 25, 2003)
5. Wooldridge, M.: An Introduction to MultiAgent Systems, 2nd edn. John Wiley & Sons (May 2009)
6. Müller, J.P.: The Design of Intelligent Agents. Springer, Heidelberg (1996)
7. `http://www.onesaf.net/community`
8. Chi, S.D., Zeigler, B.P.: Hierarchical Model-based Designs for High Autonomy Systems. J. Intelligent and Robotic Systems 9, 193–203 (1994)
9. Russell, S., Norvig, P.: Artificial Intelligence: A Modern Approach, 3rd edn. Pearson (2010)
10. Zeigler, B.P.: Object-oriented Simulation with Hierarchical, Modular Models: Intelligent Agent and Endomorphic Systems. Academic Press (1990)
11. Zeigler, B.P., et al.: Theory of Modeling and Simulation. Academic Press (2000)
12. Chi, S., You, Y., Jung, C., Lee, J., Kim, J.: Fames: Fully agent-based modeling & emergent simulation. In: Proc. 2009 Spring Simulation Multiconf., March 22-27, Society for Computer Simulation International, San Diego (2009)

Study of Multi-agent Based Combat Simulation for Orientation of OODA Loop

Aya Fusano, Hiroshi Sato, and Akira Namatame

National Defense Academy, Department of Computer Science, Hashirimizu 1 10 20, 236-8686 Yokosuka, Kanagawa, Japan
{em49036,hsato,nama}@nda.ac.jp
http://www.nda.ac.jp/cc/cs/eng/

Abstract. OODA Loop is strategic decision making theory that military operations are basically a sequence of decision making process that consists four main points as observe, orient, decide, and act. Orientation is a stage where the decision-maker adds the evaluation to Information on external situation. We think that Orientation consists of default and exception rule. In this study, we paid attention to orientation of OODA Loop, and investigated how two rules of orientation would influence the result of combat.

1 Introduction

Boyd's OODA Loop is strategic decision making theory in military forces that military operations are basically a sequence of decision processes or cycles. These cycles have four main points: observe, orient, decide, and act (OODA) as Fig. 1. Observation part is a stage where surrounding information is observed. Orientation is a stage where the decision-maker adds the evaluation to information on external situation understood at the previous stage. This stage where evaluation engine processes raw data to information including evaluation decides whether the loop rotates well. And Boyd emphasized that this stage was the most important. Next, it is a decision stage to select the means as action from information including evaluation invented by situation assessment. Actually executing it is action part. Because the situation changes by the action, The decision-maker will return first stage of the loop and observe it again[1] [2] [3].

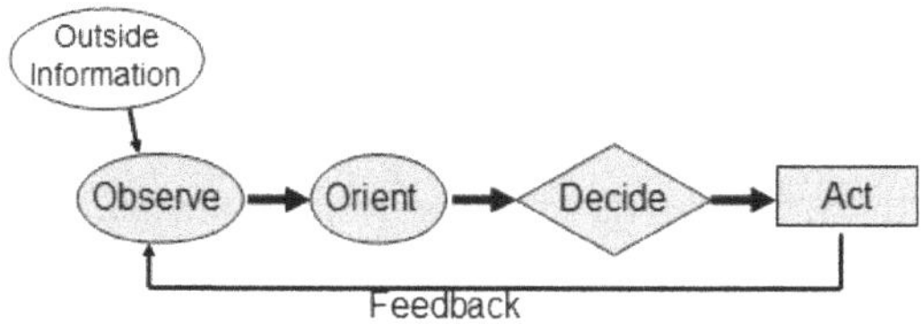

Fig. 1. Conceptual diagram of OODA Loop

J.-H. Kim et al. (Eds.): AsiaSim2011, PICT 4, pp. 432–440, 2012.

The core of OODA Loop is to obstruct enemy's OODA Loop by using OODA Loop whose performance is better than the enemy. The interaction with the enemy is a main object in OODA Loop theory unlike the PDCA cycle and other decision theory.

An initial investigation about speed of OODA Loop has already been done by Tighe. Tighe conducted the agent-based combat simulation to examine Boyd's theory, and insisted rotating fast was more effective than improvements of maximum speed, initial strength, or probability of hit. He made OODA of one force fast, other performance of another force better. The result was overwhelming superior force having fast OODA[4].

The research about other elements than speed has not been considered yet. For instance, each elements as observe and each step of OODA Loop. And OODA Loop theory pay attention to view point of 1vs1, and verification from plurals or network perspective has not been considered too. The biggest difference of 1vs1 OODA and plurals OODA is number of OODA Loop that has the influence to other OODA. For 1vs1, interacting OODA is only each OODA. On the other hand, for plurals, enemies and friend's grouped OODA is interactive.

Orientation is a stage where the decision-maker adds the evaluation to Information on external situation. We think that Orientation consists of default and exception rule. When planning next action based on information, we use a default rule and some exception rule. Not only military but anyone needs to consider default plan and exception, in order to carry out the plan smoothly. For example, when default plan about battle is the advance and attack to enemy, if there are many enemies than friends, the exception rule of changing a plan and retreating functions. At the orient part, plans based on information about surrounding are evaluated by default rule and exception rule.

We experimented especially paying attention to observation and orientation of OODA Loop of grouped combatant.

Lanchester pointed out the importance of analysis of war quantitatively. Mathematical formulae for calculating the relative strengths of a predator/prey pair called Lanchester's laws are a research method used still widely. But this model depends only two elements of initial strength and abrasion coefficient and not depends strategy of combat unit. Moreover, some assumptions are not filled at real combat.

Combat simulation using multi-agent can bring spatial elements and heterogeneity that moves of unit, information, strategy and so on. We therefore experimented on combat simulation focused on considering about the default and exception of orient part and relation between orient part and result of combat..

The paper is organized as follows: Section 2 describes agent and OODA Loop, and Section 3 reports on the results of simulation experiments. Section 4 we discuss and Section 5 summarizes this paper.

2 OODA Loop of Agent

As a subject of simulation, we use EINSTein model developed by the Center for Naval Analyses of USA [6].

2.1 Battle Field

The putative combat battlefield is represented by a two-dimensional lattice of discrete sites. Each site of the lattice may be occupied by one of two kinds of agents: red or blue. The initial state consists of user-specified formations of red and blue agents positioned anywhere within the battlefield.

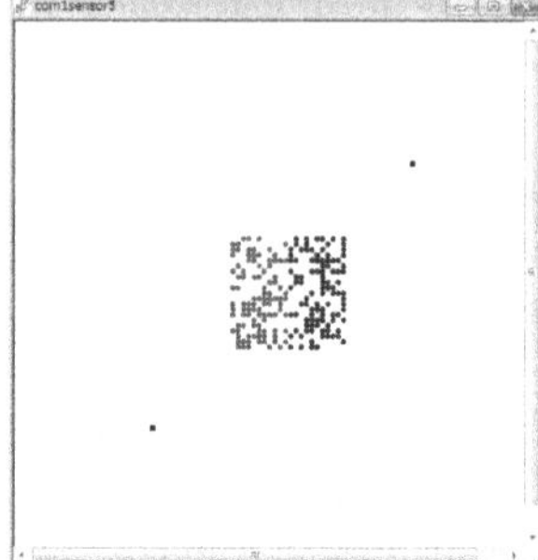

Fig. 2. Battle field of EINSTein

2.2 Agent

When the attack more than regulation is received, agent will be killed. Once killed, an agent is permanently removed from the battlefield. Basic capability contains fire range, firepower, sensor range, combat threshold rang, movement range, maximum simultaneous targets number and so on. They can see agents inside of sensor range and, and it corresponds to observation part of OODA Loop.

2.3 OODA Loop of Agent

Agents act based on the information that they obtained from inside their own sensor range.

As first step of decision, the agent gather information of their sensor range as observation of OODA Loop.

As a second step, orientation stage, the agent calculates the penalty function of all places where he can be moved. The agent's preference as how agents want to move is defined by six-component personality weight vector as default rule. Each weight shows how much agents are attracted or repulsed by other friendly and enemy agents or flags.

$$\vec{w} = (w_{AF}, w_{IF}, w_{AE}, w_{IE}, w_{FF}, w_{EF})$$
$$s.t. -1 \le w_X \le 1, \sum |w_X| = 1 \tag{1}$$

Penalty function is given by:

$$z(B_{xy}) = \frac{1}{\sqrt{2}r_s}\left[\begin{array}{l} \frac{w_{AF}}{N_{AF}}\sum_{i\in AF}^{N_{AF}} D_{i,B_{xy}} + \frac{w_{AE}}{N_{AE}}\sum_{j\in AE}^{N_{AE}} D_{j,B_{xy}} \\ + \frac{w_{IF}}{N_{IF}}\sum_{i\in IF}^{N_{IF}} D_{i,B_{xy}} + \frac{w_{IE}}{N_{IE}}\sum_{j\in IE}^{N_{IE}} D_{j,B_{xy}} \end{array}\right] + w_{FF}\frac{D^{new}_{FF,B_{xy}}}{D^{old}_{FF,B_{xy}}} + w_{EF}\frac{D^{new}_{EF,B_{xy}}}{D^{old}_{EF,B_{xy}}} \tag{2}$$

The agent changes the rule of the orientation by the situation. Exception-rule represents condition modifier. If a certain condition was satisfied, then penalty function is changed over partly.

About "Combat", if an agent senses that it has less than a threshold advantage of surrounding forces over enemy forces, then it will choose to move away from enemy agents rather than moving towards them. Combat exception-rule is implemented by:

$$\begin{array}{ll} \mathrm{N}_{\mathrm{friend}}(r_T) - \mathrm{N}_{\mathrm{enemy}}(r_S) \geq \mathrm{N}_{\mathrm{Combat}} & \rightarrow w_{3,\mathrm{default}}, w_{4,\mathrm{default}} \\ \mathrm{N}_{\mathrm{friend}}(r_T) - \mathrm{N}_{\mathrm{enemy}}(r_S) < \mathrm{N}_{\mathrm{Combat}} & \rightarrow -|w_3|, -|w_4| \end{array} \tag{3}$$

Exception-rule "Cluster" specifies the threshold number of friendly agents that must be within a given agent's threshold range beyond which that agent will no longer seek to move toward friendly agents.

Where B_{xy} is the (x, y) coordinate of battlefield B; AF, AE, IF and IE represent the sets of alive friends, alive enemies, injured friends and injured enemies within the agent's sensor range r_S ; N_X is the total number of elements of type X within the agent's sensor range; $D_{A,B}$ is the distance between elements A and B; FF and EF denote the friendly and enemy flags;

$D^{new}_{x,y}$ and $D^{old}_{x,y}$ represent new and current distance of x and y.

As a third step, decision stage, the agent decides to move to position where the penalty function is the least. Next step, as action stage, agent actually moves and attacks to enemy in threshold range.

This is one cycle of agent's OODA Loop. Afterwards, agent returns first stage of OODA Loop and begins to observe the surrounding environment again.

3 Experiment

3.1 Experimental Setting

Experiments were done as follows to evaluate influence of orientation part on the combat. The baseline scenario is that both red and blue forces have aimed to beat one another force. Experiments were conducted using the following parameters:

Number of steps in a trial run: 499
Size of battlefield: 100*100

Red and blue forces had common parameters shown in Table1 to investigate the influence that orientation gives combat. Red force has the same value as a blue force excluding the sensor range.

Table 1. Common parameter of scenarios

parameter	Red	blue
Initial number of agents	100	100
Movement range	1	1
Threshold range	1	1
Fire range	5	5
Sensor range	Varies (5,10,15)	15
Personality weight vector $(w_{AF}, w_{IF}, w_{AE}, w_{IE}, w_{FF}, w_{EF})$	Varies	
Meta personality Cluster	varies	
Meta personality Combat	varies	

Default-rule and exception rule took the value as Table2 and Table3, red and blue were changed similarly.

At scenario1,2,3 of Table2, because the vector to enemy is negative, agent doesn't to approach the enemy. Agents are very cowardly as long as meta-rule doesn't function. On the contrary, at scenario 9, 10, 11, agent doesn't approach the friends. Agents are very brave.

Table 2. Varies default rule as weight vector $(w_{AF}, w_{IF}, w_{AE}, w_{IE}, w_{FF}, w_{EF})$

Scenario	W_{AF}	W_{AE}	W_{IF}	W_{IE}	W_{FF}	W_{EF}
1	1	-0.5	1	-0.5	0	0
2	0.8	-0.3	0.8	-0.3	0	0
3	0.5	0	0.5	0	0	0
4	0.4	0.1	0.4	0.1	0	0
5	0.3	0.2	0.3	0.2	0	0
6	0.25	0.25	0.25	0.25	0	0
7	0.2	0.3	0.2	0.3	0	0
8	0.1	0.4	0.1	0.4	0	0
9	0	0.5	0	0.5	0	0
10	-0.3	0.8	-0.3	0.8	0	0
11	-0.5	1	-0.5	1	0	0

At scenario A, agent tries to run away from the enemy when the enemies are more than friends. At the scenario F, the agent doesn't try to run away even if the number of friends decreases more than 10 enemies. At the Scenario F, the agent orient as weight vector regardless of number of surrounding enemies and friends.

Table 3. Varies exception rule

scenarios	Cluster (alive)	Cluster (injured)	Combat (alive)	Combat (injured)
A	2	2	0	2
B	0	0	-2	0
C	0	0	-4	-2
D	0	0	-6	-4
E	0	0	-8	-6
F	0	0	-10	-8
G	nonuse	nonuse	nonuse	nonuse

3.2 Results

There is various kinds of evaluation method of victory or defeat in the war, such as remaining number, mortality rate, number of accomplishment of a goal people, combat entropy, and so on. In this paper, we adopt the number of survivors as a standard of victory or defeat.

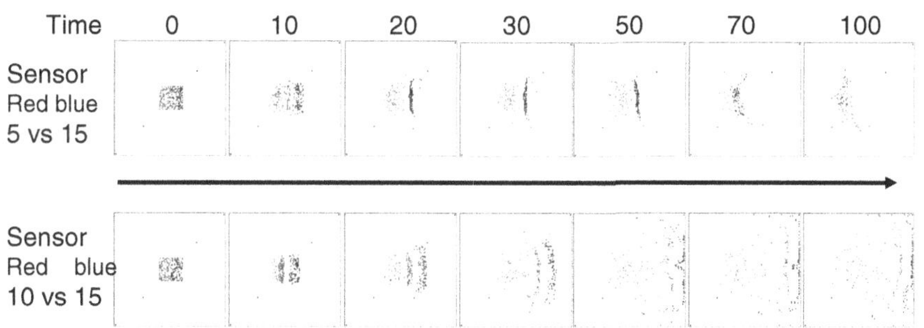

Fig. 3. Battle scene Scenario A-6

Fig.3 is the Example of combat scenarioA-6. Blue force has better performance of observation than red force at both of two case, but, blue force wins at the case of sensor 15 vs 5 and is defeated at the case of sensor 15 vs 10.

Table 4. showed results of all game's ending as the 499th step. It showed that blue force can not win though blue force had superior ability of sensor range 15 and red force had sensor range 10.

Table 4. Results of all games

cowardly ← → brave (columns 1–11); rows: 15vs5 A–G cowardly → brave, 15vs10 A–G cowardly → brave

Default Ex╲	1	2	3	4	5	6	7	8	9	10	11
15vs5 A	t	t	t	b	b	b	b	b	b	b	b
15vs5 B	t	t	t	b	b	b	b	b	b	b	b
15vs5 C	t	t	t	b	b	b	b	b	b	b	b
15vs5 D	t	t	t	b	b	b	b	b	b	b	b
15vs5 E	t	t	t	b	b	b	b	b	b	b	b
15vs5 F	t	t	t	b	b	b	b	b	b	b	b
15vs5 G	t	t	r	r	b	b	b	b	b	b	b
15vs10 A	t	t	t	r	r	r	r	r	r	r	t
15vs10 B	t	t	t	r	r	r	r	r	r	r	t
15vs10 C	t	t	t	t	t	t	t	t	t	t	t
15vs10 D	t	t	t	t	t	t	t	t	t	t	t
15vs10 E	t	t	t	t	t	t	t	t	t	t	t
15vs10 F	t	t	t	b	t	t	b	t	t	t	b
15vs10 G	t	t	t	t	t	b	b	b	b	b	b

4 Discussions

Fig. 4 is average of ten games number of alive agents of scenario A(sensor 5vs15). If the vector to the enemy is not 0 or less, it is understood that the result of the combat is not different regardless of the weight-vector.

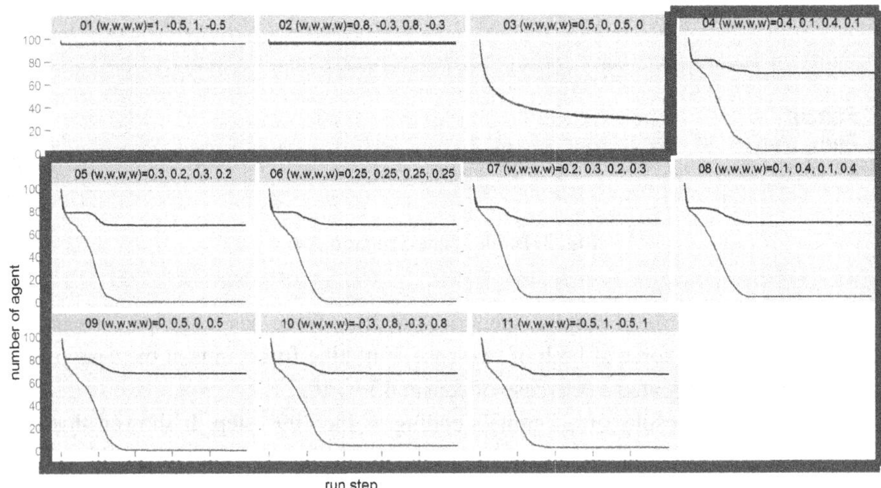

Fig. 4. Comparison concerning default rule at sensor Red 5 vs Blue 15 and scenario A ex-rule(2,2,0,2)

Figure.4. shows that change of default rule doesn't exert big influence on combat when weight vector to enemy positive.

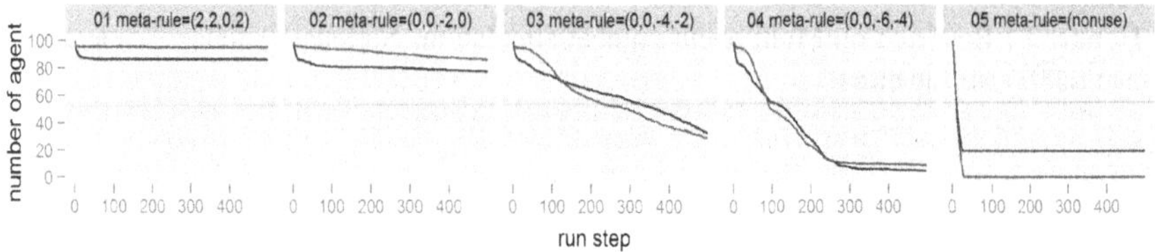

Fig. 5. Comparison concerning exception rule at sensor red10 vs blue 15 and default-rule (0.25,0.25,0.25,0.25,0,0)

Figure. 5. is results of scenario 6, and red sensor range 10 for comparison concerning exception rule. Figure.5. shows that Exception rule exert a big influence on combat.

The performance of the observation counterprductivity works when exception-rule expresses not courageous orientation even if the weight vector of default is brave.

5 Conclusion

We use the agent-based simulation in order to investigate how the orientation rule effects on the combat of grouped agent. What we have shown is that The result of a combat is influenced by not Default rule but Exception rule and The performance counterproductively works when Exception-rule is bad even if default rule is good. So, we think that When determining a guideline, we should consider the exception matter well from the basic policy.

Actually, it is difficult to know the enemy's orientation rule and sensor range, not to mention OODA Loop. Therefore, we should forecast victory or defeat from understand surely, for instance, transition of combat that friends understand, and take better measures at next decision.

We think that network structure as their crowd can be used for the index of victory or defeat, and as future works, we will analyze it and examine whether OODA Loop guessed from those data.

References

1. Osinga, F.: Science, Strategy and War: The Strategic Theory of John Boyd. Eburon Academic Publishers (2005)
2. Kitamura, J., Kitamura, A.: Doctrine of United States Marine Corps. Huyo-Syobo-Syuppan Ltd. (2009) (in Japanese)
3. Nakamura, Y.: A art of Strategy decision making of U.S. military type for business. Toyo-keiza Inc. (2006) (in Japanese)
4. Tighe, T.R.: Strategic Effects of Airpower and Complex Adaptive Agents: an Initial Investigation. Air Force Institute of Technology, USA (1999)

5. Masuda, N., Konno, N.: Complex Networks -From the base to the application. KindaiKagakusya Ltd. (2010) (in Japanese)
6. Ilachinski, A.: Artificial War: Multiagent-Based Simulation of combat. World Scienc. (2004)
7. Honda, T.: Research on Multi-agent combat simulation. National Defence Academy of Japan (2007) (in Japanese)

Clustered Parallel Coordinates with High-Speed k-Means Algorithm and Out-of-Core Feature

Chi Zhang, Naohisa Sakamoto, and Koji Koyamada

Graduate School of Engineering, Kyoto University, Japan
zhang.chi.24e@st.kyoto-u.ac.jp

Abstract. In this paper, we introduce the clustered parallel coordinates based on the progressively processing technique to solve the problems of clutter and memory limitation when visualizing and exploring large-scale data. The clustering method is based on the k-means method so that it is possible to partition large-scale datasets with relatively high speed. The progressively processing technique called out-of-core feature can enable the system processes input data part by part, which reduces the requirement of memory capacity.

1 Introduction

As an important visual representation, parallel coordinates plot is widely used for visualizing and analyzing multivariate discrete data. However, when exploring large-scale data, there is an urgent need for the solution of data cluttering in parallel coordinates and the limitation of the memory in a computer. Clutter is characterized by crowded and disordered visual entities that obscure the structure in visual displays. Clutter is certainly undesirable since it hinders viewers' understanding of the displays content. However, when the dimensions or number of data items grow high, it is inevitable for users to encounter clutter, no matter what visual method is used [8]. Out-of-core or External memory algorithms are algorithms that are designed to process data that is too large to fit into a computer's main memory at one time. In our system, we apply the clustering algorithm and the out-of-core feature to parallel coordinates to solve the problems. The clustering function partitions the input data to several clusters and each cluster is shown as a polygon. By means of clustered parallel coordinates, not only can we display much more information in a window with limited size, but also the clusters show the trend of the input data and mitigate the cluttering problem. The out-of-core algorithm makes it possible for our programme to process large-scale data even if the size of the data is larger than the memory capacity. Processing with this feature, large-scale data is divided into many parts and each part can be loaded into the main memory. When a part is finished to be processed, the memory will be cleared and the next part will be loaded. This technique is used for calculating the center points of the clusters and computing the euclidean distance in our system.

J.-H. Kim et al. (Eds.): AsiaSim2011, PICT 4, pp. 441–455, 2012.

2 Related Works

The k-means clustering algorithm was described in detail by Hartigan [1]. The most popular algorithm for k-means was developed by Lloyd [7]. Due to its algorithmic simplicity and fast computing speed, k-means clustering becomes an option for us to partition large-scale data. The traditional k-means algorithm was improved by Elkan [5] and Hamerly [6]. The distance bounds and the triangle inequality were used to accelerate the k-means method. Michaels et al. used the clustering algorithm and visualization method to analyze large-scale gene data [2]. Hierarchical parallel coordinates and binned parallel coordinates has been used for exploring large-scale datasets as well [3,4]. The binning parallel coordinates plot partitions input data to several bins to avoid the cluttering problem and simplifies the input data. Therefore, after the binning processing, the input data can be rendered very fast. However, since the width of every bin is definitely the same, to avoid the loss of the details about the distribution of data points, we use clustered parallel coordinates instead of binned parallel coordinates.

3 Proposed System

Our proposed system is used for partitioning and visualizing large-scale data. First, the supported input data is loaded as table object, in the structure of which a row indicates a multi-dimensional date point with columns indicating the dimensions. After loading the input data file, the clustering function is called to partition the original data. Since the number of clusters needs to be specified in advance as the value of k, it is set as an optional parameter in the command-line programme. Clusters are shown as polygons with different colours, or an additional axis is added to show the cluster indices. Therefore, users can select a particular cluster and hide other clusters if needed. Although other algorithms can also manage to partition data, we use k-means algorithm because it is relatively simple and fast, and the accuracy is enough for us to visualize and explore large-scale data.

3.1 Clustering Algorithm

The clustering function based on k-means algorithm is used for partitioning input data to a number of clusters. In Lloyds algorithm, first, it assigns a cluster ID to each row of the input matrix randomly. Then the center of every cluster with the same ID is calculated. After calculating the euclidean distance between the center of each cluster and data points, a new cluster ID will be assigned to a data point if the euclidean distance between this point and another center of a cluster with relevant ID is shorter. The loop of the calculation of the euclidean distance and the reassignment of cluster IDs will not stop until there is no change in the cluster ID of every data point. This algorithm is simple and easy to be implemented, however, when partitioning large-scale data, it takes a lot of time because the distance between each point and cluster center has to be calculated. In order to ulteriorly shorten the computation time, a hign-speed k-means algorithm is presented by Elkan and improved by Hamerly, which is applied to our system.

Based on Elkan's algorithm, Hamerly's algorithm uses the lower bound to eliminate the unnecessary innermost loop computation in Lloyd's algorithm so as to save time. The main steps of Hamerly's algorithm can be described briefly as follows:

- Assign cluster center, calculate the upper bound and lower bound.
- For a data point, if the upper bound is no more than the lower bound, the center assignment cannot change.
- Otherwise, calculate the distance between the center and its second closest other center. If the half of this distance is larger than the upper bound, the triangle inequality guarantees this center is the closest center of this data point.
- If the above two conditions cannot be satisfied, calculate the distance with the innermost loop to determine the correct assignment.

The Hamerly's algorithm takes advantage of these two conditions to avoid some of the innermost loops of Lloyd's algorithm, hence accelerate the computation. Besides, an alternative method called "delta updates" is used for updating the movement of cluster centers, which also reduces computation and makes the algorithm faster. In this method, if the assignment has changed, the algorithm adjusts the cluster sums and counts for the two affected centers. It subtracts the influence of the point from its old assigned center, and adds its influence to the new assigned center [6]. When the number of clusters k is small, this optimization is expected to be useful.

3.2 The Progressively Processing Technique

Since what we analyze is large-scale data, the file size is usually larger than a computer's main memory capacity. Therefore, it is impossible to read all of the data into memory at once. As a solution, a class named "`OutOfCore`" is used for processing large-scale data. During computing process, only the necessary part of the input data is loaded into the memory. When a calculation is complete, the memory will be cleared and another part of the input data will be loaded for the next computation.

In our system, a data set is imported as a table object, for example, with n rows and m columns, as shown in Figure 1(a). In a table object, a data point is represented as a row, in which the dimensions of the point are represented as the columns. Therefore, a point can be treated as a row in a table object, which is illustrated in Figure 1(b). Data sets in some other formats such as the GrADS format can be converted to table data in advance and then be processed by our programme. Some libraries which are used for converting data format and generating supported table data have been added into our system to integrate the conversion function and improve user experience.

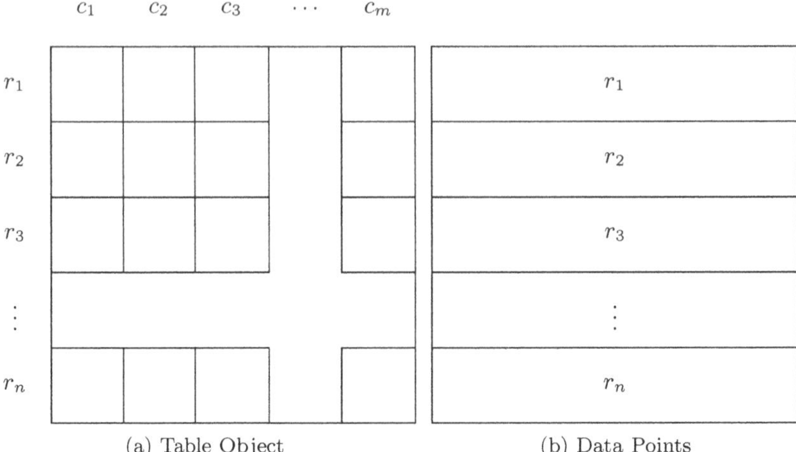

(a) Table Object (b) Data Points

Fig. 1. Input Data

For table object, the input data is stored as a matrix and a data point is stored as an array in the programme. When it is processed with out-of-core feature, a function is used to get the index of the data point which needs to be processed. Then the function checks that if there is a corresponding array in the main memory. If the array is not in the memory, the existing data point is removed and required data point is loaded into the memory.

In Lloyd's algorithm, the random assignment of the center points can be implemented by the random assignment of cluster ID. Then the center points of the data points with the same cluster IDs are calculated, which are the cluster centers. Similarly, the movement of center points can also be implemented by reassigning cluster IDs. When we partition the table object with the out-of-core feature, the progressively processing technique is used when the center point of each cluster is calculated and the euclidean distance is computed. The flow chart about the partitioning process in our system is illustrated as Figure 2. When the system calculate the center points, the data points with the same cluster ID is loaded into the memory one by one and their center point is computed. In this processing, only the data point required for the current calculation is loaded into the main memory. After the calculation is finished, the memory will be cleared and another data point with the same cluster ID will be loaded into the memory. In this way, the programme loads the data points one by one and calculates their summation. The loop for loading, calculating and clearing will not be ended until the summation of all data points is figured out. Then the center point of this cluster is calculated based on the summation. This procedure will be executed k times, where k is the number of cluster centers. Then, all of the center points are figured out. Similarly, when the euclidean distance is calculated, instead of loading all points data into the memory at once, only one data point is loaded each time. After the distance between the center points and this data points being calculated, the memory will be cleared and the next data point will be

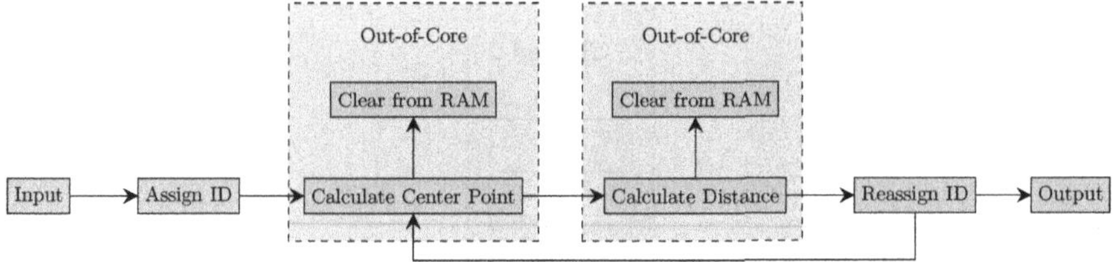

Fig. 2. Partitioning Computation with Out-of-Core Feature

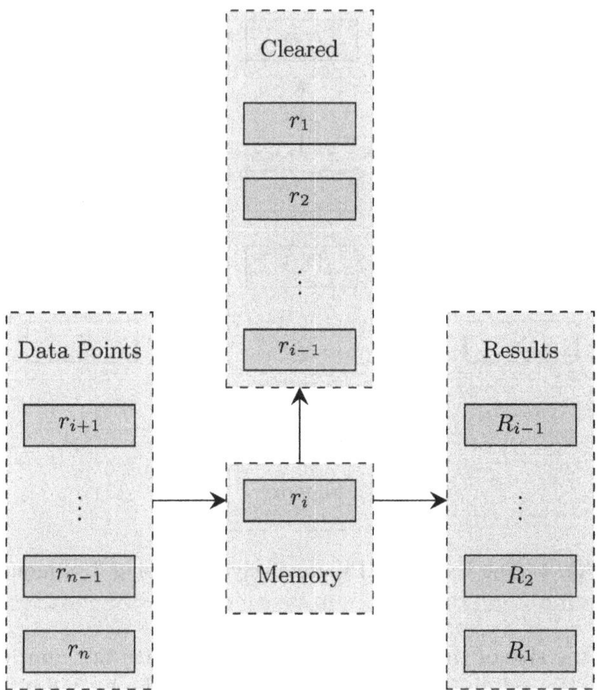

Fig. 3. The Progressively Processing Technique

loaded. The loop will be repeated until all of the necessary euclidean distances are figured out. The progressively processing technique is illustrated as Figure 3. In this way, the system can partition and render large-scale data despite the limitation of the memory capacity.

The processing mentioned above is a kind of extreme case, as the progressively processing technique calculates only one data point every time. However, in most conditions, the size of main memory is not so small that only one point is allowed to be loaded at a time. To extend the algorithm and make it possible to load more data points into the memory at a time, we set a threshold in advance to specify how much memory capacity is allowed to be used, which is the "cache" option in our system. With this option, users can specify a suitable cache size

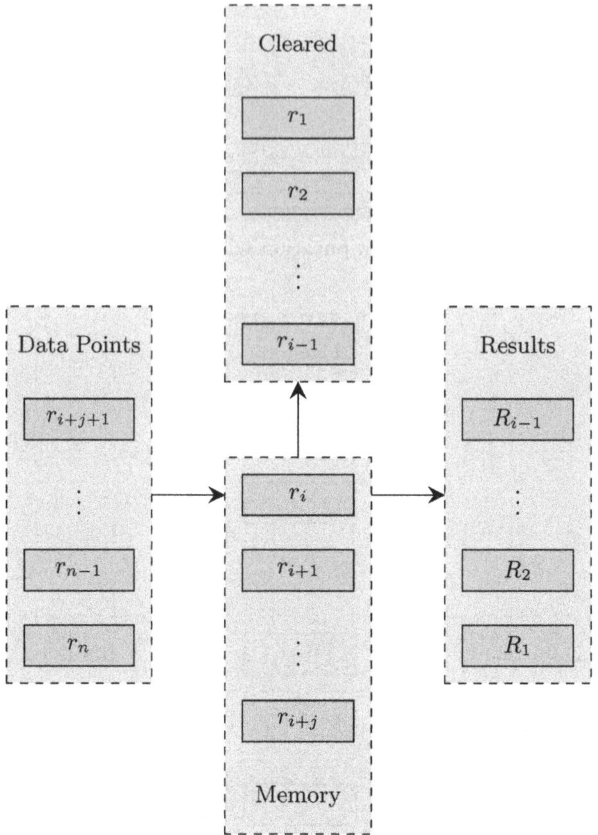

Fig. 4. The Extended Progressively Processing Technique

depending on the size of the main memory to accelerate the computation. If the "cache" option is used, more than one data point will be loaded into the memory and processed, which can be illustrated as Figure 4.

3.3 Clustered Parallel Coordinates Plot

Parallel coordinates plot is considered to be an effective tool to represent N-dimensional data in 2-dimensional space. However, for large-scale datasets, too many polylines are shown in a window with limited size, which causes the clutter problem. Thus, the effectiveness of parallel coordinates on large-scale data is reduced. Clustering is one of the most popular method used for solving this problem. Clustering algorithm partitions the input data to a number of clusters before the data is plotted. As what is introduced, we use k-means algorithm to partition the large-scale input data, which is a type of data-centric clustering.

After the input data is partitioned to a number of clusters, it needs to be plotted in a way to represent the clustering feature and avoid the clutter problem.

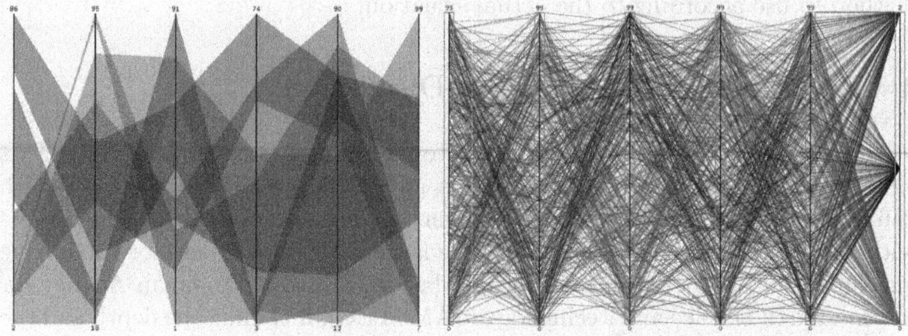

(a) Clusters are plotted as polygons (b) Another axis is added to show cluster ID

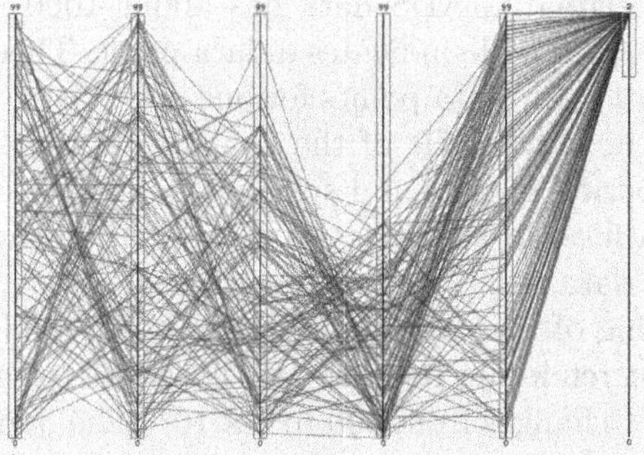

(c) Explore One of the Clusters

Fig. 5. Clustered Parallel Coordinates

In our system, there are two ways to plot the clustered parallel coordinates. In the first method, each cluster is plotted as a polygon in the parallel coordinates plot. The colour of a cluster is decided by the predefined colour map in the programme, which is illustrated in Figure 5(a). There is also another way to show the clustered parallel coordinates. In the second way, an additional axis is added to show the cluster ID, as illustrated in Figure 5(b). Therefore, users can explore each cluster simply by selecting the relevant part on the axis of cluster ID, as shown in Figure 5(c). The number of clusters is set as an optional value k with the default value of 3. In case that users just want to show a parallel coordinates plot without the clustering computation, the value of k can be set to 0 to disable the clustering feature.

Both of the two methods have their advantages and disadvantages. Plotting the clustered data points as polygons reduces the computation burden of the GPU. The region between the maximum value and the minimum value on each axis in a cluster is represented as a polygon. Therefore, there is no need for GPU to render all of the data points as polylines. However, on the other hand, some details about the data points in a cluster are ignored. If we add another axis to show the cluster indices, the details are mostly preserved, but it takes more time for GPU to render these data points as polylines. Users can choose which method to use according to the actual situation.

4 Experimental Results and Discussion

We implement our experiments on a desktop with Intel Core 2 Duo E8400 CPUs and 4 GB memory. The graphic card is an NVIDIA GeForce 8600 GT with 256 MB memory. The software environment is based on openSUSE 11.4 x86_64 with Linux kernel version of 2.6.37.6. The large-scale data we use for visualizing and analyzing is the ocean information dataset provided by Japan Agency for Marine-Earth Science and Technology (JAMSTEC), including the depth and the current velocity data about the ocean in the world from the year 2000 to 2009. The original data is in the format of GrADS and we convert it to the format of table data. The original GrADS data has about 1000×1000 grids for the data on one day, and each node indicates a data point. Therefore, the converted table data has about 1×10^6 data points for one day's data. The number of days can be specified in the control file of the GrADS dataset. Hence it is possible for us to select the period of time and specify the number of the data points in converted table data files. In this way, we can test the converted datasets with different numbers of data points and compare the results.

First, the processing of the two rendering methods, rendering clustered data points as polygons or rendering them as polylines, are tested when the number of plotted data points changes from 300 to 3×10^5. The dimensions of the input data is set to 6 and only the rendering time is recorded. The results are listed as Table 1. From the table we can know that when the number of data points

Table 1. Processing Time of Two Rendering Methods

	Average Time (msec)	
Number of Points	Polygons	Polylines
300	42.13	0.3283
600	42.76	0.7082
1000	46.76	1.140
1500	45.38	1.425
3000	44.89	2.740
5000	46.29	4.551
8000	45.68	7.272
10000	45.51	8.120
15000	46.30	12.04
30000	46.90	22.92
50000	45.28	37.40
80000	46.78	57.17
100000	46.14	74.20
150000	45.85	108.1
300000	47.45	226.8

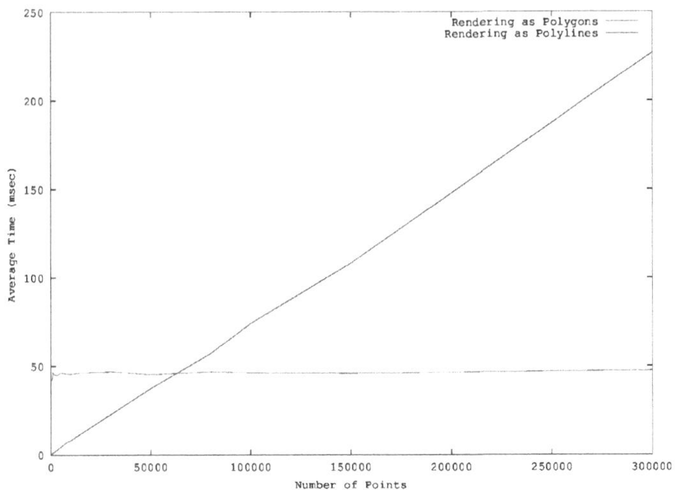

Fig. 6. Processing Time of Two Rendering Methods

is small, the processing speed is relatively fast if we choose to render the data points as clustered polylines. While if the number of data points is very large, rendering data points as polylines will cost much more time because the GPU has to render and plot each data point as a polyline. Whereas it will take relatively less time to render the data points as polygons in this condition because the number of polygons equals to the number of clusters. Therefore, the time cost is nearly unchanged when the number of points increases from 300 to 3×10^5.

According to the data from Table 1, the relationship between the number of data points and the processing time of the two rendering method can be illustrated as Figure 6. In this figure, the red line indicates the method of rendering data points as polygons and the blue line indicates the method of rendering data points as polylines. From the figure we can see that the red line is horizontal, which shows a relatively stable time cost. Whereas the blue line shows a linear relationship between the number of the points and the processing time. Therefore, the point of intersection shows a balance point, at which the two rendering methods spend the same time on rendering data points.

The average processing time of the two kinds of k-means clustering algorithm, Lloyd's algorithm and Hamerly's algorithm, is tested and recorded. Two experiments are done to compare the time cost of two algorithms. In the first one, we change the number of input data points and keep the number of clusters unchanged. In the second one, the number of clusters is changed while the number of input data points is a constant value. The results are listed as Table 2 and Table 3 respectively.

The figures of the results about Table 2 and Table 3 are shown in Figure 7 and Figure 8 respectively. Comparing with Lloyd's algorithm, Hamerly's algorithm improves the efficiency and saves the computation time by about 50% - 80%.

Table 2. Processing Time of Conventional and High-Speed k-Means Algorithm (8 dimensions, 5 clusters)

	Average Time (msec)	
Number of Points	Conventional	Improved
300	0.9145	0.2392
600	3.379	1.044
1000	7.391	2.307
1500	15.81	4.400
2000	22,72	4.858
2500	25.21	11.71
3000	46.43	16.27
3500	52.66	15.84
4000	103.8	19.12
5000	124.5	26.03
6000	127.2	32.30
7000	177.8	69.90
8000	182.7	84.57
9000	196.2	101.1
10000	294.8	118.1

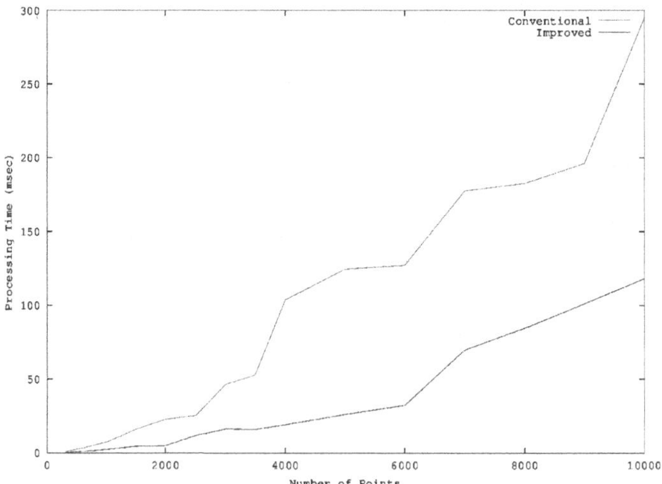

Fig. 7. Processing Time of Conventional and High-Speed k-Means Algorithm (8 dimensions, 5 clusters)

Besides, the larger the number of clusters is, the better the acceleration effect of Hamerly's method will be.

In order to test the progressively processing technique, an option is set in our programme to force enable the out-of-core feature. Therefore, even if the

Table 3. Processing Time of Conventional and High-Speed k-Means Algorithm (10000 points, 8 dimensions)

	Average Time (msec)	
Number of Clusters	Conventional	Improved
3	92.54	56.58
6	362.4	95.68
10	448.5	171.8
15	480.5	224.7
20	624.5	236.9
25	909.0	264.1
30	1005	273.6
35	1112	279.8
40	1229	290.7
45	1398	316.9
50	1512	334.8
55	1511	350.3
60	1573	352.1
65	1964	397.1
70	1990	397.5

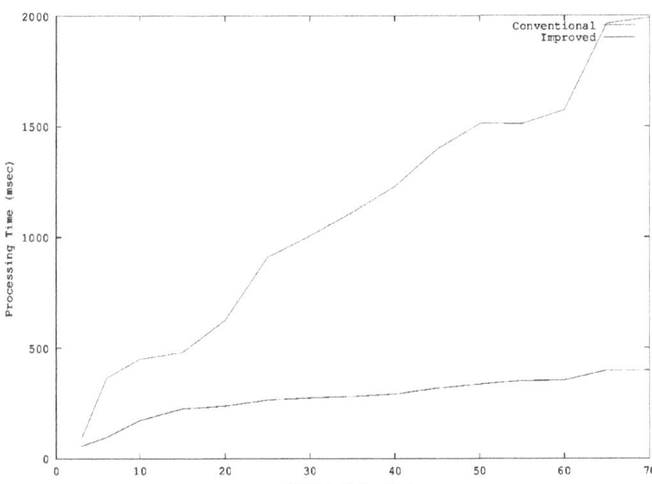

Fig. 8. Processing Time of Conventional and High-Speed k-Means Algorithm (10000 points, 8 dimensions)

size of the input data is smaller than the computer's main memory, it can be processed with the out-of-core feature as well, taking advantage of which, it is possible to compare the processing time of the out-of-core processing with the in-core processing. The results are listed as Table 4. In this table we can find that

Table 4. Processing Time of In-Core and Out-of-Core

	Processing Time (sec)	
Number of Points	In-Core	Out-of-Core
1×10^6	1.538	3.166
2×10^6	3.607	6.934
3×10^6	8.332	13.54
4×10^6	14.94	21.98
5×10^6	23.38	32.19
6×10^6	33.61	44.75
7×10^6	45.69	58.28
8×10^6	60.63	75.18
9×10^6	77.16	93.32
1×10^7	96.12	113.4
1.2×10^7	139.9	159.6
1.4×10^7	193.0	215.2
1.6×10^7	252.9	276.8
1.8×10^7	319.9	357.3
2×10^7	397.1	437.6

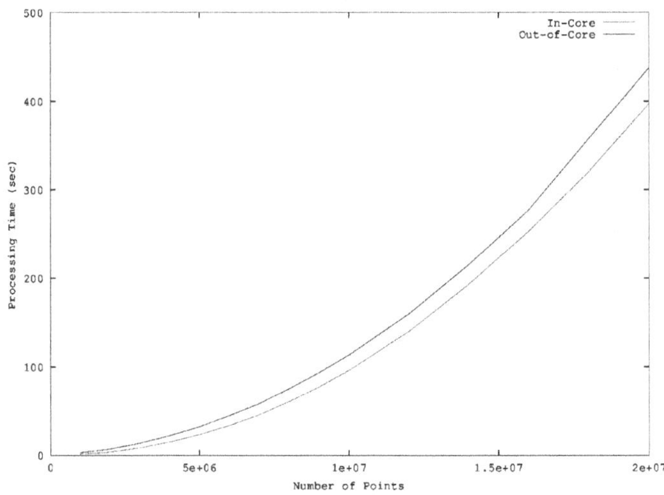

Fig. 9. Processing Time of In-Core and Out-of-Core

comparing with the in-core processing, the time cost of out-of-core processing is acceptable if the algorithm is not very complex. Furthermore, if we plot the results as Figure 9, where the red line indicates the in-core processing while the blue line indicates the out-of-core processing, we can see clearly that as the number of data points increases from 1×10^6 to 2×10^7, the proportion of the extra time spent by out-of-core processing in the total duration becomes less and

Table 5. Processing Time of Out-of-Core in Different Cache Sizes (1×10^7 points)

Cache Size (MB)	Processing Time (sec)
0	60.73
0.1	60.34
0.2	60.12
0.3	60.02
0.4	60.40
0.5	60.78
1	44.31
2	42.78
3	41.67
4	43.00
5	41.67
10	43.18
20	42.87
30	44.75
50	43.12

less, which shows that the progressively processing technique has more apparent advantages when the number of points is relatively large.

We have also tested the performance of the progressively processing technique under different cache size and listed the results as Table 5. The results show that a suitable cache size can shorten the computation time. For different numbers of the input data points, we can find their optimal cache sizes which cost relatively less memory and get a relatively significant acceleration.

5 Conclusions and Future Work

Visualizing large-scale data as clustered parallel coordinates is very challenging due to its clutter problem and the problem of the limitation of computer's main memory capacity. The clutter problem can be mitigated effectively by applying clustering algorithm to parallel coordinates. However, for large-scale data, it takes a lot of time to complete the partitioning computation. Therefore, to find a high-speed algorithm is of great significance. We have used k-means method to partition the large-scale ocean data because the algorithm is simple and fast. But even though k-means algorithm is used for partitioning the input data, the processing time will increase dramatically as the processed number of data points becomes larger and larger. Considering about the requirement of visualizing large-scale data, a further optimized clustering algorithm may be found to replace k-means algorithm if we do not need a high-level of precision. Besides, a suitable colour map also helps to mitigate the clutter problem. Rainbow color map is considered to be ineffective because it obscures data and introduce artifacts [9]. It confuses viewers through its lack of perceptual ordering [10]. Instead, a continuous diverging color map such as cool to warm color map is proposed to satisfy the general requirement [9].

After the input data is partitioned to a number of clusters, two methods are used for rendering the data points in these clusters. In our system, clusters can be plotted as polygons, or its indices can be plotted on an additional axis. The former takes nearly constant time to render the data points whatever the number of points is if the number of clusters is unchanged, because the number of polygons equals to the number of clusters. When the maximum value and minimum value of all clusters on each axis is figured out, what the GPU need to do is to plot the region between the maximum value and minimum value as polygons. Hence the time cost is relatively stable. On the other hand, the rendering time of the latter represents a linear relation between itself and the number of points. The positive linear relationship indicates that the method will cost much more time as the number of data points grows to a high level, although it has higher accuracy and flexible interactive feature. The two methods can be switched through an option in the programme so that users can make their own choice. Besides, other method can also used for rendering and plotting clusters. For example, when rendering clusters as polygons, if some areas of a polygon include more data points and others include less data points, we can use different transparency to represent the density of data points. A higher transparency value indicates a higher data density, while a lower transparency value indicates a lower data density. In this way, more information can be included in the polygons and the rendering time is expected to remain at a low level.

The progressively processing technique using external memory algorithm is applied to k-means algorithm when partitioning input data. The experimental results prove that the extra time spent by this technique is acceptable comparing with that of the in-core processing. Therefore, it is very useful when the size of the input data exceeds the computer's main memory. Taking advantage of the "cache" option, users can flexibly specify the size of the memory which is able to be used in the progressively processing. It will increase the speed of the out-of-core processing ulteriorly under the condition of the low memory consumption.

References

1. Hartigan, J.A., Wong, M.A.: A K-Means Clustering Algorithm. Journal of the Royal Statistical Society. Series C (Applied Statistics), 100–108 (1979)
2. Michaels, G.S., Carr, D.B., Askenazi, M., Fuhrman, S., Wen, X., Somogyi, R.: Cluster Analysis and Data Visualization of Large-Scale Gene Expression Data. In: Pacific Symposium on Biocomputing (PSB 1998), vol. 3, pp. 42–53 (1998)
3. Fua, Y.H., Ward, M.O., Rundensteiner, A.: Navigating Hierarchies with Structure-Based Brushes. In: Information Visualization 1999, pp. 58–64 (1999)
4. Sakamoto, N., Tanaka, T., Koyamada, K.: Multi-dimensional Binning for Parallel Coordinates and Application to Large-Scale CFD Results. In: 11th Asian Symposium on Visualization (2011)
5. Elkan, C.: Using the Triangle Inequality to Accelerate k-Means. In: International Conference on Machine Learning, pp. 147–153 (2003)

6. Hamerly, G.: Making k-means Even Faster. In: SIAM International Conference on Data Mining, pp. 130–140 (2010)
7. Lloyd, S.: Least squares quantization in PCM. IEEE Transactions on Information Theory 28(2), 129–137 (1982)
8. Peng, W., Ward, M.O., Rundensteiner, E.A.: Clutter Reduction in Multi-Dimensional Data Visualization Using Dimension Reordering. In: Proceedings of the IEEE Symposium on Information Visualization, pp. 89–96 (2004)
9. Moreland, K.: Diverging Color Maps for Scientific Visualization. In: Bebis, G., Boyle, R., Parvin, B., Koracin, D., Kuno, Y., Wang, J., Pajarola, R., Lindstrom, P., Hinkenjann, A., Encarnação, M.L., Silva, C.T., Coming, D. (eds.) ISVC 2009, Part II. LNCS, vol. 5876, pp. 92–103. Springer, Heidelberg (2009)
10. Borland, D., Taylor II., R.M.: Rainbow Color Map (Still) Considered Harmful. IEEE Computer Graphics and Applications, 14–17 (2007)

A Bug Reproducing Method for the Debugging of Component-Based Parallel Discrete Event Simulation*

Zhu Feng and Yao Yiping

School of Computer Science, National University of Defense Technology, Changsha, P.R. China
{zhufeng,ypyao}@nudt.edu.cn
http://yhsim.nudt.edu.cn

Abstract. Component debugging is critically important for diagnosing failures of programs. In component-based Parallel Discrete Event Simulation (PDES), the bug probably not to be reproduced for the different orders of event processing in different simulation runs, so there is a great challenge in the debugging for components. To solve the problem, this paper proposes a bug reproducing method based on checkpoint/restart mechanism, which schedules a simultaneous interaction event after each event processing to request the checkpoint operation. Moreover, a module called CheckpointRestartMgr (CRM) has been designed to provide checkpoint/restart service for simulation objects. The cost brought from cooperation between Logical Processes when performing checkpoint operation can be reduced through the simultaneous interaction event and CRM. The usage of our method in the debug framework for component-based PDES demonstrates it can be feasible.

Keywords: Component debug, PDES, checkpoint/restart mechanism, CRM.

1 Introduction

Complex system simulation such as large-scale ecological simulation, computational biology simulation and complicated war simulation is usually composed of a large number of entities [1]. Each entity is composed of one or more components and intricate interactions exist among these components. This will be a great challenge for domain experts to apply these entities to buildup simulation applications. Component-based PDES supports distributed developing and hierarchical composing for models, which provides an efficient way to build large-scale complex system simulation. However, the current major PDES platforms based on event scheduling, such as SPEEDES (Synchronous Parallel Environment for Emulation and Discrete-Event Simulation) [2, 3] and GTW (Georgia tech Time Warp) [4, 5] do not provide a correlative module of debugging to support the development of components. Therefore, the research on the debugging of component-based PDES becomes a very important issue gradually.

* This work has been funded by the National Science Foundation of China (No. 61170048).

J.-H. Kim et al. (Eds.): AsiaSim2011, PICT 4, pp. 456–465, 2012.

In component-based PDES, the simulation object composed of one or several components performs in parallel on different processors. This parallel paradigm will lead the nondeterministic program behavior for the components running, and the indeterminism of a parallel program adds to the complexity of the debugging process. Certain bugs manifest themselves only due to a specific ordering of events and may not show up in a rerun of the program due to a different ordering of events in the rerun [6]. Therefore, How to reproduce the bugs is a very challenging issue in the debugging of component-based PDES.

With checkpoint/restart mechanism, parallel debugging returns the developer to an intermediate state closer to the bug [7]. Checkpoint/restart mechanism provides for bug reproducing capability as well as other benefits for component-based PDES [8]. Component developers can save hours or days of time spent in debugging by checkpoint and restarting the parallel debugging session at intermediate points in the debugging cycle [9]. Therefore, this paper proposes a checkpoint/restart based approach triggered by event-processing. In another word, after processing an event, SimObj will schedule a simultaneous interaction event to request checkpoint operation. And a simulation object CheckpointRestartMgr (CRM) has been constructed on each processor, which provides checkpoint/restarting service for simulation program through processing interaction events. Then the checkpoint image will be saved into disk, which will be used in the resuming process later. The implementation based on YH-SUPE (YinHe Simulation Utilities for Parallel Environment) [10] is conducted to investigate the applicability of the proposed method in a realistic system environment. And the usage of our method in the event-driven debug framework for PDES base on components demonstrates this method can be feasible.

The remainder of this paper is structured as follows: in section 2 we analyze the motivation of our work. In section 3 we explain our method in detail for reproducing bugs in the components running. In section 4 we describe our implementation based on YH-SUPE. Finally, our conclusion will be made with an indication of the future work.

2 Motivation

In the execution of a parallel program, a bug can manifest itself because of an unusual ordering of events. The bug may not recur if the experiment is repeated, because the processor may alter the original ordering of events [11]. In such cases it would be helpful to be able to deterministically reproduce the bug.

In a PDES program, SimObjs (simulation object) are distributed to several processers for parallel execution. SimObj processes the events with different speeds in different running course, which will alter the ordering of event-processing. There is a probable scenario where a PDES program is composed of two SimObjs named SimObj1 and SimObj2, which are distributed to two processers for parallel execution. SimObj1 composed of a component named ComponentX processes four events. They are event1, event2, event3 and event4 with time stamp 100, 130, 160 and 190. SimObj2 composed of a component named ComponentY processes four events. They are eventA, eventB, eventC and eventD with time stamp 100, 160, 210 and 240. The event-scheduling process in the two components is shown in figure 1.

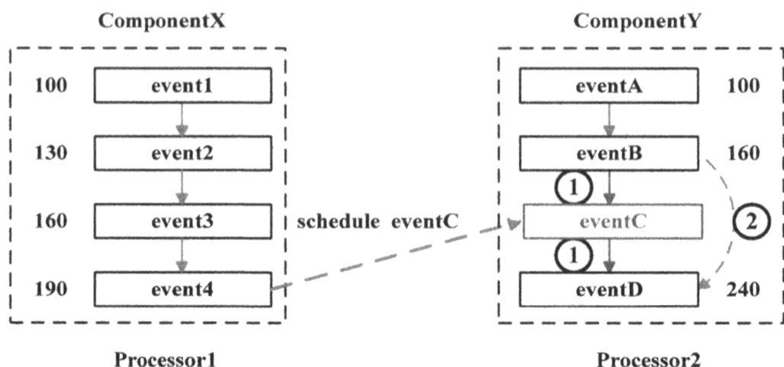

Fig. 1. 1. Processor2 processes the events faster than Processor1. 2. Processor1 processes the events faster than Processor2. This shows the event-scheduling process in the two components.

In the first running process, providing Processor2 processes the events faster than Processor1. There will be the possible situation that after eventB has been processed in ComponentY, ComponentX hasn't scheduled eventC. At that time, ComponentY proceses eventD, and then the program will be error.

In the second running process, providing Processor1 processes the events faster than Processor2. Before ComponentY processes eventD, ComponentX has processed event4 and scheduled eventC. At that time, eventD will be processed correct, because the state of ComponentY has been changed through processing eventC. Therefore, the bug will not appear.

The indetermination of the event processing order is a serious problem for components debugging. Although in PDES with optimistic time management algorithm, the error behavior of the program can be rectified through rollback mechanism which uses incremental state saving method [12, 13]. However, incremental state saving becomes more expensive than copy state saving if the state of the SimObjs is modified by each event [14]. And it will reduce the performance of the program, because the state of the SimObj modified by an event will be restored when the simulation program perform a rollback, one after the other, until the point of rollback is reached. Moreover, the rollback mechanism does not always maintain the consistency of a parallel simulation, especially the program with wrong behavior.

Checkpoint/restart mechanism makes debugging easier because it enables cyclic debugging. Cyclic debugging is possible with checkpoint/restart mechanism because order of execution is repeatable. In the further, checkpoint/restart mechanism recovers the state of the program from the latest checkpoint, which avoids starting from scratch. However, for traditional periodic checkpoint/restart mechanism, the checkpoint operation is requested periodically, which will bring a problem that how to choose the periodical time. Especially in PDES, the time cost of event-processing is usually different. If the periodical time chosen is shorter, the times for requesting checkpoint operation will be more, which will lead the bigger memory cost to store the checkpoint file. Whereas the time is longer, we must record the event processing order between the checkpoint and the error position. Otherwise, the bug appeared in the components running will be hard to reproduce.

In summary, the bug reproduction needs to be solved in the debugging of component-based PDES, but the current methods didn't provide an efficient support for this. So we need to propose another method for component debugging to promote the development of component-based PDES.

3 Checkpoint/Restart Based Bug Reproducing Method

YH-SUPE is a common simulation environment for PDES, which provides services for parallel simulation application such as time management, memory management, persistent mechanism and event scheduling strategy etc. In YH-SUPE, the simulation system can be viewed as a collection of Logical Processes that interacts in some fashion, and each Logical Process is assigned to a different processor. So we can refer a Logical Process to a processor in the following sections. The checkpoint/restart technique is utilized to save the state of the simulation application program though persistent mechanism, which will be the beginning point of recovering this program. The challenge for using checkpoint/restart technique to reproduce a bug includes the following two aspects: (1) How to store the state of Logical Process to reduce the time cost? (2) How to guarantee the order of event processing between the checkpoint and the error position the same as previous?

3.1 Persistent Mechanism in YH-SUPE

YH-SUPE provides the persistent mechanism which serves for checkpoint/restart with persistent memory and persistent pointer [10]. Our method will utilize the persistent mechanism to store the states and events to be processed to a checkpoint file. The persistent memory is used to track persistent memory allocation and reclamation, and the persistent pointer to maintain the address of the persistent memory. YH-SUPE provides a persistent database which is utilized to store persistent memory and persistent pointer. Figure 2 shows the structure of the persistent database.

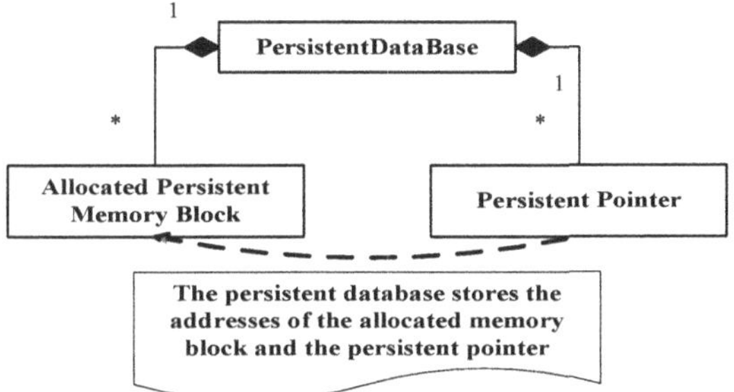

Fig. 2. The structure of the persistent database

When requesting checkpoint operation, the persistent mechanism packages the states and the events to be processed of SimObjs to a buffer using the copy state saving approach, and then compresses the buffer to a checkpoint file in the disk. When restarting from the checkpoint file, the memory block will be constructed through the persistent mechanism. Then all of the persistent pointer will be adjusted to the new address of the memory block.

3.2 Event-Driven Checkpoint/Restart Method

PDES program advances time by processing events, thus the error behavior of the program most appears in the function of event-processing. Our method triggers a checkpoint operation by event-processing. In another word, after processing an event, the SimObj will schedule a simultaneous interaction event for requesting checkpoint operation. For the purpose of guaranteeing the simultaneous interaction event processed before other simultaneous events, it must be set higher priority.

In order to maintain the consistency of global checkpoint and reduce the impact of the operation of checkpoint on execution time, the simulation object CheckpointRestartMgr (CRM) is constructed on each processor, which provides checkpoint/restarting service for simulation application, such as requesting checkpoint interaction, restoring the state of simulation objects and the events to be processed, and recovering the state of SimObjs and the event queue, etc. CRM on each processor works cooperative communication though YH-SUPE engine. The figure 3 shows the operation flow of checkpoint.

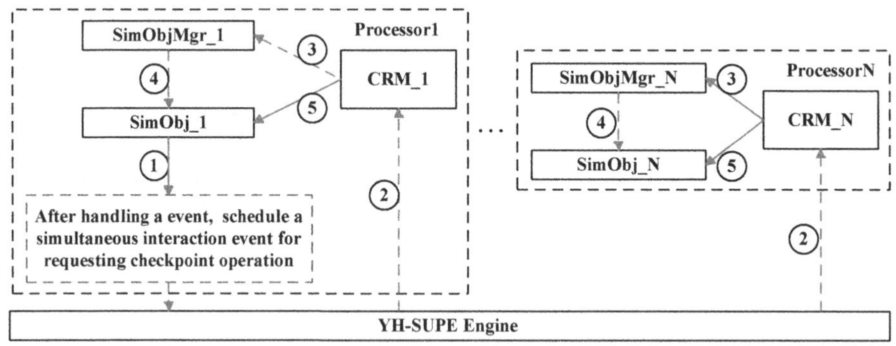

Fig. 3. The operation flow of requesting checkpoint

1. The SimObj_1 schedules a simultaneous interaction event to request a checkpoint operation after processing an event.
2. YH-SUPE engine receives a request about saving interaction, and then sends this request to all of the CRM on the other processors.
3. Each CRM notifies the simulation object manager SimObjMgr, which creates the simulation object dynamically, manages the subscribing of interaction events and event handlers on the same processor etc.
4. Each SimObjMgr notifies all of the SimObjs on the same processor.

5. Each SimObjs recalls the function of CRM to save the persistent state and event message with which the stamp time greater than the checkpoint time to a parameter set. Then this parameter is compressed to a buffer which is written to a checkpoint file in the disk.

When restarting simulation application to reproduce the error in components running, it needs to input command to restore the states of SimObjs packed in the latest checkpoint file on each processer. In the course of restarting checkpoint, the operation routine on each processor is as follows: First, the user input ProgramName –restart on the command line to restart from checkpoint. Second, each CRM searches for the latest checkpoint file from the error position occurring in the components running, and then unpack the checkpoint file to reconstruct the local buffer. Third, the parameter set of each SimObj is recreated recording to the local buffer. Fourth, the state of each SimObj and the events to be processed are reconstructed though the persistent data in the parameter set. Last, CRM sends ReconstructComplete message to YH-SUPE engine to restart running the program. Figure 4 shows the workflow of restarting checkpoint under YH-SUPE.

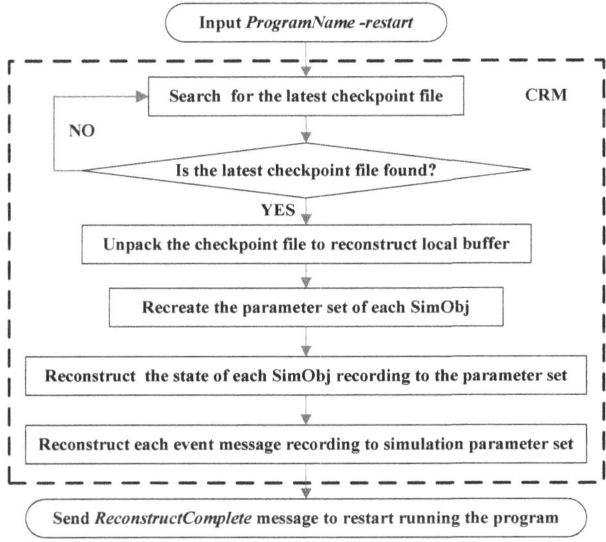

Fig. 4. The workflow of restarting checkpoint under YH-SUPE

3.3 Analysis

Our method applies the persistent mechanism which utilizing copy state saving approach to store the state of SimObjs. For some simulations that contain a large number of events and in which the event processing cost less time, making a copy of the state of all SimObjs before each event may consume large amounts of time and memory. So the effect of utilizing our method is worse. But for complex computation system simulation which contains fewer events, the function of event-processing often needs more time. And copy state saving can perform block moves to save and restore state. Therefore, the copy state saving probably becomes less expensive than incremental state saving because most of the SimObjs state is modified by each event. Furthermore, our method can reproduce the bug in component running instead of avoiding errors through rollback mechanism. It can help users to explore the impact of event-scheduling time which causes the error program behavior. This exploration is helpful to decrease the times of rollback.

Comparing traditional periodic checkpoint technique, in event-driven checkpoint operation, the times of checkpoint operation will be fewer owing to the events to be processed are fewer. So the operation of checkpoint driven by event - namely, to request the operation of checkpoint after event-processing will not increase the memory cost of checkpoint file. As a result of using copy state saving, the time cost by restarting from a checkpoint file in our method is approximately the same as that in periodic checkpoint operation.

In event-driven checkpoint/restart method, after processing an event, SimObj will schedule a simultaneous interaction event to request checkpoint operation. Then CRM will process this interaction event to store the states of Logical Process almost at the same time because of the higher priority. Thus there is an obvious benefit that it needn't to store the orders between the checkpoint and the error position in the components running because there is only one event. So the issue of storing the order of event processing does not exist. And in this way, Logical Processes will not be suspended when requesting checkpoint in the components running. Thereby, the cost of process switching will be reduced.

4 Implementation

YH-SUPE provides interaction event, which support communication for SimObjs through the subscribe approach. Macro DEFINE_INTERACTION(ClassName, MethodName) is used to define interaction event, which will introduce some functions to support register/unregister and subscribe/unsubscribe operation for interaction event. ClassName specifies the name of the interaction class and MethodName specifies the name of the interaction event processing function in this interaction class. Macro SCHEDULE_INTERACTION(const SimTime simTime, const char* interactionName, ParameterSet ¶meterSet) is utilized to schedule interaction event. simTime specifies simulation time, interactionName specifies interaction name and parameterSet is a container restoring data, which is utilized to transfer parameter for interaction event.

CRM as a SimObj defines the event-processing function Checkpoing_Requested for checkpoint operation in our method. The function of Checkpoing_Requested is to restore the state and events to be processed of each SimObj through persistent mechanism in YH-SUPE. Before the components assembled to a simulation application, macro SCHEDULE_INTERACTION will be inserted in the end of each event-processing function to schedule the simultaneous interaction event. The events are processed in the order of time stamp, therefore as soon as the event of components is processed, the interaction event with the same time stamp will be processed. That is calling function Checkpoing _Requested to restore the state and event queue of each SimObj.

CRM notifies each SimObjMgr to request checkpoint operation through the following codes. The handler of each SimObjMgr is fetched through function getSimObjMgrId. Macro SCHEDULE_CheckpointSimObjMgr schedules interaction event for each SimObjMgr. This interaction event has the same time stamp with the event to be processed. In order to maintain the causal order, this interaction event is set priority before scheduled.

```
for (int i = 0; i < numSimObjMgrs; i++){
  obj = getSimObjMgrId(i);
  //set priority for simultaneous interaction event
  checkpointTime.SetPriority(0x7ffffffe, 0x7ffffffe);
  SCHEDULE_CheckpointSimObjMgr(checkpointTime, obj);
}
```

CRM generates checkpoint file through the following codes. The parameter set of each SimObj is initialed from SimObjDataList, and then loop to fetch the state and events to be processed to a buffer, which is compressed to the checkpoint file.

```
ParameterSet *ps = SimObjDataList->GetTopElement();
   while (ps) {
      SimObjDataList->PopTop();
      //save the state and events to checkpoint file
      Fetch the state and events to a parameter set
      Compress the parameter set to checkpoint file
      ps = SimObjDataList->GetTopElement();
   }
```

In the course of restarting checkpoint, CRM recovers the state and event queue of each SimObj through the following codes. At first CRM read the content of the checkpoint file latest from the error position to a buffer, and then call InitFromBuffer to construct local parameter set, which is used to construct parameter set for each SimObj. With that recover the state and event queue of each SimObj.

```
Read the content of checkpoint to a buffer
ParameterSet *localParameter = InitFromBuffer(buffer);
Bool status = localParameter->GetSimObj(name, buffer);
while (status) {
   ParameterSet *simObjParameter= Generate_SimObj(buffer);
   //reconstruct the parameter set of SimObjs
   Recover the state of SimObj
   Recover the event queue of SimObj
   status = localParameter->GetNextSimObj(name, buffer);
   delete simObjParameter;
}
```

The implementation based on YH-SUPE is conducted to investigate the applicability of our method. We have used it in event-driven debug framework for EDEVS based component (details can be found in [15]). For complex computation system which contains fewer events and whose functions of event-processing often needs more time, this method can work better.

5 Conclusion and Future Work

Component-based PDES supports distributed developing and hierarchical composing for models, which provides an efficient way to build large-scale complex system simulation. But the indeterminism of a parallel simulation program composed of one or several components makes the bug reproduction become an increasingly important issue. In this paper, we present a bug reproducing approach based on checkpoint/restart mechanism. And a module called CheckpointRestartMgr is constructed on each processor to provide checkpoint/restarting service. In this way, our method is not to increase the memory cost of checkpoint file, but also not to increase the time of restarting from the checkpoint file, and there is no need to store the order of event processing between the checkpoint and the error position. For complex computation system simulation which contains fewer events and whose function of event-processing often needs more time, our method will work better.

As for our future work, we plan to study the fault location in which we will utilize the proposed bug reproducing technique in this paper. Another interesting line of investigation would be the usage of the checkpoint/restart mechanism to implement fault tolerance.

Acknowledgment. I would like to thank Liu Gang, the member of High Performance Simulation Group, for his research of the EDEVS component based YH-SUPE debug technology.

References

1. Li, B.H.: Some Focusing Points in Development of Modern Modeling and Simulation Technology. In: Baik, D.-K. (ed.) AsiaSim 2004. LNCS (LNAI), vol. 3398, pp. 12–22. Springer, Heidelberg (2005)
2. Jeff, S.: SPEEDES: Synchronous Parallel Environment for Emulation and Discrete Event Simulation. In: Proceedings of Advances in Parallel and Distributed Simulation, pp. 95–103 (1991)
3. Chris, B., Robert, M., Jeff, S., Jennifer, W.: SPEEDES: A Brief Overview. In: Proceedings of SPIE, Enabling Technologies for Simulation Science V, pp. 190–201 (2001)
4. Fujimoto, R.M., Das, S., Panesar, K.: Georgia Tech Time Warp programmer's manual. Technical report, College of Computing, Georgia Institute of Techmology, Atlanta, GA (July 1994)
5. Perumalla, K.S., Fujimoto, R.M.: GTW++ An Object-oriented Interface in C++ to the Georgia Tech Time Warp System. Georgia Institute of Technology (1996)

6. May, J., Berman, F.: Designing a Parallel Debugger for Portability. In: Proceedings of the International Conference on Parallel Processing, pp. 909–914 (1994)
7. Wang, Y.-M., Huang, Y., Vo, K.-P., Chung, P.-Y., Kintala, C.: Checkpoint and Its Application. In: IEEE Fault-Tolerant Computing Symp., pp. 22–31 (June 1995)
8. Bouguerra, M.-S., Gautier, T., Trystram, D., Vincent, J.-M.: A Flexible Checkpoint/Restart Model in Distributed Systems. In: Wyrzykowski, R., Dongarra, J., Karczewski, K., Wasniewski, J. (eds.) PPAM 2009, Part I. LNCS, vol. 6067, pp. 206–215. Springer, Heidelberg (2010)
9. Hursey, J., January, C., O'Connor, M., Hargrove, P.H., Lecomber, D., Squyres, J.M., Lumsdaine, A.: Checkpoint/Restart-Enabled Parallel Debugging. In: Keller, R., Gabriel, E., Resch, M., Dongarra, J. (eds.) EuroMPI 2010. LNCS, vol. 6305, pp. 219–228. Springer, Heidelberg (2010)
10. Yao, Y.-p., Zhang, Y.-x.: Solution for Analytic Simulation Based on Parallel Processing. Journal of System Simulation 20(24), 6617–6621 (2008)
11. Netzer, M.: Optimal tracing and replay for debugging message-passing parallel programs. In: Proceedings of the 1992 ACM/IEEE Conference on Supercomputing, pp. 502–511 (1992)
12. Avril, H., Tropper, C.: On Rolling Back and Checkpointing in Time Warp. IEEE Transactions on Parallel and Distributed Systems 12(11), 1105–1120 (2001)
13. Steinman, J.S.: Incremental State Saving in SPEEDES Using C++. In: Proceeding of the 1993 Winter Simulation Conference, pp. 687–696 (1993)
14. Fujimoto, R.M.: Parallel and Distributed Simulation Systems. John Wiley & Sons, Inc. (2000)
15. Liu, G., Yao, Y.: Event-Driven Debug Framework for EDEVS based Components. In: 2011 International Symposium on Computer Science and Society (ISCCS 2011), pp. 411–414 (2011)

Diagnosis of Flood Risk in South Bay Area Using Hydrology Analysis

Carine Joungyeon Yi[1] and Roy Soonchan Park[2]

[1] Graduate School of Systems and Information Engineering, University of Tsukuba, 1-1-1 Tennodai, Tsukuba, Japan
carineyi@gmail.com
[2] R.Park and Associates Inc., 555 Burnhamthorpe Rd. Toronto, Canada
rpark.associates@gmail.com

Abstract. The south part of the San Francisco Bay Area is home to millions of people. Climate change is one of the main environmental issues in California. Sea level rise has been observed globally, forcing local municipalities and regions to develop adaptive strategies. In this study, sea level rise inundation areas were diagnosed on several levels over the next 50 years using topographic and geographical information, including hydrology analysis. A high rate of precipitation has been observed in the southwest Bay Area for the past five years where rainfall, as an extreme natural event, caused flash flooding. The flow direction is towards the City of San Jose which faces a high risk of urban flooding. In this study, a series of thematic maps were created to represent the flooding risk.

1 Introduction

The San Francisco Bay Area consists of nine counties[1] including the mega-size metropolitan areas of San Francisco, Oakland and San Jose. The Bay Area, with a population of more than 7.5 million, has the core IT industry in California State with most developed urban areas spread out in the estuary region. Its abundant natural resources are attractive for human life and climate change in the ecosystem is often discussed scientifically.

This study aims to understand the risk of flooding by sea level rise and potential unexpected rainfall runoff, and help build appropriate management strategies for the future.

1.1 Impact of Climate Change in the San Francisco Bay Area

Climate change is one of the main issues in California. Sea level rise has been observed globally, and regions and local municipalities have been working to develop their own adaptation strategies.

[1] San Francisco Bay Area counties: Alameda, Contra Costa, Marin, Napa, San Francisco, San Mateo, Santa Clara, Solano, and Sonoma.

J.-H. Kim et al. (Eds.): AsiaSim2011, PICT 4, pp. 466–474, 2012.

For past decades, the State of California is concerned with the sea level rise by global warming. The state government discussed the mitigation and adaptation strategies in several reports [3], [4], [5], [7], [8]. California's annual temperature has risen 0.33ºF (0.56ºC) per decade in night time, daytime 0.1ºF (0.17ºC) since 1920 [11]. For some states - such as California, Florida, New Jersey, and South Carolina – the vulnerability of the area facing highest damage risk will be when sea level rise is 1m and 1.4m using several scenarios by the year 2100 [2], [11], [13]. By end of this century the sea level rise of 55 inches (140 cm) is estimated in the bay area.

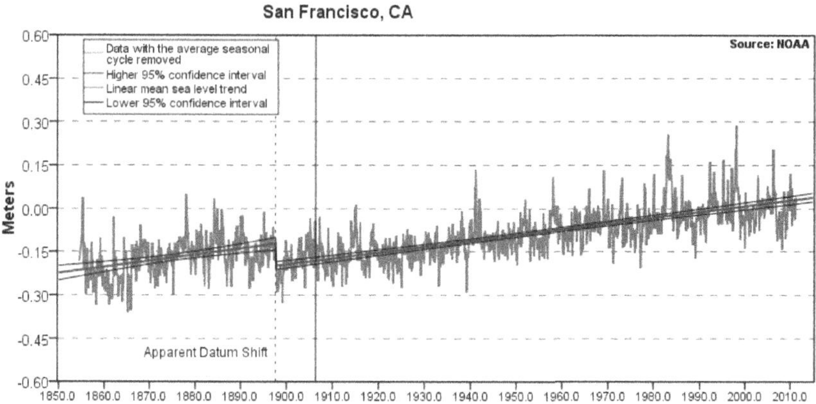

Fig. 1. Mean Sea Level Trend[2], San Francisco, California (Source: NOAA)

Among serious climate change impacts, sea level rise in California is chosen by USGS for the project, CASCaDE (Computational Assessments of Scenarios of Change for the Delta Ecosystem)[3] that was launched in 2006.

Sea level rise impacts in the Bay area are often observed with other weather events and those events bring huge damage to urban infrastructures, economy and human life. Public assistance is eligible to request the emergency works and repairs of facilities damaged by severe winter storms, flooding, mud flow and debris in some areas in California (FEMA-1952-DR, 2011)[4]. The residents of the mid-south in California are

[2] The mean sea level trend is 2.01 millimeters/year with a 95% confidence interval of +/- 0.21 mm/yr based on monthly mean sea level data from 1897 to 2006 which is equivalent to a change of 0.66 feet in 100 years. `http://tidesandcurrents.noaa.gov/sltrends/sltrends_station.shtml?stnid=9414290`, Accessed 2011-08-20

[3] USGS CASCaDE; Computational Assessments of Scenarios of Change for the Delta Ecosystem. CASCaDE I (2006-2010) developed a set of linked models to assess Delta ecosystem response to climate change. CASCaDE II (2010-present) is proposed to refine and extend those modeling capabilities to assess Delta ecosystem response to changes in climate and physical configuration (USGS CASCaDE Project).

[4] California – Severe Winter Storms, Flooding, and Debris and Mud Flows, FEMA-1952-DR (2011) : Total Individual Assistance cost estimate: $1,909,557, Total Public Assistance cost estimate: $75,414,223.

experienced heavy rain and flash flooding (CHA: Natural Disaster: Flood). Altered frequencies and intensities of extreme weather, together with sea level rise, are expected to have mostly adverse effects on natural and human systems. The impacts projected in industry, settlement and society are: distribution of vulnerable areas by flood and high winds; cost of coastal protection versus costs of land use relocation; potential movement of population and infrastructures; loss of property [9].

2 Data and Method

To figure out topography of the study area, the National Elevation Dataset of Seamless Data Warehouse (NED), 1/3 arc-second (approximately 10 meter) resolution that is provided by United States Geological Survey (USGS) and it was rectified with the North American Datum of 1983 (NAD 83). For surface water analysis, a series of National Hydrography Dataset (NHD) was obtained from USGS. The file used and a series of terrain processing and basin processing implemented for hydrology analysis using HEC Geospatial Hydrology Modeling Simulation software (HEC-GeoHMS) 5.0 that is developed by the Hydrologic Engineering Center (HEC), an organization within the Institute for Water Resources, is the designated Center for the US Army Corps of Engineers.

Monthly precipitation data from National Weather Service (NWS) and date data are used to understand the South Bay's early spring season (Jan – March, between 1895 -2010) flood trend. The 5-year mean precipitation (2007-2011) data are interpreted after kriging interpolation method.

Layers of potential inundation due to rising sea levels in the San Francisco Bay Region are the results of CASCaDE project and provided by USGS. These layers correspond to varying amounts of long-term period sea level rise in conjunction with various return heights of sea levels corresponding to shorter-term period variability. A 50 year inundation level means the 50-year recurrence interval inundation map corresponding to a 50cm or 140cm sea level rise. The 50cm-50year layer and 140cm-50year layers are selected for this study because 50cm is the first interval value and 140cm is the value that calculated as maximum sea level rise by end of this century. The horizontal resolution of these data layers is 2 meters and their projection is UTM Zone 10, NAD 83.

Last format of the Land Use Data and Agricultural Economic Data that was provided by United States Department of Agriculture (USDA) was used to estimate urban enlargement. The raster data of land use reclassified; agricultural classification are removed; water, several levels of developed area are considered as urban area; and open space area and grass area can be considered as potential urban expand areas.

Census data was obtained from U.S. Census Bureau. TIGER/Line ® data was used to understand the population distribution in this area.

3 Results

3.1 Impacts by Climate Change

The South Bay area affected by flooding by a 50cm sea level rise will be 749,336.24km^2; with 140cm, the flooded area will be 874,925.69km^2 by the middle of this century. In this study, the risks of vulnerable areas are interpreted using a series of GIS analysis processes. Figure 2 shows the inundation area at a different inundation levels. Figure 2 c) shows the inundation area is wider than 2b).

Fig. 2. Predicted inundation areas by sea level rise in The South Bay: a) 50cm and 140cm_mean sea level, b) 50cm and 140cm_50-year, c) 50cm and 140cm_100-year

By the middle of this century, 180,000 acres near Bay shoreline and, by the end of the century, 213,000 acres are vulnerable to flooding. Moreover, the area that will be vulnerable to inundation with a 16-inch sea level rise at the mid-century corresponds to today's 100-year floodplain [14]. A reconstruction of global sea level using tide-gauge data from 1950 to 2000 indicates a higher rate of rise after 1993 and other periods of rapid sea level rise acceleration over this period [1], [6].

Flooding is one of the main issues in California's climatology and hydrology for a long term period. California's aging water supply and flood protection infrastructure, including more than a thousand kilometers of levees, is being challenged by pushing floods and increased standards for urban flood protection. The South Bay has a high population and the major city, San Jose, is in the heart of IT industry facing unacceptable risks from flooding. Figure 3 shows monthly mean precipitation during 1895 to 2010 water year data. There is high rate of precipitation between December and March.

Figure 4 shows the precipitation distribution of the study area; a) shows higher rate precipitation in west of the city of San Jose and the east side. Figure 4 b), c) and d) show the precipitation in South Bay where flooding was experienced caused by heavy rainfall during 20-22 March 2011. Heavy rain caused flash flooding downstream where population is dense. It shows a tendency similar to that of the previous five years, although there are exceptions.

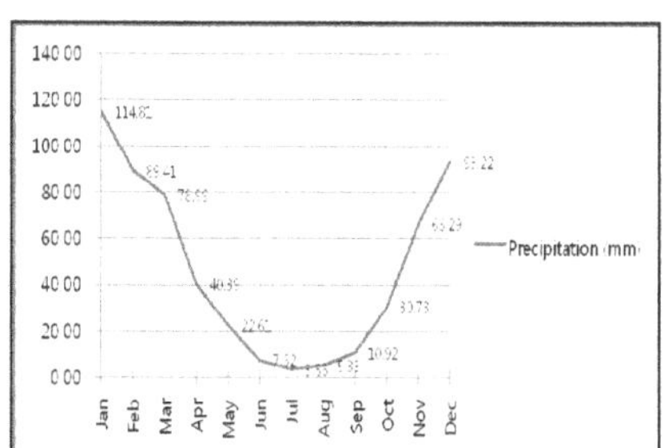

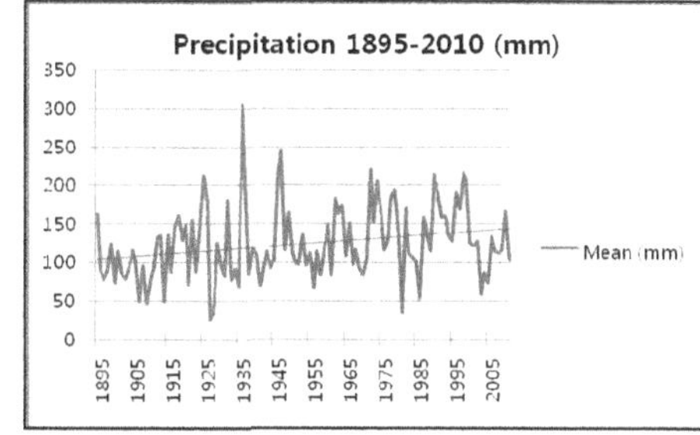

Fig. 3. Mean precipitation in California between 1895 and 2010

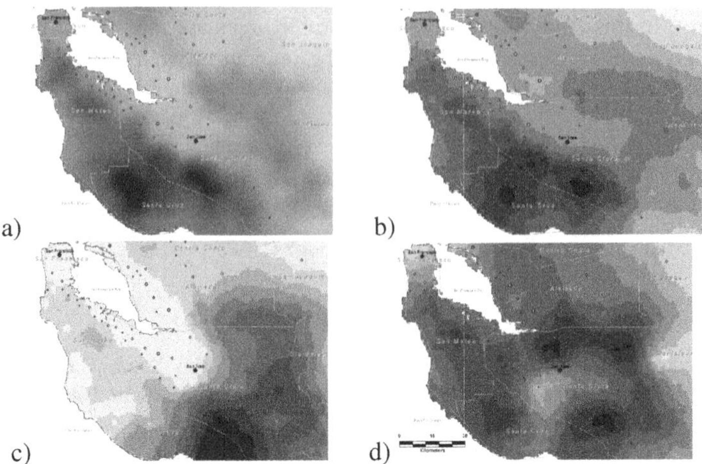

Fig. 4. a) 5-year mean precipitation (2007-2011); Flash flooding experienced in March 20-22, 2011, b) March 20, 2011 c) March 21, 2011 and d) March 22, 2011

3.2 Hydrology Analysis

Through hydrology analysis, the characteristics of the river basin in South Bay Area can be explained.

Figure 5 shows the topology and river basin in South Bay Area. The City of San Jose is located in the middle of the flat area and two main rivers runs from east and west: Guadalupe River and Coyote Creek.

Through analyzing digital terrain data, HEC-GeoHMS transforms the drainage path and watershed boundaries into a hydrologic data and represents the drainage network in South Bay Area. A series of terrain processing and Basin processing such as flow direction, flow accumulation, stream segment and more are represented in Figure 6. Figure 6 a) shows flow direction and the direction of north and northeast of the area showing a high rate of precipitation heading to the City of San Jose. Figure 6 b) represents flow accumulation: that function is created using flow direction data set where each cell assigned a value equal to the number of cells that flow to it. Figure 6 c) shows the distributary segment. Figure 6 d), e) and f) shows adjoining upstream catchment south of the City of San Jose.

Figure 7 a) shows the longest river basin of Coyote Creek that runs just east of the City of San Jose, length approximately 121.63 km, highest elevation level 660.66 m and lowest elevation level 18.24 m. Created HMS schematic can be imported and its HEC-HMS legend represents on the map of Figure 7 b) and c). Figure7 d) represents the flow length after grid cells intersect processed. The darker area indicates downstream in each catchment. The water flow direction is from darker area to the light area.

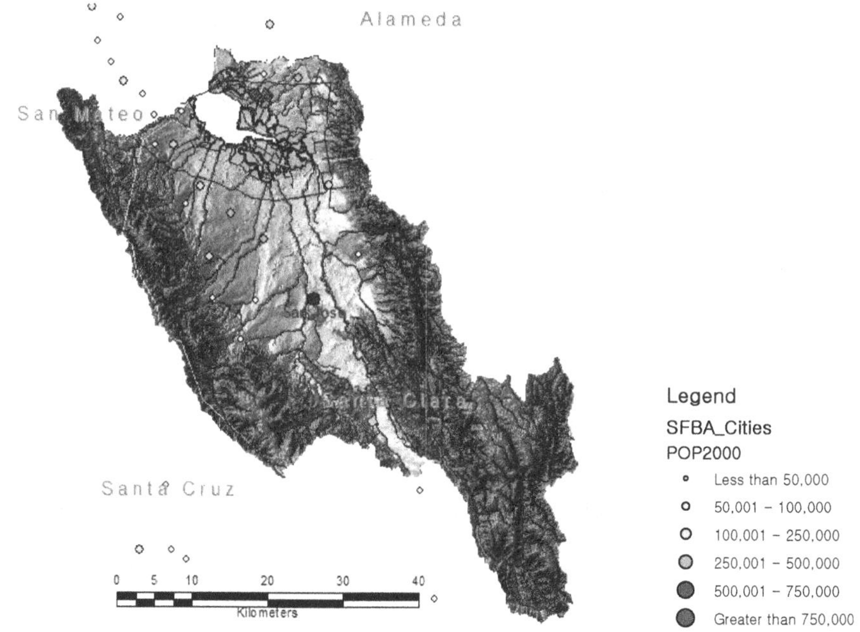

Fig. 5. Hydrology network and topology in the South Bay Area

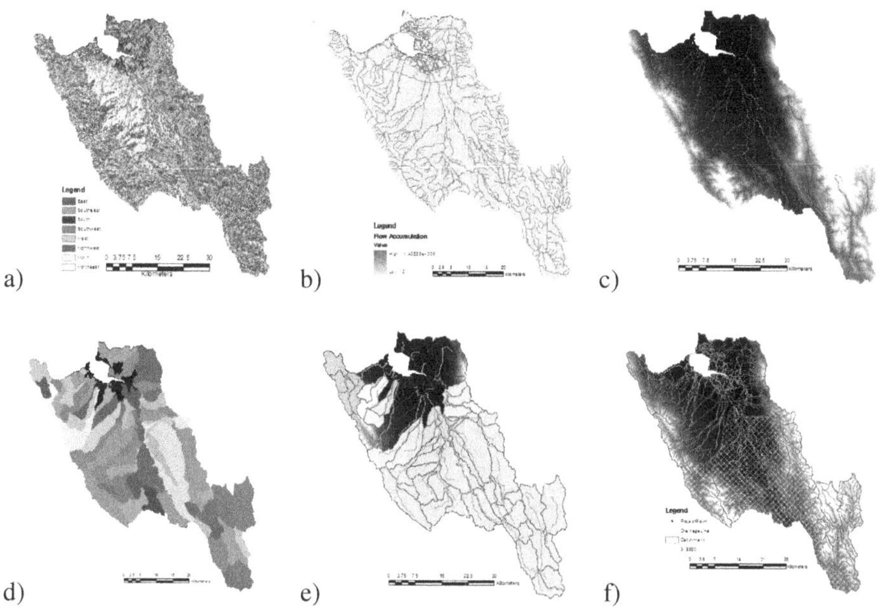

Fig. 6. a) flow direction, b) flow accumulation, c) stream segment, d) adjoining catchment, e) catchment, and f) downstream catchment of San Jose

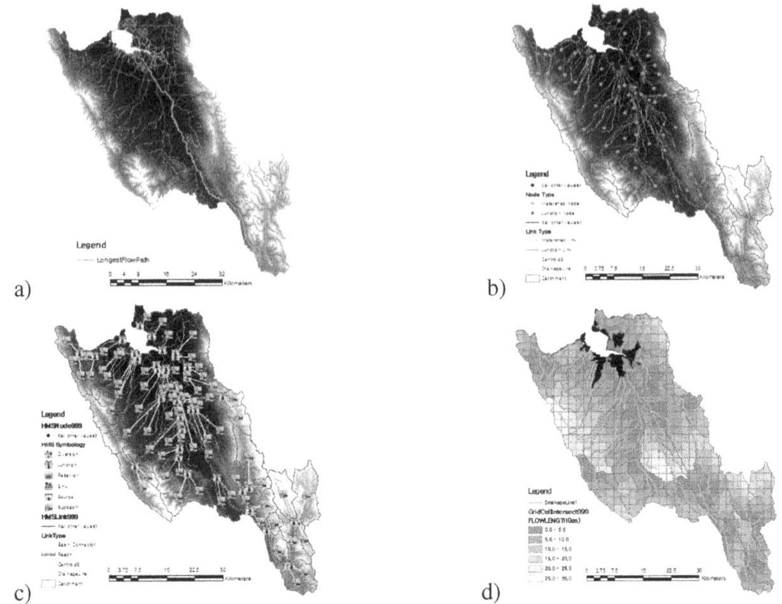

Fig. 7. a) Longest pathway and basin centroid, b) HMS schematic c) HMS-legend, and d) grid cell Intersect – flow length

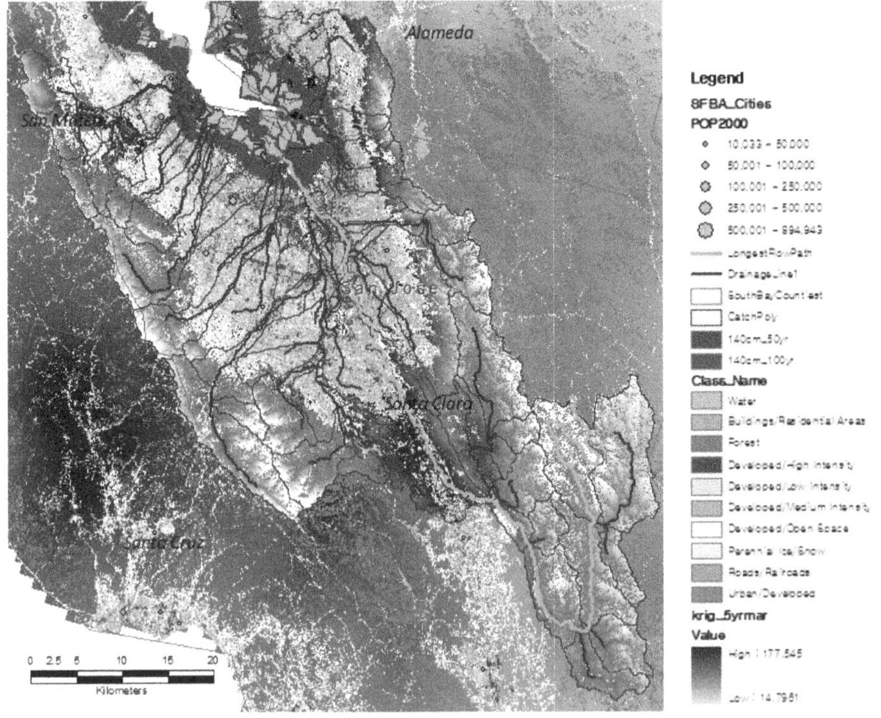

Fig. 8. Flood risk map in the South Bay Area

4 Conclusion

The South Bay Area population of 270,000 are put at risk of flooding/disaster due to global warming. About $62 billion worth of economic resources would be damaged due to risk of flooding. This area experiences high precipitation during early spring season. The huge urban areas including Silicon Valley are at risk from rising sea level and inland flooding.

Figure 8 depicts overlapped results that help to understand the South Bay Area's risks by potential flooding. The area in color of pink indicates developed urban areas and light yellow areas are open space which can be considered future expansion urban areas. For the past 5 years precipitation laid as background and darker blue color as high rate of precipitation observed in the southwest of the City of San Jose.

Flash flooding often leads to extremely dangerous and sometimes catastrophic conditions in rivers due to characteristics such as: short timescales, the limited opportunity for issuing warnings, and the frequent high average mortality. Many past extreme flood events have been accompanied by flash floods, and they have been one of the main sources of serious loss of human life among the world's worst natural disasters [15]. Considering the size and population of the South Bay Area and the impervious materials of paved road, intensive rainfall can be one of the significant factors of urban flooding because of the basin catchments head to the City of San Jose. Water management and local municipality's collaborative integrated works are required in the future.

Acknowledgement. The author would like to thank Yasuaki Hijioka, Kiyoshi Takahashi, and Yong Hee Shin who of the National Institute for Environmental Studies (NIES), Japan for their valuable professional advice. I would like to special thank to Dr. Samuel Lee who is a civil engineer at U.S. federal energy Regulatory Commission for encouraging me to complete this work.

References

1. Church, J.A., White, N.J.: A 20th century accelerate in global sea-level rise. Geo-physical Research Letters, 33, L01602, 4 (2006)
2. Curtis, K.J., Schneider, A.: Understanding the demographic implications of climate change: estimates of localized population predictions under future scenarios of sea-level rise. Population and Environment, Online First™ (2007)
3. Florsheim, J.L., Dettinger, M.D.: Climate and floods still govern California levee breaks. Geohpys. Res. Lett. 34(22) (2007)
4. Ganju, N.K., Schoellharmer, D.H.: Calibration of an estuarine sediment transport model to sediment fluxes as an intermediate step for simulation of geomorphic evolution. Continental Shelf Research, 29, 148–158 (2009)
5. Hidalgo, H.G., Dettinger, M., Cayan, D.: Downscaling with constructed analogues: Daily precipitation and temperature fields over the United States CEC-500-2007-123, California Energy Commission. PIER Energy-Related Environmental Research (2008)
6. Kenneth, K., Grubler, A., Jung, T.Y., Kram, T., La Rovere, E.L., Michaelis, L., Mori, S., Morita, T., Pepper, W., Pitcher, H.M., Price, L., Riahi, K., Roehrl, A., Rogner, H., Sankovski, A., Schlesinger, M., Shukla, P., Smith, S.J., Swart, R., van Rooijen, S., Victor, N., Dadi, Z.: California Climate Adaptation Strategy (Full report). California Natural Resources Agency. Cambridge University Press, New York (2009)

7. Knowles, N.: Potential Inundation Due to Rising Sea Levels in the San Francisco Bay Region. California Climate Change Center. CEC-500-2009-023-F, California Energy Commission, PIER Energy-Related Environmental Research (2008)
8. Knowles, N.: Potential Inundation Due to Rising Sea Levels in the San Francisco Bay Region. California Climate Change Center. San Francisco Estuary and Watershed Science 8(1) (2010)
9. Parry, M.L., Canziani, O.F., Palutikof, J.P., van der Linden P.J., and Hanson, C.E. (eds.): Climate Change 2007: Impacts, Adaptation and Vulnerability, Contribution of Working Group II to the Fourth Assessment Report of the Intergovernmental Panel on Climate Change (AR4), 976 p. Cambridge University Press, Cambridge (2007)
10. Ramstorf, S.: A Semi-empirical approach to projecting future sea-level-rise. Science 315, 368–370 (2007)
11. Ramstorf, S.: A Semi-empirical approach to projecting future sea-level-rise. Science 315, 368–370 (2007)
12. Solomon, S., Qin, D., Manning, M., Chen, Z., Marquis, M., Averyt, K.B., Tignor, M., Miller, H.L. (eds.): Climate Change 2007: Contribution of Working Group I to the Fourth Assessment Report of the Intergovernmental Panel on Climate Change, 996 p. Cambridge University Press, Cambridge (2007)
13. van Raalten, D., van der Laan, T., Wijsman, P., Boeije, L., Schellekens, E., Dircke, P., Pyke, B., Moors, E., van Pelt, S., Elias, E., Dijkman, J., Travis, W., McCrea, B., LaClair, J., Goldbeck, S.: San Francisco Bay: Preparing for the next step. San Francisco Bay Conservation and Development Commission, BCDC (2009)
14. van Raalten, D., van der Laan, T., Wijsman, P., Boeije, L., Schellekens, E., Dircke, P., Pyke, B., Moors, E., van Pelt, S., Elias, E., Dijkman, J., Travis, W., McCrea, B., LaClair, J., Goldbeck, S.: Living with a Rising Bay: Vulnerability and Adaptation in San Frnacisco Bay on its Shoreline, San Francisco Bay Conservation and Development Commission (BCDC). Shoreline (Draft Staff Report), 175 p. (2009)
15. Xia, J., Falconer, R.A., Lin, B., Tan, G.: Numerical assessment of flood hazard risk to people and vehicles in flash floods. Environmental Modelling & Software 26(8), 987–998 (2011)

Data Acquisition

16. Data: CA.gov: Cal-Atlas Geospatial Clearinghouse, `http://atlas.ca.gov/` (accessed March 19, 2011)
17. Data, `http://water.weather.gov/precip/download.php` (accessed July 29, 2011)
18. Data: Natural Resources Conservation Service, `http://www.nrcs.usda.gov/wps/portal/nrcs/home` (accessed August 15, 2011)
19. Data: U.S. Census Bureau; TIGER, TIGER/Line® and TIGER – Related Products, `http://www.census.gov/geo/www/tiger/` (accessed Mach 29, 2011)
20. USGS CASCaDE, `http://cascade.wr.usgs.gov/index.shtm` (accessed July 29, 2011)
21. Data: United States Department of Agriculture (USDA) 2007 Census of Agriculture, `http://www.agcensus.usda.gov/Surveys/index.asp` (accessed July 11, 2011)
22. Data: USGS; Seamless Data Warehouse, 1/3 arc-second NED, `http://seamless.usgs.gov` (accessed December 10, 2010)

Extendable Simulation Framework for Virtual World Environment Based on the DEVS Formalism

Changbeom Choi[1], Se Jung Kwon[1], Tag Gon Kim[1],
Jae Hyun Lim[2], Dong-Hyun Baek[3], and Soohan Kim[4]

[1] KAIST, 291 Daehak-ro, Yuseong-gu, Daejeon 305-701, Republic of Korea
{cbchoi,sjkwon}@smslab.kaist.ac.kr, tkim@ee.kaist.ac.kr
[2] Network Customizing Technology, 3-1 Doryung-dong,
Yuseong-gu, Daejeon 305-733, Republic of Korea
jhlim@netcus.com
[3] Kyungwon Univ., 69, Seongnam-Daero, Bundang-gu, Seongnam-si, Gyeonggi-do,
Korea Youngtong-gu, Suwon, Kyungqi-do, Republic of Korea
dhbaek@kyungwon.ac.kr
[4] SAMSUNG Electronics HQ 416, Youngtong-gu, Suwon, Kyunggi-do, Republic of Korea
ksoohan@samsung.com

Abstract. A virtual world is an interactive virtual environment in which users interact with each other with computers, and can be used as a platform for virtual training activities. In order to enhance the trainee's immersive experience, domain-specific simulation models are required for virtual world services. For this reason, we propose an extendable simulation framework for the virtual world. The simulation framework is composed of Core Simulation Framework and Virtual Level Architecture. By utilizing the Core Simulation Framework and the Virtual Level Architecture, the content creator of the virtual world can create extendable simulations for domain-specific content.

1 Introduction

Generally, a virtual world is an interactive virtual environment in which users interact using their own graphical representations, known as avatars. In the virtual world, users interact with each other using computers, and they can collaboratively create and edit objects in the virtual world without meeting each other and interacting in the real world. Therefore, the virtual world has been effectively utilized in education, especially in virtual training.

Virtual training is a training method in a computer-simulated environment. The virtual world allows users to access educational content from their homes or work places. In order to enhance the effectiveness of education via the virtual world, the virtual world services utilize physics engines. Usually, the physical simulation modules are embedded in the game engine. However, there are several limitations of supporting simulation models in the virtual world environment. To support immersive experiences for the trainee, domain-specific simulation models are required, such as the weather simulation model, intelligent virtual agents, and war-game simulation

J.-H. Kim et al. (Eds.): AsiaSim2011, PICT 4, pp. 475–483, 2012.

models. In order to provide domain-specific simulation content to the virtual world, the virtual world must embrace the simulation engine as a component. As the simulation content evolves, the simulation engine should embrace various levels of simulation models.

To give an immersive experience to users of the virtual world, we propose a simulation framework based on the discrete event system specification (DEVS) formalism [1]. The DEVS formalism specifies discrete event models in a hierarchical and modular manner. Therefore, the user may model the target system by decomposing the large system into smaller components by applying coupling schema to them. The proposed simulation framework consists of Core Simulation Framework, which manages and executes the DEVS simulation model, and Virtual Level Architecture, which support discrete event simulation and distributed simulation.

The rest of this paper is organized as follows: Section 2 introduces existing virtual worlds and simulation models of the virtual worlds. In Section 3, we propose the extendable simulation framework. Then, we introduce the case study. In the case study, we introduce the prototype of the firefighting training system. Finally, we conclude the paper.

2 Related Works

2.1 Second Life

Second Life® is an online virtual world that enables users to interact with each other via avatars [2]. Because Second Life supports in-world object creation, users can build various objects with primitive 3D polygonal shapes. In addition, Second Life supports the procedural script language so that users can add interactive behaviors to the object and the avatar.

Second Life was utilized as an educational platform at various institutions. For example, Loyalist College utilized Second Life to help Customs and Immigration students experience the daily routine of their future jobs [3]. Moreover, in the scientific community, Second Life is used for collaboration and data visualization [4]. Since the virtual world can overcome time and regional restrictions, Second Life was utilized as a suitable platform for real-time meetings with other people to share knowledge. Second Life utilizes the Linden Scripting Language [5] and Havok physics [6] to support the simulation of various machines and devices. However, Second Life has limited features for the domain-specific simulation. First, memory for a single script is restricted to 64 KiB. Therefore, the domain-specific simulation model may not be loaded in Second Life. Second, Havok physics is not designed as a simulation engine; it is sufficient to handle real-time collision and dynamics of rigid bodies, but not sufficient to handle domain-specific simulation in the virtual world.

2.2 OpenSim

The objective of OpenSim is to develop the full open-source server to host virtual world services using existing Second Life clients [7]. One characteristic of OpenSim is that the server can easily expand through plug-in modules, and it has been utilized in various researches.

OpenSim supports several physics simulation engines such as Open Dynamics Engine [8] and BulletX [9]. By supporting several physics simulation engines, OpenSim can handle physics simulation in the virtual world, such as detecting the collision of objects and solving the dynamics of rigid bodies. However, OpenSim does not consider domain-specific simulation in the virtual world. Therefore, OpenSim does not contain the simulation engine necessary to handle the domain-specific simulation model.

3 Extendable Simulation Framework for Virtual World

When the content creator of the virtual world wants to utilize realistic simulation models in the virtual world, the virtual world service framework should find the simulation models and send them to the virtual world servers. In order to support the virtual world service framework to handle various level of the simulation model, the virtual world service framework must provide simulation model management service. The simulation model management service is composed of storing the simulation model, searching the simulation model, transferring the simulation model, and controlling the simulation model. The store and search services assist the authors with registering the simulation model and searching the proper simulation model for the authors' intent. When the simulation models are decided, the virtual world service framework initializes the virtual world server with simulation models by transferring the simulation models from the simulation model storage to the server. After the initialization of the simulation models, the authors of the virtual world control the simulation using the simulation model control service.

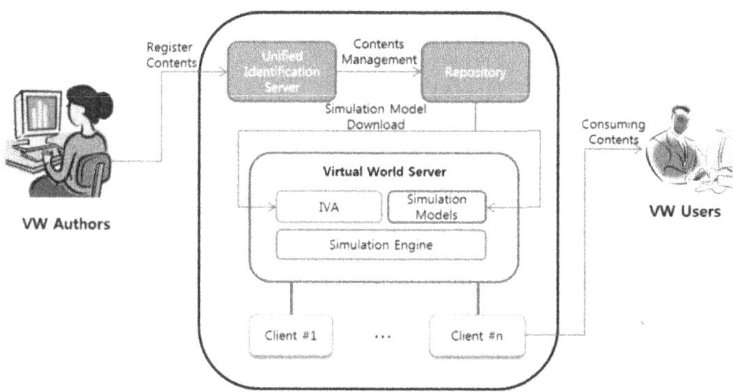

Fig. 1. Conceptual figure of the virtual world service framework

Figure 1 conceptualizes the virtual world service framework. The role of the Unified Identification Server (UIS) is to handle account information, manage content for the virtual environment, and search the account information and content. The Repository server is used to manage assets of the virtual world, such as item information for avatars, objects in the virtual world environment, and simulation models. If the content creator wants to create virtual worlds with domain-specific simulation models, the virtual world server must contain a simulation engine that supports simulations.

Moreover, the server should manage the virtual world's users, the simulation models, and the simulation engine. In this virtual world service framework, when the authors of the virtual world request the simulation models, the UIS sends the simulation model request to the Repository server. Then the Repository server transfers the corresponding simulation models to the virtual world server and the virtual world server initializes the domain-specific simulation models. However, if the virtual world server contains the simulation engine as a module, supporting various simulation engines may be limited, and the managing simulation engine and the executing simulation model may influence the overhead of the virtual world server.

Therefore, we propose an extendable simulation framework for the virtual world. Figure 2 shows the conceptual figure of the extendable simulation framework used in the virtual world service framework. As shown in the figure, the simulation engines are separated from the virtual world server. The benefits of separating the simulation engine from the server are twofold. First, various simulation engines can be applied in the virtual world without modifying the virtual world server. Second, when the content creator wants to utilize more than two domain-specific simulations in the virtual world, the extendable simulation framework can support more than two simulation engines to generate simulation results.

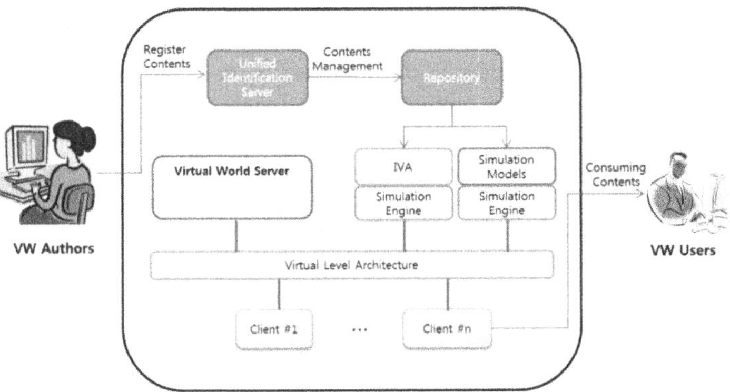

Fig. 2. Conceptual figure of the extendable simulation framework

In order to support domain-specific simulation models, we propose the Core Simulation Framework (CSF) and Virtual Level Architecture (VLA). The CSF supports the domain-specific simulation, and the VLA supports the interoperation between the CSF and the virtual world server.

3.1 Core Simulation Framework

The CSF consists of the Model API layer, the simulation engine, and the system layer, and supports the extendable simulation environment based on the plug-in feature. Figure 3 shows the conceptual architecture of the CSF.

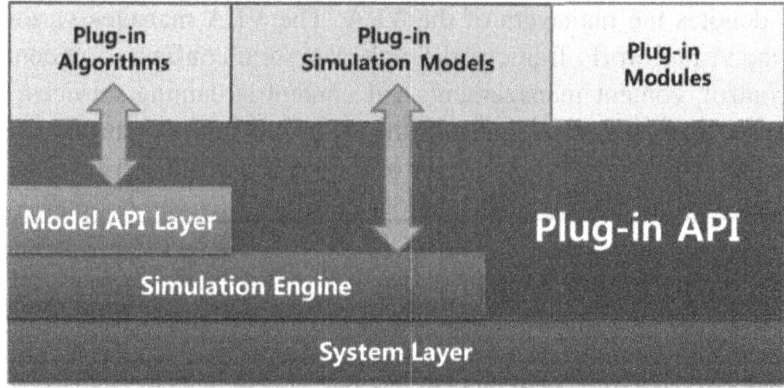

Fig. 3. Core Simulation Framework

The system layer of the CSF controls the management modules, such as network communication, simulation model initialization, and simulation engine control modules. The simulation engine of the framework may depend on the simulation domain. As a result, the system layer of the CSF supports the plug-in API to change the simulation engine based on the simulation domains [10].

The CSF's simulation engine should support the plug-in feature for the simulation models. In this research, we adopted the simulation engine for DEVS formalism. By adopting the formalism, the authors of the simulation model can reuse the domain-specific simulation models, or change the domain-specific algorithms for the simulation model by applying a layered approach [11].

3.2 Interoperation of Simulation

Interoperation between the CSF and the virtual world server could be achieved by using the Virtual Level Architecture (VLA). The VLA supports interoperation among simulation servers and the virtual world server by providing core services, such as data synchronization and time synchronization.

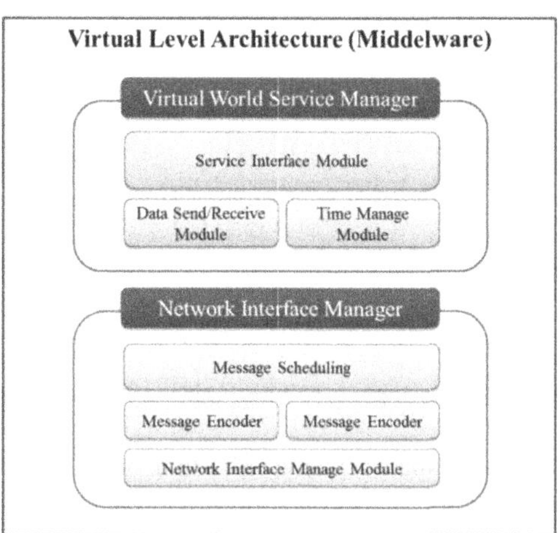

Fig. 4. Conceptual figure of the Virtual Level Architecture

Figure 4 denotes the managers of the VLA. The VLA manages various levels of services in the virtual world framework, such as system configuration control, system execution control, content management, and content streaming services. In order to handle these services, the VLA provides the API with time synchronization and data synchronization. Therefore, the developer of the virtual world server or the simulation model can develop each component without considering time and data synchronization.

In order to support interoperation between the simulation engine and the virtual world server, the service interface module in the virtual world service manager provides the interface to handle data synchronization and time synchronization. In order to support the time synchronization between the virtual world server and the simulation engine, the time synchronization algorithms are implemented in time management modules. For data synchronization, the data send/receive module manages the data based on the time. In order to support virtual world service manager, the network manager handles message scheduling and encoding/decoding the network packets.

4 Case Study

The purpose of this case study is to build a prototype of the firefighting training simulator. Modeling and simulation experts and the domain experts collaborated with each other to develop the firefighting training simulator [11]. The modeling and simulation experts built the simulation models and implemented the simulation framework using VLA, and the domain experts consulted on the development of the fire simulation model. Figure 5 shows the conceptual design of the simulator. Each simulation server has fire simulation models and engines, and every simulation server is interconnected with its neighboring simulator through VLA.

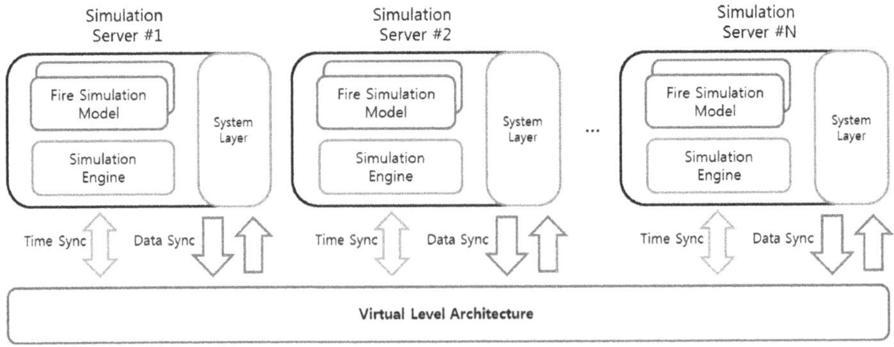

Fig. 5. Conceptual architecture of the Fire-Fighting Training Simulator Prototype

The prototype simulates spaces inside a certain building virtually. A trainer can make scenarios with flammable materials, non-flammable materials, different starting points for the fire, and so on. As the fire starts at some point and spreads, several trainees should move and extinguish the spreading fire. From the simulator, the firefighter can experience these diverse scenarios virtually. The design of the overall simulation models including the fire models is described in Figure 6. The coupled model consists of the *UserController* model, the Transducer model, and the grid of Fire models.

The *UserController* model is controlled by user interface, and has the positions of the users (trainee). The positions (i,j) that users locate means that the user tries to extinguish the flames in the *Fire model* (i,j). Hence, the *UserController* model sends extinguishing messages to proper *Fire models* through the *toFireModel* ports. The *Fire models* are interconnected to adjacent Fire models through *in_east*, *in_west*, *out_east*, *out_west*, and so on. The flames on the models spread to other models by sending messages of the fire model to adjacent to the other fire models.

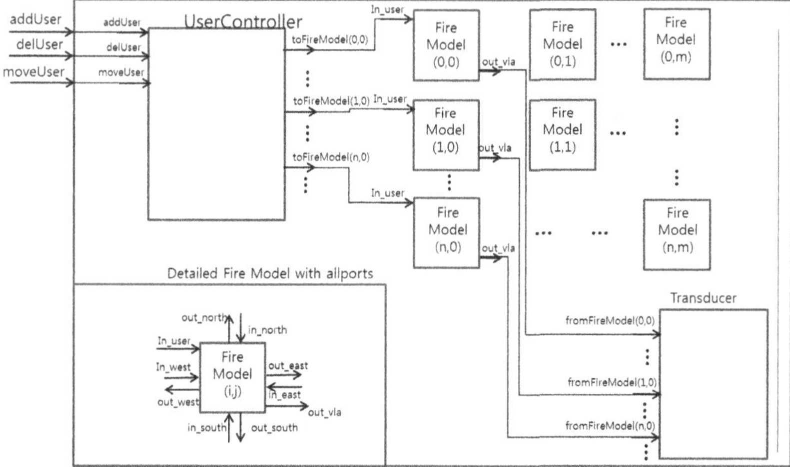

Fig. 6. Overall Coupled Model for Firefighting Simulation

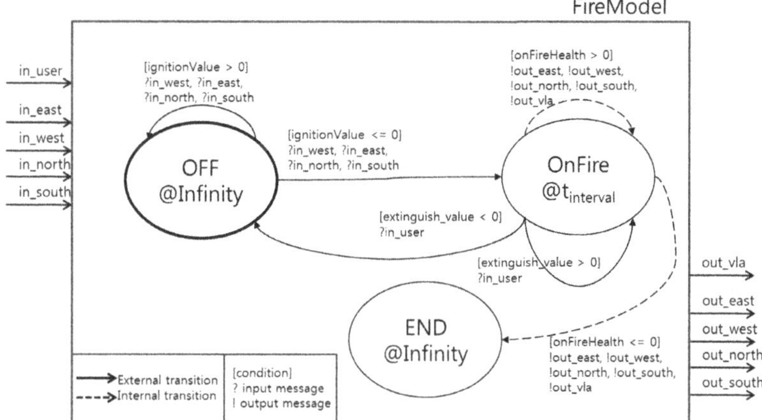

Fig. 7. Design of a Fire Model

The design of the fire model as an atomic model is described in Figure 7. The atomic model has three states: *OFF*, *OnFire*, and *END*. *OFF* is an initial state or a state with an extinct fire. *OnFire* is that the space is burning. *END* state is when the flammables in the space are burned down. There are three variables of state transitions: *ignitionValue*, *extinguish_*value, and *onFireHealth*.

- *ignitionValue*: A value representing the level for ignition of the cell. Reduced by the message from the other fire models.
- *extinguish_*value: A value representing the level for extinguishing the flames completely in the cell. Reduced by the extinguishing message from the *in_user* port.
- *onFireHealth*: Duration of the flames. Reduced as the flames are held on the space.

When the fire model on the edges of the grid send messages, some messages of output ports are sent to another simulation server. Through the *out_vla* port, the message is sent to the *Transducer* model. The *Transducer* model collects the messages and sends them to other servers through VLA.

Figure 8 is a captured picture of the four-simulation server. On the left side of the figure, there are four GUI windows for each server. Black squares indicate non-flammable materials such as walls, and green squares represent flammable materials. In this figure, red flames are spreading in several places. If the red flames arrive at the edge of the server, we can see the red flames spreading to other server.

On the right side of the figure, there are GUI windows for the user, the trainee. Users' avatars are depicted by simple yellow circles. As trainees move their avatars on the GUI windows, they can increase their experience in extinguishing the building's flames.

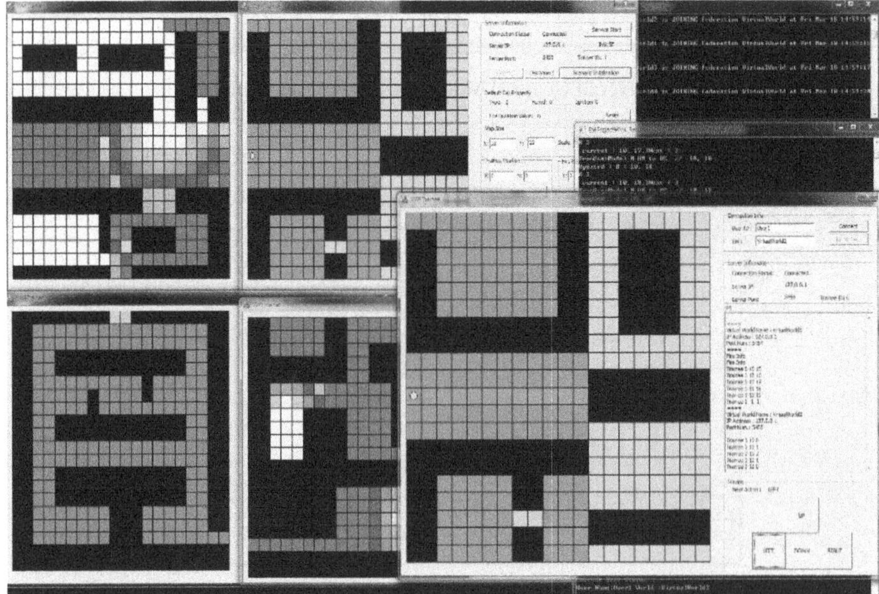

Fig. 8. The captured picture of execution in the four-simulation server

5 Conclusion

This paper introduces an extendable simulation framework for the virtual world. The virtual world is a computer-based interactive virtual environment where users interact with each other through computers. In order to give a trainee an immersive experience, the virtual world's content may be base on the domain-specific simulation models. In order to support the extendibility of the simulation framework, we propose a simulation framework based on the DEVS formalism. Therefore, the creator can utilize the existing simulation models by utilizing the DEVS-based simulation models. The simulation framework, which consists of the Core Simulation Framework and the Virtual Level Architecture, allows the user to reuse the simulation models.

The Core Simulation Framework embraces plug-in features so that the simulation framework can plug-in various modules during run-time. The Virtual Level Architecture handles time synchronization and data synchronization among components of the virtual world services. Regarding future research, we will extend the simulation framework to support extendable virtual world environments.

Acknowledgement. This research is supported by Ministry of Culture, Sports and Tourism (MCST) and Korea Creative Content Agency (KOCCA) in the Culture Technology (CT) Research & Development Program 2009.

References

1. Zeigler, B.P., Kim, T.G., Praehofer, H.: Theory of Modeling and Simulation. Academic Press, Orlando (2000)
2. Second Life, `http://www.secondlife.com`
3. Rippin, P.: Virtual World Simulation Training Prepares Real Guards on the US-Canadian Border: Loyalist College in Second Life. Linden Lab. (2009), Source, `http://secondlifegrid.net.s3.amazonaws.com/docs/Second_Life_Case_Loyalist_EN.pdf`
4. Lang, A.S., Bradley, J.C.: Chemistry in Second Life. Chemistry Central Journal 3, 14 (2009), doi:10.1186/1752-153X-3-14.
5. Linden Scripts Language, `http://wiki.secondlife.com/wiki/List`
6. Havok Phsyics, `http://www.havok.com/`
7. OpenSim, `http://www.opensimulator.org`
8. Open Dynamics Engine, `http://www.ode.org/`
9. BulletX, `http://xnadevru.codeplex.com/`
10. Bae, J.W., Kim, T.G.: DEVS Based Plug-in Framework. In: Spring Simulation Multiconference 2010, Orlando, FL, pp. 147–153 (2010)
11. Sung, C.H., Hong, S.Y., Kim, T.G.: Layered Structure to Development of OO War Game Models Using DEVS Framework. In: Proceedings of Summer Computer Simulation Conference, Philadelphia, pp. 65–70 (2005)

Visualization of Plasma Plume Collisions Using Fused Volume Data

Naoya Kawamoto[1,*], Yuiti Hatanaka[1], Takashi Yamamoto[1], Kyoko Hasegawa[2], Susumu Nakata[3], Satoshi Tanaka[3], Naohisa Sakamoto[4], Koji Koyamada[4], Seigo Misaki[5], and Kazuo Tanaka[5]

[1] Graduate School of Science and Engineering, Ritsumeikan University, Japan
tyamamoto@cg.is.ritsumei.ac.jp
[2] Kinugasa Research Organization, Ritsumeikan University, Japan
hasegawa@media.ritsumei.ac.jp
[3] College of Information Science and Engineering, Ritsumeikan University, Japan
{snakata,stanaka}@media.ritsumei.ac.jp
[4] Institute for the Promotion of Excellence in Higher Education, Kyoto University, Japan
{naohisas,koyamada}@viz.media.kyoto-u.ac.jp
[5] Graduate School of Engineering, Osaka University, Japan
misaki-s@ile.osaka-u.ac.jp, katanaka@eei.eng.osaka-u.ac.jp

Abstract. We report our study to visualize plasma plume collision based on XYT-space (space and time) volume data, not depending on the conventional animation.This aims at supporting reseraches in plasma physics. We create 3D volume data in the XYT space by piling up a time series of XY-plane photo images taken in the experiment. Then We execute isosurface/volume rendering to visualize the created data. We adopt the following "fused" visualization based on the particle-based rendering technique: (1) multiple-isosurface rendering of different temperatures, and (2) fused volume rendering of different materials and temparatures, i.e., carbon and tungsten.We can easily compare physics of different elements in flexible ways by applying the particle-based rendering technique.

Keywords: plasma plume collision, XYT space, fused visualization, particle-based rendering.

1 Introduction

Laser nuclear fusion is a promising future technology that uses high-energy lasers to activate the fusion [1]. When the laser is irradiated to a reactor wall, high-temperature clouds of plasma arise from the wall. Below we call the clouds "plasma plumes". The plasma plumes collisions occurring in the reactor wall become an obstruction of the chain reaction of the nuclear fusion. Therefore it is important to understanding their dynamical properties. Researches are going on for this issue [2]. This paper proposes visualization to support scientific researches in the field of plasma physics.

J.-H. Kim et al. (Eds.): AsiaSim2011, PICT 4, pp. 484–491, 2012.

On the other hand, visualizing volume data is a key technology to understand phenomena in physics [3]. For visualizing dynamical phenomena, we usually use the animation, i.e. movies. In fact, the animation is suitable for intuitive understanding of dynamical phenomena. But the animation is not suitable for focusing on a specific time region of interest. It is not suitable for overviewing the whole time region at one glance, either. Therefore we adopt static visualization for the dynamical phenomena of the plasma plume collision with a single image. Such visualization of ours is inspired by the recent work [4], where the whole process of the cell death is visualized in a higher-dimensional static image.

We seek for a new visualization suitable for dynamical phenomena in one image by executing "fused" visualization of different elements. In this paper, we execute two kinds of the fused visualization: (1) multiple-isosurface rendering of different temperatures, and (2) fused volume rendering of different materials, i.e. carbon and tungsten. For such fused visualization, the particle-based rendering technique [5], works well.

2 Visualized Data and Visualization Method

Our visualizing target is a time series of photo images. It is obtained by taking the experiment. The standard way to visualize these photo images is the animations, displaying each image in order with small time intervals. But the animation is not always convenient for scientific analysis as mentioned in Section 1.

2.1 Visualized Data

We visualize a time series of images photographed in the experiment (see Fig.1 (left)). In the experiment, carbon/tungsten specimens are irradiated with high-energy laser beams. The specimens are placed in two positions on a cylindrical wall such that the laser beam to each specimen runs at right angles to one another. Then the plasma plume arisen from each specimen also runs at right angles to another. The two plasma plumes collide in the center of the cylindrical.

In the above experiment, a time series of photo images are taken with a fast frame is shifted in time with 50 nsec time step at the camera trigger. Thus we obtain a time series of the photo images that show the process of the plasma plume collision (see Fig.1 (middle)). Fig.2 shows a time series of photo images displaying that the carbon plasma plumes rises (see Fig.2 (a)), collides (see Fig.2 (b), (c) and (d)) and diffuses (see Fig.2 (e) and (f)) from two positions in the appearance of the experiment with carbon specimen. For visualizing these photo images which show dynamical phenomena of plasma plumes, we usually use the animation, rendering each image at regular small time intervals. In fact, animation can visualize time development of dynamical phenomena very well. But the animation is not always convenient for scientific analysis. To observe the whole process in one glance, we create 3D volume data in the XYT space by piling up the time series of the XY-plane photo images (see Fig.1 (right)). The created volume data are normalized to values between 0 and 255 before visualization.

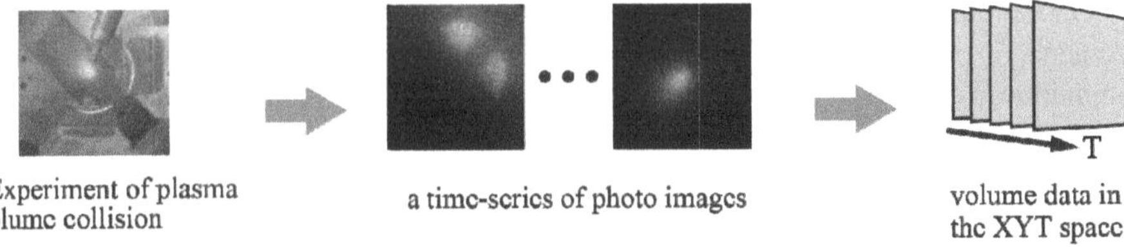

Fig. 1. Creation of volume data from a time series of photo images taken in the experiment

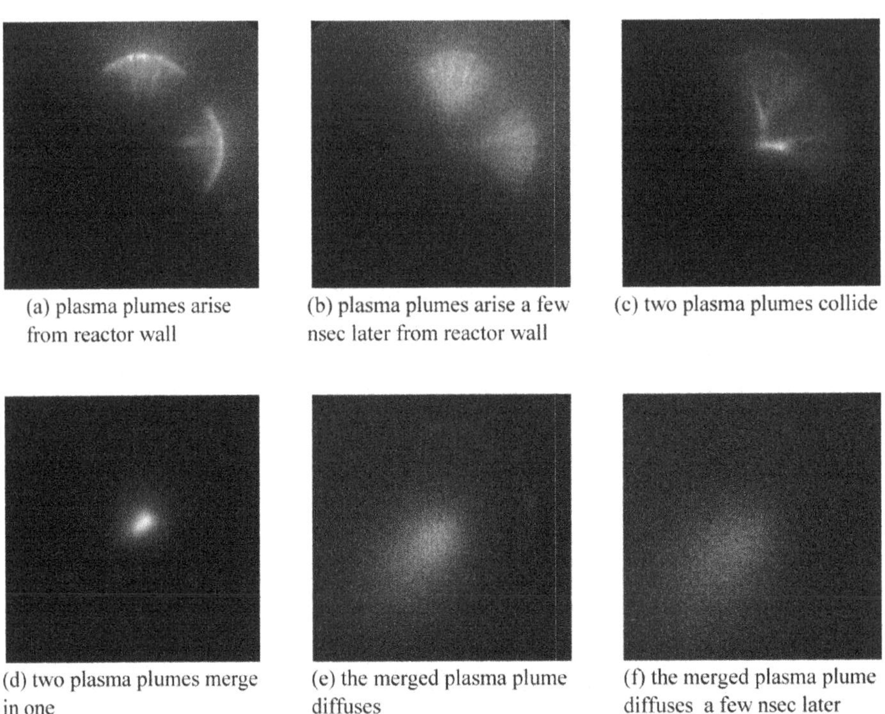

(a) plasma plumes arise from reactor wall

(b) plasma plumes arise a few nsec later from reactor wall

(c) two plasma plumes collide

(d) two plasma plumes merge in one

(e) the merged plasma plume diffuses

(f) the merged plasma plume diffuses a few nsec later

Fig. 2. A time series of photo images in appearance of experiment with carbon specimen

2.2 How to Visualize Created XYT-Space Volume Data

We visualize dynamical plasma plume collision in the XYT space. We execute isosurface/volume rendering in the XYT space and time. It can visualize dynamical behaviour of the plasma plume collision very well. So we adopt the static visualization for dynamical phenomena.

To visualize the created XYT space volume data, we adopt the particle-based rendering technique [5]. The particle-based rendering technique is suitable for "fused visualization" [6], superposing multiple images 3-dimensionally and realizes easy-to-understand 3D transparent visualization. In this paper, we execute two types of the fused visualization, i.e. multiple-isosurface rendering and fused volume rendering of different materials (carbon and tungsten). The multiple-isosurface rendering shows temperature dependence of physics. The fused volume rendering visualizes material dependence of physics. The fused visualization visualizes dynamical behaviour of plasma plume collision.

2.3 Particle-Based Rendering

In volume graphics, we adopt translucent visualization which displays inside of object. It is indispensable to sort many sampling points by an eye direction, for example, ray casting rendering. If we don't sort them, it doesn't display the frame of inner structures correctly and such. But it is galloping amount of calculation which is needed to sort that the visualized data is larger.

We developed a particle-based rendering which don't need to sort many sampling points by volume rendering. Particle-based rendering is a kind of point rendering [5]. Fig.3 shows that we compare particle-based rendering with conventional polygon rendering. We can see that complex inner structures are visualized more clearly. Besides, there is a transfer function that makes scalar value responsive to gray or RGB color. It needs to fix the transfer function in according to a visualized purpose of the volume data.

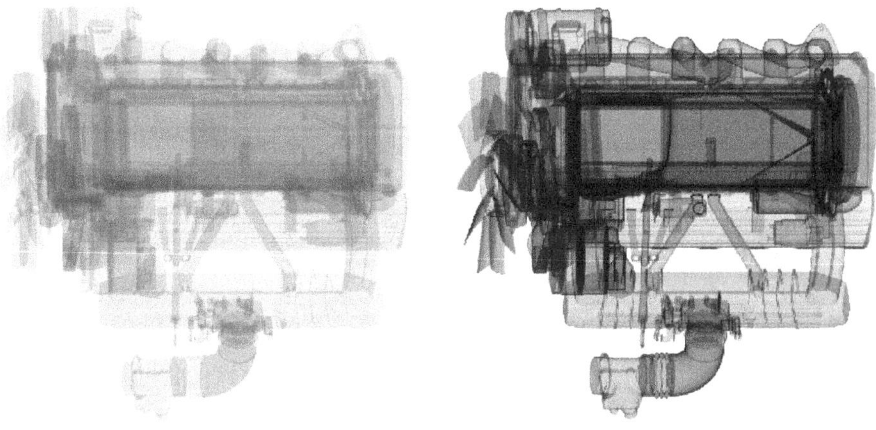

Polygon rendering Particle-based rendering

Fig. 3. Comparison of particle-based rendering with conventional polygon rendering

3 Visualization Result

We execute two kinds of XYT-space rendering: (1) multiple-isosurface rendering, and (2) fused volume rendering. The former rendering visualizes temperature dependence of physics. In the latter rendering, a material dependence of physics is visualized. The visualization is made based on our particle-based rendering.

3.1 Multiple-Isosurface Rendering of Carbon Plasma Plume Collision

Fig.4 visualizes the plasma plume collision of carbon in the XYT space. We visualize three transparent isosurfaces of hot (red; scalar value 220), medium-temperature (green; scalar value 127), cool (scalar value 64) regions for each temperature. Time flows from back to front (see Fig.4 (upper)) and from bottom to top (see Fig.4 (lower)). For each temperature, the static visualization in the XYT space clearly shows dynamical features of the plasma plume collision.

We superpose and visualize three transparent isosurfaces of hot, medium-temperature and cool regions within a single image, using particle-based rendering technique [5] (see Fig.5). Time flows in the upward direction, i.e. from bottom to top. From Fig.5, we can easily understand the dynamical features of the plasma plume collision as follows. First two medium-temperature plasma plumes collide (green parts at the bottom). Then they merge in one and become hotter (red parts in the middle). Finally, the merged larger plasma plume diffuses and cools down (blue parts at the top).

Multiple-isosurfaces rendering in the XYT space as Fig.5 enables us to compare physics of different temperatures or energy scales easily. Besides, the visualization in the XYT space makes information of the time development included. In these senses, Fig.5 tells us lots of information with a single static image.

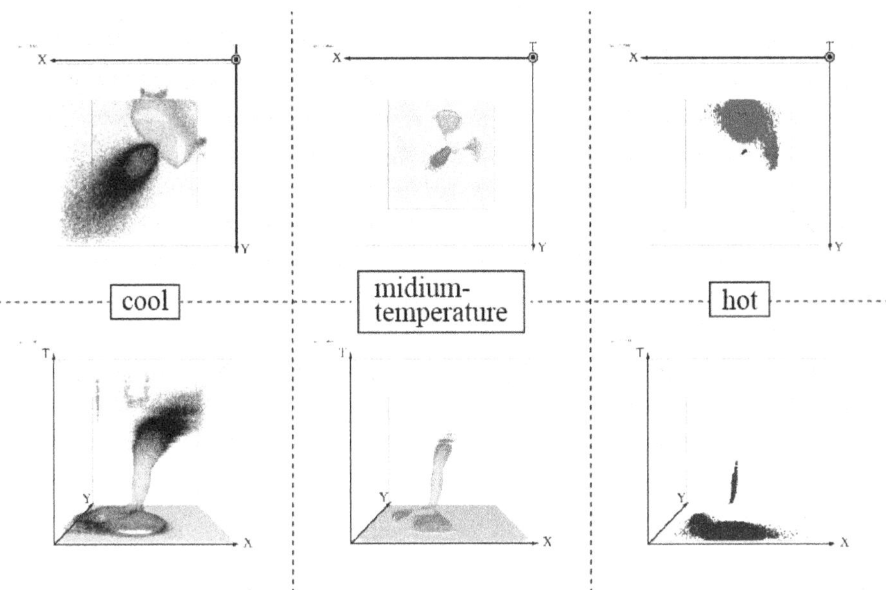

Fig. 4. Isosurface of hot (red), middle-temperature (green) and cool (blue) regions in plasma collision of carbon. In the upper figure, time flows back to front. In the lower figures, time flows from bottom to top.

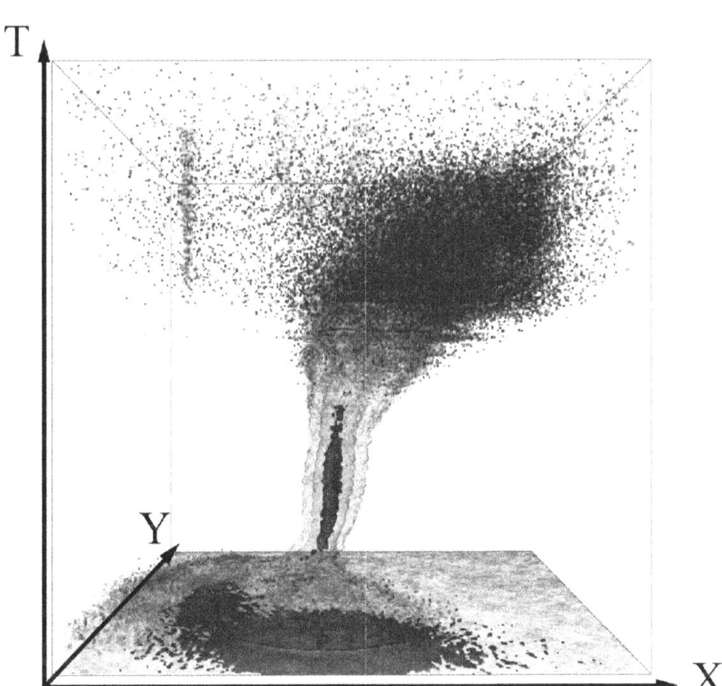

Fig. 5. Multiple-isosurface rendering of plasma plume collision of carbon in the XYT-space

3.2 Fused Volume Rendering of Different Materials

We visualize the plasma plume collisions of different materials. In executing the volume rendering, an opacity transfer function with a sharp peak around the scalar value 64 is adopted. The yellow cloud shows cool region in the plasma plume collision of carbon (see Fig.6 (a)). The other purple one is that of tungsten (see Fig.6 (b)).

Fig.7 shows that cool regions in the plasma plume collisions for two materials are superposed in a single visualization, using the particle-based rendering technique [5]. Time flows from back to front in Fig.7. We can clearly see tungsten behaves very differently from carbon. The plasma plumes of tungsten never merge in the collision, but just through. This is an evidence to prove that types of interaction between plasma particles are different, depending on materials.

Besides, we adopt the fused volume rendering of multiple temperatures and multiple volumes. Fig.5 is a similar fused volume rendering that visualizes three temperature regions for carbon, superposing only the cool region for tungsten. It visualizes hot (red), middle-temperature (green) and cool (blue) regions for carbon and the cool region for tungsten (purple). Such flexible superposing is an advantage of the particle-based rendering.

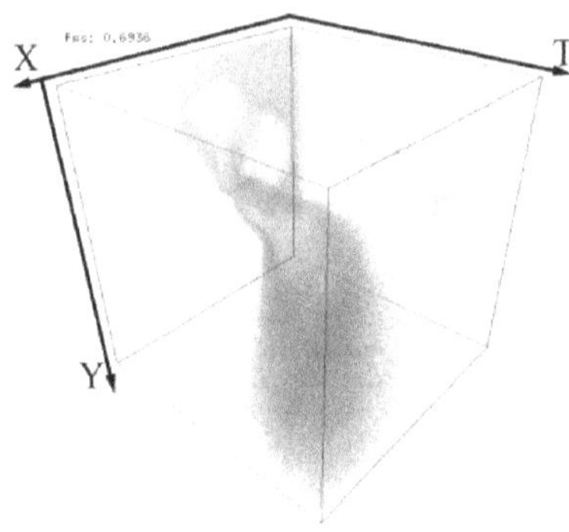

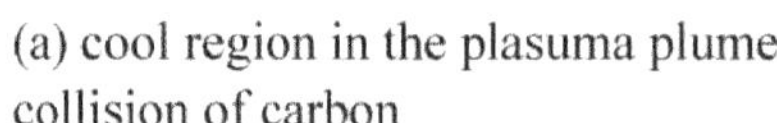
(a) cool region in the plasuma plume collision of carbon

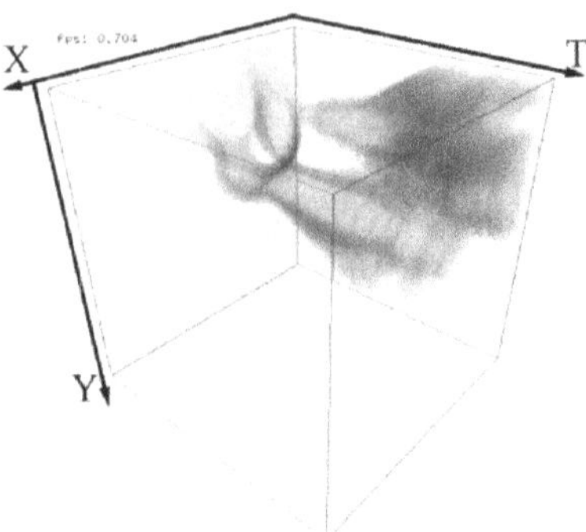

(b) cool region in the plasuma plume collision of tungsten

Fig. 6. Volume rendering that cool regions in the plasma plume collision for each materials

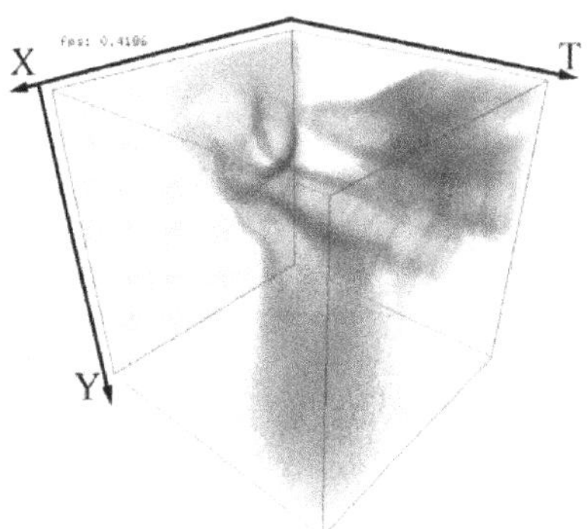

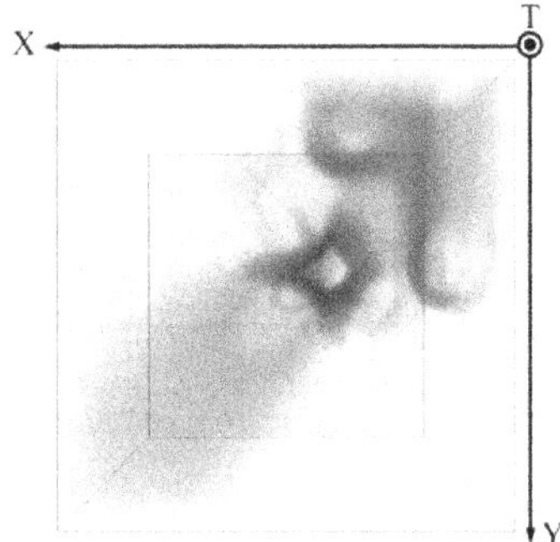

Fig. 7. Fused volume rendering of cool regions in carbon (yellow) and tungsten (purple) plasma plume collisions.

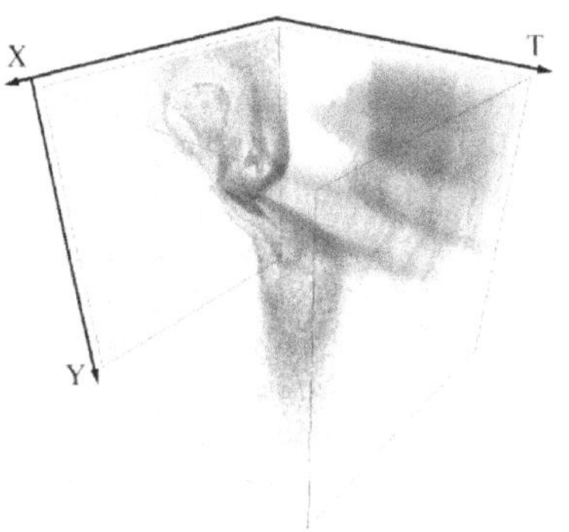

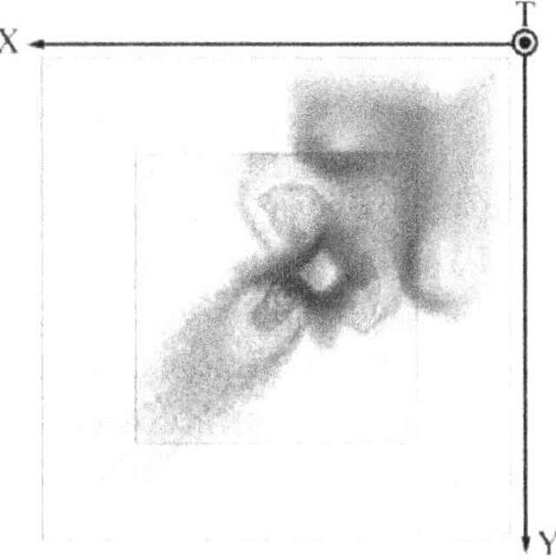

Fig. 8. Fused volume rendering of hot(red), middle-temperature (green) and cool (blue) regions in carbon and tungsten (purple) plasma plume collisions.

4 Conclusions

In this paper, we have studied the visualization of plasma plume collision based on the XYT-space volume data, not depending on animation. It can be done by piling up a time series of photo images to create the XYT-space volume data. We also executed two types of visualization based on the particle-based rendering technique, i.e. multiple-isosurface rendering of different temperatures and fused volume rendering of different materials (carbon and tungsten). The former visualization clearly shows us time development of hot, medium-temperature and cool regions in the process of plasma plume collision with a single static image. The latter visualization gives us a comprehensible way to compare plasma dynamics of different materials.

We are now trying to do surface-volume fusion visualization and make fused visualization of multiple views from different camera angles. Besides, we plan to do fusion of simulation and experimental results.

References

1. Sunahara, A., Tanaka, K.A.: Fusion Engineering and Design 85, 935 (2010)
2. Hirooka, Y., Tanaka, K.A., Sato, H., Ishihara, K., Sunahara, A.: Laboratory experiments on cluster/aerosol formation by colliding ablation plumes. Journal of Physics 244, 32–37 (2010)
3. Lichtenbelt, B., Naqui, S., Co., H.-P., Crane, R., Naqvi, S.: Introduction to Volume Rendering. Prentice Hall (1998)
4. Sakamoto, N., Kawamura, T., Kioka, M., Koyamada, K., Sakamaki, K.: Spatiotemporal Analysis of Morphological Cahnges in Cell Death Using Multiple Volume Visualization. In: IEEE Visualization Proceedings Compendium, pp. 52–53 (2008)
5. Koyamada, K., Sakamoto, N., Tanaka, S.: A Particle Modeling for Rendering Irregular Volumes. In: Proceedings of theInternational Conference on Computer Modeling and Simulation, Cambridge, England, pp. 372–377 (2008)
6. Sakamoto, N., Koyamada, K., Saito, A., Kimura, A., Tanaka, S.: Multi-Volume Rendering Using Particle Fusion. In: IEEE VGTC Pacific Visualization Symposium 2008, Kyoto, Japan (2008)

Tiled Display Visualization System with Multi-touch Control

Jun Nishimura[1], Naohisa Sakamoto[2], and Koji Koyamada[2]

[1] Graduate School of Engineering, Kyoto University, Japan
nishimura@viz.media.kyoto-u.ac.jp
http://www.viz.media.kyoto-u.ac.jp
[2] Institute for the Promotion of Excellence in Higher Education,
Kyoto University, Japan
naohisas@viz.media.kyoto-u.ac.jp,
koyamada.koji.3w@kyoto-u.ac.jp
http://www.viz.media.kyoto-u.ac.jp

Abstract. This paper presents a visualization system for tiled display wall (TDW) environments and user interface to control objects on TDW by using multi-touch devices. This system communicates messages between a master node and renderer nodes by using TCP/IP protocol for control of objects and frame synchronization. Renderer nodes receive messages from a master node at each frame, apply messages, and then render images of each frustum. It also enables multi-touch control by creating a thread for communication with multi-touch device on the master node. In this paper, visualization results of a volume rendering technique called stochastic projected tetrahedra are shown on TDW as an application.

1 Introduction

Visualization in a high-resolution display environment is crucial not to miss scientific discoveries from large-scale simulation dataset. One of the solutions to achieve this environment is tiled display wall (TDW). TDW is composed of a number of LCD displays arranged in tile on wall. These display are connected to several computers, and these computers are connected with each other in a network. By using this environment, high-resolution visualization results can be observed simultaneously by many people.

Recently, the spread of multi-touch devices is remarkable; these devices are smart phones and tablet devices equipped with a multi-touch screen. The advantages of those devices are the intuitive control by using multi-touch gestures and the easy development of its applications by using SDKs.

In this paper, we propose an original visualization system for TDW environments with multi-touch user interface. As an application, a volume rendering application is executed on a TDW environment composed of 18 LCD displays.

J.-H. Kim et al. (Eds.): AsiaSim2011, PICT 4, pp. 492–497, 2012.

2 Related Works

There are some libraries to display visualization results on TDW, and also some user interfaces to control TDW applications by using multi-touch devices.

2.1 CGLX

CGLX (Cross-Platform Cluster Graphics Library) is a library for scalable, distributed, high-performance visualization environments like TDW. This system executes OpenGL commands on each renderer node, so it can obtain high-resolution visualization results. Ponto et al. [1] proposed user interface for CGLX by using some multi-touch devices and a multi-touch table. The same image as TDW screen are shown in each device and synchronized by network.

2.2 SAGE

SAGE (Scalable Adaptive Graphics Environment) is also a library for TDW, but this library sends only pixels of client applications to TDW renderer nodes, so the image on TDW is stretched. This stretched image is not suitable to high-resolution of TDW. Carillo et al. [2] proposed an application to control 2D images for TDW environments based on SAGE library by using a multi-touch device.

3 Proposed Technique

In this paper, we assume TDW environments like Fig.1. Each renderer node mounts some directories that contain source codes and execution files by using NFS.

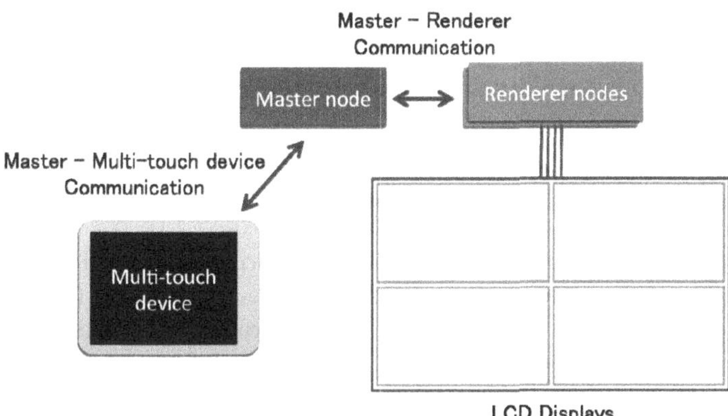

Fig. 1. TDW environment assumed in this paper. TDW is composed of one master node and some renderer nodes. Multi-touch devices can access the same network as TDW clusters via wire-less LAN.

3.1 Frame Synchronization with Renderer Nodes

It is important to synchronize each frame of renderer nodes in order not to miss some important information when controlling objects at data visualization. A frame synchronization system is embedded in this visualization system.

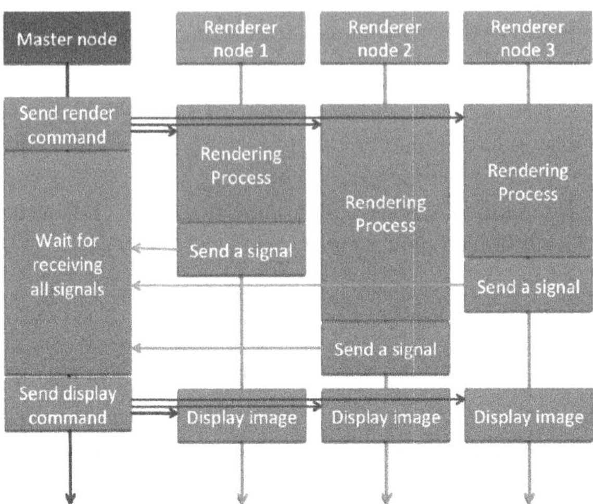

Fig. 2. Frame synchronization system. This process is done in Master – Renderer Communication of Fig.1.

To synchronize all renderer frames, several communications between a master node and renderer nodes are required (Fig.2). First, a master node sends a render command to renderer nodes. Each renderer node needs to send a signal to master node after rendering process. When a master node receives the signals from all renderer nodes, a master node sends back a command to order all renderer nodes to display the rendered back frame at the same time. TCP/IP is used to ensure these communications. The images rendered in each node can be displayed at the same timing by using this system.

3.2 Communication with Multi-touch Devices

To communicate information with multi-touch devices firmly, TCP/IP is also used. When a touch event is detected in multi-touch device, a calculation correspond to the touch gesture is done and a communication for controlling objects on TDW starts. In this connection, at first, only master node receives the information from multi-touch devices via a wireless network, and then the received data of master node is sent to each renderer node via a wired network. After receiving data in renderer nodes, the message is analyzed and reflected to the visualization result on a TDW application (Fig.3).

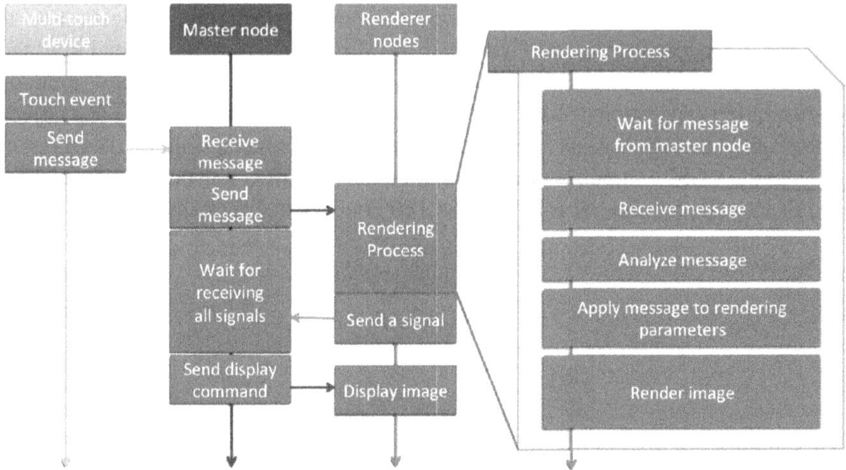

Fig. 3. Communication in this system. A master node manages messages from multi-touch devices.

A common rule for communication between the TDW application and the multi-touch device application is necessary. This rule is free to decide, provided that it is the same on the receiving side and the sending side. In our implementation, the following rule is decided. First, a message has the byte size of the whole message, and then has 4 characters represented as message type. These are common header information. Data following the header information is different by the message type. For example, rotation presents a quaternion (4 float values), scaling presents a float value and translation presents a vector (3 float values).

3.3 Multi-touch Application

In multi-touch devices, information is sent to TDW application under the rules described in the chapter 3.2. After launching a multi-touch application, users first input an IP address of the master node to the device. When users do some gestures on multi-touch devices, the difference of the rotation, scaling or translation is calculated in the device, and only the difference is sent to the TDW application. When an isosurface visualization application runs on TDW, we can also change iso value in multi-touch devices and reflect to the TDW application.

4 Application and Discussion

Our porosed technique was applied to a TDW environment composed of one master node and 18 renderer nodes connected each other with 1G Ethernet. Each renderer has a GPU (NVIDIA Quadro 4000 2GB), and connected one LCD display (1366 x 768). The total resolution of this TDW is 8196 x 2304.

In this paper, we applied our proposal to a volume rendering application on this TDW. SPT (Stochastic Projected Tetrahedra) [3] is used as a volume rendering technique. SPT is effective technique for irregular volume data, and easily applied to high-resolution environment like TDW because memory cost of SPT is less affected by the render resolution. Data used in this paper is an irregular volume dataset generated from the CSM computation by a finite element method solver. This data is composed of 282,128 tetrahedral cells with 68,170 nodes called "V6 engine".

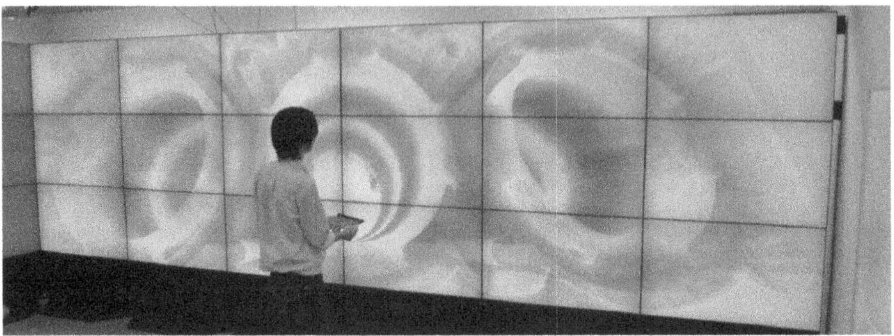

Fig. 4. Volume rendering application running on a TDW composed of 18 LCD displays. This is a photo which changing parameters of volume rendering by using a multi-touch tablet device.

Fig.4 shows a photo running our application. When the object is controlled, coarse rendering is used to render images fast. In this environment, the average time for frame synchronization was about 0.452 msec. This time is a time after the renderer node where drawing takes time most sends a synchronization signal to a master node until the display signal comes back. This result shows that the time concerning frame synchronization is very small.

5 Conclusion

This paper presented a visualization system for TDW to analyze the irregular volume datasets by using SPT, and a user-interface to control the rendered object on TDW by using the multi-touch devices. In order to verify the effectiveness of our system, we applied it to CSM simulation results composed of tetrahedral cells, and we confirmed the effectiveness by rendering these dataset on TDW with low processing cost of the frame synchronization and controlling its results with our multi-touch interface intuitively. In future work, we plan to develop a parallel visualization system on TDW for a time-varying irregular volume dataset.

References

1. Ponto, K., Doerr, K., Wypych, T., Kooker, J., Kuester, F.: CGLXTouch: A Multi-user Multi-touch Appoarch for Ultra-high-resolution Collaborative Workspaces. Future Generation Computer Systems 27(6), 649–656 (2011)

2. Carillo, E., Kadobayashi, R., Chikama, M., Shimojo, S., Rodriguez-Echavarria, K., Arnold, D.: Maiking History Alive and Interactive: Designing an iPhone App to Present the Summer War of Osaka Byobu. In: 16th International Conference on Virtual Systems and Multimedia, pp. 365–368 (2010)
3. Sakamoto, N., Kawamura, T., Kuwano, H., Koyamada, K.: Sorting-free Pre-Integrated Projected Tetrahedra. In: Proceedings of the 2009 Workshop on Ultrascale Visualization 2009, pp. 11–18 (2009)

Field Approximation Using Piecewise Polynomials for Fast Volume Rendering on GPU

Hidetaka Takagi[1,*], Shuuhei Aoyama[1], Ryousuke. Makino[1], Takenori Hatsuda[1], Susumu Nakata[2], and Satoshi Tanaka[2]

[1] Graduate School of Science and Engineering, Ritsumeikan University, Japan
{htakagi,s-aoyama,rmakino,thatsuda}@cg.is.ritsumei.ac.jp
[2] College of Information Science and Engineering, Ritsumeikan University, Japan
{snakata,stanaka}@media.ritsumei.ac.jp

Abstract. Volume data are stored as different data types depending on the application and the discrete data needed to be interpolated for visualization. In this study, we present a parallel volume rendering algorithm that accelerates the direct rendering of scalar fields obtained as interpolations of unorganized scattered data. The rendering process consists of two phases: (1) the preprocessing phase where the given scalar field is converted to a grid of polynomials for fast field evaluation and efficient parallel computation, and (2) the rendering phase where the direct rendering of the polynomials' grid is performed in parallel on the graphics processing unit GPU. Our performance tests show that the method enables interactive volume rendering of smooth scalar fields.

Keywords: Volume Rendering, Local Polynomial, GPU Parallel Processing.

1 Introduction

Volume visualization is a set of techniques used to visualize fields in three-dimensional space. In practical applications, the fields are given as sets of scattered data and need to be interpolated as continuous fields in order to apply visualization techniques. For the interpolation, different interpolation techniques are used depending on the data types such as regular grid data, tetrahedral volume data, or unorganized scattered data.

The volumetric multilevel partition of unity (MPU) method is an efficient scattered data interpolation technique [1, 2]. It generates a smooth field while interpolating a given set of scattered data in three-dimensional space. Furthermore, it was proposed as an extended version of the original MPU method [3], which is a surface reconstruction method for three-dimensional shape modeling. The field is defined using adaptive domain decomposition, which contributes to effective data compression of the original scattered data, and the value is given as a weighted linear combination of locally defined polynomial fields. The evaluation process requires tree search for detecting the local fields that correspond to the evaluation point and for computing the weighted sum. Although a direct volume rendering result of the smooth field can be obtained by applying a rendering scheme to the field of volumetric MPU, the computational cost in time can be high depending on the fineness of domain decomposition.

J.-H. Kim et al. (Eds.): AsiaSim2011, PICT 4, pp. 498–505, 2012.

In this study, we present a novel efficient rendering scheme for smooth fields that performs direct volume rendering while preserving the smoothness of the original fields. Our method accelerates the rendering process by using the combination of the following two steps: conversion to a polynomial' grid and parallel volume rendering on the GPU.

The first step, conversion to a grid of polynomials, is a method for polynomial approximation [4], which contributes in reducing the cost for field evaluation. The conversion is performed using quadratic B-spline interpolation and the approximated fields inherit the smoothness of the original fields because of the smoothness of the quadratic B-spline basis functions.

The second step, parallel volume rendering on GPU, is incorporated into the rendering scheme. Our rendering is based on a ray casting method [5] and its parallel computation is performed by dividing the process into threads corresponding to the pixels. The GPU supporting the programming model called CUDA [6] works as a single-instruction multiple-data (SIMD) parallel processor, and our parallel algorithm is suitable for the device because the cost for evaluating the field, a six degree polynomial, is constant in the domain.

2 Conversion to a Grid of Polynomials

We assume that a scalar field $f(\boldsymbol{x})$ is given in the rectangular domain, $\boldsymbol{x} \in [x_{\min}, x_{\max}] \times [y_{\min}, y_{\max}] \times [z_{\min}, z_{\max}]$. Such a scalar field is obtained as a result of data interpolation using tri linear interpolation, spline interpolation, or the volumetric MPU method. Note that the volumetric MPU method can be applied not only to regular grid data but also to irregularly scattered data and produces smooth fields by interpolating the original data. Moreover, the field is effectively generated in the sense that the rectangular domain is adaptively divided into subdomains depending on the local complexity of the original discrete data, and then, the interpolating field is locally determined for each local area.

2.1 Approximation of Fields with B-Spline Functions

The evaluation process of the field $f(\boldsymbol{x})$ generated by the volumetric MPU requires the following two steps: the tree-search algorithm to determine the locally defined fields containing the evaluation point and the computation of the weighted sum of the local fields. In the volume rendering process, the field is evaluated at a number of sampling points along the rays cast through the domain and the process requires high computational cost in time depending on the complexity of the data because the cost for field evaluation increases with the increase of the local subdomains. In addition, the evaluation process is not appropriate to efficient parallel computation on the GPU because the process contains the tree-search the depth of which varies depending on the position of the evaluation point. In addition, the cost for evaluation of each local field changes with the distribution of the scattered data.

In order to eliminate the computation of the tree-search and the weighted sum for the field evaluation, we convert the field to a set of polynomials of degree six defined at each cell of the uniform rectangular grid of the domain. In our method, the conversion is performed using the quadratic B-spline interpolation technique which leads to piecewise polynomials with 27 terms. Note that the quadratic B-spline can be replaced with a higher-order B-spline in general. In that case, the order of the B-spline determines the trade-off between the smoothness of the approximated field and the cost for field evaluation.

Given a scalar field in a rectangular domain, the B-spline interpolation is applied to grid data, f_{ijk}, which are obtained as a set of scalar values sampled at uniform grid points in the domain. The domain can be assumed to be $[0,L]\times[0,M]\times[0,N]$ for simplicity without loss of generality, where L, M and N are the number of sampling points in x, y and z direction, respectively. Under this assumption, the result of the B-spline interpolation is given in the following form:

$$g(\boldsymbol{x}) = \sum_{i=-1}^{L}\sum_{j=-1}^{M}\sum_{k=-1}^{N} c_{ijk}\varphi_i(x)\varphi_j(y)\varphi_k(z) \tag{1}$$

where $\varphi_i(t)$ are the B-spline basis functions defined as:

$$\varphi_i(t) = \varphi(t-i),$$
$$\varphi(t) = \frac{1}{2}\begin{cases}(t+1)^2 & (-1\le t<0)\\ -2t^2+2t+1 & (0\le t<1)\\ (t-2)^2 & (1\le t<2)\\ 0 & (\text{otherwise})\end{cases} \tag{2}$$

and c_{ijk} are the coefficients determined to satisfy the interpolating condition, $g(\boldsymbol{x}_{ijk}) = f_{ijk}$.

2.2 Extraction of Piecewise Polynomials

Assume the domain $[0,L]\times[0,M]\times[0,N]$ be uniformly divided into $L\times M\times N$ cells and denote the interpolated field in the (i,j,k)-th cell by $g_{ijk}(\boldsymbol{x})$. If $\boldsymbol{x}$ is in the (i,j,k)-th cell, i.e. $\boldsymbol{x}\in[i,i+1[\times[j,j+1[\times[k,k+1[$, the interpolated function is expressed using the adjacent 27 bases as:

$$g_{ijk}(\boldsymbol{x}) = \sum_{I=-1}^{1}\sum_{J=-1}^{1}\sum_{K=-1}^{1} c_{i+I,j+J,k+K}\varphi_{i+I}(x)\varphi_{j+J}(y)\varphi_{k+K}(z) \tag{3}$$

The basis function $\varphi_i(t)$ can be expressed using quadratic polynomials defined by

$$\varphi_i(t) = \begin{cases}\alpha_1^{(2)}(i)t^2+\alpha_1^{(1)}(i)t+\alpha_1^{(0)}(i) & (i-1\le t<i)\\ \alpha_0^{(2)}(i)t^2+\alpha_0^{(1)}(i)t+\alpha_0^{(0)}(i) & (i\le t<i+1)\\ \alpha_{-1}^{(2)}(i)t^2+\alpha_{-1}^{(1)}(i)t+\alpha_{-1}^{(0)}(i) & (i+1\le t<i+2)\\ 0 & (\text{otherwise})\end{cases} \tag{4}$$

where

$$\alpha_1^{(2)}(i)=1/2,\ \alpha_1^{(1)}(i)=-i+1,\ \alpha_1^{(0)}(i)=(i-1)^2/2,$$
$$\alpha_0^{(2)}(i)=-1,\ \alpha_0^{(1)}(i)=2i+1,\ \alpha_0^{(0)}(i)=-i^2-i+1/2,$$
$$\alpha_{-1}^{(2)}(i)=1/2,\ \alpha_{-1}^{(1)}(i)=-i-2,\ \alpha_{-1}^{(0)}(i)=(i+2)^2/2.$$

and, if $t \in [i, i+1[$, the equation (4) leads to the following expression:

$$\begin{aligned}
\varphi_{i-1}(t) &= \alpha_{-1}^{(2)}(i-1)t^2 + \alpha_{-1}^{(1)}(i-1)t + \alpha_{-1}^{(0)}(i-1), \\
\varphi_i(t) &= \alpha_0^{(2)}(i)t^2 + \alpha_0^{(1)}(i)t + \alpha_0^{(0)}(i), \\
\varphi_{i+1}(t) &= \alpha_1^{(2)}(i+1)t^2 + \alpha_1^{(1)}(i+1)t + \alpha_1^{(0)}(i+1)
\end{aligned}$$

or equivalently, we have

$$\varphi_{i+I}(t) = \sum_{p=0}^{2} \alpha_i^{(p)}(i+I)t^p \ (i \le t < i+1, I = -1,0,1). \tag{5}$$

The polynomial expression of the interpolated field is obtained by substituting (5) into (1) as

$$g_{ijk}(\boldsymbol{x}) = \sum_{p=0}^{2}\sum_{q=0}^{2}\sum_{r=0}^{2} \beta_{ijk}^{(p,q,r)} x^p y^q z^r, \tag{6}$$

where $\beta_{ijk}^{(p,q,r)}$ are the coefficients determined by

$$\begin{aligned}
\beta_{ijk}^{(p,q,r)} &= \sum_{I=-1}^{1}\sum_{J=-1}^{1}\sum_{K=-1}^{1} c_{i+I,j+J,k+K} \\
&\cdot \alpha_i^{(p)}(i+I)\alpha_j^{(q)}(j+J)\alpha_k^{(r)}(k+K).
\end{aligned} \tag{7}$$

The local polynomial (6) is evaluated independently of the other cells as shown in Figure 1 (left). Note that the field inherits the continuity and the smoothness of (3) at the interface between neighboring cells. The coefficients can be computed in advance using (7) and 27 coefficients are required for the piecewise polynomial of a cell, i.e. the number of coefficients to be stored is $27LMN$ in total.

2.3 Computational Cost for Field Evaluation

In this section, we compare the cost for field evaluation of the polynomial grid with the volumetric MPU. The field of the volumetric MPU is based on the cells obtained as a recursive subdivision of the domain. The value of the field is defined as a weighted sum of locally determined fields that are defined on overlapping spheres located corresponding to all the cells as shown in Figure 1 (right). The definition of the field is

$$f(\boldsymbol{x}) = \sum_{k=1}^{K} \varphi_i(\boldsymbol{x}) Q_i(\boldsymbol{x}),$$

where K is the number of the cells, $Q_i(\boldsymbol{x})$ are the local fields and $\varphi_i(\boldsymbol{x})$ are the weights. Each of the local fields is constructed as a polynomial the degree of which is selected from 0, 1, 2 or 3 appropriately depending on the local frequency of the field. The weights are determined depending on $\boldsymbol{x}$ so that the weights satisfy the partition of unity property (see [2] for detail).

The evaluation process consists of three steps: find all the local fields containing the evaluation point using the tree structure for fast search (see Figure 2), evaluate the weights and the local polynomials and compute the weighted sum in (6). The cost for field evaluation of the volumetric MPU varies depending on the position because of the three characteristics: the depth of the tree-search is not constant, the number of local fields required for the weighted sum changes depending on the position and the degree of the local polynomials changes depending on the local frequency. The varying computational cost leads to ineffective parallel computation on the GPU.

By replacing the original field with the gird of polynomials, the tree-search algorithm can be eliminated because the index of the cell containing the evaluation point can be determined the grid structure. In addition, the process of weighted sum is not required because the polynomial of each cell is independently defined preserving the smoothness of the field. The computational complexity of (1) is constant (52 floating-point operations for one evaluation point if the Horner scheme is applied) which contributes to efficient parallel computation on GPU. Note that a smoother field can be generated by applying a higher order B-spline interpolation although the computational and memory costs increase with the order.

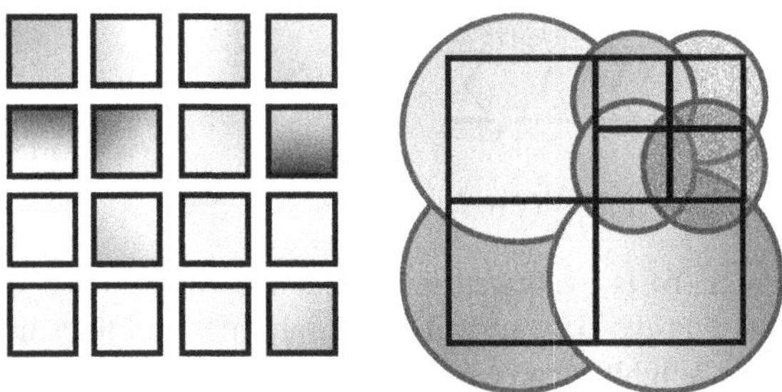

Fig. 1. Domain decomposition and locally determined fields used in the grid of polynomials (left) and the volumetric MPU (right)

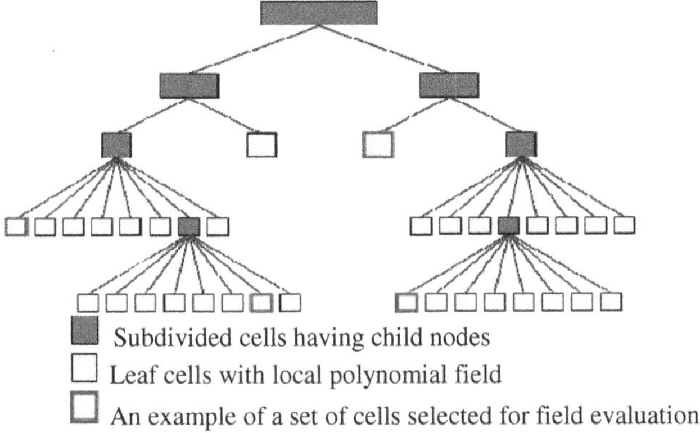

Fig. 2. Data structure of volumetric MPU

3 Volume Ray-Casting on GPU

In this study, we use the volume ray-casting algorithm for the visualization of fields. Assuming a camera and a screen, the algorithm requires field evaluation at sampling points located along all the rays corresponding to the pixels on the screen. For parallel computation, we assign the process on one pixel to one thread which is performed on one processor on the GPU. For the ideal computation on GPU, the computational complexity for all the thread needs to be constant. In our implementation, the cost for one thread is proportional to the number of sampling points along the corresponding ray because the cost for field evaluation is constant. Although the costs are not constant as the number of sampling point for different pixels can differ depending on the geometry of the rays and the domain, fast parallel computation on GPU is still expected because the difference of the number of sampling points is small in the vicinity of a pixel.

The algorithm of the volume ray-casting consists of the following two steps:

Step 1. (Data transfer from CPU to GPU) Send the coefficients of the polynomials to the global memory on GPU.

Step 2. (Rendering on GPU) For each rendering frame, send the camera position to GPU, determine the pixel values in parallel and display the image. Each of the pixel values is determined as follows:

2-1 Determine the intersection between the ray and the domain.

2-2 Compute the volume rendering integral discretely by evaluating the polynomial field at the uniform sampling points along the interval.

The data transfer needs to be transferred in advance of the rendering process and is not required in the rendering process except for the parameters related to the camera position. Note that any feedback from GPU to CPU is not required if the rendering results are directly displayed on the monitor.

4 Results

We use four volume data sets which are called (a) hydrogen, (b) neghip and (c) fuel. The original volume data are given at (a) 128^3, (b) 128^3 and (c)64^3 grid points. We used the volumetric MPU for the construction of fields approximating the given volume sets and converted the fields to grids of polynomials using the quadratic B-spline interpolation, where the size of the grid is 100^3. The process of the volumetric MPU interpolation and the polynomial approximation are performed on CPU and the coefficients of the polynomials, the number of which is 27×100^3 in total, are transferred to the GPU global memory as preprocessing. The size of the screen is set to500×500 throughout this section and all the tests are performed on CPU: Intel Core i7 (2.67GHz) and GPU: NVIDIA GeForce GTX 580 with 512 streaming processers.

Figure 3 (a), (c) and (e) are the results of volume rendering using the volumetric MPU generated from the volume data and (b), (d) and (f) are the results using the grids of polynomial. The results show that the polynomial approximation gives accurate rendering under the rendering parameters described above. The computational costs in time for rendering of volumetric MPU on CPU (1 core is used) and of grid of polynomials on GPU are listed in Table 1. These results show that the rendering algorithm is suitable for parallel computation on GPU and we obtain significant acceleration with a factor of over 1380. In addition, the rendering time of grid of polynomials on GPU is constant even if the original field is different. Table 2 shows the difference between the rendering results using the original fields and the approximated fields. The differences are small enough on average although the field is approximated by the grid of polynomials.

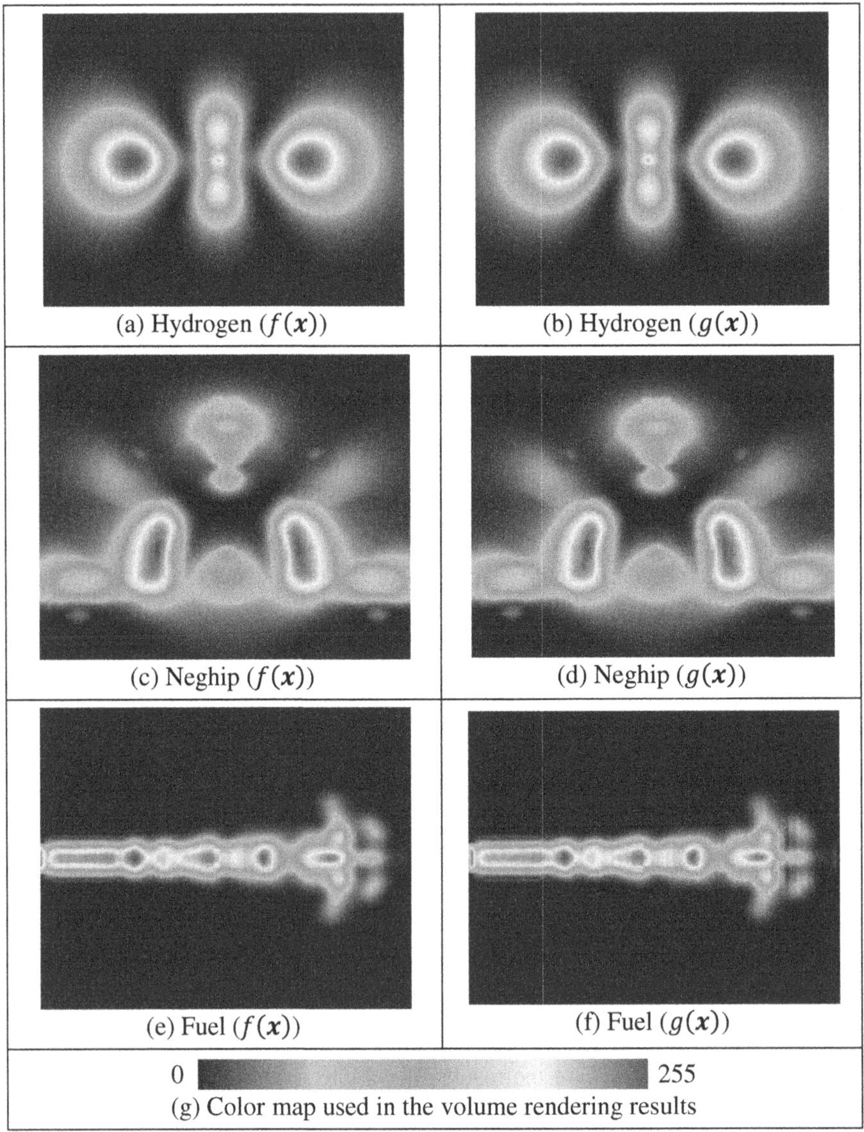

Fig. 3. Images of the volume rendering. Left: rendering results of the volumetric MPU field performed on CPU, Right: rendering results of the piecewise polynomial approximation performed on GPU. All the images are illustrated using the color map in (g) after normalization.

Table 1. Computational time for volume rendering

Field	Volumetric MPU (CPU, 1 core)	Grid of polynomials (GPU, 512 SPs)
Hydrogen	415.67 [sec]	0.30 [sec]
Neghip	582.67 [sec]	0.30 [sec]
Fuel	387.10 [sec]	0.30 [sec]

5 Conclusion

In this study, we have proposed a volume rendering algorithm for scalar fields. The rendering process in the algorithm is accelerated using the two steps: conversion of the fields to grids of polynomials and parallel computation on GPU.

In the first step, given scalar fields can be converted to a set of polynomials using the idea of B-spline interpolation and the conversion is accurate enough if the fineness of the grids is appropriately chosen. The property that the field of a cell can be evaluated independently of the others contributes to fast field evaluation without using weight functions and to efficient parallel computation.

The second step, parallel computation on GPU, is performed by simply dividing the rendering process with respect to the pixels on the screen. The simple parallel algorithm is suitable for the computation on GPU because the cost for evaluating the grid of polynomials is constant and independent of position or complexity of the field.

References

1. Tsukamoto, Y., Kataoka, S., Hasegawa, K., Nakata, S., Tanaka, S.: Data Interpolation Independent of Grid Structure Using a Volumic Version of MPU. In: Asia Simulation Conference 2009, Ritsumeikan University, Shiga, Japan (2009)
2. Tsukamoto, Y., Kawashima, S., Inoue, S., Ito, S., Kataoka, S., Kojima, K., Hasegawa, K., Nakata, S., Tanaka, S.: Data Fitting Independent of Grid Structure Using a Volumic Version of MPU. Journal of Visualization 4(3), 161–170 (2011)
3. Ohtake, Y., Belyaev, A., Alexa, M., Turk, G., Seidel, H.-P.: Multi-Level Partition of Unity Implicits. ACM Transaction on Graphics 22(3), 463–470 (2003)
4. Nakata, S., Aoyama, S., Makino, R., Hasegawa, K., Tanaka, S.: Real-Time Isosurface Rendering of Smooth Fields. In: Proceedings of 11th Asian Symposium on Visualization (2011)
5. Hansen, D.C., Johnson, R.C.: Visualization Handbook. Academic Press, New York (2004)
6. Kirk, D.B., Hwu, W.W.: Programming Massively Parallel Processors: A Hands on Approach. Morgan Kaufmann (2010)

A New Framework for Visualizing a Time-Varying Unstructured Grid Dataset with PBVR

Kun Zhao[1], Jun Nishimura[1],
Naohisa Sakamoto[2], and Koji Koyamada[2]

[1] Graduate School of Engineering, Kyoto University, Japan
{zhao.kun,nishimura}@viz.media.kyoto-u.ac.jp
http://www.viz.media.kyoto-u.ac.jp
[2] Institute for the Promotion of Excellence in Higher Education, Kyoto University, Japan
naohisas@viz.media.kyoto-u.ac.jp,
koyamada.koji.3w@kyoto-u.ac.jp

Abstract. In this paper, we propose a new framework for a large-scale time-varying unstructured grid dataset visualization based on the Particle-Based Volume Rendering (PBVR). A previous rendering system can handle a time-varying grid dataset which is composed of only hundreds of thousands of unstructured grid cells. To overcome the limitation, we employed the PBVR in which the computational complexity of rendering is dominated not by the number of cells but by that of particles. Regardless of the number of cells, the PBVR can efficiently render a large-scale grid dataset by reducing particles in the location or the time frame which is not so important. In this system, we develop a hierarchical framework for managing the particles for efficient rendering. Moreover, by compressing the particle data, we also succeed in reducing the loading time of the whole time step particle data from the hard disk. We applied the system to rendering of 465 time steps of 3,676,828 hexahedral cell grid dataset to confirm the effectiveness.

1 Introduction

Time-dependent simulations and time-varying data can be found in almost every major scientific discipline. The time-varying data usually involves a certain phenomenon that grows, persists, and declines in several distinct stages. The rate of change at each stage could vary dramatically in space and time. As a result, the data from these simulations are always highly complex, large-scale.

For these time-varying data, effective visualization is extremely difficult to achieve due to the complexity of illustrating multiple time-steps and clearly showing changes and variations over time. Since the data is always organized into unstructured grid dataset which is hard to be directly rendered, a good rendering method is also needed to render the unstructured dataset effectively.

In the paper, we present a new framework for effective time-varying unstructured grid visualization. In this system, we use the Particle-Based Volume Rendering (PBVR)[1] as the rendering method. With this method, we first generate particles from the original unstructured volume data. And then, the particles are projected onto

J.-H. Kim et al. (Eds.): AsiaSim2011, PICT 4, pp. 506–516, 2012.

the image plane, so that the visibility sorting is not necessary. Also, along with the changes of view point, there is no need to do the resort operation. Furthermore, when rendering a large-scale grid dataset, PBVR allows user to control the level of detail (LOD) dynamically by changing the number of particles projected to the image plane. With these features, our system can render time-varying unstructured grid dataset very effectively. Moreover, for the large-scale time-varying volume data, as the generated particle data size is always large, the loading of these data from the hard disk can always cost much time. As a result, the compression of the particle data is necessary to reduce the loading time. The stored compressed data can be read and expanded in the CPU memory, and then rendered by graphics processing unit (GPU). Since the loading time is reduced for the smaller data size, the rendering animation of the time-varying particle data can be launched faster than loading the original particle data.

In this paper, we apply our system to large-scale simulation results of oral airflow data composed of 465 time steps in order to confirm its effectiveness.

2 Related Research

Bernardon et al. use the technique of HAVS (Hardware Assisted Visibility Sorting) to render the time-varying unstructured grid dataset[2]. Since they have to store the volume data to the memory of GPU, it works well when there are just hundreds of thousands grids. However, when the dataset have millions of grids, this framework just doesn't work since there is not enough GPU memory. Also, HAVS only supports data grid consisting of tetrahedral cell. For the data grid consisting of hexahedral cell such as the data grid used in this paper, the HAVS system does not work.

There are also many other methods of visualizing the time-varying volume dataset[3,4]. However, for the large-scale unstructured dataset like the data used by this paper, effective approaches have not been proposed so far.

3 Particle-Based Volume Rendering

In our rendering system, we use PBVR to render the time-varying unstructured dataset. PBVR is based on the density emitter model of Sabella's theory[5], which uses grainier particles than the splatting algorithms and assumes that the particles are fully opaque. With this proposed particle model, our system requires neither alpha blending nor visibility ordering when rendering.

PBVR can be generally divided into three steps: particle generation, particle projection and ensemble averaging (Fig. 1).

Firstly, the particle generation is to generate particles from the volume data by using the given transfer function (Fig. 1.A). The particle is generated in cell-by-cell manner. As the pre-process, this process is done by CPU. And then the generated particles are projected to the image plane in the particle projection process (Fig. 1.B). This projection process is performed by GPU. At the same time, the calculating of the particle size and shadow processing is also performed by using normal vector. The third step ensemble averaging process means that, for the group of particles generated by using different random numbers, particle projection process is performed multiple times, and the generated image superimposes each other averagely (Fig. 1.C). The multiple times is called repeat level. As a result, a high-quality rendering results can be obtained by setting a high repeat number. However, with the increasing of the particle number, the particle data size is also increasing which would exert pressure on the CPU memory and may lead to a deceleration of the rendering speed. To solve this problem, PBVR provide a LOD control which allows user to control the level of detail dynamically by changing the number of particles projected to the image plane. By reducing the repeat level, we can obtain a high-speed rendering result.

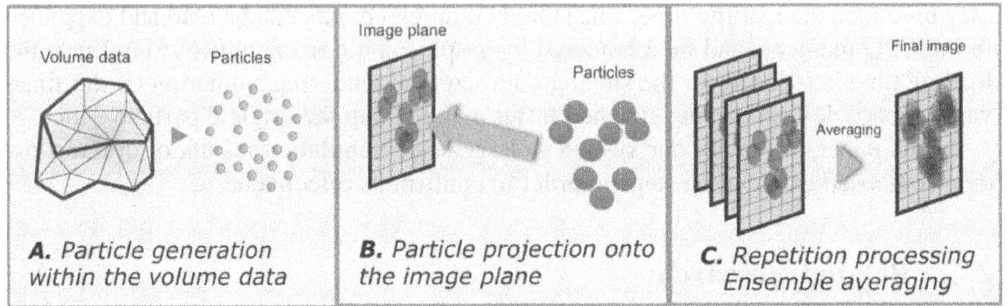

Fig. 1. The Three Step of PBVR: Particle Generation, Particle Projection, Ensemble Averaging

4 Proposed System

In this paper, we use the method of PBVR to perform the time-varying unstructured grid data visualization. By using the LOD control of PBVR, we can obtain a high-speed performance by setting a low repeat level when we rendering the time-varying particle as animation. And if we stop the animation, we can obtain a high-quality rendering by setting a high repeat level. Furthermore, the compression of the particle data is conducted to reduce the loading time from the hard disk. As a result, the rendering animation can be launched faster than loading the original particle data.

4.1 Particle Generation

As the pre-process, we first need to generate particle data from original volume data. Using the given transfer function, we generate particles from the time-varying volume data in cell-by-cell manner (the left figure of Fig. 2). In every time step, particles are repeatedly generated at a high repeat level (repetition number) and saved as one file (the right figure of Fig. 2).

Since the generated particle dataset of the whole time steps are too large to be stored in the CPU memory, we store these data to the hard disk. During the rendering step, only the particles amounting to repeat level, which can be assigned by the user, are loaded and projected to the screen. Therefore, the user can control the number of particles projected to the image plane by changing the repeat level. The lower repeat level can provide a higher speed rendering with a lower quality. The repeat level provides a LOD control to our system, so that this system can be applied in any environment. This operation is processed by CPU.

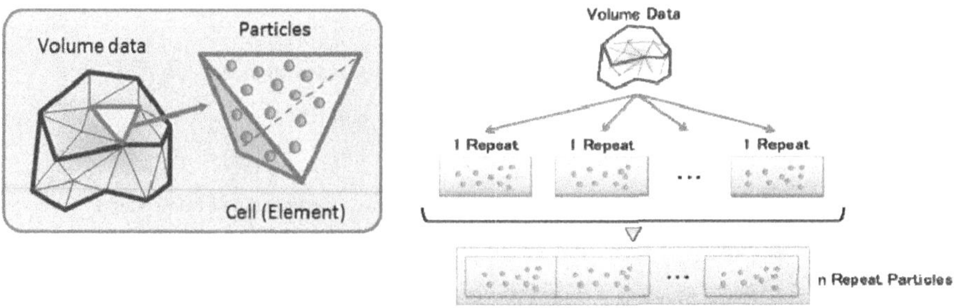

Fig. 2. The left figure shows the generation of Particles in cell-by-cell manner. The right figure shows the process of repeatedly generation of particles. For one time step, the repeatedly generated particles are saved in one file. And during the rendering step, only the particles amounting to repeat level are projected to the screen.

4.2 Particle Data Visualization

Before we launch the rendering animation, we first need to load the particle data amounting to a low repeat level set by user for all time steps. These data are loaded to CPU memory so that there is no need to read data from the hard disk during the animation. In the rendering process, GPU projects these particles to the image plane depending on the number of repeat levels. For one time step, one repeat level's data can be projected as one image, and since there are many repeat level's data of one time step, the generated images averagely superimpose each other (Fig.1.C). This operation is ensemble averaging. And the higher repeat level can provide a more high-quality image.

To get the visualization result, the process flow of our system shows in the Fig. 3. At first, the amount of particle data at a low repeat level has been loaded to the CPU memory before the animation is launched. And the particle data of time step 0 also has been rendered to the image plane, so that the user can get an initial image of the time-varying data. As we begin the animation, our system transfers the particle data at a low repeat level to GPU memory and these data are rendered step by step in regular intervals. At certain time step ***t***, the particle data of previous step ***t-1*** that has been rendered by GPU would be deleted from GPU memory. And the particle data of current time step ***t*** is transferred to the GPU memory and rendered to the image plane. After the rendering of the time step ***t*** ended, the same process would be performed to time step ***t+1***. By repeating this until the last time step ***n***, we can obtain the animation of visualization result.

And when we stop the animation at any time step, the system load the particle data generated at a high repeat level of this time step from the hard disk and renders it to the image plane. As a result, we can obtain a high-quality rendering image when the animation is stopped. In this case, the pre-loaded low repeat level particle data are still stored in the CPU memory; therefore the animation can be continued without reading data from hard disk (Fig. 4).

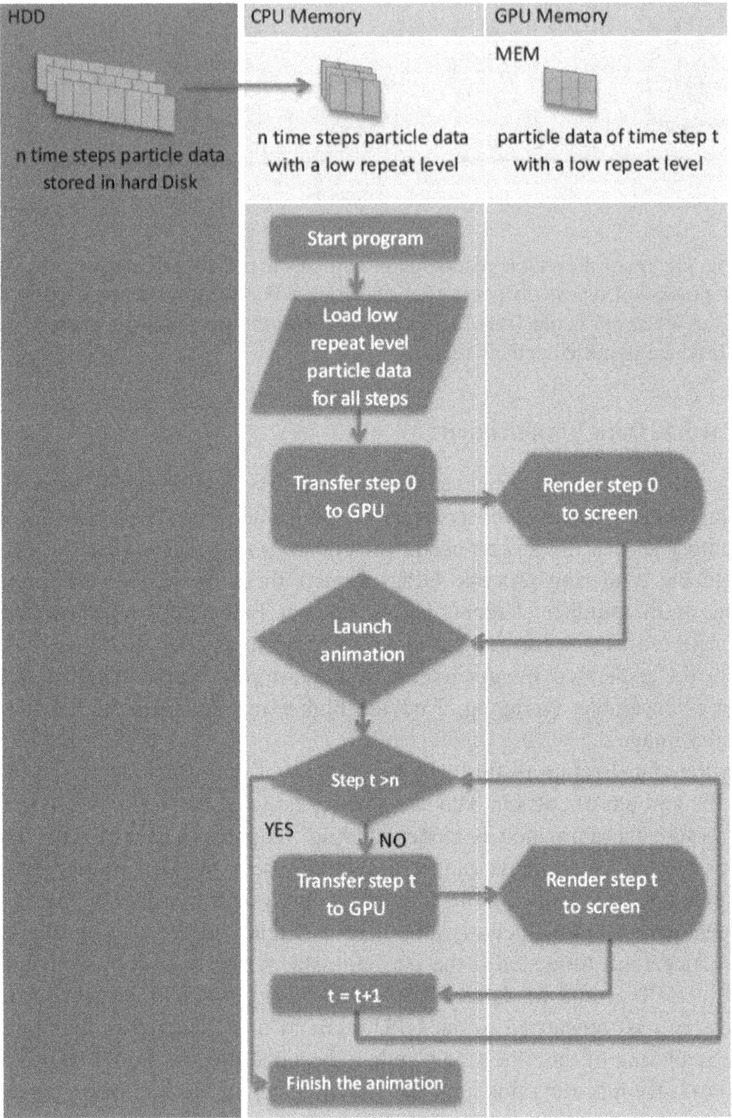

Fig. 3. Process Flow Diagram of the Animation

As the user interface, user can stop or resume the animation by pressing keys on the keyboard. Moreover, our system also provides a time slider that shows which time step that is currently displayed. As a result, by moving the slider using the mouse, user is able to quickly view the particle data of the time step interested in.

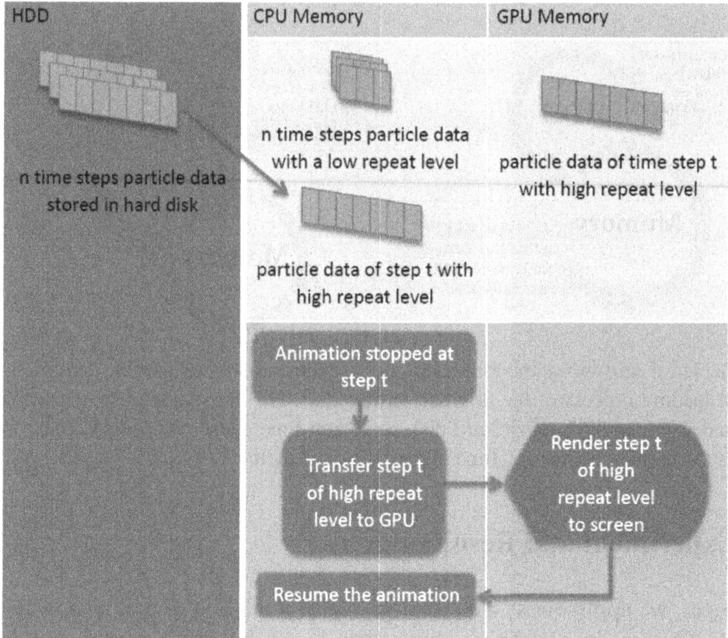

Fig. 4. Process flow diagram when the animation is stopped

4.3 Compression of Particle Data

As mentioned above, before the animation begins, we need to pre-load the particle data for the whole time step. This can always cost much time to access the hard disk. And if we can compress the particle data into a smaller size, the accessing time would be reduced. With the compressed particle data stored in the hard disk, the reading would be faster than the original data. Then the memory-loaded compressed data need to be expanded in to another buffer in the memory. And when the expanding is completed, the loaded compressed data would be deleted to save the memory space. Until here, the loading of the compressed data is completed (the left figure of Fig 5). For the time-varying particle data, before the animation begins, the loading of the compressed particle data would be done on the whole time step. And after the whole time step particle data are loaded, the animation rendering can be launched (when animation is launched, the process flow is the same to the figure shown in Fig. 3 and Fig. 4). Also, the loading process of the original particle data is shown in the right figure of Fig. 5 to make a comparison.

As we can see in the figures, the expanding would be very important to determine whether the loading of the compressed data would be efficient. In this paper, we choose the Zlib library as the compression method, since it can provide a lossless data-compression, and can conduct compression and expanding in memory.

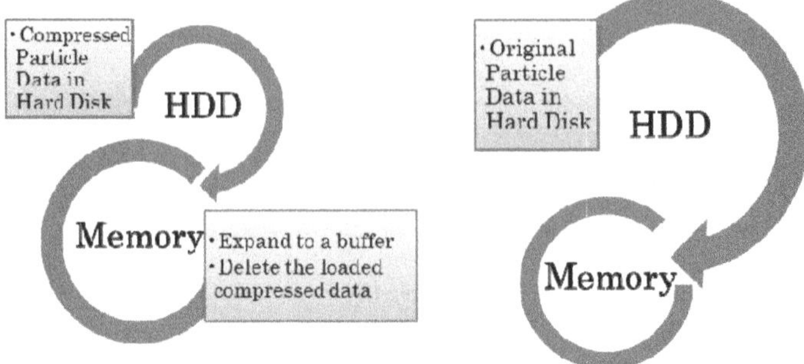

Fig. 5. The left figure shows the loading process of the compressed particle and the right figure shows the loading process of the original particle data. For the smaller size, the loading of the compressed data can save much hard disk accessing time. Here, the length of the blue circle represents the relative processing time in hard disk and memory.

5 Experiment and Result

In this paper, we apply our system to large-scale simulation of oral airflow data (composed of 465 time steps). This volume data has 3,769,092 vertexes and 3,676,828 hexahedral grids. With the time step varying, the grid of the volume data not changes, and only the value of the vertex is changing. The data size is 494.4MB on the vertex coordinates and connectivity information, and 92.2GB on the value information of each vertex that represents the scalar value of velocity vector and pressure and so on.

In this experiment, the value of each vertex represents the scalar value of velocity vector of the airflow. The transfer function used in the experiment is shown in Fig. 6. We conduct our experiment with the environment of two Quad-Core AMD Opteron Processer CPU, NVIDIA QuadroFX 4700 1GB GPU, and 32GB CPU memory. The Operation System used is SUSE Linux Enterprise Server 10 SP2.

Fig. 6. Transfer Function Used in the Particle Generation

As the pre-process, the whole time step particle data are generated from the volume data. And the reading time, generation time, generated particle number are shown in Table 1. By rendering these particle data, we can obtain the animation of the time-varying unstructured grid dataset (Fig. 7). And when we stop the animation, we can obtain a high-quality rendering (Fig. 8). The measuring result of the animation is shown in Table 2. In this experiment, the data with the repeat level of 9 is used in the animation, and the data with the repeat level of 144 is used to be drawn in the image plane when the animation is stopped.

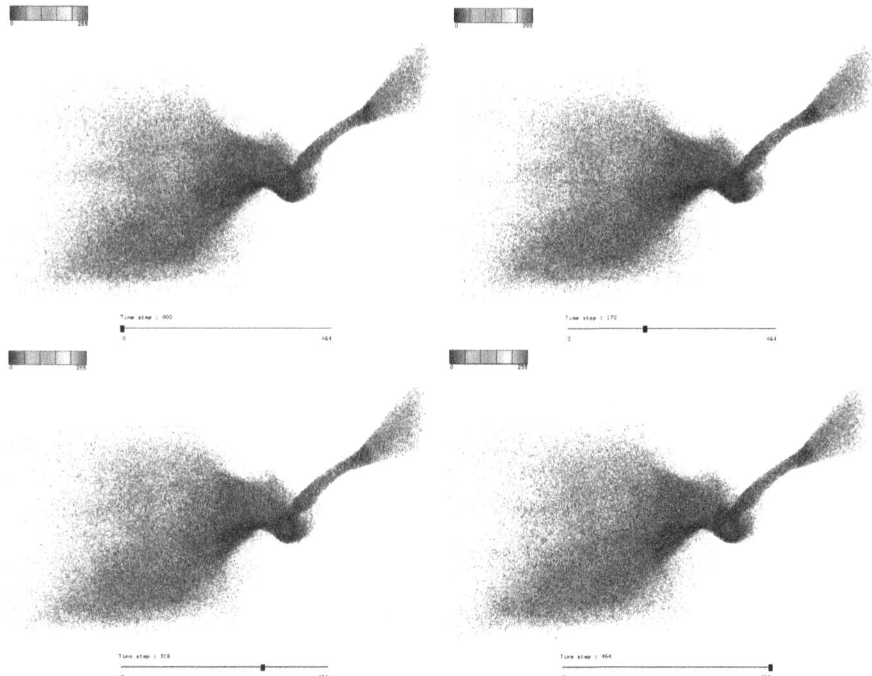

Fig. 7. This figure shows the rendering image during the animation. The figure on the upper left represents the image of time step 0, the upper right shows the time step 172, the lower left shows the time step 318 and the last one shows the image of time step 464.

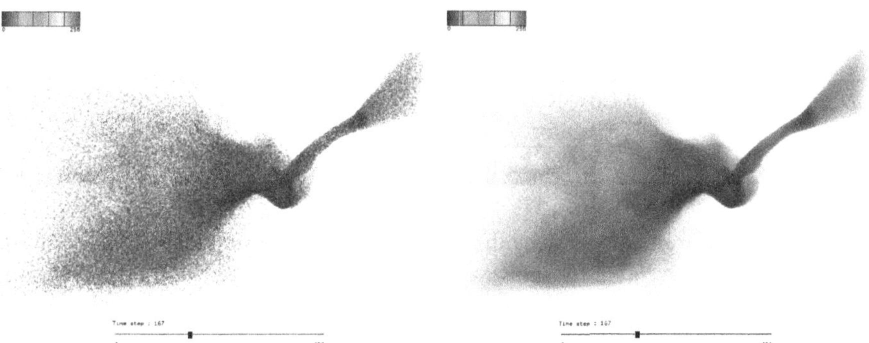

Fig. 8. The contrast rendering image of showing time and stopping time at time step 167(the right one is the image of stopping time)

As we can see from Table 2, the loading of low repeat level particle data for animation costs as much as 45.80 seconds. To reduce this time, the compression is implemented to the low repeat level particle data. For the oral airflow particle data, every time step data contains color, coordinate and normal vector information of the particles. In fact, we find the coordinate data is hardly to be compressed and the expanding always cost too much time. Since expanding speed can determine the efficiency of loading, we only conduct compression on color and normal data for every time step and the original coordinate data is saved with the compressed color and normal as one file. This can provide us a good compression ratio and a fast loading speed (Table 3).

Table 1. This table shows the measuring result of particle generation

Volume Data Reading Time	189.2 sec
Particle Generation Time	2,188.4 sec
Average Particle Number of One Time Step (Low Repeat Level=9)	220,534
Average Particle Number of One Time Step (High Repeat Level=144)	3,528,544

Table 2. This table shows the measuring result of animation. The loading of low repeat level data for all time steps is done before the animation launched.

Loading of Low Repeat Level Particle Data for All Time Step	45.80 sec
Average Rendering Time of Low Repeat Level Particle Data for One Time Step (Frames Per Second)	96.83msec (10.33FPS)
Loading and Rendering Time of High Repeat Level Particle Data for One Time Step	1.54 sec

Table 3. This table shows the totally compression ratio and the loading time of the compressed particle data before the animation begins. The ratio means the compressed data size over the original data size. The loading time includes the reading time of the compressed data from hard disk, the expanding time in memory and also the deleting time of the loaded compressed data.

Compression Ratio of Low Repeat Level Particle Data for All Time Step with Zlib	50.36%
Loading of Low Repeat Level Particle Data for All Time Step	27.89sec

From this table we can see, since the loading time for all time step is reduced from 45.80second to 27.89second for the smaller data size, we can obtain a rendering animation faster than loading the original particle data.

6 Discussion

When the particle data of whole time steps with low repeat level is rendered as an animation (Fig. 7), the average rendering time of one time step is 96.83millisecond (10.33FPS). From this result, we can know the animation can show the whole time steps smoothly. From these results, we can see our system is efficient on the rendering of the volume data with a large number of cells and time steps. However, the loading of the whole time step low repeat level particle data cost as much as 45.80second

(Table 2) before the animation can be launched. To solve this problem, we conduct compression on the particle data, and since the loading time for all time steps is reduced for the smaller data size, we succeed in reducing the loading time from 45.80second to 27.89second (Table 3).

However, as the large number of the cells and time steps, the particle generation costs too much time. As a result, when the transfer function is changed, particle data would need to be re-generated by costing about 40 minutes. In fact, we conduct the generation only using one process, and the parallel process would provide a better performance. Moreover, since the whole time step particle data with the low repeat level need to be stored in memory before animation begins, if there are too many steps, there would be not enough memory space. In this case, the whole time step particle data should be loaded partly. As the animation flows, the rendered particle data should be deleted from the memory, and the other particle data should be loaded from the hard disk simultaneously. In addition, in this paper we only conduct compression on low repeat level data, and when animation stopped, the compressed high repeat level data would also provide a fast loading speed. These works would be done in the future.

In this paper, we apply our system to large-scale simulation results of oral airflow data, the grid of which not changes. Since our system generates particles from the cells, for the data with the changing grid in each time step, our system is also applicable.

7 Conclusion

This paper presents a new framework for visualizing a time-varying unstructured grid dataset with PBVR. As the pre-process, the whole time step particle data are generated from the volume data. With these data, we can obtain a high-speed rendering animation result and a high-quality rendering image when the animation is stopped. As a result, the oral airflow data consisting of 3,676,828 hexahedral cells composed of 465 time steps could be visualized at more than 10FPS during the animation. Moreover, by compressing the particle data, we also succeed in reducing the loading of the particle data from 45.80second to 27.89second before the animation is launched. From these experimental results, we can verify that our system can provide a fast and relatively accurate rendering result in the time-varying unstructured grid visualization.

References

1. Kawamura, T., Sakamoto, N., Koyamada, K.: A Level-of-Detail Rendering of a Large-Scale Irregular Volume Dataset Using Particles. Journal of Computer Science and Technology 25(5), 905–915 (2010)
2. Bernardon, F.F., Callahan, S.P., Comba, J.L.D., Silva, C.T.: An Adaptive Framework for Visualizing Unstructured Grids with Time-varying Scalar Fields. Parallel Computing 33(6), 391–405 (2007)

3. Lum, E.B., Ma, K., Clune, J.: A Hardware-Assisted Scalable Solution for Interactive Volume Rendering of Time-Varying Data. IEEE Transactions on Visualization and Computer Graphics, 286–301 (2002)
4. Maximo, A., Marroquim, R., Farias, R.: Hardware-Assisted Projected Tetrahedra. Commputer Graphics Forum 29(3), 903–912 (2010)
5. Sabella, P.: A Rendering Algorithm for Visualizing 3D Scalar Fields. Computer Graphics 22(4), 51–58 (1988)

Volume-Surface Fused Visualization Applied to Medical Data Based on Particle-Based Rendering

Kyoko Hasegawa[1], Susumu Nakata[2], and Satoshi Tanaka[2]

[1] Kinugasa Research Organization, Ritsumeikan University,
1-1-1 Noji-higashi, Kusatsu, Shiga, Japan
hasegawa@cg.is.ritsumei.ac.jp

[2] College of Information Science and Engineering, Ritsumeikan University,
1-1-1 Noji-higashi,Kusatsu, Shiga, Japan
{snakata,stanaka}@media.ritsumei.ac.jp

Abstract. This paper proposes a method to create 3D fusion images, such as volume–volume, volume–surface, and surface–surface fusion. Our method is based on the particle-based rendering, which uses tiny particles as rendering primitives. The method can create natural and comprehensible 3D fusion images simply by merging particles prepared for each element to be fused. Moreover, the method does not require particle sorting along the line of sight to realize right depth feel.

1 Introduction

In medical, scientific, and other fields, transparent surface visualization is used to investigate inner 3D structures. Usually, this type of visualization uses polygon graphics, where the polygons must be sorted along the line of sight. The sorting, however, takes a long computation time for large-scale data. Besides, the order of polygons in the sorting often becomes indefinite, especially for intersecting surfaces.

Recently, Koyamada et al. proposed the particle-based volume rendering (PBVR) method [1, 2], which uses tiny particles as rendering primitives. This method does not require any sorting and applicable to large-scale data. It also enables natural volume fusion [3]. We extend PBVR such that it becomes applicable to surfaces as well as volumes. A great advantage of this extension is that 3D fused visualization of different volume/surface objects becomes possible simply by merging particles prepared for each element to be fused. In this paper, we demonstrate 3D fusion of volume–volume, volume–surface, and surface–surface fusions.

2 Particle-Based Rendering

PBVR renders 3D scalar fields as particle clouds and incorporates both the emission and absorption effects. The particle density depends on the transfer function and is used to evaluate the number of particles to be generated in the volume data. Because the particles can be considered opaque, no visibility sorting processing is required. PBVR has three processes: particle generation, particle projection onto the image, and

J.-H. Kim et al. (Eds.): AsiaSim2011, PICT 4, pp. 517–524, 2012.

ensemble averaging of particle luminosities. The first process stochastically generates particles according to the form of transfer function. We repeat to create such particle distribution until L_R statistically independent particle sets are prepared (see Fig. 1(a)). Below we call L_R the 'repeat level'. Such particle projection with the occlusion effect is executed for each particle set prepared in the first process. As the result, L_R similar images are created (see Fig. 1(b)). The second process projects particles onto an image plane. The third process calculates the ensemble average of the L_R images created in the second process (see Fig. 1(c)), which realizes transparent images.

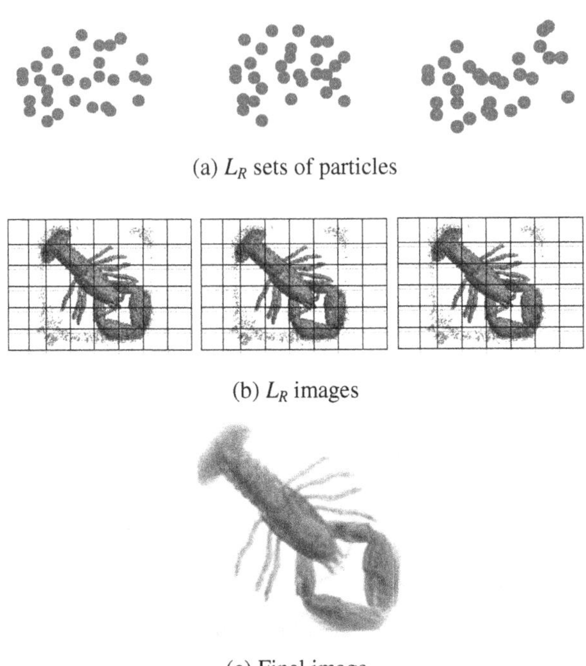

(a) L_R sets of particles

(b) L_R images

(c) Final image

Fig. 1. Schematic view of the PBVR

Using the above particle-based policy of rendering, we propose particle-based surface rendering (PBSR). For PBSR, we stochastically and uniformly generate illuminant opaque particles on a visualized surface [4]. Then, we project the particles onto the image plane and assign pixel brightness based on the ensemble averages of particle luminosities. In PBSR, the opacity depends on the generated particle number. The opacity of the surface can be decided by the surface-coverage ratio. Especially in a complex surface, we consider a small hypothetical sphere whose center is on the surface. Below we call this sphere the 'counting sphere'. During the sampling we count the number of particles generated inside the counting sphere. The number of particles is defined by the surface-coverage ratio in the counting sphere, and it defined as follows:

$$n = \frac{\pi r^2}{s} L_R \alpha , \tag{1}$$

where r and s are the radius of the counting sphere and the sectional area of a particle, respectively. And α is opacity of the surface. For statistical accuracy we should prepare plural counting spheres, and use the average number of particles as n.

To execute the 3D fused visualization, the particles generated by PBVR and PBSR techniques are combined, and the particle projection and ensemble averages are executed. Because the particle projection and ensemble averages have been quickly rendered by GPU computing, the 3D fusion images are quickly created, once the particles are generated.

2.1 Creating Particles on the Surface

PBSR may have applicability to defined drawing surface (for example, a cross-section surface, line). In the case of a polygonized surface, a cross-section surface, the total number of particles can be calculated by the replacement of the counting sphere with the total area size. In this sub-section, we describe the method of creating particles in a polygonized surface, a cross-section surface, and a contour line.

First, we explain how to execute the uniform sampling of a polygonized surface, i.e., a polygon mesh (see Fig. 2 (a)). For a polygonized surface, it is easy to calculate the total areas of the constituent polygons, which we use as area size S in Eq. (1). Correspondingly, we regard the total number of particles generated on all the constituent polygons as n_{all}.

The sampling is made such that the generated particles form the 2D square grid with inter-particle distance d in each polygon (see Fig.3). Then the particle density becomes $1/d^2$, i.e., one particle per one square. On the other hand, the particle density should be n_{all}/S. Therefore an equation $1/d^2 = n_{all}/S$ holds, and d is determined as:

$$d = \sqrt{\frac{S}{n_{all}}}. \tag{2}$$

The sampling consists of three steps:

Step 1. Sampling polygon edges,
Step 2. Sampling inside of polygons,
Step 3. Particle shuffling.

By executing step 1 first, we can avoid duplicative sampling of edges shared by neighboring polygons. In step 1, we simply put particles with distance d on each edge of the whole polygon mesh (see Fig.2 (b)). In step 2, for each polygon, we put particles with distance d on scan lines, which are placed parallelly to an edge at intervals of d. In step 3, we shuffle the generated particles that are stored in an array. Step 3 ensures randomness that is required to apply the probabilistic theory of opacity developed in the PBSR. Note that computation time for the particle shuffling is much shorter than the particle sorting. The former is proportional to n_{all}, while the latter is proportional to $n_{all} \log n_{all}$. After completion of the above three steps, we obtain a particle set that is uniformly distributed and has well a statistical property.

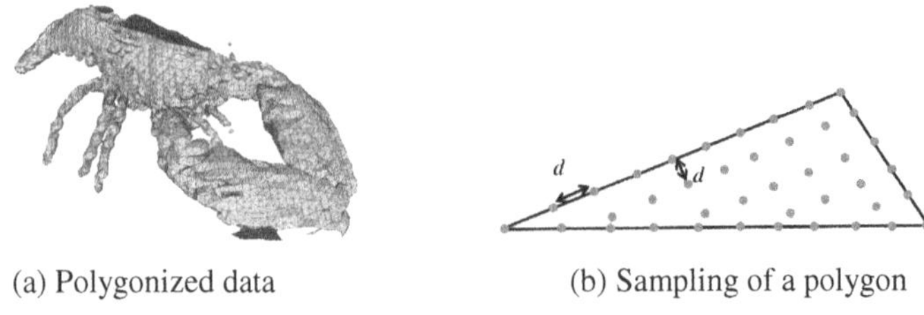

(a) Polygonized data (b) Sampling of a polygon

Fig. 2. Creating particles on the polygon

Next, we describe how to execute the random sampling of the cross-section surface. We defined as any plane in the volume data, and the particles are stochastically generated on the plane (see Fig. 3). The total number of particles can be evaluated by the area size of the plane that can easily calculate from number of voxels of the volume data. Here, the field value of the particles is evaluated by the volume date using tri-linear interpolation. In addition, the color of particles are decided by the transfer function same as volume rendering.

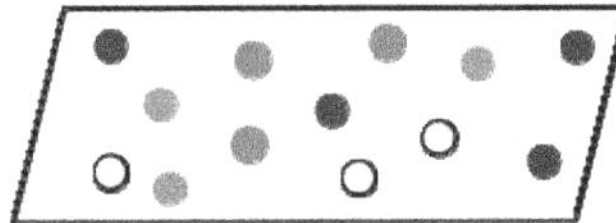

Fig. 3. Creating particles on the cross-section plane

Finally, we explain how to execute the uniform sampling of the contour line. We define as the contour line using marching squares method on the cross-section surface (see Fig. 4(a)). The total area size S is calculated by: S = (total length) × s where the total length is obtained from the defined line in each grid (see Fig. 4(b)). Thus, n_{all} and d is evaluated form Eq. (1) and Eq. (2), respectively. In the line, we simply put particles with distance d on each line, after that we shuffle the generated particles that are stored in an array same as polygon.

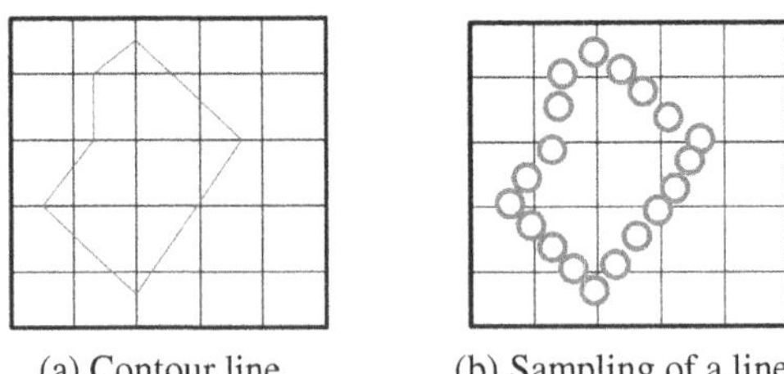

(a) Contour line (b) Sampling of a line

Fig. 4. Creating particles on the contour line

3 Experiments

The experiments in this paper were conducted on a computer that uses a Linux operating system, an Intel Core i7 processor with 2.80 GHz clock speed, 8 GB of memory, and a NVIDIA GeForce GTX470 graphics card. The created images are fixes image resolution 1024^2.

3.1 Creating Particles on the Surface

In this sub-section, we show the fused visualization from two volume data sets, the human-breast data and its dose distribution (see Fig. 5). A PBVR result similar to ray-casting rendering can be obtained, as shown in Fig. 5. Fig. 6 is an example of visualizing polygonized isosurfaces with different isovalues (human-breast's isovalues are -141.255, 300.118, 741.49, and dose distribution's isovalues are 3921.18, 588.76, 7842.35). All isosurface is defined as $\alpha = 0.1$, $L_R = 500$, the number of particles of the human-breast and the dose distribution are 91 million and 4 million, respectively. These particles combines and fused visualizes shown in Fig. 6(c). It can be seen from this figure that the dose distribution exists in the upper part of the human-breast and that the crossed isosurface is exactly visualized. In Fig. 7, we show the cross-section surface with $z = 25$, $L_R = 500$. We omit unnecessary parts from the cross-section, which is field value less than -1024 in the human-breast data and is field value equal to 0 in the dose distribution data. The number of particles of human-breast and dose distribution are 44 million and 15 million with $\alpha = 0.5$ and $\alpha = 0.3$, respectively, and these image fused visualize shown in Fig. 7(c). Fig. 8 shows the contour line on the cross-section surface same as Fig. 7 with $\alpha = 1.0$, $L_R = 500$. The contour vales are same as Fig.6's isovalues. The numbers of particles of human-breast and dose distribution are 27 million and 3 million, respectively. The line colors are decided from the transfer function same as Fig. 5. These particles combines and fused visualizes shown in Fig. 8(c). We can confirm that natural fusion is realized by simply merging the particle sets from the different volume data.

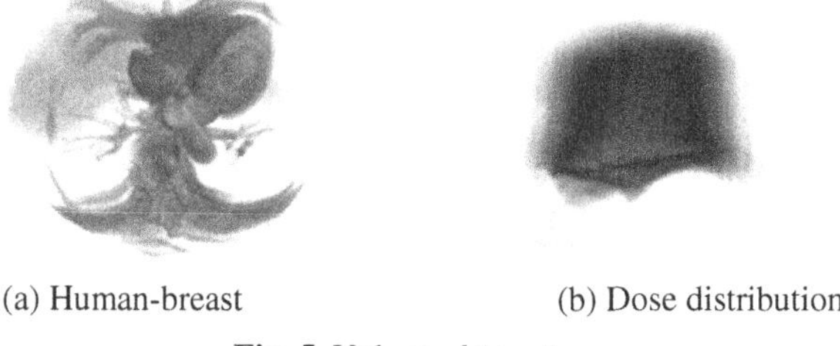

(a) Human-breast (b) Dose distribution

Fig. 5. Volume data sets

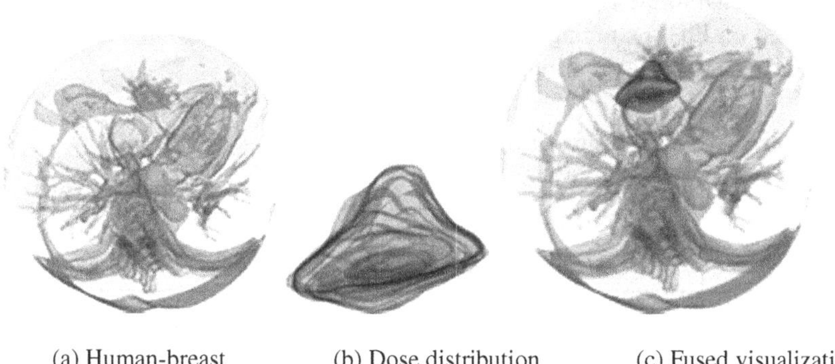

(a) Human-breast (b) Dose distribution (c) Fused visualization

Fig. 6. Fused visualization of surface rendering

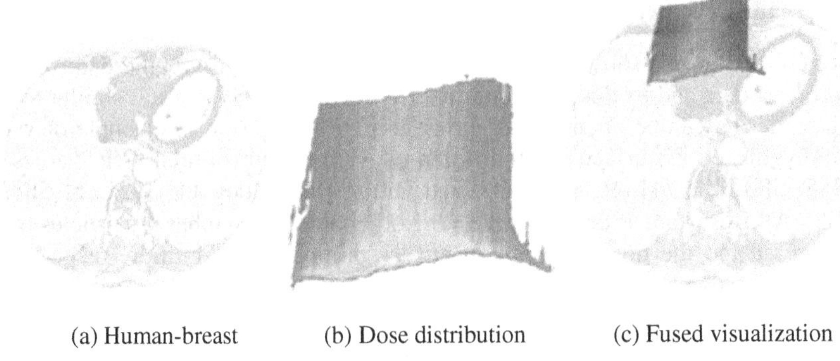

(a) Human-breast (b) Dose distribution (c) Fused visualization

Fig. 7. Fused visualization of slice surface rendering

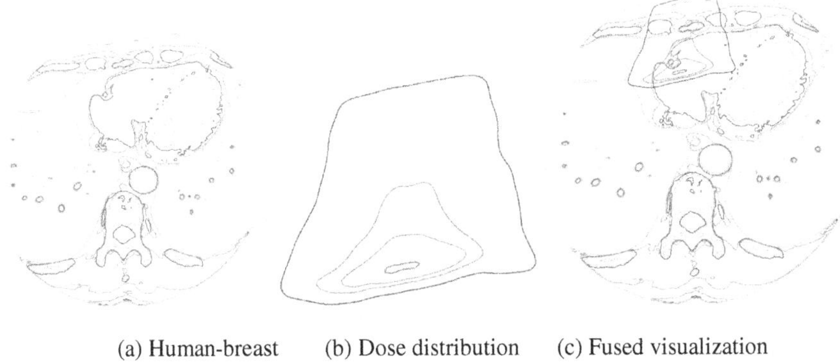

(a) Human-breast (b) Dose distribution (c) Fused visualization

Fig. 8. Fused visualization of contour lime rendering

Table 1 and 2 compares the time of creating particles between different rendering types. It can be seen from the tables that the sampling speed is in the same range of all rendering type and independent on the volume data. Note that the sampling time is clearly proportional to the number of generated particles, and that the number is controllable based on Eq. (1), tuning the image-quality parameters L_R and s. The particle cross-section s is controllable through the image resolution.

Table 1. Time of creating particles in Human-breast

Rendering Type	Number of particles (A)	Creating Time (B)[s]	(B)/(A)
Volume	5.61×10^7	11.61	2.06×10^{-7}
Surface	9.10×10^7	12.12	1.33×10^{-7}
Slice	4.43×10^7	6.51	1.47×10^{-7}
Line	2.79×10^7	2.60	9.63×10^{-8}

Table 2. Time of creating particles in dose distribution

Rendering Type	Number of particles (A)	Creating Time (B)[s]	(B)/(A)
Volume	1.00×10^{7}	2.09	2.09×10^{-7}
Surface	4.30×10^{6}	0.58	1.34×10^{-7}
Slice	1.51×10^{7}	2.04	1.35×10^{-7}
Line	2.74×10^{6}	0.21	7.66×10^{-8}

3.2 Fused Visualization of Different Rendering Type

In this sub-section, we show the fused visualization from different rendering type. Fig. 9 is an example of the fused visualization consisted of the volume, surface and cross-section surface. To fuse these images in three dimensions, only the created

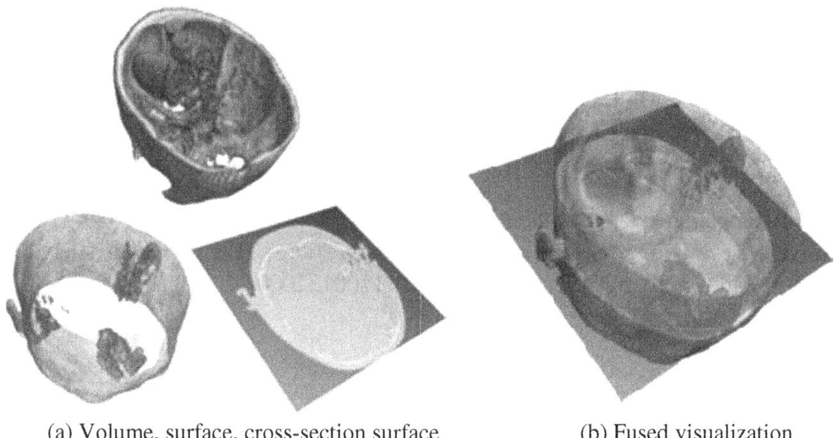

(a) Volume, surface, cross-section surface (b) Fused visualization

Fig. 9. Fused visualization from volume, surface and cross-section surface

(a) Surface, cross-section surface, contour line (b) Fused visualization

Fig. 10. Fused visualization from surface, cross-section surface and contour line

particles have to be combined. Fig. 9 is the result of fused visualization, where the number of particles used is about 76 million (including 16 million particles in the volume, 30 million particles on the isosurface, and 30 million particles on the slice plane). Fig. 10 is an example of the fused visualization from three rendering type, surface, cross-section surface and contour line. We can confirm that natural fusion is realized by simply merging the particle sets from the different rendering types.

4 Conclusion

In this paper, we described a method to create fusion images using particle-based volume rendering (PBVR) and particle-based surface rendering (PBSR). The PBSR creates transparent images with right depth feel without necessity of particle sorting. The nonnecessity of the particle sorting originates from opaqueness of each particle and probabilistic determination of surface opacity. Our probabilistic opacity theory can be controlled on the surface opacity. The sampling speed is in the same range of all rendering type and independent on the input volume data. In addition, we demonstrated that 3D fusion images with right depth feel can be created only by combining particles prepared for each element to be fused. The element can be either volume, surface, cross-section or contour line.

Acknowledgements. The authors wish to thank Koji Koyamada, Hiromi T. Tanaka, and Naohisa Sakommoto for their valuable suggestions. This work was supported by Grant-in-Aid for Young Scientists (B) (No. 217001220004).

References

1. Koyamada, K., Sakamoto, N., Tanaka, S.: A Particle Modeling for Rendering Irregular Volumes. In: Proceedings of the International Conference on Computer Modeling and Simulation (UKSIM 2008), Cambridge, England, pp. 372–377 (2008)
2. Sakamoto, N., Kawamura, T., Koyamada, K.: Improvement of particle-based volume rendering for visualizing irregular volume data sets. Computers & Graphics 34(1), 34–42 (2010)
3. Sakamoto, N., Koyamada, K., Saito, A., Kimura, A., Tanaka, S.: Multi-Volume Rendering Using Particle Fusion. In: IEEE VGTC Pacific Visualization Symposium 2008, Kyoto, Japan (2008)
4. Satoshi, T., Akio, M., Satoru, N., Yasushi, F., Hiroaki, Y.: Sampling Implicit Surfaces Based on Stochastic Differential Equations with Converging Constraint. Computers & Graphics 24(3), 419–431 (2000)

Author Index